Fundamentals of Actuarial Mathematics

Fundamentals of Actuarial Mathematics

Third Edition

S. David Promislow

York University, Toronto, Canada

This edition first published 2015

Registered office
John Wiley & Sons Ltd, The Atrium, Southern Gate, Chichester, West Sussex, PO19 8SQ, United Kingdom

For details of our global editorial offices, for customer services and for information about how to apply for permission to reuse the copyright material in this book please see our website at www.wiley.com.

Library of Congress Cataloging-in-Publication Data

Promislow, S. David.
Fundamentals of actuarial mathematics / S. David Promislow. – Third edition.
pages cm
Includes bibliographical references and index.
ISBN 978-1-118-78246-0 (hardback)
1. Insurance–Mathematics. 2. Business mathematics. I. Title.
HG8781.P76 2014
368′.01–dc23

2014027082

A catalogue record for this book is available from the British Library.

ISBN: 9781118782460

Set in 10/12pt Times by Aptara Inc., New Delhi, India

1 2015

To Georgia and Griffith

Contents

Preface xix

Acknowledgements xxiii

About the companion website xxiv

Part I THE DETERMINISTIC LIFE CONTINGENCIES MODEL 1

1 Introduction and motivation 3
- 1.1 Risk and insurance 3
- 1.2 Deterministic versus stochastic models 4
- 1.3 Finance and investments 5
- 1.4 Adequacy and equity 5
- 1.5 Reassessment 6
- 1.6 Conclusion 6

2 The basic deterministic model 7
- 2.1 Cash flows 7
- 2.2 An analogy with currencies 8
- 2.3 Discount functions 9
- 2.4 Calculating the discount function 11
- 2.5 Interest and discount rates 12
- 2.6 Constant interest 12
- 2.7 Values and actuarial equivalence 13
- 2.8 Vector notation 17
- 2.9 Regular pattern cash flows 18
- 2.10 Balances and reserves 20
  - 2.10.1 Basic concepts 20
  - 2.10.2 Relation between balances and reserves 22
  - 2.10.3 Prospective versus retrospective methods 23
  - 2.10.4 Recursion formulas 24
- 2.11 Time shifting and the splitting identity 26

*2.11 Change of discount function 27
2.12 Internal rates of return 28
*2.13 Forward prices and term structure 30
2.14 Standard notation and terminology 33
2.14.1 Standard notation for cash flows discounted with interest 33
2.14.2 New notation 34
2.15 Spreadsheet calculations 34
Notes and references 35
Exercises 35

3 The life table 39
3.1 Basic definitions 39
3.2 Probabilities 40
3.3 Constructing the life table from the values of q_x 41
3.4 Life expectancy 42
3.5 Choice of life tables 44
3.6 Standard notation and terminology 44
3.7 A sample table 45
Notes and references 45
Exercises 45

4 Life annuities 47
4.1 Introduction 47
4.2 Calculating annuity premiums 48
4.3 The interest and survivorship discount function 50
4.3.1 The basic definition 50
4.3.2 Relations between y_x for various values of x 52
4.4 Guaranteed payments 53
4.5 Deferred annuities with annual premiums 55
4.6 Some practical considerations 56
4.6.1 Gross premiums 56
4.6.2 Gender aspects 56
4.7 Standard notation and terminology 57
4.8 Spreadsheet calculations 58
Exercises 59

5 Life insurance 61
5.1 Introduction 61
5.2 Calculating life insurance premiums 61
5.3 Types of life insurance 64
5.4 Combined insurance–annuity benefits 64
5.5 Insurances viewed as annuities 69
5.6 Summary of formulas 70
5.7 A general insurance–annuity identity 70
5.7.1 The general identity 70
5.7.2 The endowment identity 71

5.8 Standard notation and terminology 72
5.8.1 Single-premium notation 72
5.8.2 Annual-premium notation 73
5.8.3 Identities 74
5.9 Spreadsheet applications 74
Exercises 74

6 Insurance and annuity reserves 78
6.1 Introduction to reserves 78
6.2 The general pattern of reserves 81
6.3 Recursion 82
6.4 Detailed analysis of an insurance or annuity contract 83
6.4.1 Gains and losses 83
6.4.2 The risk–savings decomposition 85
6.5 Bases for reserves 87
6.6 Nonforfeiture values 88
6.7 Policies involving a return of the reserve 88
6.8 Premium difference and paid-up formulas 90
6.8.1 Premium difference formulas 90
6.8.2 Paid-up formulas 90
6.8.3 Level endowment reserves 91
6.9 Standard notation and terminology 91
6.10 Spreadsheet applications 93
Exercises 94

7 Fractional durations 98
7.1 Introduction 98
7.2 Cash flows discounted with interest only 99
7.3 Life annuities paid *m*thly 101
7.3.1 Uniform distribution of deaths 101
7.3.2 Present value formulas 102
7.4 Immediate annuities 104
7.5 Approximation and computation 105
*7.6 Fractional period premiums and reserves 106
7.7 Reserves at fractional durations 107
7.8 Standard notation and terminology 109
Exercises 109

8 Continuous payments 112
8.1 Introduction to continuous annuities 112
8.2 The force of discount 113
8.3 The constant interest case 114
8.4 Continuous life annuities 115
8.4.1 Basic definition 115
8.4.2 Evaluation 116
8.4.3 Life expectancy revisited 117

8.5 The force of mortality 118
8.6 Insurances payable at the moment of death 119
8.6.1 Basic definitions 119
8.6.2 Evaluation 120
8.7 Premiums and reserves 122
8.8 The general insurance–annuity identity in the continuous case 123
8.9 Differential equations for reserves 124
8.10 Some examples of exact calculation 125
8.10.1 Constant force of mortality 126
8.10.2 Demoivre's law 127
8.10.3 An example of the splitting identity 128
8.11 Further approximations from the life table 129
8.12 Standard actuarial notation and terminology 131
Notes and references 132
Exercises 132

9 Select mortality 137
9.1 Introduction 137
9.2 Select and ultimate tables 138
9.3 Changes in formulas 139
9.4 Projections in annuity tables 141
9.5 Further remarks 142
Exercises 142

10 Multiple-life contracts 144
10.1 Introduction 144
10.2 The joint-life status 144
10.3 Joint-life annuities and insurances 146
10.4 Last-survivor annuities and insurances 147
10.4.1 Basic results 147
10.4.2 Reserves on second-death insurances 148
10.5 Moment of death insurances 149
10.6 The general two-life annuity contract 150
10.7 The general two-life insurance contract 152
10.8 Contingent insurances 153
10.8.1 First-death contingent insurances 153
10.8.2 Second-death contingent insurances 154
10.8.3 Moment-of-death contingent insurances 155
10.8.4 General contingent probabilities 155
10.9 Duration problems 156
*10.10 Applications to annuity credit risk 159
10.11 Standard notation and terminology 160
10.12 Spreadsheet applications 161
Notes and references 161
Exercises 161

11 Multiple-decrement theory 166
11.1 Introduction 166
11.2 The basic model 166
11.2.1 The multiple-decrement table 167
11.2.2 Quantities calculated from the multiple-decrement table 168
11.3 Insurances 169
11.4 Determining the model from the forces of decrement 170
11.5 The analogy with joint-life statuses 171
11.6 A machine analogy 171
11.6.1 Method 1 172
11.6.2 Method 2 173
11.7 Associated single-decrement tables 175
11.7.1 The main methods 175
11.7.2 Forces of decrement in the associated single-decrement tables 176
11.7.3 Conditions justifying the two methods 177
11.7.4 Other approaches 180
Notes and references 181
Exercises 181

12 Expenses and profits 184
12.1 Introduction 184
12.2 Effect on reserves 186
12.3 Realistic reserve and balance calculations 187
12.4 Profit measurement 189
12.4.1 Advanced gain and loss analysis 189
12.4.2 Gains by source 191
12.4.3 Profit testing 193
Notes and references 196
Exercises 196

*13 Specialized topics 199
13.1 Universal life 199
13.1.1 Description of the contract 199
13.1.2 Calculating account values 201
13.2 Variable annuities 203
13.3 Pension plans 204
13.3.1 DB plans 204
13.3.2 DC plans 206
Exercises 207

Part II THE STOCHASTIC LIFE CONTINGENCIES MODEL 209

14 Survival distributions and failure times 211
14.1 Introduction to survival distributions 211
14.2 The discrete case 212

14.3 The continuous case 213
14.3.1 The basic functions 214
14.3.2 Properties of μ 214
14.3.3 Modes 215
14.4 Examples 215
14.5 Shifted distributions 216
14.6 The standard approximation 217
14.7 The stochastic life table 219
14.8 Life expectancy in the stochastic model 220
14.9 Stochastic interest rates 221
Notes and references 222
Exercises 222

15 The stochastic approach to insurance and annuities 224
15.1 Introduction 224
15.2 The stochastic approach to insurance benefits 225
15.2.1 The discrete case 225
15.2.2 The continuous case 226
15.2.3 Approximation 226
15.2.4 Endowment insurances 227
15.3 The stochastic approach to annuity benefits 229
15.3.1 Discrete annuities 229
15.3.2 Continuous annuities 231
*15.4 Deferred contracts 233
15.5 The stochastic approach to reserves 233
15.6 The stochastic approach to premiums 235
15.6.1 The equivalence principle 235
15.6.2 Percentile premiums 236
15.6.3 Aggregate premiums 237
15.6.4 General premium principles 240
15.7 The variance of ${}_rL$ 241
15.8 Standard notation and terminology 243
Notes and references 244
Exercises 244

16 Simplifications under level benefit contracts 248
16.1 Introduction 248
16.2 Variance calculations in the continuous case 248
16.2.1 Insurances 249
16.2.2 Annuities 249
16.2.3 Prospective losses 249
16.2.4 Using equivalence principle premiums 249
16.3 Variance calculations in the discrete case 250
16.4 Exact distributions 252
16.4.1 The distribution of $\bar{Z}$ 252
16.4.2 The distribution of $\bar{Y}$ 252

16.4.3 The distribution of L 252
16.4.4 The case where T is exponentially distributed 253
16.5 Some non-level benefit examples 254
16.5.1 Term insurance 254
16.5.2 Deferred insurance 254
16.5.3 An annual premium policy 255
Exercises 256

17 The minimum failure time 259
17.1 Introduction 259
17.2 Joint distributions 259
17.3 The distribution of T 261
17.3.1 The general case 261
17.3.2 The independent case 261
17.4 The joint distribution of (T, J) 261
17.4.1 The distribution function for (T, J) 261
17.4.2 Density and survival functions for (T, J) 264
17.4.3 The distribution of J 265
17.4.4 Hazard functions for (T, J) 266
17.4.5 The independent case 266
17.4.6 Nonidentifiability 268
17.4.7 Conditions for the independence of T and J 269
17.5 Other problems 270
17.6 The common shock model 271
17.7 Copulas 273
Notes and references 276
Exercises 276

Part III ADVANCED STOCHASTIC MODELS 279

18 An introduction to stochastic processes 281
18.1 Introduction 281
18.2 Markov chains 283
18.2.1 Definitions 283
18.2.2 Examples 284
18.3 Martingales 286
18.4 Finite-state Markov chains 287
18.4.1 The transition matrix 287
18.4.2 Multi-period transitions 288
18.4.3 Distributions 288
*18.4.4 Limiting distributions 289
*18.4.5 Recurrent and transient states 290
18.5 Introduction to continuous time processes 293
18.6 Poisson processes 293
18.6.1 Waiting times 295
18.6.2 Nonhomogeneous Poisson processes 295

18.7 Brownian motion 295
18.7.1 The main definition 295
18.7.2 Connection with random walks 296
*18.7.3 Hitting times 297
*18.7.4 Conditional distributions 298
18.7.5 Brownian motion with drift 299
18.7.6 Geometric Brownian motion 299
Notes and references 299
Exercises 300

19 Multi-state models 304
19.1 Introduction 304
19.2 The discrete-time model 305
19.2.1 Non-stationary Markov Chains 305
19.2.2 Discrete-time multi-state insurances 307
19.2.3 Multi-state annuities 310
19.3 The continuous-time model 311
19.3.1 Forces of transition 311
19.3.2 Path-by-path analysis 316
19.3.3 Numerical approximation 317
19.3.4 Stationary continuous time processes 318
19.3.5 Some methods for non-stationary processes 320
19.3.6 Extension of the common shock model 321
19.3.7 Insurance and annuity applications in continuous time 322
19.4 Recursion and differential equations for multi-state reserves 324
19.5 Profit testing in multi-state models 327
19.6 Semi-Markov models 328
Notes and references 328
Exercises 329

20 Introduction to the Mathematics of Financial Markets 333
20.1 Introduction 333
20.2 Modelling prices in financial markets 333
20.3 Arbitrage 334
20.4 Option contracts 337
20.5 Option prices in the one-period binomial model 339
20.6 The multi-period binomial model 342
20.7 American options 346
20.8 A general financial market 348
20.9 Arbitrage-free condition 351
20.10 Existence and uniqueness of risk-neutral measures 353
20.10.1 Linear algebra background 353
20.10.2 The space of contingent claims 353
20.10.3 The Fundamental theorem of asset pricing completed 357
20.11 Completeness of markets 359
20.12 The Black–Scholes–Merton formula 361

20.13 Bond markets 364
20.13.1 Introduction 364
20.13.2 Extending the notion of conditional expectation 366
20.13.3 The arbitrage-free condition in the bond market 367
20.13.4 Short-rate modelling 368
20.13.5 Forward prices and rates 370
20.13.6 Observations on the continuous time bond market 371
Notes and references 372
Exercises 373

Part IV RISK THEORY 375

21 Compound distributions 377
21.1 Introduction 377
21.2 The mean and variance of S 379
21.3 Generating functions 380
21.4 Exact distribution of S 381
21.5 Choosing a frequency distribution 381
21.6 Choosing a severity distribution 383
21.7 Handling the point mass at 0 384
21.8 Counting claims of a particular type 385
21.8.1 One special class 385
21.8.2 Special classes in the Poisson case 386
21.9 The sum of two compound Poisson distributions 387
21.10 Deductibles and other modifications 388
21.10.1 The nature of a deductible 388
21.10.2 Some calculations in the discrete case 389
21.10.3 Some calculations in the continuous case 390
21.10.4 The effect on aggregate claims 392
21.10.5 Other modifications 393
21.11 A recursion formula for S 393
21.11.1 The positive-valued case 393
21.11.2 The case with claims of zero amount 397
Notes and references 398
Exercises 398

22 Risk assessment 403
22.1 Introduction 403
22.2 Utility theory 403
22.3 Convex and concave functions: Jensen's inequality 406
22.3.1 Basic definitions 406
22.3.2 Jensen's inequality 407
22.4 A general comparison method 408
22.5 Risk measures for capital adequacy 412
22.5.1 The general notion of a risk measure 412
22.5.2 Value-at-risk 413

22.5.3 Tail value-at-risk 413
22.5.4 Distortion risk measures 417
Notes and references 417
Exercises 417

23 Ruin models 420
23.1 Introduction 420
23.2 A functional equation approach 422
23.3 The martingale approach to ruin theory 424
23.3.1 Stopping times 424
23.3.2 The optional stopping theorem and its consequences 426
23.3.3 The adjustment coefficient 429
23.3.4 The main conclusions 431
23.4 Distribution of the deficit at ruin 433
23.5 Recursion formulas 434
23.5.1 Calculating ruin probabilities 434
23.5.2 The distribution of $D(u)$ 436
23.6 The compound Poisson surplus process 438
23.6.1 Description of the process 438
23.6.2 The probability of eventual ruin 440
23.6.3 The value of $\psi(0)$ 440
23.6.4 The distribution of $D(0)$ 440
23.6.5 The case when $\boldsymbol{X}$ is exponentially distributed 441
23.7 The maximal aggregate loss 441
Notes and references 445
Exercises 445

24 Credibility theory 449
24.1 Introductory material 449
24.1.1 The nature of credibility theory 449
24.1.2 Information assessment 449
24.2 Conditional expectation and variance with respect to another random variable 453
24.2.1 The random variable $E(X|Y)$ 453
24.2.2 Conditional variance 455
24.3 General framework for Bayesian credibility 457
24.4 Classical examples 459
24.5 Approximations 462
24.5.1 A general case 462
24.5.2 The Bühlman model 463
24.5.3 Bühlman–Straub Model 464
24.6 Conditions for exactness 465
24.7 Estimation 469
24.7.1 Unbiased estimators 469
24.7.2 Calculating $\mathrm{Var}(\bar{X})$ in the credibility model 470

24.7.3 Estimation of the Bülhman parameters 470
24.7.4 Estimation in the Bülhman–Straub model 472
Notes and references 473
Exercises 473

Answers to exercises 477

Appendix A review of probability theory 493
A.1 Sample spaces and probability measures 493
A.2 Conditioning and independence 495
A.3 Random variables 495
A.4 Distributions 496
A.5 Expectations and moments 497
A.6 Expectation in terms of the distribution function 498
A.7 Joint distributions 499
A.8 Conditioning and independence for random variables 501
A.9 Moment generating functions 502
A.10 Probability generating functions 503
A.11 Some standard distributions 505
A.11.1 The binomial distribution 505
A.11.2 The Poisson distribution 505
A.11.3 The negative binomial and geometric distributions 506
A.11.4 The continuous uniform distribution 507
A.11.5 The normal distribution 507
A.11.6 The gamma and exponential distributions 509
A.11.7 The lognormal distribution 510
A.11.8 The Pareto distribution 511
A.12 Convolution 511
A.12.1 The discrete case 511
A.12.2 The continuous case 513
A.12.3 Notation and remarks 515
A.13 Mixtures 516

References **517**

Notation index **519**

Index **523**

Preface

The third edition of this book continues the objective of providing coverage of actuarial mathematics in a flexible manner that meets the needs of several audiences. These range from those who want only a basic knowledge of the subject, to those preparing for careers as professional actuaries. All this is carried out with a streamlined system of notation, and a modern approach to computation involving spreadsheets.

The text is divided into four parts. The first two cover the subject of life contingencies. The modern approach towards this subject is through a stochastic model, as opposed to the older deterministic viewpoint. I certainly agree that mastering the stochastic model is the desirable goal. However, my classroom experience has convinced me that this is not the right place to begin the instruction. I find that students are much better able to learn the new ideas, the new notation, the new ways of thinking involved in this subject, when done first in the simplest possible setting, namely a deterministic discrete model. After the main ideas are presented in this fashion, continuous models are introduced. In Part II of the book, the full stochastic model of life contingencies can be dealt with in a reasonably quick fashion.

Another innovation in Part II is to depart from the conventional treatment of life contingencies as dealing essentially with patterns of mortality or disability in a group of human lives. Throughout Part II, we deal with general *failure times* which makes the theory more widely adaptable.

Part III deals with more advanced stochastic models. Following an introduction to stochastic processes, there is a chapter covering multi-state theory, an approach which unifies many of the ideas in Parts I and II. The final chapter in Part III is an introduction to modern financial mathematics.

Part IV deals with the subject of risk theory, sometime referred to as loss models. It includes an extensive coverage of classical ruin theory, a topic that originated in actuarial science but recently has found many applications in financial economics. It also includes credibility theory, which will appeal to the reader interested more in the casualty side of actuarial mathematics.

This book will meet the needs of those preparing for the examinations of many of the major professional actuarial organizations. Parts I to III of this new third edition covers all of the material on the current syllabuses of Exam MLC of the Society of Actuaries and Canadian Institute of Actuaries and Exam LC of the Casualty Actuarial Society, and covers most of the topics on the current syllabus of Exam CT5 of the British Institute of Actuaries.

In addition, Part IV of the book covers a great deal of the material on Exam C of the Society of Actuaries and Canadian Institute of Actuaries, including the topics of Frequency, Severity and Aggregate Models, Risk Measures, and Credibility Theory.

The mathematical prerequisites for Part 1 are relatively modest. comprising elementary linear algebra and probability theory, and, beginning in Chapter 8, some basic calculus. A more advanced knowledge of probability theory is needed from Chapter 13 onward, and this material summarized in Appendix A. A usual prerequisite for actuarial mathematics is a course in the theory of interest. Although this may be useful, it is not strictly required. All the interest theory that is needed is presented as a particular case of the general deterministic actuarial model in Chapter 2.

A major source of difficulty for many students in learning actuarial mathematics is to master the rather complex system of actuarial notation. We have introduced some notational innovations, which tie in well with modern calculation procedures as well as allow us to greatly simplify the notation that is required. We have, however, included all the standard notation in separate sections, at the end of the relevant chapters, which can be read by those readers who desire this material.

Keeping in mind the nature of the book and its intended audience, we have avoided excessive mathematical rigour. Nonetheless, careful proofs are given in all cases where these are thought to be accessible to the typical senior undergraduate mathematics student. For the few proofs not given in their entirety, mainly those involving continuous-time stochastic processes, we have tried at least to provide some motivation and intuitive reasoning for the results.

Exercises appear at the end of each chapter. In Parts I and II these are divided up into different types. Type A exercises generally are those which involve direct calculation from the formulas in the book. Type B involve problems where more thought is involved. Derivations and problems which involve symbols rather than numeric calculation are normally included in Type B problems. A third type is spreadsheet exercises which themselves are divided into two subtypes. The first of these asks the reader to solve problems using a spreadsheet. Detailed descriptions of applicable Microsoft Excel® spreadsheets are given at the end of the relevant chapters. Readers of course are free to modify these or construct their own. The second subtype does not ask specific questions but instead asks the reader to modify the given spreadsheets to handle additional tasks. Answers to most of the calculation-type exercises appear at the end of the book.

Sections marked with an asterisk * deal with more advanced material, or with special topics that are not used elsewhere in the book. They can be omitted on first reading. The exercises dealing with such sections are likewise marked with *, as are a few other exercises which are of above average difficulty.

There are various ways of using the text for university courses geared to third or fourth year undergraduates, or beginning graduate students. Chapters 1 to 8 could form the basis of a one-semester introductory course. Part IV is for the most part independent of the first three parts, except for the background material on stochastic processes given in Chapter 18 and would constitute another one-semester course. The rest of the book constitutes roughly another two semesters worth of material, with possibly some omissions; Chapter 13 is not needed for the rest of the book. Chapters 7 (except for Section 7.3.1), 9 and 12 deal with topics that are important in applications, but which are used minimally in other parts of the text. They could be omitted without loss of continuity.

CHANGES IN THE THIRD EDITION

There are several additions and changes to the third edition.

The most notable is a new Chapter 20 providing an introduction to the mathematics of financial markets. It has been long recognized that knowledge of this subject is essential to the management of financial risk that faces the actuary of today.

Other additions include the following:

- Chapter 12, on expenses, has been considerably enlarged to include the topic of profit testing.
- The chapter on multi-state models has been expanded to include discussion of reserves and profit testing in such models, as well as several additional techniques for continuous-time problems.
- Some extra numerical procedures have been included, such as Euler's method for differential equations, and the three-term Woolhouse formulas for fractional annuity approximations.
- An introduction to Brownian motion has been added to the material on continuous-time stochastic processes.
- The previous material on universal life and variable annuities has been rewritten and included in a new chapter dealing with miscellaneous topics. A brief discussion of pension plans is included here as well.
- Additional examples, exercises, and clarification have been added to various chapters.

As well as the changes there has been a reorganization in the material The previous two chapters on stochastic processes have been combined into one and now appear earlier in the book as background for the multi-state and financial markets chapters. In the current Part IV, the detailed descriptions of the various distributions have been removed and added as a section to the Appendix on probability theory.

Acknowledgements

Several individuals have assisted in the various editions of this book. I am particularly indebted to two people who have made a significant contribution by providing a number of helpful comments, corrections, and suggestions. They are Virginia Young for her work on the first edition, and Elias Shiu for his help with the third edition.

There are many others who deserve thanks. Moshe Milevsky provided enlightening comments on annuities and it was his ideas that motivated the credit risk applications in Chapter 10, as well as some of the material on generational annuity tables in Chapter 9. Several people found misprints in the first edition and earlier drafts. These include Valerie Michkine, Jacques Labelle, Karen Antonio, Kristen Moore, as well as students at York University and the University of Michigan. Christian Hess asked some questions which led to the inclusion of Example 21.10 to clear up an ambiguous point. Exercise 18.13 was motivated by Bob Jewett's progressive practice routines for pool. My son Michael, a life insurance actuary, provided valuable advice on several practical aspects of the material. Thanks go to the editorial and production teams at Wiley for their much appreciated assistance. Finally, I thank my wife Shirley who provided support and encouragement throughout the writing of all three editions.

About the companion website

This book is accompanied by a companion website:

www.wiley.com/go/promislow/actuarial

The website includes:

- A variety of exercises, both computational and theoretical
- Answers, enabling use for self-study.

Part I

THE DETERMINISTIC LIFE CONTINGENCIES MODEL

1

Introduction and motivation

1.1 Risk and insurance

In this book we deal with certain mathematical models. This opening chapter, however, is a nontechnical introduction, designed to provide background and motivation. In particular, we are concerned with models used by actuaries, so we might first try to describe exactly what it is that actuaries do. This can be difficult, because a typical actuary is concerned with many issues, but we can identify two major themes dealt with by this profession.

The first is *risk*, a word that itself can be defined in different ways. A commonly accepted definition in our context is that risk is the possibility that *something bad* happens. Of course, many bad things can happen, but in particular we are interested in occurrences that result in *financial loss*. A person dies, depriving family of earned income or business partners of expertise. Someone becomes ill, necessitating large medical expenses. A home is destroyed by fire or an automobile is damaged in an accident. No matter what precautions you take, you cannot rid yourself completely of the possibility of such unfortunate events, but what you can do is take steps to mitigate the financial loss involved. One of the most commonly used measures is to purchase insurance.

Insurance involves a sharing or pooling of risks among a large group of people. The origins go back many years and can be traced to members of a community helping out others who suffered loss in some form or other. For example, people would help out neighbours who had suffered a death or illness in the family. While such aid was in many cases no doubt due to altruistic feelings, there was also a motivation of self-interest. You should be prepared to help out a neighbour who suffered some calamity, since you or your family could similarly be aided by others when you required such assistance. This eventually became more formalized, giving rise to the insurance companies we know today.

With the institution of insurance companies, sharing is no longer confined to the scope of neighbours or community members one knows, but it could be among all those who chose to purchase insurance from a particular company. Although there are many different types

Fundamentals of Actuarial Mathematics, Third Edition. S. David Promislow.

Companion website: http://www.wiley.com/go/actuarial

of insurance, the basic principle is similar. A company known as the *insurer* agrees to pay out money, which we will refer to as *benefits*, at specified times, upon the occurrence of specified events causing financial loss. In return, the person purchasing insurance, known as the *insured*, agrees to make payments of prescribed amounts to the company. These payments are typically known as *premiums*. The contract between the insurer and the insured is often referred to as the *insurance policy*.

The risk is thereby transferred from the individuals facing the loss to the insurer. The insurer in turn reduces its risk by insuring a sufficiently large number of individuals, so that the losses can be accurately predicted. Consider the following example, which is admittedly vastly oversimplified but designed to illustrate the basic idea.

Suppose that a certain type of event is unlikely to occur but if so, causes a financial loss of 100 000. The insurer estimates that about 1 out of every 100 individuals who face the possibility of such loss will actually experience it. If it insures 1000 people, it can then expect 10 losses. Based on this model, the insurer would charge each person a premium of 1000. (We are ignoring certain factors such as expenses and profits.) It would collect a total of 1 000 000 and have precisely enough to cover the 100 000 loss for each of the 10 individuals who experience this. Each individual has eliminated his or her risk, and in so far as the estimate of 10 losses is correct, the insurer has likewise eliminated its own risk. (We comment further on this statement in the next section.)

We conclude this section with a few words on the connection between insurance and gambling. Many people believe that insurance is really a form of the latter, but in fact it is exactly the opposite. Gambling trades certainty for uncertainty. The amount of money you have in your pocket is there with certainty if you do not gamble, but it is subject to uncertainty if you decide to place a bet. On the other hand, insurance trades uncertainty for certainty. The uncertain drain on your wealth, due to the possibility of a financial loss, is converted to the certainty of the much smaller drain of the premium payments if you insure against the loss.

1.2 Deterministic versus stochastic models

The example in Section 1.1 illustrates what is known as a *deterministic* model. The insurer in effect pretends it will know exactly how much it will pay out in benefits and then charges premiums to match this amount. Of course, the insurer knows that it cannot really predict these amounts precisely. By selling a large number of policies they hope to benefit from the diversification effect. They are really relying on the statistical concept known as the 'law of large numbers', which in this context intuitively says that if a sufficiently large number of individuals are insured, then the total number of losses will likely be close to the predicted figure.

To look at this idea in more detail, it may help to give an analogy with flipping coins. If we flip 100 fair coins, we cannot predict exactly the number of them that will come up heads, but we expect that most of the time this number should be close to 50. But 'most of the time' does not mean always. It is possible for example, that we may get only 37 heads, or as many as 63, or even more extreme outcomes. In the example given in the last section, the number of losses may well turn out to be more than the expected number of 10. We would like to know just how unlikely these rare events are. In other words, we would like to quantify more precisely just what the words 'most of the time' mean. To achieve this greater sophistication a stochastic model for insurance claims is needed, which will assign probabilities to the occurrence of various numbers of losses. This will allow adjustment of premiums in order to allow for the risk that the actual number of losses will deviate from that expected. We will however begin the study of actuarial mathematics by first developing a deterministic approach, as this seems to be the best way of learning the basic concepts. After mastering this, it is not difficult to turn to the more realistic stochastic setting.

We will not get into all the complications that can arise. In actual coin flipping it seems clear that the results of each toss are independent of the others. The fact that one coin comes up heads, is not going to affect the outcomes of the others. It is this independence which is behind the law of large numbers, and which results in outcomes that are usually close to what is expected. There are some risks, often referred to as *systematic* or *non-diversifiable,* where the independence assumption fails, and which can adversely affect all or a large number of members of a group at the same time. For example, a spreading epidemic could cause life or health insurers to pay more in claims than they expected. Selling more policies in order to diversify would not help their financial situation. It could in fact make it worse, if the premiums were not sufficient to cover the extra losses. Severe climatic disturbances causing storms could impact property insurance in the same way. In 2008, falling real estate prices in the United States affected mortgage lenders and those who insured mortgage lenders against bad debts, to the extent that this helped trigger a global financial crisis. A detailed discussion of these matters is not within the scope of this work, and for the most part, the stochastic model we present will confine attention to the usual insurance model where the risks are considered as independent. It should be kept in mind however that the detection and avoidance of systematic risk are matters that the actuary must always be aware of.

1.3 Finance and investments

The second theme involved in an actuary's work is finance and investments. In most of the types of insurance that we focus on in this book, an additional complicating factor is the long-term nature of the contracts. Benefits may not be paid until several years after premiums are collected. This is certainly true in life insurance, where the loss is occasioned by the death of an individual. Premiums received are invested and the resulting earnings can be used to help provide the benefits. Consider the simple example given above, and suppose further that the benefits do not have to be paid until 1 year after the premiums are collected. If the insurer can invest the money at, say, 5% interest for the year, then it does not need to charge the full 1000 in premium, but can collect only 1000/1.05 from each person. When invested, this amount will provide the necessary 1000 to cover the losses. Again, this example is oversimplified and there are many more complications. We will, in the next chapter, consider a mathematical model that deals with the consequences of the payments of money at various times. A much more elaborate treatment of financial matters, incorporating randomness, is presented in Chapter 20.

1.4 Adequacy and equity

We can now give a general description of the responsibilities of an actuary. The overriding task is to ensure that the premiums, together with investment earnings, are *adequate* to provide for the payment of the benefits. If this is not true, then it will not be possible for the insurer to meet its obligations and some of the insureds will necessarily not receive compensation for their losses. The challenge in meeting this goal arises from the several areas of uncertainty. The amount and timing of the benefits that will have to be paid, as well as the investment earnings, are unknown and subject to random fluctuations. The actuary makes substantial use of probabilistic methods to handle this uncertainty.

Another goal is to achieve *equity* in setting premiums. If an insurer is to attract purchasers, it must charge rates that are perceived as being fair. Here also, the randomness means that it is not obvious how to define equity in this context. It cannot mean that two individuals who are charged the same amount in premiums will receive exactly the same back in benefits, for that would negate the sharing arrangement inherent in the insurance idea. While there are different possible viewpoints, equity in insurance is generally expected to mean that the mathematical expectation of these two individuals should be the same.

1.5 Reassessment

Actuaries design insurance contracts and must initially calculate premiums that will fulfill the goals of adequacy and equity, but this is not the end of the story. No matter how carefully one makes an initial assessment of risks, there are too many variables to be able to achieve complete accuracy. Such assessments must be continually re-evaluated, and herein lies the real expertise of the actuary. This work may be compared to sailing a ship in a stormy sea. It is impossible to avoid being blown off course occasionally. The skill is to detect when this occurs and to take the necessary steps to continue in the right direction. This continual monitoring and reassessing is an important part of the actuary's work. A large part of this involves calculating quantities known as *reserves*. We introduce this concept in Chapter 2 and then develop it more fully in Chapter 6.

1.6 Conclusion

We can now summarize the material found in the subsequent chapters of the book. We will describe the mathematical models used by the actuary to ensure that an insurer will be able to meet its promised benefits payments and that the respective purchasers of its contracts are treated equitably. In Part I, we deal with a strictly deterministic model. This enables us to focus on the main principles while keeping the required mathematics reasonably simple. In Part II, we look at the stochastic model for an individual insurance contract. In Part III, we look at more advanced stochastic models and introduce the mathematics of financial markets. In Part IV, we consider models that encompass an entire portfolio of insurance contracts.

2

The basic deterministic model

2.1 Cash flows

As indicated in the previous chapter, a basic application of actuarial mathematics is to model the transfer of money. Insurance companies, banks and other financial institutions engage in transactions that involve accepting sums of money at certain times, and paying out sums of money at other times.

To construct a model for describing this situation, we will first fix a time unit. This can be arbitrary, but in most applications it will be taken as some familiar interval of time. For convenience we will assume that time is measured in years, unless we indicate otherwise. We will let time 0 refer to the present time, and time t will then denote t time units in the future. We also select an arbitrary unit of capital. In this chapter, we assume that all funds are paid out or received at integer time points, that is, at time $0, 1, 2, \ldots$. The amount of money received or paid out at time k will be called the *net cash flow* at time k and denoted by c_k. A positive value of c_k denotes that a sum is to be received, whereas a negative value indicates that a sum is paid out. The entire transaction is then described by listing the sequence of cash flows. We will refer to this as a *cash flow vector*,

$$\mathbf{c} = (c_0, c_1, \ldots, c_N),$$

where N is the final duration for which a payment is made.

For example, suppose I lend you 10 units of capital now and a further 5 units a year from now. You repay the loan by making three yearly payments of 7 units each, beginning 3 years from now. The resulting cash flow vector from my point of view is

$$\mathbf{c} = (-10, -5, 0, 7, 7, 7).$$

From your point of view, the transaction is represented by $-\mathbf{c} = (10, 5, 0, -7, -7, -7)$.

Fundamentals of Actuarial Mathematics, Third Edition. S. David Promislow.

Companion website: http://www.wiley.com/go/actuarial

One of our main goals in this chapter is to provide methods for analyzing transactions in terms of their cash flow vectors. There are several basic questions that could be asked:

- When is a transaction worthwhile undertaking?
- How much should one pay in order to receive a certain sequence of cash flows?
- How much should one charge in order to provide a certain sequence of cash flows?
- How does one compare two transactions to decide which one is preferable?

All of these questions are related, and we could answer all of them if we could find a method to put a value on a sequence of future cash flows. If all cash flows were paid at the same time, or if the value of money did not depend on the time that a payment was made, the problem would reduce to one of simple addition. We could simply value a cash flow sequence by adding up all the payments. We cannot proceed in this naive way, however and must consider the *time value of money*. It is a basic economic fact that we prefer present to future consumption. We want to eat the chocolate bar now, rather than tomorrow. We want to enjoy the new car today, rather than next month. This means of course that money paid to us today is worth more than money paid in the future. We are no doubt all very familiar with this fact. We pay interest for the privilege of borrowing money today, which lets us consume now, or we advance money to others, giving up our present consumption and expecting to be compensated with interest earnings. In addition, there is the effect of risk. If we are given a unit of money today, we have it. If we forego it now in return for future payments, there could be a chance that the party who is supposed to make remittance to us may be unable or unwilling to do, and we expect to be compensated for the possible loss. A major step in answering the above questions is to quantify this dependence of value on time.

Readers who have taken courses on the theory of compound interest will be familiar with many of the ideas. However, our treatment will be somewhat different than that usually given. One reason for this is that we want to develop the concepts in such a way that they are applicable to more general situations, as given in Chapters 3–5. A second reason is that our approach is designed to be compatible with modern-day computing methods such as spreadsheets.

To conclude this section, we remark that many complications arise when the cash flows are not exactly known in advance. They may depend on several factors, including random elements. There may be complicated interrelationships between the various cash flows. These matters involve advanced topics in finance and actuarial mathematics and for the most part will not be dealt with in this book. In Part I we deal mainly with a simplified model, where all cash flows are fixed and known in advance. In later parts of the book we will consider certain aspects of randomness, but will not get into the full extent of complications that can arise.

2.2 An analogy with currencies

To motivate the basic ideas, we will consider first a completely different problem, which is nonetheless related to that introduced above. Suppose that I give you 300 Canadian dollars, 200 US dollars and 100 Australian dollars. How much money did I give you? It would be naive indeed to claim that you received 600 dollars, for clearly the currencies are of different value. To answer the question we will need conversion factors that allow us to deduce the value of each type of dollar in terms of others. Let $v(c, u)$ denote the value in Canadian dollars of one US dollar. Assume that $v(c, u) = 1.05$, which means that a US dollar is worth 1.05 Canadian dollars. (Our numbers here are for purposes of illustration only. They are close to the conversion rates at the time of writing, but they may well have changed considerably by the time you are reading this.) Similarly, letting a stand for Australian, we will assume that $v(c, a)$ equals 0.95, which means 95 cents Canadian will buy one Australian dollar. The convention we are using here, which should be noted for later use, is that the v function returns the value of *one* unit of the *second* coordinate currency in terms of the *first* coordinate currency.

There are four more conversion factors of interest, but an important fact is that they can all be deduced from just these two (or indeed from any two that have a common first or common second coordinate). We note first that if it takes 1.05 Canadian dollars to buy 1 US dollar, then a single Canadian dollar is worth $1/1.05 = 0.9524$ US dollars. That is,

$$v(u, c) = v(c, u)^{-1} = 0.9524, \qquad v(a, c) = v(c, a)^{-1} = 1.0526,$$

where we use similar reasoning for the Australian dollar.

Next consider $v(u, a)$. We want the amount of US dollars needed to buy one Australian dollar. We could conceivably effect this purchase in two stages, first using US money to buy Canadian, and then using Canadian to buy Australian. Working backwards, it will take 0.95 Canadian to buy 1 Australian, and it will take $v(u, c)$ 0.95 US dollars to buy the 0.95 Canadian. To summarize,

$$v(u, a) = v(u, c)v(c, a) = 0.9048.$$

Our calculations are completed with

$$v(a, u) = v(u, a)^{-1} = 1.1052.$$

The reader may notice, given a typical real-life listing of currency prices, that the relationships we state here do not hold exactly, but that is due to commissions and other charges. In the absence of these, they must necessarily hold.

Let us now return to the original problem of determining of how much I paid you. We must first select a currency to express the answer in. For example, we could say that the total was equivalent to $300 + 200v(c, u) + 100v(c, a) = 605$ Canadian dollars. We could also say that the total was equivalent to $300v(u, c) + 200 + 100(u, a) = 576.20$ US dollars. Notice as a shortcut, that we did not need to compute the latter sum (which could be a significant saving in calculation if we had several rather than just three currencies). If the total amount is equivalent to 605 Canadian, then it must also be equivalent to $605v(u, c) = 576.20$ US dollars. Similarly, the total in Australian dollars can be computed as $605v(a, c)$ or alternatively as $576.20v(a, u)$, both of which are equal to 637 (approximately as there are some rounding differences).

2.3 Discount functions

We now go back to the original situation. We want to value a sequence of cash flows, which are all in the same currency, but which are paid at different times. Conversion factors are needed to convert the value of money paid at one time to that paid at another. The principles involved are exactly the same as in Section 2.2. Let $v(s, t)$ denote the value at time s, of 1 unit paid at time t. (Note again that our convention is that the 1 unit goes with the second coordinate. In other words, 1 unit paid at time t is equivalent to $v(s, t)$ paid at time s.) In the case where $s < t$, you can interpret $v(s, t)$ as the amount you must invest at time s in order to accumulate 1 at time t. In the case where $s > t$ you can interpret $v(s, t)$ as the amount that you will have accumulated at time s from an investment of 1 at time t. The fundamental relationship that must be satisfied is the same as we noted with currencies, namely

$$v(s, t)v(t, u) = v(s, u), \quad \text{for all } s, t, u. \tag{2.1}$$

Due to its importance, we will repeat the reasoning for this fundamental fact in the current context. It is simply that 1 unit at time u is equivalent to $v(t, u)$ at time t, and this $v(t, u)$ at time t is equivalent to $v(s, t)v(t, u)$ at time s, showing that 1 unit at time u is indeed equivalent to $v(s, t)v(t, u)$ at time s.

We now make a formal definition.

Definition 2.1 A *discount function* is a positive valued function v, of two nonnegative variables, satisfying (2.1) for all values of s, t, u.

Other desirable features follow immediately from (2.1). Taking $s = t = u$, we deduce that $v(s, s)v(s, s) = v(s, s)$ and, since $v(s, s)$ is nonzero, we verify the obvious relationship

$$v(s, s) = 1, \quad \text{for all } s. \tag{2.2}$$

From this we deduce that $v(s, t)v(t, s) = v(s, s) = 1$ so that we recover the relationship, noted in the currency case, that

$$v(s, t) = v(t, s)^{-1}. \tag{2.3}$$

Although we have called v a 'discount' function, the common English usage of the word really applies to the case where $s < t$. In that case, $v(s, t)$ will be normally less than 1, and the function is returning the *discounted amount* of 1 unit paid at a later date. Some authors would prefer to define the discount function to apply only to this case and then define another function, called an *accumulation function*, to cover the case where s is greater than t. In that case, the function returns the accumulated amount from an investment of 1 unit at an earlier date, which will normally be greater than or equal to 1. We find it more convenient to use only the one function as given above, since the ideas, and the key relationship (2.1), are the same regardless of the ordering on s, t and u.

The concept of a discount function will be a key ingredient in what follows. We will suppose that, given any financial transaction, there is a suitable discount function that governs the investment of all funds. We will deal only briefly with the important problem of choosing the discount function. There are many factors governing this choice and it will depend on the nature of the transaction. It may be chosen to simply reflect the preferences of the parties for present as opposed to future consumption. It may reflect the desired return that an investor wishes to achieve. In many cases it is based on a prediction of market conditions that will determine what returns can be expected on invested capital.

A possible complication that we will not deal with to any great extent is that which arises when two parties to a transaction have different choices of a suitable discount function. We will assume, unless indicated otherwise, that the same function applies to both parties.

2.4 Calculating the discount function

We now provide a procedure for calculating the values of v in a systematic way. The currency example indicated that we can calculate all the values of a discount function just by knowing those at points with a fixed first coordinate, that is, with a common comparison point. In most applications it is convenient to take this as time 0. To simplify notation, we drop the first coordinate in this case and define

$$v(t) = v(0, t).$$

It follows from (2.1) that $v(s, t) = v(s, 0)v(0, t)$ and then from (2.3) that

$$v(s, t) = \frac{v(t)}{v(s)}. \tag{2.4}$$

In the first few chapters of this book, we will need to know the value of $v(s, t)$ only for integral values of s and t. From (2.4), it will be sufficient to know $v(n)$ where n is any nonnegative integer.

To calculate $v(n)$, note first that the key relationship (2.1) can be extended from one involving three terms to an arbitrary number. That is, given times $t_1, t_2, \ldots, t_n$,

$$v(t_1, t_2)v(t_2, t_3) \cdots v(t_{n-1}, t_n) = v(t_1, t_n). \tag{2.5}$$

To see this, take for example, $n = 4$. The quantity $v(t_1, t_2)v(t_2, t_3)v(t_3, t_4)$ is equal to $v(t_1, t_3)v(t_3, t_4)$ by applying (2.1) to the first two terms. By another application of (2.1) it is equal to $v(t_1, t_4)$. We have extended (2.1) to a formula involving four terms. A similar step extends from four to five, and so on. Formally, we are using mathematical induction.

It follows from (2.5) that

$$v(n) = v(0, 1)v(1, 2) \cdots v(n - 1, n), \tag{2.6}$$

so we need only know $v(n - 1, n)$ for all positive integers n. Given such values, we can use the recursion formula

$$v(n) = v(n - 1)v(n - 1, n), \qquad v(0) = 1, \tag{2.7}$$

to calculate all values of $v(n)$. The information we need is then summarized by the vector

$$\mathbf{v} = \big(v(0), v(1), v(2), \ldots, v(N)\big),$$

where N is the final duration at which a nonzero cash flow occurs.

2.5 Interest and discount rates

In practice, rather than specifying $v(k-1,k)$ directly, it is more common to deduce this quantity from the corresponding rates of interest or discount. Given any discount function v and a nonnegative integer k, these are defined as follows.

Definition 2.2 The *rate of interest* for the time interval k to $k+1$ is the quantity

$$i_k = v(k+1,k) - 1.$$

Definition 2.3 The *rate of discount* for the time interval k to $k+1$ is the quantity

$$d_k = 1 - v(k,k+1).$$

Note that an investment of 1 unit at time k will produce $v(k+1,k) = 1 + i_k$ units at time $k+1$. Similarly, an investment of $1 - d_k = v(k,k+1)$ units at time k will accumulate to 1 unit at time $k+1$. So d_k is the amount you must take off from each unit paid at the end of the period, to get the equivalent amount at the beginning of the period.

Given any of the three quantities $v(k,k+1)$, i_k or d_k we can easily obtain the other two. For example, using the definitions and (2.3), it is straightforward to deduce that

$$d_k = i_k v(k,k+1) = \frac{i_k}{1+i_k}, \quad i_k = d_k v(k+1,k) = \frac{d_k}{1-d_k}. \tag{2.8}$$

Remark In most cases we expect i_k will be a nonnegative number, although from the definition it could in theory take on any value greater than -1. Indeed, people who make investments yielding less than the rate of inflation are in effect receiving a negative interest rate. We will not be concerned with inflation in this text however, and the reader can generally assume that interest and discount rates are nonnegative, unless otherwise specified.

Remark The reader is cautioned that some other authors use a different convention for the subscript k in interest and discount rates. They would refer to our i_0 as i_1, since it is the interest rate for the *first* time interval, and in general would use i_{k+1} for our i_k. We find it more convenient to start all indexing at 0.

2.6 Constant interest

Readers who have previously studied compound interest will be familiar with one particular family of discount functions. Suppose we believe that accumulation of invested funds depends only on the length of time for which the capital is invested, and not on the particular starting time. That is, we postulate that for all nonnegative s, t, h,

$$v(s,s+h) = v(t,t+h). \tag{2.9}$$

If this holds, then

$$v(s+t) = v(0, s+t) = v(0, s)v(s, s+t) = v(0, s)v(0, t) = v(s)v(t).$$

We have a familiar functional equation, and it is well known that if we assume some regularity condition, for example, that v be continuous, we must have $v(t) = v^t$ and therefore that

$$v(s, t) = v^{t-s}$$

for some constant v.

For such a discount function, the rate of interest i_k is a constant $i = v^{-1} - 1$, and the rate of discount d_k is a constant $d = 1 - v$. The discount function is therefore conveniently given by simply stating a single parameter, which is usually taken as i, the constant rate of interest. So, for example, if we want to know how much we will accumulate at time n from an investment of 1 at time 0, this is just $v(n, 0) = (1 + i)^n$, which is the usual starting point for the subject of compound interest in elementary texts. In the pre-calculator, pre-computer age, this constant interest family of discount functions was almost always used, mainly to facilitate the computation. Many current textbooks on this subject still deal largely with this constant interest case. However, with modern computing methods, such as spreadsheets, the extra effort involved in using a general discount function is negligible, and there is no reason to restrict the flexibility that one can achieve. Throughout this book we will use arbitrary discount functions, although occasionally we will restrict discussion to the constant interest case in order to simplify the notation. One of the main advantages of not restricting ourselves to constant interest will be apparent in Chapters 4 and 5 where we will be able to incorporate the contingencies of life and death into the discount function. A key point in these chapters is that the calculation of premiums for life insurance and life annuities can be expressed as a special case of interest theory, using a general discount function.

2.7 Values and actuarial equivalence

We now put together the two key concepts of *cash flow vector* and *discount function*. Suppose we are given a cash flow vector $\mathbf{c} = (c_0, c_1, \ldots, c_N)$ and a discount function v. We want to calculate the single payment at time zero that is equivalent to all the cash flows, assuming that the time value of money is modeled by the given discount function v. This amount is commonly referred to as the *present value* of the sequence of cash flows, and sometimes abbreviated as P.V. We can think of it as the amount we would pay at time zero in order to receive all of the cash flows, or equivalently as the single payment that we would be willing to accept now in lieu of all these future cash flows. The cash flow at time k has a present value of $c_k\, v(k)$ by definition of the discount function. When $c_k > 0$ this is what we have to pay now in order to receive c_k at time k. When $c_k < 0$, receiving $-c_k v(k)$ now will let us pay out $-c_k$ at time k.

We then add up the individual present values to get

$$\text{Present value of all cash flows} = \sum_{k=0}^{N} c_k v(k). \tag{2.10}$$

Example 2.1 To make sure this is understood, take a very simple example. Let $v(k) = 2^{-k}$ for all k. This is a constant interest rate per period of 100 %. In other words, money doubles itself every period. This is of course not very realistic for a period of a year, but it could hold for a sufficiently long interval. Suppose we are to receive 12 units at time 2, but will be required to pay out 8 units at time 3. Find the present value, and verify that it makes sense.

Solution. From (2.10) the present value is $12v(2) - 8v(3) = 12(1/4) - 8(1/8) = 2$. To verify this, we note that the 12 units received at time 2 will accumulate to 24 at time 3. We then have to pay out 8, leaving an accumulation of 16 units by time 3. Contrast this with receiving instead a single payment of 2 at time 0. This will accumulate to 4 at time 1, 8 at time 2 and at 16 at time 3. Therefore assuming (as we do throughout) that all money accumulates according to the given discount function, we are in exactly the same position in both cases.

Remark Unrealistic interest rates will be frequently used in examples and exercises throughout the book, in order to simplify the numerical computation, and allow the reader to concentrate on the underlying concepts. So for example, we will often take $i_k = 20\%$, 25%, 50%, 100%, which correspond respectively to $v(k, k+1)$ equal to $5/6$, $4/5$, $2/3$, $1/2$. Calculation is even easier when we have an interest rate of 0, in which case $v(k) = 1$ for all k.

Note that it was convenient in the above example to compare the amounts accumulated at the time of the last payment. This is known as the *accumulated value* and in general is given by

$$\sum_{k=0}^{N} c_k v(N, k).$$

It represents the amount we will have at time N, resulting from all the cash flows.

More generally, we can calculate a value at *any time* between 0 and N. In this chapter we concentrate on integer times. We take the present value of the future cash flows at that time, plus the accumulated values of the past cash flows. The following definition formulates this precisely.

Definition 2.4 For any time $n = 0, 1, \dots, N$, the *value at time n* of the cash flow vector $\mathbf{c}$ with respect to the discount function v is given by

$$\mathrm{Val}_n(\mathbf{c}; v) = \sum_{k=0}^{N} c_k v(n, k).$$

It represents that single amount that we would accept at time n in place of all the other cash flows, assuming that everything accumulates according to the discount function v.

The values at various times are related in a simple way. Since $v(m, k) = v(m, n)v(n, k)$, it follows immediately that

$$\mathrm{Val}_m(\mathbf{c}; v) = \mathrm{Val}_n(\mathbf{c}; v)v(m, n). \tag{2.11‡}$$

(Formulas marked with ‡ denote key facts, of particular importance.)

Formula (2.11) should be intuitively clear. The single amount we would accept at time m in place of all the cash flows must be the value at time m of the single payment that we would accept at time n, in place of all the cash flows. We can then easily deduce all values from the value at a particular time. Normally it will be convenient to take this as time 0. The same point was illustrated in the currency example of Section 2.2, where we pointed out that the total amount in any one currency was easily converted to the total in any other by a single multiplication.

Notation For the particular case of values at time 0 we will use a special symbol. Let

$$\ddot{a}(\mathbf{c}; v) = \mathrm{Val}_0(\mathbf{c}; v).$$

The letter a is a standard actuarial symbol that is used to stand for *annuity*, another name for a sequence of periodic payments. See Section 2.14 for an explanation of the two dots.

When there is only one discount function under consideration we often suppress the v and just write $\mathrm{Val}_n(\mathbf{c})$ or $\ddot{a}(\mathbf{c})$.

We can express and calculate $\ddot{a}$ conveniently by expressing it in vector form:

$$\ddot{a}(\mathbf{c}) = \mathbf{v} \cdot \mathbf{c} = \mathbf{v}\mathbf{c}^{\mathrm{T}}. \tag{2.12}$$

The second term is the (scalar) inner product of the two vectors. The third views the vectors $\mathbf{v}$ and $\mathbf{c}$ as $1 \times N$ matrices, with the superscript T denoting a matrix transpose.

Formulas (2.11) and (2.12) make it clear that calculating values is a linear operation, a fact we will often exploit. That is,

$$\mathrm{Val}_k(\mathbf{c} + \mathbf{d}) = \mathrm{Val}_k(\mathbf{c}) + \mathrm{Val}_k(\mathbf{d}), \qquad \mathrm{Val}_k(\alpha\mathbf{c}) = \alpha\mathrm{Val}_k(\mathbf{c}), \tag{2.13}$$

for any cash flow vectors $\mathbf{c}$ and $\mathbf{d}$, scalar α and duration k.

We have been comparing the values of a sequence of cash flows with a single payment at a particular time. We often wish to compare the values of two sequences of cash flows. For this we have the following definition.

Definition 2.5 Two cash flow vectors $\mathbf{c}$ and $\mathbf{e}$ are said to be *actuarially equivalent* with respect to the discount function v if, for some nonnegative integer n,

$$\mathrm{Val}_n(\mathbf{c}; v) = \mathrm{Val}_n(\mathbf{e}; v).$$

From (2.11), if the above holds for some n, it holds for all n.

Take, for example, $n = N$. We see that a person taking the payments given by $\mathbf{c}$ and letting them accumulate according to the given discount function v will eventually be in exactly the same financial position as one taking the payments given by $\mathbf{e}$. This is the meaning of actuarial equivalence.

Many problems in actuarial mathematics reduce to the following. We are given a cash flow vector $\mathbf{c}$, and another cash flow vector $\mathbf{e}$ that depends on some unknown parameters. We have to solve for the unknown parameters in order to make the two vectors actuarially equivalent.

We have already done a simple example of this in calculating $\text{Val}_n(\mathbf{c})$. In that case we wanted a single payment at time n that is actuarially equivalent to $\mathbf{c}$. In another common application, a lender advances payments to a borrower, and the borrower must return the money by loan repayments. The lender is therefore trading one sequence of payments (the advances) for another (the repayments) and wants the two to be actuarially equivalent. We look at a simple example.

Example 2.2 A lends B 20 units now and another 10 units at time 1. B promises to repay the loan by two payments, made at times 2 and time 3. The repayment at time 3 is to be twice as much as that at time 2. If A wishes to earn interest of 25% per period, what should these repayments be?

Solution. Let K be the unknown payment at time 2. We want to find K so that the vectors $\mathbf{c} = (20, 10, 0, 0)$ and $\mathbf{e} = (0, 0, 2K, 3K)$ are actuarially equivalent. There are many possible calculation methods. We could determine the vector $\mathbf{v}$ and use formula (2.12), which is essentially the best approach for a spreadsheet method, as we describe later in Section 2.14. For small problems to be done by hand calculation, it is convenient to make use of a time diagram, where we indicate the payments and the 1 year discount factors $v(k, k+1)$ which are all equal to $(1.25)^{-1} = 0.8$. See Figure 2.1. Readers may find to useful to write down their own time diagrams for the examples in the book, if they are not given.

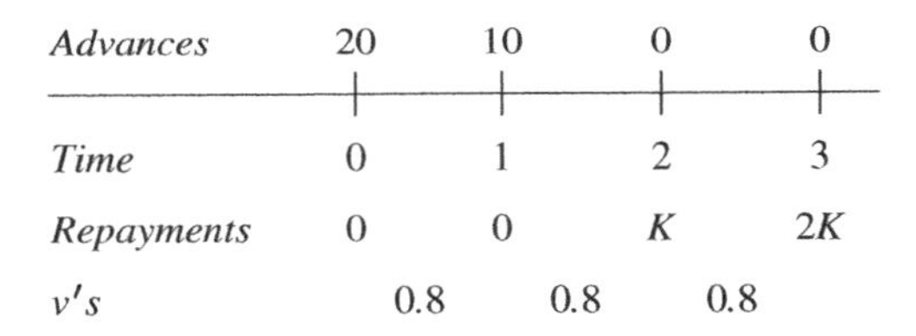

Figure 2.1 Example 2.2

Making use of formula (2.6), we calculate values at time 0 by multiplying each payment c_k by $v(k)$, which we calculate as the product of all the preceding discount factors $v(i, i+1)$. So,

$$\text{P.V. of advances} = 20 + 10\,(0.8) = 28$$
$$\text{P.V. of repayments} = K[(0.8^2 + 2(0.8^3)] = 1.664K.$$

Equating values to make the advances and repayments actuarially equivalent, $K = 28/1.664 = 16.83$. The borrower pays 16.83 at time 2 and 33.66 at time 3.

We conclude this section by describing a useful technique that we will call the *replacement principle*. Suppose we are given a cash flow vector and some subset of the entries $(0, 1, \ldots, N)$. Take the value at time k of just those cash flows in the subset and then replace all entries in the subset by a single payment at time k equal to that value. This leaves a vector that is actuarially equivalent to the original. A formal derivation can be given by writing the vector as the sum of two vectors and using linearity. We will leave the details to the interested reader.

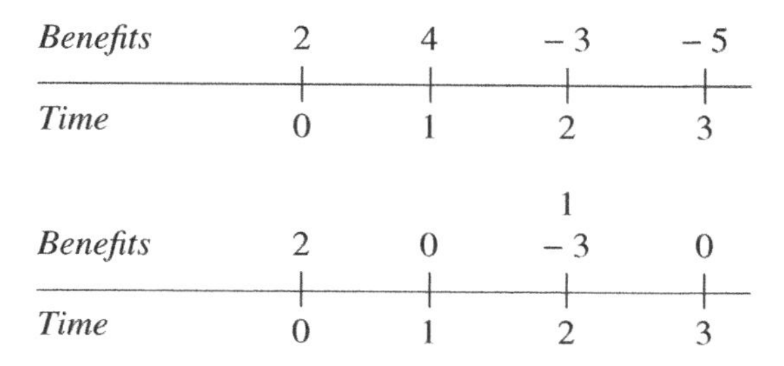

Figure 2.2 Example 2.3

The following is a simple, although artificial example. We will give more relevant uses of this principle later. See Figure 2.2.

Example 2.3 Let

$$\mathbf{c} = (2, 4, -3, -5).$$

Assume a constant interest rate of 0.25. Find the actuarially equivalent vector by applying the replacement principle with time $k = 2$ and the subset $\{1,3\}$.

Solution. The value of c_1 and c_3 at time 2 is $4(1.25) - 5(1.25)^{-1} = 1$. Making the replacement, we obtain the vector $(2, 0, -2, 0)$ that is actuarially equivalent to $\mathbf{c}$, as can be verified by direct calculation.

2.8 Vector notation

We now introduce some convenient notation for vectors. In all examples, we will have a maximum duration N, and all of our vectors will be $(N + 1)$-dimensional, with entries indexed from 0 to N. However, we will often write a vector of lower dimension with the understanding that all subsequent entries will be zero. For example if $N = 8$, then (1, 3, 2) will denote the vector (1, 3, 2, 0, 0, 0, 0, 0, 0). For vectors $\mathbf{c}$ and $\mathbf{d}$, written as above, we will write ($\mathbf{c}$, $\mathbf{d}$) to denote the vector consisting of the entries of $\mathbf{c}$ followed by those of $\mathbf{d}$. For example if $\mathbf{c} = (2, 2)$, and $\mathbf{d} = (3, 7, 4, 1)$ then $(\mathbf{c}, \mathbf{d}) = (2, 2, 3, 7, 4, 1, 0, \ldots, 0)$. (Note that the juxtaposition must come *before* filling in the ending zero entries.) For any number r and integer k we will write (r_k) to denote the vector consisting of k entries of r. For example, $(1_3, 2_5) = (1, 1, 1, 2, 2, 2, 2, 2, 0, \ldots, 0)$. We let $\mathbf{e}^i$ denote the standard ith basic vector, $i = 0, 1, 2, \ldots, N$, that is, a vector with an entry of 1 in position i and zeros elsewhere.

For a given vector $\mathbf{b} = (b_0, b_1, \ldots, b_n)$, let

$$\Delta\mathbf{b} = (b_0, b_1 - b_0, b_2 - b_1, \ldots, b_n - b_{n-1}, -b_n).$$

That is, $\Delta\mathbf{b}$ is obtained from $\mathbf{b}$ by subtracting from each entry the immediate preceding entry, except for the entry in position 0, which remains the same. (Think of the entry in position -1

as 0). Note also that by our notational convention, $b_j = 0$ for $j > n$, which accounts for the final entry of $-b_n$. As a check, the sum of all the entries in $\Delta\mathbf{b}$ must be 0. (The reader should also note that our definition differs from some authors' usage, whereby $\Delta\mathbf{b}(k)$ would equal $b_{k+1} - b_k$, rather than $b_k - b_{k-1}$ as we have defined it.)

We use the symbol * to denote pointwise multiplication of vectors, that is,

$$(a_1, a_2, \ldots, a_n) * (b_1, b_2, \ldots, b_n) = (a_1 b_1, a_2 b_2, \ldots, a_n b_n).$$

The remainder of this section can be omitted on first reading. We develop an identity which is somewhat technical but which will be very useful in later chapters.

For any vector $\mathbf{b} = (b_0, b_1, \ldots, b_n)$ and a discount function v, define a new vector $\nabla\mathbf{b}$ whose entry of index k is given by

$$\nabla b_k = b_k - v(k, k+1) b_{k+1}. \tag{2.14}$$

So ∇ is something like Δ except we subtract in the reverse order and discount the second term. Note that ∇ depends on v and it could be denoted as ∇_v if there is any confusion.

Our main result is that for any other vector $\mathbf{c} = (c_0, c_1, \ldots)$

$$\ddot{a}(\nabla\mathbf{b} * \mathbf{c}) = \ddot{a}(\mathbf{b} * \Delta\mathbf{c}). \tag{2.15}$$

To see this, we just note that the left hand side is

$$\big(b_0 - v(1)b_1\big)c_0 + \big(b_1 - v(1,2)b_2\big)c_1 v(1) + \big(b_2 - v(2,3)b(3)\big)c_2 v(2) + \cdots.$$

Expanding and using the fact that $v(k)v(k, k+1) = v(k+1)$, this equals

$$b_0 c_0 + b_1(c_1 - c_0)v(1) + b_2(c_2 - c_1)v(2) + \cdots$$

which is the right hand side. We can remember this formula as saying that when computing the present value of a pointwise product of one vector multiplied by Δ applied to a second vector, the Δ can slide off the second vector and become a ∇ on the first.

2.9 Regular pattern cash flows

Before the introduction of computers and calculators, finding values of cash flows could involve a lengthy computation. Accordingly, the emphasis was not only on constant interest rates, but also on vectors where all nonzero cash flows were of the same amount, usually referred to as the case of *level* cash flows, or where the cash flows followed some regular pattern, such as entries increasing in arithmetic progression. Values in this case were fairly easy to compute by algebraic means. One would then try to handle more general cash flow vectors by expressing them in terms of those with regular patterns. Readers who have had previous experience with compound interest courses have no doubt seen such techniques. With modern computing methods there is little need for these methods for computing purposes, although they are sometimes useful for theoretical matters. We give a brief illustration of some of the more useful formulas of this type. We assume throughout the section that the discount function is given by $v(n) = v^n$ for some constant v.

Consider the vectors (1_n) and $\mathbf{j}^n = (1, 2, \ldots, n-1, n)$. Then

$$\ddot{a}(1_n) = 1 + v + v^2 + \cdots + v^{n-1}.$$

Multiplying by v,

$$v\ddot{a}(1_n) = v + v^2 + \cdots + v^n.$$

Subtracting the second equation from the first and dividing by $1 - v$ gives

$$\ddot{a}(1_n) = \frac{1 - v^n}{1 - v}. \tag{2.16}$$

Many readers will have seen this technique for summing a geometric progression. A similar trick handles the vector $\mathbf{j}$ by reducing to the level payment case.

$$\ddot{a}(\mathbf{j}^n) = 1 + 2v + 3v^2 + \cdots + nv^{n-1}.$$

Multiplying by v,

$$v\ddot{a}(\mathbf{j}^n) = v + 2v^2 + \cdots + nv^n.$$

Subtracting the second equation from the first gives

$$(1 - v)\ddot{a}(\mathbf{j}^n) = (1 + v + v^2 + \cdots + v^{n-1}) - nv^n,$$

and dividing by $(1 - v)$,

$$\ddot{a}(\mathbf{j}^n) = \frac{\ddot{a}(1_n) - nv^n}{1 - v}. \tag{2.17}$$

There is an alternate way to derive (2.16) and (2.17), as well as many other similar formulas, which does not involve any series summation. It is based on the fact that a loan may be paid off by paying interest each period on the prior amounts advanced, and then paying off the total principal at the end. This is intuitively clear. A formal derivation will be given in Section 2.11.

Suppose you receive a loan of 1 unit. You could repay i at the end of each period for n periods and then eventually repay the principal at time n. The repayment vector is therefore $(0, i, i, \ldots, i + 1)$, which by the replacement principle is actuarially equivalent to $(d, d, \ldots, d, 1) = d(1_n) + \mathbf{e}^n$, since a cash flow of d at any time k is actuarially equivalent to $d(1 + i) = i$ at time $k + 1$. (In other words, if you pay interest at the beginning of the year, the appropriate rate is d rather than i.) Equating the present value of the advances and repayments, $1 = d\ddot{a}(1_n) + v^n$, which leads to (2.16).

For the second formula, suppose you are to receive loan advances of 1 unit at the beginning of each year for n years. According to our scheme you will pay 1 unit of interest at time 1,

2 units of interest at time 2, etc. You then will repay the total principal of n at time n. The repayment vector is $(0, i, 2i, \dots (n-1)i, ni+n)$. By replacing each i by a d one period earlier, this is actuarially equivalent to $d\mathbf{j}^n + n\mathbf{e}^n$. Equating present values of this with the vector of loan advances gives $\ddot{a}(1_n) = d\ddot{a}(\mathbf{j}^n) + nv^n$, which leads to (2.17).

2.10 Balances and reserves

2.10.1 Basic concepts

This section will introduce one of the most fundamental actuarial concepts, that of a reserve.

Suppose we enter into a financial transaction, represented by the cash flow vector $\mathbf{c}$. At any future time k there are two fundamental quantities to compute. First, we would like to know the total amount accumulated from all payments of the transaction up to this point. Second, we would like to know how much money we will *need* at that time in order to discharge our future obligations under the transaction. We illustrate with a simple example.

Example 2.4 Let

$$\mathbf{c} = (3, 6, 1, 2, -20), \quad v(0,1) = 0.6, \quad v(1,2) = 0.5, \quad v(2,3) = 0.4, \quad v(3,4) = 0.5.$$

How much do we have, and how much will we need at time 2, just *before* the 1-unit payment due at that time?

See Figure 2.3, where we have inserted an arrow to indicate the time that values are taken.

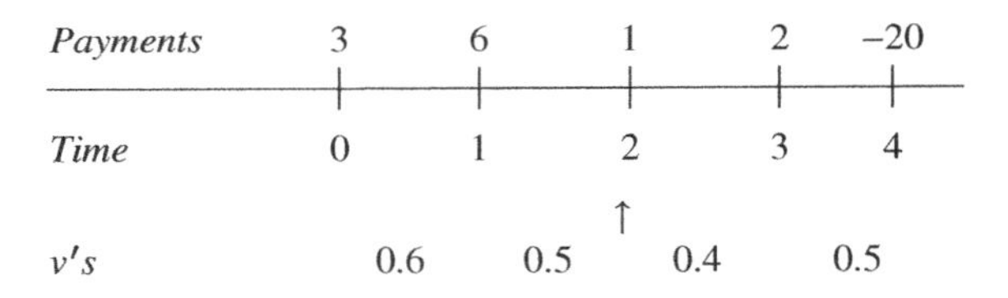

Figure 2.3 Example 2.4

Solution. The amount we have is clearly the 3 units paid to us at time 0, accumulated for two periods, and the 6 units paid at time 1 accumulated for one period, for a total amount of

$$3v(2,0) + 6v(2,1) = 3v(2,1)v(1,0) + 6v(2,1) = \frac{3}{0.6 \times 0.5} + \frac{6}{0.5} = 22.$$

For the second question, note first that we have an obligation to pay out 20 units at time 4, and the amount we need at time 2 to provide for this is $20v(2,4) = 20v(2,3)v(3,4) = 20 \times 0.4 \times 0.5 = 4$. We can offset this with the positive cash flows that we will acquire after time 2. The value at time 2 of the 1 unit due immediately at time 2 is just 1, and the value at time 2 of the 2 units payable at time 3 is just $2v(2,3) = 0.8$. The total needed to ensure we can meet our obligations is $4 - 1 - 0.8 = 2.2$.

Let us verify this directly. Suppose we have 2.2 units at time 2, just before the payment due at that time. The payment of 1 will come in to give us 3.2. This will accumulate to

$(3.2/0.4) = 8$ units at time 3. We will receive another 2 units at time 2 for a total of 10, and this will accumulate to the 20 units that we need at time 4 in order to meet our obligation at that time.

We now introduce some notation and terminology to express these concepts in a general case.

Notation Given any cash flow vector $\mathbf{c}$ and nonnegative integer k, let

$$_k\mathbf{c} = (c_0, c_1, \ldots, c_{k-1}, 0, \ldots, 0), \qquad ^k\mathbf{c} = (0, 0, \ldots, 0, c_k, c_{k+1}, \ldots, c_N)$$

so that

$$\mathbf{c} = {}_k\mathbf{c} + {}^k\mathbf{c}. \tag{2.18}$$

For example, for the vector $\mathbf{c}$ in Example 2.4, ${}_2\mathbf{c} = (3, 6, 0, 0, 0)$ while ${}^2\mathbf{c} = (0, 0, 1, 2, -20)$.

The idea is that ${}_k\mathbf{c}$ represents the *past* cash flows and ${}^k\mathbf{c}$ the *future* cash flows when measured from time k. It is important to note that the payment at exact time k is by our convention taken as future. (We will see in Chapter 6 that this fits in with the usual treatment for life insurance contracts.) Note also that ${}_0\mathbf{c}$ is the zero vector while ${}^0\mathbf{c}$ is just $\mathbf{c}$.

Definition 2.6 For $k = 0, 1, \ldots, N$, the *balance* at time k with respect to $\mathbf{c}$ and v, is defined by

$$B_k(\mathbf{c}; v) = \mathrm{Val}_k({}_k\mathbf{c}; v) = \sum_{j=0}^{k-1} c_j v(k, j).$$

(We will, as before, suppress the v when there is no confusion.) The balance at time k is simply the accumulated amount at time k resulting from all the payments received up to that time, and answers the question of how much we will have. Note again that by our conventional treatment, balances are computed just before the payment at exact time k is made, so that the time k payment is *not* included in B_k.

Definition 2.7 For $k = 0, 1, \ldots, N$, the *reserve* at time k with respect to $\mathbf{c}$ and v is defined by

$${}_kV(\mathbf{c}; v) = -\mathrm{Val}_k({}^k\mathbf{c}; v) = -\sum_{j=k}^{N} c_j v(k, j).$$

The reserve at time k is the *negative* of the value at time k of the future payments, and is equal to the amount we will need in order to meet future obligations. It is important to remember the negative sign, to reflect the fact that we are measuring the value of net obligations, that is, amounts to be paid out of the fund.

Referring again to Example 2.5, we calculated that $B_2(\mathbf{c}; v) = 22$ and ${}_2V(\mathbf{c}; v) = 2.2$.

The terminology and notation for reserves were borrowed from a particular life insurance application (which we will discuss in detail in Chapter 6). In keeping with this standard reserve notation, we have placed duration as a left rather than right subscript. We introduce the reserve concept now, in order to stress that its use is more general than just this particular life insurance application and can apply to any sequence of cash flows. One can think of this quantity as representing capital that must be set aside or 'reserved' for future use. For those familiar with accounting terminology, reserves, if positive, represent *liabilities,* which are amounts that we owe to other parties.

It is possible of course for the reserve to be negative, which indicates that at time k the present value of the future amounts coming in for this transaction will exceed the present value of the future amounts to be paid out. In such a case, the absolute value of the reserve is a 'receivable', that is, an amount that is owed to us and that will be received through future payments.

2.10.2 Relation between balances and reserves

From (2.16) and the linearity of Val_k,

$$\text{Val}_k(\mathbf{c}) = B_k(\mathbf{c}) - {}_kV(\mathbf{c}), \tag{2.19‡}$$

which says that the value of the transaction at any point is equal to the difference between what you actually have accumulated and what you need to set aside from that accumulation to meet future obligations. An important consequence of this is that

$$\text{for a } \textit{zero-value} \text{ cash flow vector } \mathbf{c}, \qquad B_k(\mathbf{c}) = {}_kV(\mathbf{c}), \tag{2.20‡}$$

where a zero-value vector is one whose value is zero at all durations. Such vectors arise frequently. If $\mathbf{c}$ and $\mathbf{d}$ are actuarially equivalent, then for any duration k,

$$\text{Val}_k(\mathbf{c} - \mathbf{d}) = \text{Val}_k(\mathbf{c}) - \text{Val}_k(\mathbf{d}) = 0,$$

so that $\mathbf{c} - \mathbf{d}$ is a zero-value vector.

For a typical example, suppose that an individual borrows money, and agrees to repay with a sequence of repayments. Consider the net cash flows from the *borrower's* point of view, that is, advances less repayments. If the advances and repayments are actuarially equivalent, these net cash flows form a zero-value vector. What is the outstanding balance on the loan at time k? This is the value of all the payments yet to be made, which by definition is just ${}_kV(\mathbf{c})$. In view of (2.18) it is also equal to $B_k(\mathbf{c})$. This is perfectly logical. It says that the amount still owing at time k is the value of all the money that been received, offset by the value of the repayments that have been made. Calculating the outstanding balance by means of the reserve is often called the *prospective method* since it looks to the future, while calculating by means of the balance is often called the *retrospective method,* since it looks to the past. For this reason, many authors refer to what we have called a balance as a *retrospective reserve.*

Example 2.5 (Figure 2.4). An individual borrows 1000 now and another 2000 at the end of 1 year. This loan will be repaid by yearly payments for 10 years, beginning 5 years from the present. The yearly payment doubles after 5 years. The interest rate is 0.06 for the first

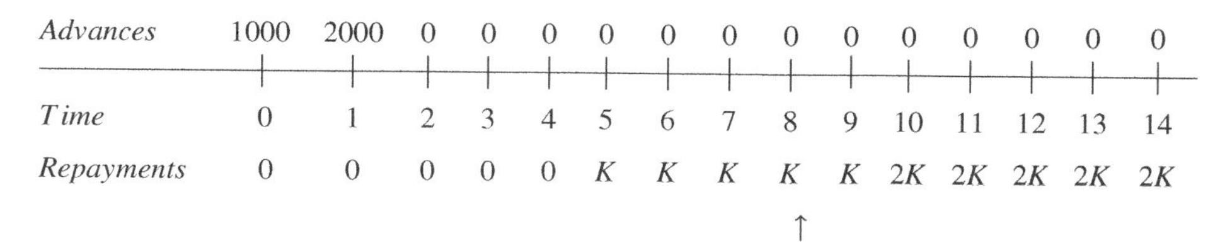

Figure 2.4 Example 2.5

5 years and 0.07 thereafter. How much is still owing 8 years from now, prior to the payment at that time? How much is still owing after the payment at time 8?

Solution. We must first determine the repayment amounts. Let K be the initial repayment. If $\mathbf{c}$ denotes the advances and $\mathbf{d}$ denotes repayments, then

$$\mathbf{c} = (1000, 2000),$$
$$\mathbf{d} = (0, 0, 0, 0, 0, K, K, K, K, K, 2K, 2K, 2K, 2K, 2K) = K(0_5, 1_5, 2_5),$$
$$\ddot{a}(\mathbf{c}) = 2886.79, \qquad \ddot{a}(\mathbf{d}) = 7.953\,26K,$$

and equating, $K = 362.97$ (see Section 2.15 for the calculation details). So the first five instalments are each 362.97, and the second five are each 725.94. The amount owing at time 8, just prior to the payment due on that date, can be calculated either as

$$B_8(\mathbf{c} - \mathbf{d}) = \mathrm{Val}_8(\mathbf{e}), \quad \text{where } \mathbf{e} = (1000, 2000, 0_3, -362.97_3)$$

or as

$${}_8V(\mathbf{c} - \mathbf{d}) = \mathrm{Val}_8(\mathbf{f}), \quad \text{where } \mathbf{f} = (0_8, 362.97_2, 725.94_5).$$

In either case the amount is 3483.97. The outstanding balance at time 8, *after* the payment due at that time, is just $3483.97 - 362.97 = 3121.00$.

It is of interest to note that the lower payments at the beginning have the effect that the borrower is not repaying enough to handle the interest on the loan, so the amount owing is greater than the total amount borrowed.

2.10.3 Prospective versus retrospective methods

For the model that we have described so far, either the prospective or retrospective approach can be used to determine the amount owing on a loan. In pre-computer days, one often chose the one that allowed easier calculation, but with modern methods it makes little difference.

There are, however, situations where one approach or the other dominates. The retrospective approach must be used in cases where the time and amounts of loan repayments are not scheduled in advance, but can be made at the option of the borrower.

Consider a variation on Example 2.5. Suppose the borrower had made payments of 300 at time 5, 400 at time 6 and 500 at time 7, and the payments after that were not yet determined. To obtain the outstanding balance at time 8 we would necessarily use the retrospective method and calculate the balance. This is $\text{Val}_8(\mathbf{b})$, where $\mathbf{b} = (1000, 2000, 0, 0, 0, -300, -400, -500, 0)$.

On the other hand, the prospective approach figures prominently in cases where the choice of an appropriate discount function may have changed since the transaction was originally entered into. In that case the reserve calculated with respect to a *new* discount function that is realistic at that time, could well give an amount that differs from that given by the balance as calculated retrospectively. After all, the equality of these two depended on the fact that the same discount function was used for both. We will not dwell too much on this complication, but would like to describe two very familiar situations.

Suppose the borrower wishes to discharge a loan at some point prior to the scheduled completion. The natural amount to pay would be the outstanding balance on the loan at that date, as calculated by either method. Suppose, however, that the loan contract requires that payments be made as scheduled and does not permit you to repay early. This is often the case with home mortgages. Suppose, further, that interest rates have gone down since the loan was originally taken out. The lender will not welcome early repayment as these amounts will have to be reinvested at a lower rate. Borrowers may however be permitted to repay, if they include an extra amount as a penalty. What is really happening is that the lender is revaluing the reserve, by applying the new discount function to the future repayments. If interest rates decrease, the value of the function v increases. Assuming the normal case where entries in the vector ${}^k\mathbf{c}$ are nonpositive as they represent repayments, the negative of these values are nonnegative, and multiplying by the higher values of v and summing will lead to a higher reserve. The penalty represents the excess value of the reserve as calculated by an up-to-date discount function, over the original value. (It should be noted that, in practice, the penalty amounts are often determined by approximate formulas rather than an exact recalculation of the reserve.)

As a second example, suppose you purchase a bond, which means that you are in the position of the lender. The nature of a bond is that this debt is assignable to another party by trading in the bond market. You sell the bond to someone who will then collect the repayments by the issuer. What is a fair price? The answer is obviously the outstanding balance, not as calculated originally, but rather with respect to the rates of interest applicable at the time of sale. As in the case above, lower interest rates will cause this outstanding balance to increase. Traders who buy bonds always hope that interest rates will fall, so the outstanding balances increase, and they can sell the bonds for a profit in the market.

2.10.4 Recursion formulas

If f is a function defined on the nonnegative integers, it is often useful to derive a recursion formula that expresses $f(k+1)$ in terms of $f(k)$. Given an initial value $f(0)$, one can then successively calculate values of $f(k)$ for all k. This is known as a *recursion* formula. A recursion formula often leads to a *difference formula* that gives an expression for $f(k+1) - f(k)$.

There are many recursion formulas in actuarial mathematics. For the most part, they can be derived from the basic recursions for balances and reserves that we present in this section.

We will first express B_{k+1} in terms of B_k. For any cash flow vector $\mathbf{c}$ and duration k, it is clear from the definitions that

$$ {}_{k+1}\mathbf{c} = {}_k\mathbf{c} + c_k\,\mathbf{e}^k. $$

The value at time k of the second term on the right is c_k and the value at time k of the first term on the right is $B_k(\mathbf{c})$. Invoking linearity and multiplying by $v(k+1,k)$ to get the value at time $k+1$, gives the recursion

$$B_{k+1}(\mathbf{c}) = (B_k(\mathbf{c}) + c_k)\, v(k+1,k), \tag{2.21}$$

where we start the recursion with the initial value, $B_k = 0$.

Subtracting B_k from each side and expressing $v(k+1,k)$ as $1+i_k$ yields the difference formula

$$B_{k+1}(\mathbf{c}) - B_k(\mathbf{c}) = i_k(B_k(\mathbf{c}) + c_k) + c_k. \tag{2.22}$$

Difference equations are not normally efficient for calculating numbers. Rather, they are useful for analyzing how quantities change from one period to another. For example, the above formula expresses the increase (or decrease if negative) in the balance over a period as a sum of two quantities, the interest earned plus the additional cash flow.

Recall that our convention was to treat cash flows at time k as future with respect to k. This ties in well with insurance contracts, as we will see later. However, the other convention is normally used when dealing with loans. Accordingly, we define

$$\tilde{B}_k(\mathbf{c}) = B_k(\mathbf{c}) + c_k,$$

the accumulated amount at time k *after* the cash flow at time k is paid. We leave it to the reader to verify the corresponding recursion and difference equations as

$$\tilde{B}_{k+1}(\mathbf{c}) = \tilde{B}_k(\mathbf{c})v(k+1,k) + c_{k+1}, \tag{2.23}$$

$$\tilde{B}_{k+1}(\mathbf{c}) - \tilde{B}_k(\mathbf{c}) = i_k\tilde{B}_k(\mathbf{c}) + c_{k+1}. \tag{2.24}$$

Rewriting (2.22), we get

$$-c_{k+1} = i_k\tilde{B}_k(\mathbf{c}) + [\tilde{B}_k(\mathbf{c}) - \tilde{B}_{k+1}(\mathbf{c})], \tag{2.25}$$

which forms the basis for the usual loan amortization schedules. (These are schedules that show the balances at each duration and how they change.) The left hand side above is the amount repaid on the loan at time $k+1$ (the negative of the negative cash flow) and the formula gives the split of this repayment into two parts. The first is the interest on the outstanding balance, and the second is the amount of principal reduction.

Let us now turn to reserves. They satisfy exactly the same recursion relation as balances. That is,

$$_{k+1}V(\mathbf{c}) = (_kV(\mathbf{c}) + c_k)\, v(k+1,k), \tag{2.26}‡$$

where we start the recursion with the initial value $_0V(\mathbf{c}) = -\ddot{a}(\mathbf{c})$. This is obvious in the case of zero-value vectors where reserves equal balances, but it holds in general, as seen for example from the relation, $^{k+1}c = {}^kc - c_k\mathbf{e}^k$, or alternatively derived from (2.19), substituting from (2.17) and using (2.11).

2.11 Time shifting and the splitting identity

In this section we provide a background for following some classical formulas that appear in the actuarial literature.

From (2.16), taking values at time 0 and applying (2.11), we derive

$$\ddot{a}(\mathbf{c}) = \ddot{a}({}_k\mathbf{c}) + v(k)\mathrm{Val}_k({}^k\mathbf{c}). \tag{2.27}$$

We call this the *splitting identity*, since it splits the calculation of a present value into two parts, considering first those cash flows before time k, and then those after time k. It can also be obtained as a special case of the replacement principle. Replace all the cash flows after time k by the single payment of $\mathrm{Val}_k({}^k\mathbf{c})$ at time k. In classical actuarial mathematics, the spitting identity was often used as a calculation tool when there was distinct change in the cash flow sequence or discount function occurring at time k. (It can still be useful in this regard in the continuous case discussed in Chapter 8.)

The remainder of this section deals mainly with notational issues, and could be deferred until we apply it later in Section 4.3.2. The splitting identity often appears in a somewhat different form, since the symbol Val_k is not standard. In traditional actuarial notation, it is common to express all formulas in terms of $\ddot{a}$ (or similar symbols which we introduce later). So the question is then, how do we write Val_k in terms of $\ddot{a}$? This is quite simple when payments and interest rates are constant. In general case we must introduce some new notation for *time shifting*.

Suppose we are at time k and we wish to consider this as the 'new' time 0. Given a cash flow vector $\mathbf{c}$, define the cash flow vector $\mathbf{c} \circ k$ by

$$(\mathbf{c} \circ k)_n = \mathbf{c}_{k+n}.$$

For example, if $\mathbf{c} = (2, 2, 3, 4, 5)$, then $\mathbf{c} \circ 2 = (3, 4, 5)$. In other words, $\mathbf{c} \circ k$ simply gives the cash flows in order, but starting with the one at time k. Another way of looking at it is that $\mathbf{c} \circ k$ is just ${}^k\mathbf{c}$ with the first k zeros removed. This notation was not needed with constant payment vectors, since when $\mathbf{c}$ is (1_n), then $\mathbf{c} \circ k = (1_{n-k})$.

Similarly, given a discount function v, define a new discount function $v \circ k$ by

$$v \circ k(n, m) = v(n + k, m + k)$$

so that

$$v \circ k(n) = v(k, k + n).$$

Once again, the idea is simply that we are treating time k as time 0 and measuring time from that point.

It is quite simple to calculate $v \circ k$ from the factors for periods of length 1 as we did in (2.6). We multiply as before, but start with $v(k, k + 1)$ rather than $v(0, 1)$.

An important point to notice is that, from (2.9)

$$v \circ k = v \quad \text{if interest is constant,}$$

which explains why this concept does not appear in classical works where interest is almost always taken as constant.

We can now write

$$\mathrm{Val}_k({}^k\mathbf{c}) = \sum_{i=k}^{N} v(k,i)c_i = \sum_{i=k}^{N} v \circ k(i-k)(\mathbf{c} \circ k)_{i-k} = \sum_{j=0}^{N-k} v \circ k(j)(\mathbf{c} \circ k)_j = \ddot{a}(c \circ k, v \circ k), \tag{2.28}$$

which is intuitively clear since we are starting at time k, treating it as a new time 0 and valuing the future payments. We can then write the splitting identity in its usual form

$$\ddot{a}(\mathbf{c}; v) = \ddot{a}({}_k\mathbf{c}; v) + v(k)\ddot{a}(\mathbf{c} \circ k; v \circ k). \tag{2.29}$$

Under constant interest, we can write this in an easier fashion as

$$\ddot{a}(\mathbf{c}) = \ddot{a}({}_k\mathbf{c}) + v^k\ddot{a}(\mathbf{c} \circ k). \tag{2.30}$$

The splitting identity can sometimes be useful in determining the effect of changes in certain quantities.

Example 2.6 Suppose that $\ddot{a}(\mathbf{c}) = 19.6$ and $\ddot{a}({}_{10}\mathbf{c}) = 10$. If $v(9, 10)$ increases from 0.8 to 0.81 while all other values of $v(k, k+1)$ remain the same what is the new value of $\ddot{a}(\mathbf{c})$?

Solution. From (2.29), with $k = 10$, we know that $v(0,9)(0.8)\ \ddot{a}(\mathbf{c} \circ 10, v \circ 10) = 9.6$, so $v(0,9)(0.81)\ \ddot{a}(\mathbf{c} \circ 10, v \circ 10) = 9.72$ and the new value of $\ddot{a}(\mathbf{c})$ is 19.72.

*2.11 Change of discount function

This section is somewhat technical. Its purpose is to derive a useful relationship that allows us to replace one discount function with another, which could be more convenient for the task at hand.

Suppose we are given a cash flow vector $\mathbf{c}$ and a discount function v. Let v' be another discount function. Define a new cash flow vector $\mathbf{c}'$ whose kth entry is given by

$$\mathbf{c}'_k = \mathbf{c}_k + [v'(k,k+1) - v(k,k+1)]B_{k+1}(\mathbf{c}; v). \tag{2.31}$$

It follows that

$$B_k(\mathbf{c}; v) = B_k(\mathbf{c}'; v'), \quad \text{for all } k. \tag{2.32}$$

We prove (2.30) inductively. To simplify the notation, denote $B_k(\mathbf{c}; v)$ by B_k, $B'_k(\mathbf{c}'; v')$ by B'_k, $v(k, k+1)$ by v and $v'(k, k+1)$ by v'. At $k = 0$, the balances are both equal to 0. We will assume that $B_k = B'_k$ and show that $B_{k+1} = B'_{k+1}$:

$$B_{k+1} = \frac{B_k + \mathbf{c}_k}{v} = \frac{B'_k + \mathbf{c}'_k}{v} + \frac{\mathbf{c}_k - \mathbf{c}'_k}{v} = B'_{k+1}\frac{v'}{v} + \frac{v - v'}{v}B_{k+1}.$$

Formula (2.30) for $k + 1$ follows after some minor algebraic manipulation.

A corollary of this is that if $\mathbf{c}$ has zero value with respect to v, then $\mathbf{c}'$ has zero value with respect to v'. This follows since a cash flow vector has zero value if and only if its balances eventually equal to 0.

We will later apply this result to a key idea in life insurance. In this chapter we will give a formal proof of the assertion used in Section 2.9.

Consider a loan for N periods made according to a discount function, which we will denote by v' to tie in with the notation above. We let d' and i' refer to the corresponding discount and interest rates. Suppose the loan advances are given by vector $\mathbf{c} = (c_0, c_1, \dots, c_{N-1})$. Let

$$s = \sum_{k=0}^{N-1} c_k, \qquad \mathbf{r} = \left(d'_0(c_0), d'_1(c_0 + c_1), d'_2(c_0 + c_1 + c_2), \dots, d'_{N-1}s\right).$$

We want to show that $\mathbf{c}$ is actuarially equivalent to $(\mathbf{r} + s\mathbf{e}^N)$ with respect to v'. This will mean that we can repay any loan by paying interest each year on the prior advances, and then repaying the principal at the end. While this may be intuitively clear, it is important to verify it formally to ensure that our model captures the idea of interest the way one normally perceives it. Let $v(k) = 1$ for all k, the discount function for a constant interest rate of zero. It follows that

$$B_k(\mathbf{c}, v) = \sum_{i=0}^{k-1} c_k$$

so that the vector $\mathbf{c}'$ as given by (2.31) is just $\mathbf{c} - \mathbf{r}$. Invoking (2.32), this gives

$$\mathrm{Val}_N(\mathbf{c} - \mathbf{r} - s\mathbf{e}^N; v') = B_N(\mathbf{c}'; v') - s = B_N(\mathbf{c}; v) - s = 0,$$

showing the desired actuarial equivalence.

2.12 Internal rates of return

In this section we take the point of view of an investor or lender, who is undergoing a transaction that involves investing funds, hoping to get back amounts that are greater in value than those put in. Let $\mathbf{c} = (c_0, c_1, \dots, c_N)$ denote the net cash flow vector of the transaction. We assume throughout that $c_N \neq 0$. We are not given a discount function but rather want to find the constant interest rate that will make $\mathbf{c}$ a zero-value vector. In this section it is convenient to deal with balances that include the payment due at the time the balance is computed. That is, we want to consider $\tilde{B}$ as defined above.

Definition 2.8 An *internal rate of return* (often abbreviated as *i.r.r.*) of the transaction is a number i in the interval $(-1, \infty)$ such that, for the discount function $v(n) = (1+i)^{-n}$,

(i) $\tilde{B}_N(\mathbf{c}; v) = 0$, and

(ii) $\tilde{B}_k(\mathbf{c}; v) \leq 0$ for $k = 0, 1, \ldots, N-1$.

An i.r.r. is sometimes referred to as a *yield rate*.

Remark The definition is unchanged if we use B in place of $\tilde{B}$, since $B_{k+1} = \tilde{B}_k(1+i_k)$.

Theorem 2.1 *If an i.r.r. of a transaction exists, it is unique.*

Proof. Suppose, to the contrary, that we have two such rates, i and i', with $i < i'$. Let $\tilde{B}_k$ and $\tilde{B}'_k$ denote the resulting balances. Suppose that c_j is the first nonzero entry in $\mathbf{c}$. We must have $j < N$ since $j = N$ would imply that $\tilde{B}_N = c_N \neq 0$. We will now show by induction that, for $k \geq j+1$, we have

$$\tilde{B}_k > \tilde{B}'_k. \tag{2.33}$$

If (2.33) holds for some index k then it will hold for $k+1$ since the nonpositivity of balances implies that

$$\tilde{B}_{k+1} = \tilde{B}_k(1+i) + c_{k+1} > \tilde{B}'_k(1+i') + c_{k+1} = \tilde{B}'_{k+1}.$$

Applying the above for $k = j$, and noting that $c_j = \tilde{B}_j = \tilde{B}'_j < 0$, verifies the initial step. So we have shown that $\tilde{B}_N > \tilde{B}'_N$, contradicting the fact that both rates satisfy (i). □

The reader familiar with previous literature on this subject may well be surprised at this theorem, since a great deal has been written on the non-uniqueness of the i.r.r. In our opinion this occurs from a faulty definition. The standard way of defining the i.r.r is to require only point (i) of Definition 2.8 and not (ii). The problem then reduces to finding the roots of the polynomial $\sum_{k=0}^{N} c_k(1+i)^{N-k}$, and there may indeed be several. However, when balances become positive, it means that the status of the individual has changed from that of a lender to that of a borrower. The transaction therefore involves both borrowing and lending, and it is difficult to give any interpretation to the roots. After all, a lender is seeking a high i.r.r. whereas a borrower wants it to be low. We feel that the best approach is to give the definition above and concede that other transactions just do not possess an i.r.r. under this definition. Other means are needed for their analysis.

Note however that no problem is presented by the usual type of transaction, whereby the loan advances, represented by negative entries, all precede the loan repayments, as given by positive entries. The balances will begin as negative. If they ever become positive, it could only be when the entries change to positive. But as the entries then remain positive, the balance could then never become 0. Therefore, in this case, if an interest rate i satisfies (i) it will necessarily be an i.r.r.

The remaining material in this section can be omitted as it is not used in the remainder of the book. To handle cases where the i.r.r. does not exist by the definition given above, a generalization of the definition was given by Teichroew *et al.* (1965a, 1965b), often referred to as the TRM method. Their approach was to recognize that when the status of a lender or an investor reverts to that of a borrower, as indicated by a positive balance, then that balance should accumulate at some rate r fixed in advance and not at the unknown, presumably higher rate that one is trying to solve for. The rate r could be, for example, the rate at which the individual could obtain financing. They then define a generalization of balances with respect to this rate r, building on the recursive formula (2.21). Let $\tilde{B}_0(\mathbf{c}; v; r) = c_0$ and define inductively

$$\tilde{B}_{k+1}(\mathbf{c}; v; r) = \begin{cases} \tilde{B}_k(\mathbf{c}; v; r)(1 + i_k) + c_{k+1}, & \text{if } \tilde{B}_k(\mathbf{c}; v; r) \leq 0, \\ \tilde{B}_k(\mathbf{c}; v, r)(1 + r) + c_{k+1}, & \text{if } \tilde{B}_k(\mathbf{c}; v; r) > 0. \end{cases}$$

Define the i.r.r. of $\mathbf{c}$ relative to r as the value of i for which $\tilde{B}_N(\mathbf{c}; v; r) = 0$ when $v(k) = (1 + i)^{-k}$. We will denote this quantity as i_r. Note that if i is an i.r.r. by Definition 2.8, then it will be an i.r.r. with respect to r for *any* r. In this case balances are never positive and $\tilde{B}_k(\mathbf{c}; v; r)$ will simply equal $\tilde{B}_k(\mathbf{c}; v)$.

The same type of inductive calculation as in the proof of Theorem 2.1 shows that as i increases, while k, r and a nonzero vector $\mathbf{c}$ are held fixed, the value of $\tilde{B}_k(\mathbf{c}; r; v)$ is strictly decreasing. There is therefore at most one i.r.r. relative to r. Moreover, there will always be one except when either $\tilde{B}_N(\mathbf{c}; v; r) < 0$ or $\tilde{B}_N(\mathbf{c}; v; r) > 0$ for all i in $(-1, \infty)$. Defining i_r to be -1 in the first case and ∞ in the second case, we can state the following theorem.

Theorem 2.2 *For any nonzero vector* $\mathbf{c}$ *and any* $r > -1$, *there is a unique value of* i_r *in the interval* $[0, \infty]$.

This shows that we do obtain a unique i.r.r. for any transaction after we first postulate the deposit rate r. This gives rise to a simple criterion to decide if it is worthwhile to enter into a transaction. The general rule is that the transaction is worthwhile if $i_r > r$, and not worthwhile if $i_r < r$.

*2.13 Forward prices and term structure

An interesting example of a discount function is furnished by the *forward prices* on *risk-free zero-coupon bonds*. A zero-coupon bond is a financial instrument that promises to pay a fixed sum at some future date, known as the maturity date, and which makes no intervening payments (coupons) before that time. By risk-free we mean something like a government bond, where we can safely assume that the bond is sure to be redeemed with no chance of default.

We first discuss the general idea of a *forward contract*. This is an agreement between two parties, whereby one party, the seller, agrees to deliver a certain specified asset at a specified future time to the other party, the buyer. At the time of delivery, the buyer, will remit a sum of money that is agreed upon at the time when the contract is entered into and is independent of the prevailing price at the time of exchange. The price agreed upon at the contract date is known as the *forward price* of the contract. The buyer is often referred to as taking a *long position* and the seller is referred to as taking *a short position*.

There are various reasons for such an arrangement. The buyer may be someone who needs the particular item at some future date, and wishes to lock in a price now, to protect themselves from a future increase in the price. Alternatively, the buyer may be a speculator with no need of the item, but who predicts that prices will rise, thereby giving them a profit as they can buy the asset at the delivery date and immediately sell it for a higher price. Similarly, the seller may be one who owns the asset, wishes to dispose of it at a future date, and wants to lock in the amount they will receive as protection against falling prices. Alternatively, the seller may be a speculator who predicts that prices will fall, and hopes to profit by buying the item at the delivery date for less than they have agreed to sell it for.

Let $\tilde{v}(0, t) = \tilde{v}(t)$ be the price at time 0 of a 1-unit, zero-coupon, risk-free bond maturing at time t. For any $s \leq t$, let $\tilde{v}(s, t)$ be the forward price for a 1-unit zero-coupon bond, maturing at time t, where the delivery date is time s. Now what will these forward prices be? It turns out that under a certain natural assumption they are determined from the time zero prices by the rule

$$\tilde{v}(s, t) = \tilde{v}(t)/\tilde{v}(s) \tag{2.34}$$

which implies immediately that $\tilde{v}$ is a discount function. (Note that in the notation of Section 2.11 forward prices as determined from time s are given by the time shifted discount function $\tilde{v} \circ s$.)

The assumption made is the *no arbitrage* hypothesis, which is a major concept in modern day financial economics. An *arbitrage* opportunity is one where a party can make a sure profit by buying and selling certain financial assets with no risk of a loss. The hypothesis in question says that in an environment when all parties have perfect information, arbitrage opportunities cannot persist. This is a simple result of supply and demand laws in economics. The arbitrage opportunities occur when some assets are overvalued and others are undervalued. The argument is that if such an opportunity should ever rise, individuals will, in an attempt to take advantage of this, rush to buy the undervalued assets, thereby raising their prices, and rush to sell the overvalued assets, thereby lowering their prices, which restores equilibrium and eliminates the arbitrage.

We also make some other idealized assumptions. One is that an arbitrary number of units of a bond can be bought or sold at any time. This includes the possibility of *short selling* whereby an individual can sell an asset that they do not own by borrowing the asset from another party. They plan to acquire the asset at a later date (hopefully at a lower price than they sold it for) for return to the lender. Another assumption, is that buying or selling does not involve any transaction costs such as commissions.

To verify our claim above we will show by an example, that in the absence of Equation (2.34) an arbitrage opportunity would arise. Suppose, for example, that $\tilde{v}(10) = 0.75$ and $\tilde{v}(5) = 0.90$ but $\tilde{v}(5, 10) = 0.8 < \tilde{v}(10)/\tilde{v}(5)$. An individual could sell a 1-unit bond maturing at time 10, receiving 0.75 and then (i) use the proceeds to buy 5/6 of a 1-unit bond maturing at time 5, and (ii) take a long position on a forward contract for a 1-unit bond maturing at time 10 with a delivery date of time 5. At time 5, the individual receives 5/6 from the maturing 5-year bond, uses 4/5 of that to settle the forward contract, and now owns a 1-unit bond maturing at time 10, which they use to settle the short sale. A sure profit of 5/6 − 4/5 is made at time 5.

The reader is invited to find a corresponding example of an arbitrage opportunity in the case that $\tilde{v}(s, t) > \tilde{v}(t)/\tilde{v}(s)$ and then to provide a general proof.

The analysis above indicates that for an individual or corporation whose investment environment consists of risk-free zero-coupon bonds, the forward prices constitute a reasonable choice of discount function. For example, the individual could ensure (under our assumptions) that a payment received at time s would accumulate to $\tilde{v}(s, t)$ at a subsequent time t by entering into a suitable forward contract.

We conclude this section by introducing some conventional terminology. We have given the above analysis in terms of prices, which seems the most convenient way, but typically the economic and finance literature refers to rates instead of prices. Assume we are given prices of risk-free zero-coupon bonds for all maturities.

Definition 2.9 The *spot rate of interest* y_t is the yield rate for the bond maturing at time t. That is

$$\tilde{v}(t) = (1 + y_t)^{-t}$$

so that

$$y_t = \tilde{v}(t)^{-1/t} - 1. \tag{2.35}$$

Definition 2.10 For $s \leq t$, the *forward rate of interest* $f(s, t)$ is the yield rate earned for the bond maturing at time t and acquired on a forward contract with delivery at time s. That is

$$\tilde{v}(s, t) = [1 + f(s, t)]^{s-t}$$

so that from (2.34),

$$f(s, t) = \left(\frac{\tilde{v}(t)}{\tilde{v}(s)}\right)^{1/(s-t)} - 1. \tag{2.36}$$

The relationship between the various maturity dates and the spot rates is known as the *term structure of interest rates*. A graph which shows values of y_t for various values of t is known as a *yield curve*. Examples of yield curves can be found in the financial and business sections of many daily newspapers, as well as being freely available online. The study of the various possible shapes of yield curves is an important topic for economists which is beyond the scope of this text.

As a final word, note that although we have discussed spot and forward rates in the context of risk-free bonds, they can be defined, using formulas (2.35) and (2.36) for a general discount function.

Example 2.7 Given the spot rates $y_2 = 0.05$ and $y_5 = 0.07$, find the forward rate $f(2, 5)$.

Solution.

$$f(2, 5) = \left(\frac{1.07^{-5}}{1.05^{-2}}\right)^{-1/3} - 1 = .084.$$

2.14 Standard notation and terminology

There is a system of standard international actuarial notation that can seem quite complex at first. One goal in this book is to simplify notation as much as possible. In order to read this book, it will not be necessary to learn any of this system other than that which is introduced in the text. However, the student who wishes to read other actuarial literature, or write actuarial examinations, or pursue an actuarial career, will be expected to be familiar with the full extent of the system. Accordingly, we will, at the end of each relevant chapter, review the standard notation, indicating how it ties in with the notation in the main text.

2.14.1 Standard notation for cash flows discounted with interest

The vector notation is peculiar to this book. In the standard notation the same basic symbol a is used for the present value of a cash flow stream, but the particular vector of cash flows is indicated by embellishments of this symbol. A prime example is

$$a_{\overline{n}|} = \ddot{a}(0, 1_n).$$

A few comments are in order. The 'angle' around the n is intended to signify a duration of time, as opposed to an age (which will be encountered in the chapters on life annuities and insurances). The vector above arises frequently. It is usual for a loan contract to stipulate that the first repayment is made *one period after* receiving the loan. One does not usually make a repayment as of the loan date, since that would in effect just mean you were getting a smaller loan. Therefore, the simplest unadorned symbol was reserved for the vector with first entry equal to 0. When there is a payment at time zero, standard notation denotes this by placing two dots above the a. That is,

$$\ddot{a}_{\overline{n}|} = \ddot{a}(1_n).$$

The former case is termed an *immediate* annuity, and the latter a *due* annuity. The terminology is a bit unusual since the immediate annuity does not start immediately but after one period. The name arose as a contrast to the general *deferred annuity* where there are several zero entries at the beginning. Once we move away from level payments, the distinction between due and immediate annuities no longer applies. From our point of view, there is *always* a payment at time zero although it may be of zero amount. To avoid conflicts with standard notation we use the two dots as a general symbol.

Another traditional way of looking at the difference between due and immediate was that the former applied when payments were made at the beginning of the year, and the latter when payments were made at the end of the year. Once again, these distinctions no longer apply when payments are not level, since the beginning of 1 year is the end of the preceding one. (We will, however, have need to refer to this point again, when we discuss fractional period payments in Chapter 7.)

Some other symbols are

$$Ia_{\overline{n}|} = \ddot{a}(0, 1, 2, \ldots, n, 0, \ldots, 0),$$
$$I\ddot{a}_{\overline{n}|} = \ddot{a}(1, 2, \ldots, n, 0, \ldots, 0),$$

using the vector that we denoted by $\mathbf{j}^n$ in Section 2.9,

$$Da_{\overline{n}|} = \ddot{a}(0, n, (n-1), (n-2), \ldots, 1, 0, \ldots, 0),$$
$$D\ddot{a}_{\overline{n}|} = \ddot{a}(n, (n-1), (n-2), \ldots 1, 0, \ldots, 0).$$

These are simple enough to remember from the fact that I indicates an *increasing* sequence of payments and D indicates a *decreasing* sequence.

The symbol s is used to indicate an accumulated value. So, for example,

$$s_{\overline{n}|} = \mathrm{Val}_n(0, 1_n) = v(n, 0)a_{\overline{n}|},$$
$$\ddot{s}_{\overline{n}|} = \mathrm{Val}_n(1_n) = v(n, 0)\ddot{a}_{\overline{n}|}.$$

Note that in both cases the values are at time n. For the unadorned symbol this is at the date of the last payment. For the double-dotted symbol, time n is *one period* after the last payment, since the first payment was at time 0.

2.14.2 New notation

A reader who has already spend a great deal of effort in mastering the standard notation may have felt some dismay in having to learn yet more notation introduced in this chapter. We feel, however, that the devices introduced here are useful. There are three main innovations: the vector notation; the symbol Val; and and the time-shifting notation. We consider each in turn.

The vector notation allows us to conveniently refer to an arbitrary sequence of cash flows in a systematic way. The traditional notation uses special embellishments for each particular sequence, and even so, is restricted to a small number of cases, such as payments constant over some period, or payments in arithmetic progression.

The system of writing vectors is useful since it tells you exactly how to enter a cash flow vector into a column of a spreadsheet. For example, faced with the vector $(1_{20}, 2_{10})$, one simply enters a 1 in the first column, copies it down for 20 rows, then enters a 2, and copies it down for 10 rows.

The standard notation allows you to specify values *only* at the beginning or end of a transaction, and it is often useful to denote values at an intermediate point, through the use of Val_k.

Having introduced the vector notation, it is convenient to introduce the lower and upper subscripts to denote past and future, respectively. Such a device was not needed in the classical literature that dealt with level cash flows and interest rates. For example, the vector ${}_k1_n$ is just 1_{n-k}. Similarly, a symbol like our $\circ$ for time shifting could easily be avoided since, as noted, $v \circ k = v$. Moreover, $1_k \circ h$ is just equal to 1_{k-h}. There are, however, examples in the literature where it is necessary to refer to general time-shifted cash flows, and *ad hoc* symbols are used (see Bowers *et al.*, 1997, p. 519). We feel it is much preferable to have a consistent notation to handle these.

2.15 Spreadsheet calculations

Here and at the end of Chapters 4, 5, 6, 9 and 10 we will describe spreadsheets for doing the basic calculations pertinent to the material of the respective chapter. Our descriptions will refer in particular to Microsoft Excel®, and we assume the reader is reasonably acquainted with this application. The basic ideas should be easily adaptable to other formats as well. Readers can reproduce our spreadsheets as given, or they may wish to come up with their own.

Our general idea is to enter various functions of the duration k in columns starting with row 10 for $k = 0$. The cells above can then be used for headings or information.

We put the duration k in column A. That is, we insert 0 in cell A10, 1 in cell A11, etc. This is done by putting the formula =A10+1 in cell A11 and copying down to cell A10+N.

The usual way of specifying the discount function is through the one-period interest rates i_k. We enter these in column B, with i_k being inserted into cell B10+k. We then calculate the vector of $v(k)$ values in column C. Insert 1 in cell C10 and the recursive formula (2.7) is entered into cell C11 as =C10/(1+B10). This is copied down column C to cell C10+N.

We enter the cash flows in column D. The cash flow c_k is put into cell D10+k. The value $\ddot{a}(\mathbf{c}, v)$ is then calculated in cell D8 through the formula

$$= \text{SUMPRODUCT}(\$\text{C}10 : \$\text{C}10 + N{*}\text{D}10 : \text{D}10 + N)$$

where we substitute for the particular value of N. As a check, take a constant interest rate of 0.06, and the vector $(1_{10}, 2_5)$. The answer is 12.7883. By copying cell D8 to the right, we can do calculations for several different cash flow vectors.

Values at time k are easily obtained by dividing $\ddot{a}(\mathbf{c})$ by $v(k)$, which is in cell C(10+k).

Notes and references

This chapter is not intended to provide an exhaustive treatment of the mathematical theory of interest. The goal was to provide that portion of the subject that is needed for the remainder of the book. Some additional materials will also be given in Chapters 7 and 8. Readers interested in a more detailed account can consult Broverman (2010) or Daniel & Vaaler (2009).

Additional work on internal rates of return can be found in Promislow (1980, 1997) and Teichroew *et al.* (1965a, 1965b). The concept of the generalized i.r.r. is due to the latter.

Exercises

Type A exercises

2.1 You are given a discount function v where $v(1,3) = 0.9,\ v(3,6) = 0.8,\ v(8,6) = 1.2$

(a) How much must you invest at time 1, in order to accumulate 10 at time 8?

(b) If you invest 100 at time 3, how much will have accumulated by time 8?

2.2 If $v(t) = 2^{-t}$, and you are given cash flow vectors $\mathbf{c} = (1, 2, 3)$ and $\mathbf{e} = (2, K, 1)$, find K so that $\mathbf{c}$ and $\mathbf{e}$ are actuarially equivalent with respect to v.

2.3 You are given interest rates $i_0 = i_1 = 0.25,\ i_2 = i_3 = 1$. You have entered into business transaction where you will receive 2 at time 0, 5 at time 3 and 10 at time 4, in return for a payment by you of 3 at time 2. In place of all these cash flows you are offered a single payment made to you at time 1. What is the smallest payment you would accept?

2.4 You are given rates of discount as follows: $d_k = 1/3$, for $k = 0, 1, 2$, and $d_k = 1/4$, for $k = 3, 4$. The vector $\mathbf{c} = (6_2, 9_3, 12)$. Find (a) the discount vector $\mathbf{v}$, (b) $\ddot{a}(\mathbf{c})$ and (c) $\text{Val}_2(\mathbf{c})$.

2.5 Let $\mathbf{c} = (1, 2, 4, -3, 8, -12)$. Suppose $v(0, 1) = v(1, 2) = 0.8$, $v(2, 3) = v(3, 4) = 0.75$, $v(4, 5) = 0.5$. Find (a) ${}_3V$, (b) B_3.

2.6 A discount function satisfies

$$v(k) = 2^{-k}\left[1 - \frac{k}{6}\right], \quad k = 0, 1, 2, \ldots, 5.$$

For the vector $\mathbf{c} = (1, -2, 4, 3, -3, -5)$ find (a) ${}_3V(\mathbf{c})$, (b) $B_3(\mathbf{c})$.

2.7 Given that ${}_8V(\mathbf{c}) = 100$, $c_8 = 60$, $c_9 = 70$, $v(8, 9) = 0.8$, $v(9, 10) = 0.75$, $v(0, 10) = 0.5$, and that $\ddot{a}(c) = 40$, find (a) ${}_{10}V(\mathbf{c})$, (b) $B_{10}(\mathbf{c})$.

2.8 Given the vector $\mathbf{c} = (1, 2_2, 7_3)$ and a discount function v satisfying $v(k) = 1 - k/10, k = 0, 1, \ldots, 5$, find $v \circ 3(k)$ for $k = 0, 1, 2$; the vectors ${}_3\mathbf{c}$ and $\mathbf{c} \circ 3$; and the present values $\ddot{a}(\mathbf{c}; v)$, $\ddot{a}({}_3\mathbf{c}; v)$, $\ddot{a}(\mathbf{c} \circ 3; v \circ 3)$. Verify that formula (2.29) holds.

Type B exercises

2.9 Define a two variable function for $s, t \geq 0$ by

$$v(s, t) = \begin{cases} (1 + (t - s))^{-1}, & \text{if } s \leq t, \\ 1 + (s - t), & \text{if } s \geq t. \end{cases}$$

Show that v is *not* a discount function.

2.10 Show that, given any positive-valued function g of one variable,

$$v(s, t) = \frac{g(t)}{g(s)}$$

defines a discount function.

2.11 Suppose that v_1 and v_2 are two discount functions. Are either of the functions $v_1 v_2$ or $v_1 + v_2$ discount functions? These are defined by

$$v_1 v_2(s, t) = v_1(s, t) v_2(s, t),$$
$$[v_1 + v_2](s, t) = v_1(s, t) + v_2(s, t).$$

2.12 Assume constant interest.

(a) Show that $\ddot{a}(0, 1_n) = \ddot{a}(1_n) - 1 + v^n$.

(b) Show that $\ddot{a}(0, 1_n) = (1 - v^n)/i$. Try to do this from part (a) and also directly from (2.14).

2.13 Let $\mathbf{k}^n$ denote the vector $(n, n-1, n-2, \ldots, 2, 1)$ Show that

$$\ddot{a}(\mathbf{k}^n) = \frac{n - \ddot{a}(0, 1_n)}{d}$$

in three different ways:

(a) From first principles, as in the first derivation of (2.17);

(b) By making use of a linear relationship between the vectors (1_n), $\mathbf{k}^n$ and the vector $\mathbf{j}^n$ of Section 2.9;

(c) By using the loan-interest argument given at the end of Section 2.9.

2.14 Suppose that $v(2,7) = 0.5, v(2,8) = 0.4, v(2,9) = 0.3$. Find a vector $\mathbf{d}$ actuarially equivalent to $\mathbf{c} = (1, 3, 5, 2, 9, 10, 6, 4, 8, 3)$ such that $\mathbf{d}_i = \mathbf{c}_i$ for $i < 7$ and $\mathbf{d}_i = 0$ for $i > 7$.

2.15 A loan of 2300 is to be repaid by n yearly payments of 230 beginning at time 5. The borrower is given the option of repaying only 115 at time 5, but must then pay 240 in each of the subsequent payments. Find $v(5)$. (An exact numerical answer is required. It should not be a function of n.)

2.16 A loan of 20 000, made at an interest rate of 6%, is to be repaid by level yearly payments for 10 years, beginning 1 year after the loan is advanced. Just before making the seventh repayment, the borrower wishes to repay the entire loan.

(a) If interest rates remain unchanged, what is the outstanding balance?

(b) Suppose interest rates have dropped to 5%. How much will the borrower have to pay if the lender uses the lower interest rate to calculate the outstanding balance?

2.17 A person has 1000 now and plans to invest it for 5 years. He is trying to decide between two alternatives. The first is to buy a bond that matures in 5 years. The second is to buy a bond that matures in 10 years and sell it at the end of 5 years. Assume that in both cases the bonds have no payments before maturity and can be purchased today at an interest rate of 6%. How much better or worse off is the individual at the end of 5 years if he chooses the second alternative instead of the first, assuming that at time 5 the interest rate for this class of bonds is (a) 4%, (b) 7%?

2.18 For a certain cash flow vector $\mathbf{c}$, $\mathbf{c}_0 = 1$, $\mathbf{c}_1 = 5, \ddot{a}(\mathbf{c}) = 15$. If the discount function is changed so that $v(1,2)$ is decreased by 0.1, while all other of values of $v(n, n+1)$ remain unchanged, then $\ddot{a}(\mathbf{c})$ decreases by 2.4. One the other hand, if $v(0,1)$ is decreased by 0.1 while all other values of $v(n, n+1)$ remain unchanged, then $\ddot{a}(\mathbf{c})$ decreases by 2. Find $v(0,1)$ and $v(1,2)$.

2.19 $\mathbf{c}$ and $\mathbf{d}$ are actuarially equivalent vectors such that $\mathbf{c}$ is constant and $\mathbf{d}$ is nondecreasing. Show that ${}_kV(\mathbf{c} - \mathbf{d}) \geq 0$, for $k = 0, 1, \ldots, N$.

*2.20 For the cash flow vector $(-1, 3, -2)$,

(a) show that the i.r.r. does not exist;

(b) find i_r as a function of r.

*2.21 Suppose that for a cash flow vector **c**, the number r satisfies $\tilde{B}_N(\mathbf{c}; v) = 0$, where $v(n) = (1 + r)^{-n}$. Show that $i_r = r$.

*2.22 Show that when interest is constant at rate r, then $\ddot{a}(\mathbf{c}) \geq 0$ if and only if $i_r \geq r$. (This shows that the criteria for determining if it is worthwhile to enter into a transaction given after Theorem 2.2 are in accordance with the remarks in Section 2.1.)

*2.23 Suppose that current interest yields on risk-free bonds are 4% for a 5-year bond and 5% for a 10-year bond. Calculate the forward price to be paid at the end of 5 years for a zero-coupon bond that pays 1000 units, 10 years from today. Suppose that instead of the number you just calculated, this forward price is 800. Illustrate how you would make a sure profit (under the idealized conditions given in Section 2.12).

*2.24 The current price of a 1-unit zero-coupon bond maturing in 10 years is 0.610. The forward rate $f(6, 10) = 0.06$. Find the spot rate y_6.

Spreadsheet exercises

2.25 A loan contract involves advances of 10 000 initially, 20 000 one year later, and 30 000 one year after that. This is to repaid by 20 yearly instalments beginning at time 3. The payments reduce by 5% each year (so, for example, if the first payment was 1000, the second would be 950, the third would be 902.50). Interest rates are 6% for the first 5 years, 7% for the next 5 years, and 8% after that. Find all payments and the outstanding balances at the end of each year, until the loan is fully discharged. Balances should be calculated after the payment due at the particular time is made.

3

The life table

3.1 Basic definitions

For the actuary working in the life insurance field, a major objective is to estimate the mortality pattern which will be exhibited by a group of individuals. A basic device for accomplishing this is known as a *life table*. (It is also known as a *mortality table* – an interesting example of a word and its opposite being used interchangeably.)

Let ℓ_0 be an arbitrary number, usually taken to be a round figure such as 100 000. Suppose we start with a group of ℓ_0 newly born lives. We would like to predict how many of these individuals will still be alive at any given time in the future. Of course, we cannot expect to compute this exactly, but we can hope to arrive at a close estimate if we have sufficiently good statistics. In the first part of this book we will make the assumption that we can indeed arrive at exact figures. This is in keeping with the concept of a *deterministic model* introduced in Chapter 1. In Part II, we introduce the *stochastic model* for mortality, where we investigate the more realistic assumption that the quantities we want are random variables. Let ℓ_x be the number of those original lives aged 0 who will still be alive at age x, and let d_x be the number of those original lives aged 0 who die between the ages of x and $x + 1$. The basic relationship between these quantities is

$$\ell_{x+1} = \ell_x - d_x. \tag{3.1}$$

A *life table* is a tabulation of ℓ_x and d_x where x is a nonnegative integer. The following is an example of a portion of a life table (this is an illustration only, and the figures are not intended to be realistic):

Fundamentals of Actuarial Mathematics, Third Edition. S. David Promislow.

Companion website: http://www.wiley.com/go/actuarial

x	ℓ_x	d_x
0	100 000	2000
1	98 000	1500
2	96 500	1000
3	95 500	900
⋮	⋮	⋮
ω	0	

The table will end at some age, traditionally denoted by ω, such that $\ell_\omega = 0$. This is the limiting age of the table, and denotes the first age at which all of the original group will have died. The actual value of ω will vary with the particular life table, but it is typically taken to be around 110 or higher.

3.2 Probabilities

Although we assume we can predict ℓ_x exactly, there is still randomness in our model, since it is not known whether or not any given individual will be among the survivors at a particular point of time. It is convenient to introduce some elementary probabilistic notions. For nonnegative integers n and x, let

$$ {}_np_x = \frac{\ell_{x+n}}{\ell_x}. \tag{3.2} $$

What is the meaning of this term? Consider the ℓ_x survivors age x. Out of this group, ℓ_{x+n} will survive to age $x+n$. The quotient then gives us the probability that a person aged x, hereafter denoted just by the symbol (x), will be alive at age $x+n$. Let

$$ {}_nq_x = \frac{\ell_x - \ell_{x+n}}{\ell_x}. \tag{3.3} $$

This gives us the probability that (x) will die between the ages of x and $x+n$. It is clear that

$$ {}_nq_x = 1 - {}_np_x. \tag{3.4} $$

As an example, in the table given above we would have ${}_2p_0 = 965/1000,\ {}_2q_1 = 25/980$.

Since a left subscript of '1' occurs frequently it is omitted for notational convenience. That is, p_x denotes ${}_1p_x$, and q_x denotes ${}_1q_x$. The quantity q_x is often referred to as the *mortality rate* at age x.

What is the probability that (x) will die between the ages of $x+n$ and $x+n+k$? This is a quantity which we will use frequently. There are three main ways of expressing it:

$$ \frac{\ell_{x+n} - \ell_{x+n+k}}{\ell_x}, \tag{3.5a} $$

or

$$ {}_n p_x - {}_{n+k} p_x, \tag{3.5b} $$

or

$$ {}_n p_x \, {}_k q_{x+n}. \tag{3.5c} $$

The reader should verify, by substituting values of ℓ, that all three expressions are equal. They can each be explained intuitively. Consider the first. The numerator is the number of people living at age $x + n$, less the number living at age $x + n + k$. This difference must be the number of people who died between the two ages. Dividing by the number of people that we start with will give us the required probability. In the second expression we express this quantity as the probability that (x) will live n years, but will *not* live $n + k$ years. In the third expression we consider two stages. To die between the specified ages, (x) must first live to age $x + n$. The individual, then being age $x + n$, must die within the next k years. We will have occasion to use all three of these expressions, choosing the one which is most convenient for the purpose at hand.

Another useful identity, which we will refer to as the *multiplication rule*, is

$$ {}_{n+k} p_x = {}_n p_x \quad {}_k p_{x+n}, \tag{3.6} $$

for all nonnegative integers n, k and x. It can be verified directly from (3.2). Intuitively, it says that in order for (x) to live $n + k$ years, the individual must first live n years, and then, being age $x + n$, must live another k years.

3.3 Constructing the life table from the values of q_x

The life table is constructed in practice by first obtaining the values of q_x for $x = 0, 1, \dots, \omega - 1$. Obtaining these values is a statistical problem which we will not discuss in detail. It is basically done by carrying out a study in which we observe how long people of different ages will live. For example, if we observe a group of 1000 people of exact age 50 and 10 of them die within 1 year, then we could estimate q_{50} as 0.01. This of course is an extreme simplification and the process is much more complicated. It is not practical to gather a group of people exactly age 50 at one point of time, and then to observe them for an entire year. In practice, people will enter the study at various times, and leave for reasons other than death. In addition, we must achieve consistency between values at different ages. The statistical subject known as *survival analysis* deals with these problems.

In this book we will take the values of q_x as given. The life table can then be constructed inductively, starting with ℓ_0, from the formulas

$$ d_x = \ell_x q_x, \qquad \ell_{x+1} = \ell_x - d_x, \tag{3.7} $$

which follow immediately from (3.1) and (3.3). We will see, however, that it is not usually necessary to actually calculate ℓ_x and d_x. In practice, life tables are specified by just giving the values of q_x, which is sufficient for the necessary computations. The advantage of the traditional form lies mainly in its intuitive appeal, rather than its use as a calculating tool.

3.4 Life expectancy

Life expectancy is one of the most frequently quoted actuarial concepts. Probably for this reason it is often used incorrectly, as we will explain below. A basic question is the following. How long can a person age x expect to live? Of course there is much variety in the future lifetime of various individuals of the same age. Some will live for several years, and some will die immediately, but we can attempt to arrive at some sort of average figure. One approach would be to take a large number of people age x and observe them until all have died. We could then compute the total future lifetime of all these individuals. Dividing by the number of people in the original group would give an estimate of the desired average. For a drastically oversimplified example, take three people exactly age 60. Suppose one dies at age 62, another at age $72\frac{1}{2}$ and the third at age $91\frac{1}{4}$. The total future lifetime would be $2 + 12\frac{1}{2} + 31\frac{1}{4} = 45\frac{3}{4}$. Dividing by 3 we could estimate that a person age 60 could expect to live on average another $15\frac{1}{4}$ years. Of course to be statistically accurate we would need many more than three people. Moreover, the length of time needed for such a study makes it completely impractical. Remarkably however, once we have the life table, we can obtain the figure directly, without carrying out the observations. To see this, we look at another approach to obtaining total future lifetime. Suppose we start with ℓ_x people age x. After 1 year, there will ℓ_{x+1} survivors who would have each contributed 1 year of lifetime to this total. At the end of the 2 years there will ℓ_{x+2} survivors who would have each contributed another year to the total. Continuing in this way, we can estimate the total future lifetime of all lives as

$$\ell_{x+1} + \ell_{x+2} + \ell_{x+3} + \cdots + \ell_{\omega-1},$$

and, dividing through by ℓ_x, we obtain the quantity

$$e_x = \sum_{k=1}^{\omega-x-1} \frac{\ell_{x+k}}{\ell_x} = \sum_{k=1}^{\omega-x-1} {}_kp_x. \tag{3.8}$$

The quantity e_x is known as the *curtate life expectancy* or curtate expectation of life at age x. The word *curtate* means *reduced* or *truncated*, reflecting the fact that this is not exactly the quantity that we want. We have cheated a little in our alternate measurement scheme, for it measures only whole future years of lifetime and ignores the fraction of the year lived in the year of death. In our illustration above, for example, under the alternate counting method, the 60-year-old who died at age $72\frac{1}{2}$ would be credited with only 12 years of total lifetime rather than the actual $12\frac{1}{2}$. The person who died at $91\frac{1}{4}$ would be credited with only 31 years rather than the actual $31\frac{1}{4}$. We are undercounting between 0 and 1 years for each individual, and it seems reasonable to take this to be one-half on average. The true life expectancy, usually referred to as the *complete life expectancy* at age x and denoted by $\mathring{e}_x$ is given approximately by

$$\mathring{e}_x = e_x + \frac{1}{2}.$$

We will give a more formal treatment of $\mathring{e}_x$ in Chapter 8 and again in Chapter 15.

There is a simple recursion formula to compute e_x for all values of x. From the second formula in (3.8),

$$\begin{aligned} e_x &= p_x + {}_2p_x + {}_3p_x + \cdots + {}_{\omega-x-1}p_x \\ &= p_x(1 + p_{x+1} + {}_2p_{x+1} + \cdots + {}_{\omega-x-2}p_{x+1}) \\ &= p_x(1 + e_{x+1}). \end{aligned} \tag{3.9}$$

The second line is obtained by using the multiplication rule (3.6). Note that this is a *backward* recursion formula as it gives the value of the function in terms of the next higher argument. The recursion is then started from the initial value of $e_\omega = 0$.

It is instructive to give an intuitive explanation of (3.9). To live any whole number of years in the future, (x) must first live to age $x + 1$, as reflected by the factor p_x. The individual will then have completed 1 year of lifetime, and in addition, being now age $x + 1$, will complete, on average, the expected number of future whole years for a person of that age.

The reader should note carefully that life expectancy is a function of age. For each age x, the life expectancy at that age gives the average number of future years that (x) will live. A major source of misquoting is to express this as a single figure rather than a function. One can often find statements in newspapers or similar sources, stating something like 'life expectancy has increased from 75.3 to 75.8 years'. The writer is invariably referring to the life expectancy at age 0 only. This is certainly of interest, but it conveys somewhat limited information. A person already aged 80 who wishes to estimate how much longer he/she can expect to live is not helped by a statement which claims that newborn lives live on average to age 75.8.

The reader should also note that life expectancy is the average *duration* and not the average age a person can expect to live to. We say, for example, that the life expectancy at age 50 is 31.2, meaning that on average a person age 50 can expect to live to age 81.2. There is often confusion on this point, which is again a result of the tendency to report only the age 0 figure where duration and age are the same.

There are many other quantities obtainable from the life table which are of interest, but only one in particular that we will discuss here. Sometimes we may be interested in the average duration lived by (x) over the next n years, where n is some fixed duration. The quantity

$$\sum_{k=1}^{n} \frac{\ell_{x+k}}{\ell_x} = \sum_{k=1}^{n} {}_kp_x \tag{3.10}$$

is known as the *curtate n-year temporary life expectancy* at age x. The word *temporary* comes from an analogy with life annuities which we discuss in the next chapter. It gives us the expected *complete* number of years lived *over the next n years* by people now age x. This is what we would compute if we repeated the alternate measurement system described above, but ended the observations after n years. To adjust this for the undercounting in the year of death requires some care. Those who lived to age $x + n$ will have contributed the correct total of n years, and therefore it is only the $(\ell_x - \ell_{x+n})$ people who died during the n-year period who must be considered in the adjustment. To get a more accurate n-year temporary life expectancy at age x we add to the quantity in (3.10) not $\frac{1}{2}$, but rather $\frac{1}{2}(\ell_x - \ell_{x+n})/\ell_x$, to get an approximation to the *complete n-year temporary life expectancy* at age x of

$$\sum_{k=1}^{n} {}_kp_x + \frac{1}{2}{}_nq_x.$$

3.5 Choice of life tables

The life table reflects the fact that age is the major determinant of future mortality. There are of course several other factors that affect future lifetime, such as gender, health status, lifestyle, and geographical location. In practice, the effect of these factors is handled by producing several different life tables. One confines the mortality study that produces the table to a particular group of people, namely those with the characteristic that you want to separate out. The following are a few of the more important distinctions made in practice.

It is observed, for reasons that nobody has fully explained, that females live longer than males. For the middle range of ages, it is typical for the life expectancy of a female to be from 5 to 7 years more than that of a male of the same age. To reflect this, it is usual to produce separate male and female life tables.

In recent years, there has been overwhelming statistical evidence to show the dangers of smoking. This has led insurance companies to construct different life tables for smokers and nonsmokers.

The choice of life table will also depend heavily on the type of contract that is being sold. The life tables produced for the general population from census data are not suitable for insurance purposes. People accepted for life insurance policies are usually screened by the insurance company to make sure they are in reasonable health. They can expect to live longer than a person of the same age taken from the population as a whole. Life tables for insurance purposes are constructed by looking at insurance company data only.

Purchasers of life annuities (discussed in Chapter 4) will on the average live even longer, since a person in poor health would be unlikely to buy such a product. Separate tables are needed for annuity purposes.

Still another distinction to be made is the difference between individual contracts and group contracts. In the former case the buyers makes a definite decision to enter into the insurance or annuity contract and are presumably aware of their health conditions and acting in their best interests. In the latter cases, an employer purchases the contract to cover a large group of employees. Different mortality patterns in the two cases can be expected.

There are many other examples which we will not discuss here, although some will be pointed out briefly in succeeding chapters. The reader should be aware that selecting an appropriate table for a particular use is an important actuarial task.

Sometimes a very simple method, known as *multiples of standard mortality*, is employed to construct many different tables from a given one, known as the standard table. For example, it might be decided that for risks of a certain type, the mortality is 150% of standard mortality. The life table for such risks is constructed by multiplying each q_x in the standard table by 1.5. This method is justified more by simplicity of calculation rather than any scientific rationale.

3.6 Standard notation and terminology

We have already introduced the standard symbols ${}_np_x, {}_nq_x, e_x, \mathring{e}_x$ and ω.

The symbol ${}_{n|k}q_x$ denotes the probability that (x) will die between the ages of $x+n$ and $x+n+k$, the quantity that we already have three ways of writing as shown in (3.5). A subscript of 1 is omitted, so that ${}_{n|}q_x$ denotes ${}_{n|1}q_x$. This use of a vertical bar is a typical actuarial device to denote a 'waiting period'. In this case, the symbol is intended to indicate that the person will wait n years and then die in the following k years.

The quantity in (3.10) is denoted by $e_{x:\overline{n}|}$ and the complete temporary life expectancy is denoted by $\mathring{e}_{x:\overline{n}|}$.

3.7 A sample table

To do the spreadsheet problems in the following chapters that require a life table, we introduce a sample table. It is given by

$$q_x = \begin{cases} 1 - \mathrm{e}^{-0.00005(1.09)^x}, & x = 0, 1, \dots, 118, \\ 1, & x = 119. \end{cases} \tag{3.11}$$

We have $\omega = 120$. The formula is easily programmed into a spreadsheet, and that is the reason for giving the table in this form. Giving an existing table would necessitate entering the figures individually. Another advantage of this parametric form is that the two constants of 0.00005 and 1.09 can be changed to provide a variety of different life tables for comparison purposes. Our table is, as stated, a sample table, and it should not be taken as being a realistic picture of modern-day mortality. This is true especially at the younger or very old ages, as will be discussed later in the text. Chapter 14 will provide some motivation for the formula; see, in particular, Exercise 14.11.

Notes and references

London (1997) provides an introduction to survival analysis, and gives more details on the construction of life tables.

Exercises

Type A exercises

3.1 You are given that $q_{60} = 0.20, q_{61} = 0.25, q_{62} = 0.25, q_{63} = 0.30, q_{64} = 0.40$.

(a) Find ℓ_x for ages 60–65, beginning with $\ell_{60} = 1000$.

(b) Find the probabilities of the following:

(i) (61) will die between the ages of 62 and 64.

(ii) (62) will live to age 65.

(c) Given that $e_{65} = 0.8$, find e_x for $x = 60$–64.

3.2 You are given that ${}_5p_{40} = 0.8, {}_{10}p_{45} = 0.6, {}_{10}p_{55} = 0.4$. Find the probability that (40) will die between the ages of 55 and 65.

3.3 Suppose that out of a typical group of 100 people age 70, 10 will die in the first year, 15 will die in the second year, and 20 will die in the third year. Calculate q_{70}, q_{71}, q_{72} and ${}_3p_{70}$.

Type B exercises

3.4 Suppose that

$$\ell_x = 100 - x, \qquad x = 0, 1, \ldots, 100.$$

Find expressions for (a) ${}_np_x$, (b) ${}_nq_x$, (c) the probability that (x) will die between the ages of $x + n$ and $x + n + k$.

3.5 You are given that $e_{60} = 17$, ${}_{10}p_{50} = 0.8$, and that the 10-year curtate temporary life expectancy at age 50 is 9.2. Find e_{50}.

3.6 Prove the following statement, assuming the approximation given in the text, and give an intuitive explanation:

$$\mathring{e}_x = \frac{1}{2}q_x + p_x(1 + \mathring{e}_{x+1}).$$

3.7 Suppose that q_x is equal to a constant q for all x. (Note that in this case $\omega = \infty$). Find expressions in terms of q and n for (a) ${}_np_x$, (b) e_x. Do you think that this gives a realistic life table? Why or why not?

3.8 In constructing a life table for heavy smokers, Actuary A decides to take a standard table, and double each value of q_x, using a value of 1 if $q_x > 0.5$. Actuary B takes the same table and squares each value of p_x. Show that the resulting table of Actuary B, has lower mortality rates at all ages than that of Actuary A.

Spreadsheet exercise

3.9 Taking the sample life table as given by (3.11), use recursion to find e_x for $x = 0, 1, \ldots, \omega - 1$. Focus on e_0. How much is this reduced if the constant 0.00005 is changed to 0.00006. What happens if 0.00005 is kept the same, but 1.09 is changed to 1.092?

4

Life annuities

4.1 Introduction

The financial losses that we listed at the beginning of Chapter 1 are no doubt familiar to all readers. Another type of risk, which may not be so obvious, is that of *living too long*. Given our definition of risk as the possibility of something bad happening, the reader may wonder about this statement. After all, is not living a long life a good rather than bad occurrence? It certainly can be, but it does carry with it the possibility of financial hardship if one does not have adequate income. Imagine the decision faced by a retired individual who has accumulated savings of 1 000 000, which is invested at an annual rate of 5%, providing an annual income of 50 000. Imagine also that the individual decides that this is an insufficient return, so it is necessary to consume a portion of the capital each year, as well as the interest. Doing so, however, runs the risk that the entire capital may be depleted before death, leaving the individual with no source of income for his/her remaining lifetime.

Life annuities are a means of insuring against such a risk. A life annuity is a contract between an insurance company, known as the *insurer*, and another party, known as the *annuitant,* which provides the following. In return for the payment by the annuitant of prescribed *premiums*, the insurer will provide a sequence of payments, known as *annuity benefits* of prescribed amounts and at prescribed times. The unique provision is that the annuitant must be alive to receive each benefit payment. These will terminate upon the death of the annuitant.

People who purchase such a contract are investing capital that they have decided will not be needed for any dependents after they die. They agree to give up these funds upon their death in return for the greater yield they can achieve on their investment by sharing the amounts forfeited by those who predecease them. A good way to picture the workings of a life annuity is to imagine a room with a number of small boxes, belonging to one annuitant each. Each annuitant pays the same premium into his/her box, and interest earnings are added

---

Fundamentals of Actuarial Mathematics, Third Edition. S. David Promislow.

Companion website: http://www.wiley.com/go/actuarial

to the amounts in his/her box. Each receives the same annuity benefits, which are provided by taking money out of the box. When one participant dies, his/her box is opened, and the amount is spread evenly among all the other boxes in the room. Participants then receive not only interest earnings, but also these forfeited amounts of those who die before them. We refer to the latter as *survivorship earnings*. This is a fairly accurate account of what actually happens, except that amounts are calculated on paper. There is obviously no need for the physical rooms or boxes. More precisely, it is an account of what would happen if mortality followed the life table exactly. For most annuities sold in practice, the accumulation is guaranteed in advance, so it does not depend on the actual deaths. If fewer people die than expected, the insurer would still have to add the extra survivorship amounts to people's 'boxes', and would experience a loss, while more deaths than expected would result in a gain to the insurer.

To be equitable, all participants in any one room must have roughly the same risk of dying, and in particular must be of the same age. A 70-year-old, put in the same room as 20-year-olds, would be much more likely to die early on and would clearly be at a disadvantage. We will therefore calculate annuity premiums primarily as a function of age. There are of course many other factors to consider, such as gender, which we discuss briefly in Section 4.6.

Insurance companies are not the only source of life annuities. Many pension plans pay retirement benefits in the form of a life annuity.

4.2 Calculating annuity premiums

We now turn to the mathematical aspects of annuities. The reader should review the vector notation given in Section 2.8, which we will make frequent use of.

We assume that we have fixed throughout an appropriate life table and a discount function v to model the effect of investment earnings.

Consider a contract sold to (x) – or, more accurately, *on the life of* (x), since the purchaser could be a different party than the annuitant, such as an employer. It is assumed throughout that (x) is an integer, which allows us to conveniently use a life table if desired. Indeed, the common practice is to classify an individual by their age on their last birthday, which is just the normal way that one thinks of age. We suppose that benefit payments are made yearly. Suppose that the amount of benefit paid at time k is c_k. This could of course be zero, indicating that no payment is to be made. The contract is then described by the cash flow vector $\mathbf{c} = (c_0, c_1, \ldots, c_{\omega-x-1})$. In this case we will refer to $\mathbf{c}$ as the *annuity benefit* vector. (Note that the final cash flow is made at age $\omega - 1$, which the last age at which anyone is living, according to our model.) Suppose that the benefits are to be purchased by a single premium paid at the beginning of the contract at age x. Our first problem is to calculate the premium that the annuitant should pay. The criterion is that the total premiums from all annuitants, together with the interest earned, should be sufficient to provide all of the required benefits, assuming that invested capital accumulates according to the given discount function and that mortality follows that of the given life table.

Consider the simplest possible case, namely $\mathbf{c} = \mathbf{e}^k$. This is an annuity consisting of a single benefit payment of 1 to be made at time k if (x) is then alive. Such a contract is usually referred to as a *pure endowment*. Let E denote the single premium that should be paid to provide for this. Suppose that ℓ_x people, each age x, buy such a contract. The total amount collected in premiums at time zero would be $\ell_x E$, and this will accumulate to $\ell_x E v(k, 0)$ at time k. There will be ℓ_{x+k} survivors at time k, and each of these must receive 1 unit. This means that

$$\ell_x Ev(k, 0) = \ell_{x+k}.$$

Solving, we obtain

$$E = v(k)\frac{\ell_{x+k}}{\ell_x} = v(k)_k p_x.$$

This can be verified intuitively. If the benefit of 1 unit at time k were guaranteed, we know from Chapter 2 that we would need to invest $v(k)$. In this case, where the benefit is not guaranteed, we multiply by the probability of receiving it to arrive at the premium.

The general case easily follows. Let $\ddot{a}_x(\mathbf{c})$ denote the single premium for a life annuity on (x) with benefit vector $\mathbf{c}$. This contract can be viewed as a sequence of pure endowments, one for each value of k, where the kth pure endowment pays c_k at time k. The premium for such a pure endowment is just $c_k v(k)_k p_x$, and the total premium is obtained by summation. We have

$$\ddot{a}_x(\mathbf{c}) = \sum_{k=0}^{\omega-x-1} c_k v(k)_k p_x. \tag{4.1‡}$$

Life annuities have been classified into various types. An annuity for which benefit payments are made for a fixed number of years and then cease is known as a *temporary life annuity* (an example for those who collect oxymorons).

The most common type of annuity is one for which benefits continue for as long as the annuitant lives. This is known as a *whole life* annuity. Purchasers of whole life annuities protect themselves completely against outliving their available capital, for no matter how long they live, the income from the annuity will continue. We have no need to distinguish between temporary and whole life annuities in our mathematical model. As indicated, we consider all annuities on (x) as running from time 0 to time $\omega - 1 - x$. In the case of a temporary life annuity we simply take benefits equal to 0 after the last positive payment. In real life, however, there is a difference. The insurer can obviously not inform a person who has reached the age of ω, that according to the insurer's model they are no longer alive and annuity payments will stop. This means that as a practical matter the benefit payments on a whole life annuity must eventually be constant (or at least follow some regular pattern) so it is clear what amount to continue to pay should an annuitant live beyond ω. In most tables, ω is set sufficiently high that such an occurrence is extremely rare.

Deferred annuities are contracts where the initial payment does not commence for several years. For example, a person wishes to provide for an income beginning at retirement. Mathematically, this simply means that the initial entries in the benefit vector will be zero, so no special treatment is required. The annuity benefit vector will be of the form $\mathbf{c} = (0_k, \mathbf{d})$, and the duration k is known as the *deferred period.*

Example 4.1 A temporary life annuity on (50) provides for 1 payable at age 50, 2 payable at age 51, 3 payable at age 52, and 4 payable at age 53. Suppose $q_{50} = 0.1$, $q_{51} = 0.2$, $q_{52} = 0.25$. The interest rate is 50% for the first year and 100% thereafter. Find the single premium.

Solution. For a small problem such as this, it is convenient to make use of a time diagram as introduced in Chapter 2. See Figure 4.1.

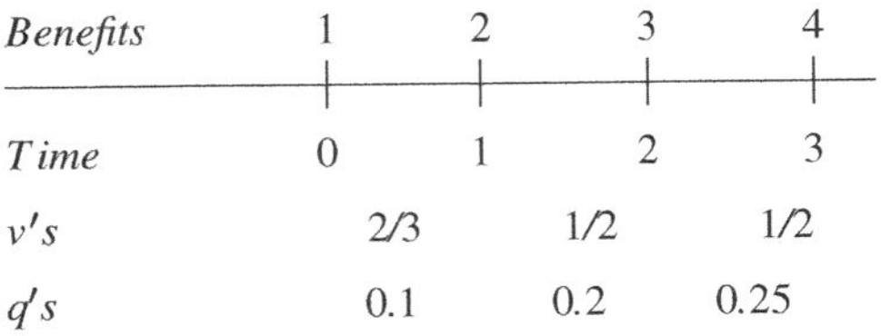

Benefits	1		2		3		4
Time	0		1		2		3
v's		2/3		1/2		1/2	
q's		0.1		0.2		0.25	

Figure 4.1 Example 4.1

Here we have inserted values of q_y, for relevant ages y, as well as the 1-year interest discount factors. We now must multiply each payment c_k by the product of all previous values of $v(i, i+1)$ as we did in Chapter 2, and *also* by ${}_kp_x$ which we calculate as the product of all the previous values of p_y. The single premium equals

$$1 + \left(2 \times \frac{2}{3} \times 0.9\right) + \left(3 \times \frac{2}{3} \times \frac{1}{2} \times 0.9 \times 0.8\right) + \left(4 \times \frac{2}{3} \times \frac{1}{2} \times \frac{1}{2} \times 0.9 \times 0.8 \times 0.75\right) = 3.28.$$

An alternate way of calculating the survival probabilities, which some may prefer, is to use (3.2). We can construct part of a life table, utilizing (3.7). In this example, starting with say $\ell_{50} = 1000$ we would get in turn $d_{50} = 100$, $\ell_{51} = 900$, $d_{51} = 180, \ell_{52} = 720$, $d_{52} = 180$, $\ell_{53} = 540$. This automatically performs all the multiplication involving the values of p_y, and gives us the factors of 0.9, 0.72, 0.54, used above.

4.3 The interest and survivorship discount function

4.3.1 The basic definition

It is important to note that the formula for $\ddot{a}_x(\mathbf{c})$ is the same as that given for $\ddot{a}(\mathbf{c})$ in Chapter 2, but with $v(k)$ replaced by $v(k)_kp_x$ for each value of k. This suggests that we are in the general situation of Chapter 2, but with a different discount function. This is indeed the case. Define

$$y_x(n) = v(n)_np_x$$

and extend to a two-variable function

$$y_x(k, n) = \frac{y_x(n)}{y_x(k)}$$

for all nonnegative integers k, and n. It is clear that y_x satisfies (2.1) and therefore is a discount function. This fact is borne out by the discussion of the pure endowment in the previous section, which showed that the value at age x of 1 paid at age $x + k$, when accumulation is by interest and survivorship, is precisely $y_x(k)$.

Using the multiplication rule we can verify that for $k \leq n$

$$y_x(k,n) = \frac{y_x(n)}{y_x(k)} = v(k,n)_{n-k}p_{x+k}, \tag{4.2}$$

and for $k > n$

$$y_x(k,n) = \frac{v(k,n)}{{}_{k-n}p_{x+n}}. \tag{4.3}$$

So for example, assuming constant interest

$$y_x(2,5) = v^3{}_3p_{x+2}, \qquad y_x(5,2) = \frac{(1+i)^3}{{}_3p_{x+2}}.$$

Note that the last quantity, giving the amount at time 5 resulting from an investment of 1 at time 2, is greater than $(1+i)^3$, the amount resulting from interest only. This reflects the extra earnings due to survivorship.

We will refer to y_x as an *interest and survivorship* discount function. From (4.1) it follows that

$$\ddot{a}_x(\mathbf{c}) = \ddot{a}(\mathbf{c}; y_x), \tag{4.4}$$

and is therefore just the present value of the payments, with resect to the discount function y_x. The significance of the above is that it shows that all the results of Chapter 2 can be carried over directly to life annuities.

As a matter of terminology we will interchangeably refer to $\ddot{a}_x(\mathbf{c})$ and similar quantities introduced later, as *net single premiums* or *present values.*

When calculating $\ddot{a}_x(\mathbf{c})$ by spreadsheets it is convenient to make use of the recursion formula (2.7) adapted to the interest and survivorship function. We have

$$y_x(k+1) = y_x(k)v(k,k+1)p_{x+k}. \tag{4.5}$$

Notation Consider the vector $1_{\omega-x}$ which refers to a 1-unit cash flow sequence continuing for the life of (x). Due to the prevalence of this symbol, it is suppressed for notational convenience. This is a standard convention of actuarial notation that will be followed throughout the book. In general, whenever a vector is omitted, it is understood to be $1_{\omega-x}$. For example, $\ddot{a}_x$ denotes the net single premium for a whole life annuity on (x), paying 1 per year for life, beginning at time 0.

In cases where payments run for life, but are not necessarily constant, it is sometimes convenient to avoid specifying ω by using the symbol 1_∞ to refer to a vector with entries of 1, running to age $\omega - 1$ For example if $\omega = 120$ we could write $\ddot{a}_{40}(2_{15}, 1_\infty)$ in place of $\ddot{a}_{40}(2_{15}, 1_{65})$.

In what follows we will keep v as the notation for the investment discount function, which reflects the discounting resulting from earnings on invested capital. We will use the letter y to denote a general discount function, which could be v or y_x or any other such function.

Life annuity symbols will be written as given with the age as a subscript. For present values involving discounting at interest only, we will then incorporate the v and write $\ddot{a}(\mathbf{c}; v)$.

4.3.2 Relations between y_x for various values of x

The reader should note that we have several interest and survivorship functions, one for each value of the initial age x. They are, however, related in the following way.

From (4.2)

$$y_x \circ k(n) = y_x(k, k+n) = v(k, k+n)_n p_{x+k} = v \circ k(n)_n p_{x+k}.$$

This tells us that

$$y_x \circ k(\text{calculated with } v) = y_{x+k}(\text{calculated with } v \circ k). \tag{4.6}$$

Therefore, if interest is constant,

$$y_{x+k} = y_x \circ k. \tag{4.7}$$

In particular, with constant interest $y_x = y_0 \circ x$, so that knowing the interest and survivorship function for initial age 0 determines it for all initial ages.

In the remainder to this section, we continue with the notational ideas first introduced in Section 2.11, with the intention of providing a background for following classical actuarial formulas. It can be safely omitted at first, as everything can be conveniently expressed by using the new notation that we introduced in Chapter 2 and discussed in Section 2.14.2. We do however occasionally refer back to this section as well as Section 2.14.2 In particular the symbols $v \circ k$, and $\mathbf{c} \circ \mathbf{k}$ prove to be convenient and will be used in various places.

In the first place, since we do not assume constant interest as was traditionally done, it is necessary to introduce a new notational device into the classical life annuity symbols in order to handle time shifting. The idea is to separate age from duration in the quantity $x+k$. We let $\ddot{a}_{[x]+k}(\mathbf{c})$ denote the present value of a life annuity on $(x+k)$, paying c_i at time i, and calculated with respect to the investment discount function $v \circ k$. Of course with constant interest, $v \circ k = v$, so that $\ddot{a}_{[x]+k}(\mathbf{c}) = \ddot{a}_{x+k}(\mathbf{c})$ for all k, and this notation is not needed. (However even with constant interest we will have need of this notation with a more refined mortality model which we discuss in Chapter 9.)

From (2.28) we can write

$$\ddot{a}_{[x]+k}(\mathbf{c} \circ k) = \text{Val}_k({}^k\mathbf{c}; \mathbf{y_x}), \tag{4.8}$$

so the symbol on the left hand side is conveniently interpreted as the value at time k, of the benefits at and *after* time k, paid on a life annuity issued to (x) with benefit vector $\mathbf{c}$.

The above equation leads to the traditional form of the life annuity splitting identity.

$$\ddot{a}_x(\mathbf{c}) = \ddot{a}_x({}_k\mathbf{c}) + y_x(k)\ddot{a}_{[x]+k}(\mathbf{c} \circ k). \tag{4.9}$$

For a particular case of (4.9), assume constant interest and let $\mathbf{c} = 1_\infty$. Then ${}_k\mathbf{c} = 1_k$ and $\mathbf{c} \circ k = 1_\infty$. Using (3.2) we can write

$$\ddot{a}_x = \ddot{a}_x(1_k) + v^k \frac{\ell_{x+k}}{\ell_x} \ddot{a}_{x+k}. \tag{4.10}$$

This was a popular formula in pre-computer days. It was used to calculate values of temporary annuities $\ddot{a}_x(1_k)$ from a table of values of whole life annuities and the life table. With modern spreadsheets, we do not need this or similar formulas for calculation purposes. They are however sometimes useful to explain relationships between various quantities.

Formula (4.9) is often applied to express the present value of a deferred annuity $\ddot{a}_x(\mathbf{c})$ where $\mathbf{c} = 0_k \circ \mathbf{d}$, directly in terms of the vector $\mathbf{d}$. Since ${}_k\mathbf{c} = 0$ and $\mathbf{c} \circ k = \mathbf{d}$, we have immediately that

$$\ddot{a}_x(\mathbf{c}) = y_x(k)\ddot{a}_{[x]+k}(\mathbf{d}). \tag{4.11}$$

Example 4.2 This example is designed to further illustrate the square bracket notation. Suppose that $q_{60} = 0.05$, $q_{61} = 0.10$, $q_{62} = 0.15$, $i_0 = i_1 = 0.04$ and $i_k = 0.08$ for $k \geq 2$. Calculate $\ddot{a}_{60}(1_4)$, $\ddot{a}_{[59]+1}(1_4)$, $\ddot{a}_{[58]+2}(1_4)$.

Solution.

$$\begin{aligned}
\ddot{a}_{60}(1_4) &= 1 + (1.04)^{-1}0.95 + (1.04)^{-2}(0.95)(0.90) \\
&\quad + (1.04)^{-2}(1.08)^{-1}(0.95)(90)(0.85) = 3.33 \\
\ddot{a}_{[59]+1}(1_4) &= 1 + (1.04)^{-1}0.95 + (1.04)^{-1}(1.08)^{-1}(0.95)(0.90) \\
&\quad + (1.04)^{-1}(1.08)^{-2}(0.95)(90)(0.85) = 3.27.
\end{aligned}$$

This could be interpreted as the amount that a person now age 59 would have to pay for a 1-unit 4-year annuity if the purchase was made at the end of 1 year (assuming no change in the life table). It is lower than the first amount, since there is only 1 year of low interest rates rather than two.

$$\begin{aligned}
\ddot{a}_{[58]+2}(1_4) &= 1 + (1.08)^{-1}0.95 + (1.08)^{-2}(0.95)(0.90) \\
&\quad + (1.08)^{-3}(0.95)(90)(0.85) = 3.19.
\end{aligned}$$

This could be interpreted as the amount that a person now age 58 would have to pay for a 1-unit 4-year annuity if the purchase was made at the end of 2 years. It is lower yet, since the buyer would avoid both years of low interest.

4.4 Guaranteed payments

Suppose an individual purchases a life annuity and dies immediately after paying the premium. The person will get nothing back, except the benefit payment at time zero if that is positive. While we have seen that this is fair, the situation is not always well understood by the dependents of the annuitant, who may well complain that the insurer has confiscated the funds.

Aside from this, many prospective purchasers are uneasy with the possibility that all or a large portion of their money could be lost. To provide additional flexibility, insurers frequently offer life annuities with a *guaranteed period.* A typical contract of this type would stipulate that for a certain duration (commonly 10 or 15 years) the benefits will be paid regardless of whether the annuitant is alive or not. After this guaranteed period the contract reverts to a life annuity and benefits are conditional on survival. The guaranteed period of course means that a higher premium must be paid for the same level of benefits. To calculate the premium, it is best to consider such an annuity as two separate contracts, one for the guaranteed payments, the other for the contingent payments, and add the respective premiums. Examples follow.

Example 4.3 A person age 40 purchases a life annuity that provides 10 000 each year for life, with the first payment starting at age 41. The first 10 payments will be paid regardless of whether the annuitant is alive or not. Find a formula for the single premium.

Solution. The premium for the guaranteed annuity is

$$10\,000\,\ddot{a}(\mathbf{c}; v), \quad \mathbf{c} = (0, 1_{10}).$$

The premium for the non-guaranteed annuity is

$$10\,000\,\ddot{a}_{40}(\mathbf{f}), \quad \mathbf{f} = (0_{11}, 1_{\infty}).$$

Thus total premium is just the sum

$$10\,000\,[\ddot{a}(\mathbf{c}; v) + \ddot{a}_{40}(\mathbf{f})].$$

Let us verify the duration in the vector $\mathbf{f}$. The first guaranteed payment is at time 1, so the last guaranteed payment is at time 10, and the first non-guaranteed payment is at time 11. As we start the indexing with 0, there will be 11 zeros in the non-guaranteed vector.

Example 4.4 A person age 40 purchases a life annuity that provides 1 per year for life with the first payment at age 65. If (40) lives to age 65 he/she will receive at least 10 payments. Nothing is paid if death occurs before age 65. Find a formula for the premium.

Solution. The premium for the non-guaranteed annuity is $\ddot{a}_{40}(0_{35}, 1_{\infty})$.

One must be careful with the 'guaranteed' portion here, which is deferred and therefore not completely guaranteed. The value of this at time 25 is $\mathrm{Val}_{25}(0_{25}, 1_{10}; v)$. To get the value at time 0, we multiply this by $y_{40}(25)$. It would be wrong to multiply by $v(25)$, since these payments are not made if (40) dies before age 65, so that accumulation is at interest and survivorship for the first 25 years. The total premium is therefore

$$\ddot{a}_{40}(0_{35}, 1_{\infty}) + y_{40}(25)\mathrm{Val}_{25}(0_{25}, 1_{10}; v).$$

The more traditional way of writing this, incorporating our time shifting notation to handle non-constant interest is

$$v(35)\,{}_{35}p_{40}\ddot{a}_{[40]+35} + v(25)\,{}_{25}p_{40}\ddot{a}(1_{10}; v \circ 25).$$

To handle the general case of guaranteed payments, let $\mathbf{u}$ denote a vector of benefit payments that are guaranteed provided the annuitant lives to time g. So in Example 4.3, $g = 0$ and $\mathbf{u} = (1_{10})$, whereas in Example 4.4, $g = 25$ and $\mathbf{u} = (0_{25}, 1_{10})$. Clearly, we can assume that the first g entries are equal to 0 so that $\mathbf{u}$ is of the form $(0_g, \mathbf{w})$. Then, reasoning as above, the present value of the guaranteed payments is

$$_g p_x \ddot{a}(\mathbf{u}; v) = v(g)\,_g p_x \ddot{a}(\mathbf{w};\ v \circ g),$$

which equals the present value with respect to the discount function v, multiplied by the probability that the payments will be made.

See Exercise 4.15 for one variation on the guaranteed payment concept. We present more modifications in the next chapter. See Example 5.11 and Exercise 5.19.

4.5 Deferred annuities with annual premiums

Deferred annuities are often purchased by a series of annual premiums rather than a single premium. The premium payment period can be any length that does not exceed the deferred period, so that premiums stop when the benefit payments commence. Deferred annuities with annual premiums are frequently used to provide pensions. An individual, together with his/her employer, will pay premiums during his/her working years in order to provide income beginning at retirement.

The sequence of annual-premium payments in the contract can be described by means of a *premium pattern vector*. This is a vector ρ with $\rho_0 = 1$. Then, ρ_k denotes the ratio of the premium payable at time k to the premium payable at time 0. If we know the premium payable at time 0, often referred to as the *initial* premium, the premium pattern vector will determine all premiums. Namely, if π_k denotes the premium payable at time k,

$$\pi_k = \pi_0 \rho_k. \tag{4.12}$$

The vector $\pi = (\pi_0, \pi_1, \ldots)$ will be called the *premium vector*.

Suppose that (x) purchases a life annuity to begin at age $x + n$. The most common pattern in practice would be $\rho = (1_n)$, signifying a level premium payable until the income commences. However, many other patterns are encountered. Some annuitants may prefer to pay a higher premium but for a shorter period, adopting a pattern of (1_m) for some $m < n$. Others may prefer to pay a lower premium for an initial period, and then more at the end. For example, the pattern $\rho = (1_k, 2_{n-k})$ would call for the premium to double after k years.

The premium payments also constitute a life annuity on (x) since they will cease upon (x)'s death. (In this case they are paid *by* (x) rather than to (x).) To achieve the goal that premiums together with investment earnings are sufficient to provide the required annuity payments, the actuary will set premiums to be actuarially equivalent to benefits, with respect to the interest and survivorship discount function.

Example 4.5 (Figure 4.2). An annuity on (40) provides for 1 annually for life, beginning at age 65. Nothing is paid for death before 65. Annual premiums are payable for 25 years beginning at age 40. The premium reduces by one-half after 15 years. Find a formula for the initial premium, assuming $\omega = 110$.

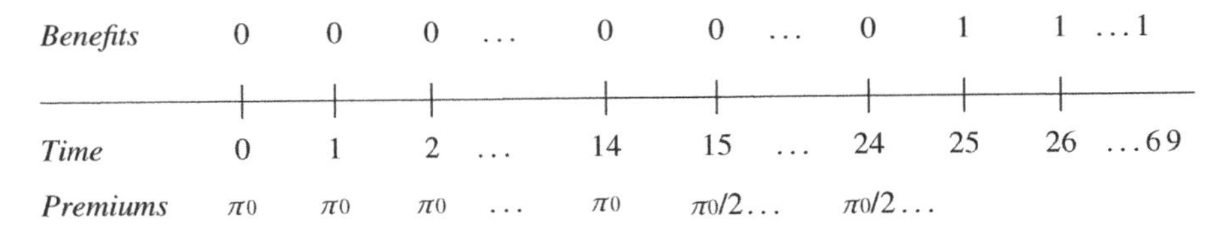

Figure 4.2 Example 4.6

Solution. Equating present values, we have

$$\ddot{a}_{40}(\mathbf{c}) = \ddot{a}_{40}(\pi) = \pi_0 \ddot{a}_{40}(\rho),$$

where

$$\mathbf{c} = (0_{25}, 1_{45}), \qquad \rho = (1_{15}, 0.5_{10}).$$

Solving, we obtain

$$\pi_0 = \frac{\ddot{a}_{40}(\mathbf{c})}{\ddot{a}_{40}(\rho)}.$$

4.6 Some practical considerations

4.6.1 Gross premiums

The annuity premiums, both single and annual, that we calculated in the previous sections are referred to as *net* premiums or *benefit* premiums. They are the premiums that are required to provide the benefits. The premiums that are actually charged in practice are known as *gross premiums*, and involve other factors in their calculation. The insurer must make provision for amounts to cover expenses and profits. In addition, a contingency charge is added to provide for adverse experience. After all, interest earnings may fall short of those predicted by the discount function or people may live longer than predicted by the given life table. Moreover, marketing considerations inevitably play a role. If an insurer wants to stay in operation it must ensure that the premiums it actually charges are competitive. The many details that go into the calculation of the actual gross premiums are beyond the scope of this book, although we will touch on this topic in Chapter 12. In any event, the initial basic step is to calculate the net premiums as done in the examples of this chapter.

4.6.2 Gender aspects

The role of gender in life annuities is a controversial issue. As noted above, the participants in a 'female age x room' can expect to live longer than those in the 'male age x room'. The females will receive less money in forfeiture due to death, and therefore must invest more to receive back the same amounts. If the insurer uses separate life tables for males and females, the result is that females must pay higher premiums to receive the same annuity benefits. This has in particular caused controversy in connection with *defined contribution* pension plans. In such a plan, the employee and employer make specified contributions into a fund and the accumulated amounts are then paid out as some form of life annuity when the employee retires. (This is as opposed to a *defined benefit* plan, where the annuity income is fixed as a function of length of service and salary.) If separate life tables are used, the same level of contributions will purchase smaller pension benefits for a female than for a male of the same age. This has resulted in charges of discrimination by gender. There have been lengthy debates on this issue and many points of view put forward. Some question the fairness of considering only gender while ignoring other factors. For example, a female smoker may actually expect to live a shorter period than a male nonsmoker of the same age, but will still receive a smaller pension if only gender is taken into account. The issue is complicated and we will not go into it further here. The trend lately has been to agree that the gender discrimination should be avoided, and many pension plans now use 'unisex life tables' that are a blend of the corresponding male and female tables. The result is that, for defined contribution plans, males receive somewhat lower pension benefits, and females receive somewhat higher pension benefits, than they would have received had separate male and female tables been used.

4.7 Standard notation and terminology

Standard notation for life annuities generally follows that for general annuities as introduced in Section 2.14, except that the age x is inserted as a right subscript, as we have done. So, for example,

$$\begin{aligned}
\ddot{a}_{x:\overline{n}|} &= \ddot{a}_x(1_n), \qquad a_{x:\overline{n}|} = \ddot{a}_x(0_1, 1_n);\\
\ddot{s}_{x:\overline{n}|} &= y_x(n)^{-1}\ddot{a}_x(1_n), \qquad s_{x:\overline{n}|} = y_x(n)^{-1}\ddot{a}_x(0, 1_n);\\
Ia_{x:\overline{n}|} &= \ddot{a}_x(0, 1, 2, \ldots, n), \qquad I\ddot{a}_{x:\overline{n}|} = \ddot{a}_x(1, 2, \ldots, n);\\
Da_{x:\overline{n}|} &= \ddot{a}_x(0, n, (n-1), (n-2), \ldots, 1), \qquad D\ddot{a}_{x:\overline{n}|} = \ddot{a}_x(n, (n-1), (n-2), \ldots, 1);\\
{}_{k|n}\ddot{a}_x &= {}_{k-1|n}a_x = \ddot{a}_x(0_k, 1_n).
\end{aligned}$$

The last item is the present value of a deferred annuity where the first payment is at time k, and there are n payments. In some cases, as here, a duration appears as a left subscript rather than under an 'angle'. To understand the second symbol, note that the unadorned a indicates a first payment at time 1, and the $k-1$ before the deferred symbol indicates that payments are $k-1$ years later than this, that is at time k, which is the same as the first symbol.

The present value of a k-year pure endowment of 1 is denoted by ${}_kE_x$, another example where duration is given by a left subscript.

We have already introduced the symbol $\ddot{a}_x$, in which the omitted duration symbol indicates a whole life annuity with payments of 1 for life. In the standard notation this carries over to the other symbols as well. That is,

$$\ddot{a}_x, \qquad a_x, \qquad I\ddot{a}_x, \qquad Ia_x, \qquad {}_{k|}\ddot{a}_x = {}_{k-1|}a_x$$

are all defined as the corresponding symbols above, with benefits continuing for as long as (x) is alive. In other words, n is taken to be $\omega - x - j$, where j is the time of the first nonzero payment.

Traditional actuarial texts include many identities relating these quantities. For the most part, they all follow from the splitting identity (4.9). For example (assuming constant interest), we get the following variation of (4.10), written in standard notation as

$$\ddot{a}_{x:\overline{n+k}|} = \ddot{a}_{x:\overline{k}|} + v^k {}_kp_x \ddot{a}_{x+k:\overline{n}|}.$$

The term *annuity certain* or *fixed term annuity* is often used to describe annuities where the payments are certain to be made, as distinguished from life annuities.

4.8 Spreadsheet calculations

In order to handle problems involving a life table, we proceed as follows. As in the Chapter 2 spreadsheet we put duration in column A, and interest rates in Column B, starting at row 10. We insert the life table in column N (columns in between are reserved for other purposes, described in later chapters), with q_y entered in cell $N(10 + y)$. Our sample table can be input as follows. Enter the two parameters 0.0005 in cell N3, 1.09 in cell N4. This allows us to change parameters when desired. Then enter the formula

$$= 1 - \text{EXP}(-N\$3 * N\$4\text{\^{}}\$A10)$$

in cell N10 and copy down to cell N128. Enter 1 in cell N129. It is important to also ensure that the remaining cells in Column N are filled with zeros up to at least N250.

In cell C1 we insert the age x. We then use the index function in Excel to select the pertinent mortality rates, putting q_{x+t} in cell C (10+t) This is done by inserting the formula

$$=\text{INDEX}(N\$10 : N\$250, C\$1 + \$A11)$$

in cell C10 and copying down. Entries of zero will appear after duration $\omega - x$ coming from the extra zeros in Column N.

We calculate $v(k)$ in column D as in Spreadsheet 2, entering 1 in cell D10, = D10/(1+B10) in column D11 and copying down.

We calculate the vector $\mathbf{y}_x$ in column E, using the recursion (4.5). The entry in E10 is 1, the entry in E11 is

$$= E10 * (1 + B10)\text{\^{}}-1 * (1 - C10)$$

and we copy down.

We insert the cash flow vector $\mathbf{c}$ in column F, and calculate the value $\ddot{a}_x(\mathbf{c})$ in cell F8 with the formula.

$$=\text{SUMPRODUCT}\ (\$E10 : \$E129, F10 : F129).$$

For example at a constant interest rate of 6%, $\ddot{a}_{50}(1_{20}) = 11.5957$.

For annual-premium deferred annuities, we enter the premium pattern vector in Column H. We copy the formula in F8–H8 (reserving column G for a later purpose). In I6, we enter = F8/H8, the initial premium. In I10, we enter =$I$6*H10 and copy down, which inserts the premium vector in column I. Copying the formula in H8–I8 provides a check. We should get the same value as in F8.

As a check, at 6% interest, for an annuity on (50) providing 1 per life beginning at age 60, the level annual premium payable for 10 years is 0.855.

We will leave it to the interested reader to modify this spreadsheet to handle guaranteed periods.

Note that this spreadsheet can be used to handle short duration problems with the mortality rates given individually, such as we have in the examples and exercises. We can ignore the given age, and simply insert the values of q_{x+t} that we want directly in column C. Remember however *not* to save changes when closing, or alternatively, copy up in column C to restore the formulas when you next use the spreadsheet.

Exercises

Type A exercises

4.1 Redo Example 4.1, only suppose now that interest rates are 50% for the first 2 years and 100% after that.

4.2 A group of individuals age 40, each invest 1000 in a fund earning interest at 5%. At the end of 20 years the fund is divided equally among the survivors. If ${}_{20}p_{40} = 0.8$, how much does each get?

4.3 You are given $q_{60} = 0.20,\ q_{61} = 0.25,\ q_{62} = 0.40,\ q_{63} = 0.50$. The interest rate is a constant 5% for the first 2 years, and 7% after that. A 5-year life annuity on (60) provides for payments of $100(1 + k)$ at time k, where $k = 0, 1, 2, 3, 4$.

(a) Find the present value.

(b) Suppose that instead of being a straight life annuity, the first three annuity payments are guaranteed regardless of whether (60) is alive or not. Find the present value now.

4.4 You are given that interest is constant and that

$$\ddot{a}_{40}(1_{30}) = 21, \qquad \ddot{a}_{60}(1_{10}) = 6, \qquad v^{20}{}_{20}p_{40} = 0.8.$$

Calculate $\ddot{a}_{40}(1_{20})$.

4.5 The present values of a 10-year life annuity on (60), and a 10-year compound interest annuity, with annual payments of 1, are given, respectively by $\ddot{a}_{60}(1_{10}) = 5$ and $\ddot{a}(1_{10}; v) = 6$. A person age 60 has 100 000 that he plans to pay as a single premium for a life annuity, beginning at age 60. This will provide a level income of 12 500 per year for life. What will the yearly income be if, instead of a straight life annuity, he purchases a life annuity with a 10-year guaranteed period?

Type B exercises

4.6 Mortality is given by $q_{52} = 0.1,\ q_{53} = 0.2$. The investment discount function v is given by $i_0 = 0.2,\ i_1 = 0.25,\ i_2 = 0.25,\ i_3 = 0.5$. Calculate $y_{50} \circ 2(2)$ and $y_{52}(2)$, using v in both cases, to show that $y_x \circ k$ is not equal to y_{x+k} when interest is not constant.

4.7 If interest is constant and q_x is a constant q for all x, find an expression for $\ddot{a}_x$ in terms of v and q.

4.8 Show that, at a constant zero rate of interest, $\ddot{a}_x(0, 1_\infty) = e_x$. Give an intuitive explanation of this fact.

4.9 A life annuity on (x) provides for annual benefit payments for life beginning at age x. The initial benefit payment is 1000 and each subsequent payment increases by 6% (i.e., the benefit payment at age $x + 1$ is 1060, the payment at age $x + 2$ is $1000(1.06)^2$, and so on). The first 10 benefit payments are *guaranteed* and will be made regardless of whether (x) is alive or not. The interest rate is a constant 6%. You are given that ${}_{10}p_x = 0.9$ and $e_{x+10} = 10$. Find the present value of this contract.

4.10 A life annuity contract on (80) has a present value of 3.14. The annuity benefits at both age 80 and 81 are 1. The interest rate in the first year is 25% and $p_{80} = 0.8$. Suppose that the value of p_{81} increases by 10% while all other mortality rates remain unchanged. What is the new present value of the contract?

4.11 Given that $\ddot{a}_{40} = 15$, $\ddot{a}_{[40]+25} = 10$, $v(25) = 0.5$, ${}_{25}p_{40} = 0.4$, find the net annual premium, payable for 25 years beginning at age 40, for a deferred annuity, paying 1000 yearly for life, with the first benefit payment at age 65.

4.12 Suppose that for all x, $q_{x+1} \geq q_x$. Show that $\ddot{a}_{x+1} \leq \ddot{a}_x$. Does this remain true if we remove the monotone condition on q_x?

Spreadsheet exercises

The following exercises are to be done using the sample life table of Section 3.7.

4.13 A deferred life annuity on (40) provides for income of 10 000 per year beginning at age 65. Nothing is paid if death occurs before age 65. This is purchased by annual premiums payable for 25 years beginning at age 40. Premiums increase each year by 10% of the initial premium, that is, $\pi_k = (1 + 0.1k)\pi_0$. Interest rates are 5% for the first 10 years and 6% thereafter. Find the initial premium π_0.

4.14 A group of individuals age 30 each agree to invest 1000 per year for 40 years beginning at age 30. At time 40, the fund is divided up among all the survivors. If interest rates are 5% for the next 10 years, 6% for the following 10 years and 8% after that, how much does each survivor receive?

4.15 A single-premium life annuity on (x) provides 1 unit per year for life, beginning at time 1, with a n-year guaranteed period, where n is the smallest integer greater than or equal to the premium. (This is known as an *instalment refund* annuity. It guarantees that at least the full amount of the premium, without interest, will be returned.) Assume a constant interest rate of 4%. Find n and the single premium if (a) $x = 40$, (b) $x = 70$. (Note that there is no direct method of calculation. A trial and error procedure is called for).

4.16 Interest rates are 8% for the first 25 years and 6% thereafter. Compare $\ddot{a}_{[40]+20}$ and $\ddot{a}_{[50]+10}$. Which one of these is smaller? Answer this before any calculation, and then do the actual calculation to verify your answer.

5

Life insurance

5.1 Introduction

A life insurance policy is a contract between the *insurer* and another party known as the *policyholder*. In return for a payment of premiums, the insurer will pay a predetermined amount of money, known as a *death benefit*, upon the death of the policyholder. The amount of the benefit can vary with the time of death. In practice, this money will be paid immediately upon death (or more realistically, a short time after, to allow for processing the claim) but for mathematical convenience we assume in this chapter that it will be paid at *the end of the year of death*. For example, if the policyholder purchases a policy on January 1 and dies a week later, our assumption means that the death benefit will not be paid until December 31. We will consider the more realistic situation of payment at the moment of death in Chapter 8.

The reader should distinguish carefully between life annuity and life insurance contracts. The life annuity provides a sequence of periodic benefit payments. The typical life insurance contract provides only a single benefit payment, paid on the occasion of death.

5.2 Calculating life insurance premiums

Consider a policy on (x). Let b_k be the amount that will be paid at time $k + 1$ for death between time k and time $k + 1$. We will refer to the vector

$$\mathbf{b} = (b_0, b_1, \ldots, b_{\omega-x-1})$$

as the *death benefit vector*.

Notation The reader is cautioned that many authors use the subscript on b to refer to the time of payment. What we call b_k, they would call b_{k+1}, since it is paid at the end of the year, which is time $k + 1$. We prefer the convention above. All our vectors are then indexed from 0

Fundamentals of Actuarial Mathematics, Third Edition. S. David Promislow.

Companion website: http://www.wiley.com/go/actuarial

to $\omega - x - 1$. In particular, this facilitates matters when dealing with contracts that combine annuity and insurance benefits.

Suppose we have fixed an investment discount function v and a life table. We want to calculate the net single premium for the above policy, which we will denote by $A_x(\mathbf{b})$. (The A is a standard symbol that probably came from the word 'assurance', an older version of the word 'insurance'.) The principle to determine this is the same as that used for annuities. The total premiums, together with all investment earnings, must be sufficient to provide all the death benefits.

To derive the formulas, we will follow the annuity procedure and begin with the case where $\mathbf{b} = \mathbf{e}^k$. This is a policy in which 1 is paid at time $k+1$ providing (x) dies between the ages of $x+k$ and $x+k+1$. All of the other death benefits are of zero amount. Suppose that we have ℓ_x people age x who each purchase this same contract. Out of these, there will be d_{x+k} individuals who die between the ages of $x+k$ and $x+k+1$, and each of them will receive 1 at time $k+1$. The total present value of all these death benefits will be $v(k+1)d_{x+k}$.

This must be equal to the total amount collected in premiums, which will be $\ell_x A_x(\mathbf{e}^k)$. Therefore

$$A_x(\mathbf{e}^k) = v(k+1)\frac{d_{x+k}}{\ell_x}.$$

The general policy can be viewed as a sequence of 1-year policies as above, one for each value of k, where the kth policy pays b_k at time $k+1$ for death in the previous year. The premium for such a 1-year policy is just $b_k v(k+1)d_{x+k}/\ell_x$ and the total premium is obtained by summation. We have

$$\begin{aligned} A_x(\mathbf{b}) &= \sum_{k=0}^{\omega-x-1} b_k v(k+1)\frac{d_{x+k}}{\ell_x} \\ &= \sum_{k=0}^{\omega-x-1} b_k v(k+1)({}_kp_x - {}_{k+1}p_x) \qquad (5.1)\ddagger \\ &= \sum_{k=0}^{\omega-x-1} b_k v(k+1)\,{}_kp_x\, q_{x+k}. \end{aligned}$$

The reader should note that the expression above follows the same pattern as the annuity single premium. It is a pattern that we will encounter many more times in subsequent material. Namely, we sum up a number of terms, each of which consists of *three* factors,

$$\text{amount} \times \text{interest discount factor} \times \text{probability that payment is made}. \qquad (5.2)$$

In the case of insurance, formulas [3.5(a)–(c)] give three different expressions for the probability that the payment will be made, giving us the three different ways of writing $A_x(\mathbf{b})$. Each will be useful in certain cases.

As with annuities, the notation will suppress the death benefit vector $1_\infty = 1_{\omega-x}$. Accordingly, A_x will be the net single premium for a policy paying 1 at the end of the year of death, whenever it occurs.

Example 5.1 Suppose that $q_{60} = 0.2$, $q_{61} = 0.4$. $q_{62} = 0.5$ and $i = 100\%$. A policy sold to (60) provides for benefits at the end of the year of death of 80 for death in the first year, 75 for death in the second year, and 100 for death in the third year. If the insured lives to age 63, the policy terminates and nothing is paid. Find the net single premium.

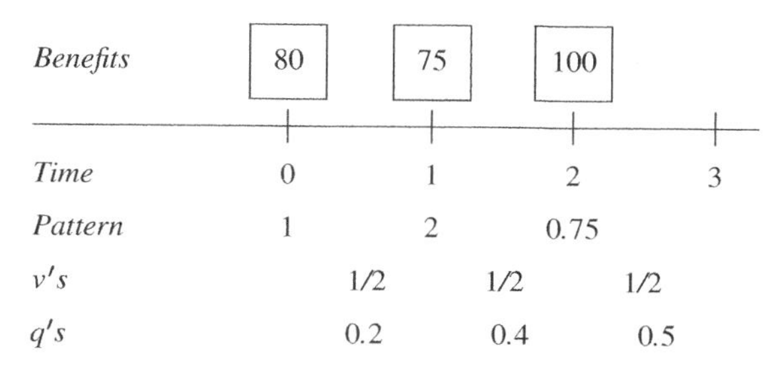

Figure 5.1 Example 5.1

Solution. We again can make use of a time diagram as shown in Figure 5.1. The death benefit amount is put at the *beginning* of the year to which it is applicable. That is, b_k is inserted above time k. Our convention here is to enclose these amounts in a box, to distinguish them from annuity benefits. The amount b_k must be multiplied by the previous interest discount factors, as well as by the interest discount factor for the year starting at time k, since it is paid at the end of the year. Moreover it is multiplied by the previous p values and as well by q_{x+k}, as indicated in the third formula in (5.1), which together give the probability of living to time k and then dying in the following year. In this case, the net single premium is

$$\left(80 \times \frac{1}{2} \times 0.2\right) + \left(75 \times \frac{1}{4} \times 0.8 \times 0.4\right) + \left(100 \times \frac{1}{8} \times 0.8 \times 0.6 \times 0.5\right) = 17$$

As with life annuities, one can as an alternative construct a partial life table, which automatically gives you the multiplication of the probabilities, and then use the first formula in (5.1). So for example in this case, starting with $\ell_{60} = 1000$, we have in turn, $d_{60} = 200$, $\ell_{61} = 800$, $d_{61} = 320$, $\ell_{62} = 480$, $d_{62} = 240$.

Normally, the policyholder does not pay for insurance by a single premium but rather by a sequence of periodic premiums. We assume in this chapter that premiums are paid annually. The premiums will be given by a premium vector $\pi = (\pi_0, \pi_1, \ldots, \pi_{\omega-x-1})$ as defined for deferred annuities in Section 4.5. Following the principle used there, we want the net single premium to be equal to the present value of the premiums, with respect to the interest and survivorship accumulation function. This means that

$$A_x(\mathbf{b}) = \ddot{a}_x(\pi) = \pi_0 \ddot{a}_x(\rho),$$

so that

$$\pi_0 = \frac{A_x(\mathbf{b})}{\ddot{a}_x(\rho)}, \qquad (5.3)\ddagger$$

where ρ is the premium pattern vector.

We follow the terminology used for single premiums and call these premiums *net annual premiums* or *annual benefit premiums*. We will later encounter examples of premiums that are different from net premiums, but our convention is that unless otherwise mentioned, all premiums will be net.

Example 5.2 Suppose that the insurance in Example 5.1 is to be purchased by three annual premiums, beginning at age 60, where the second premium is double the first, and the third premium is three quarters of the first. Find the premiums.

Solution. The premium pattern vector is given by $\rho = (1, 2, 0.75)$. We have

$$\ddot{a}_{60}(\rho) = 1 + \left(2 \times \frac{1}{2} \times 0.8\right) + \left(0.75 \times \frac{1}{4} \times 0.8 \times 0.6\right) = 1.89.$$

so that the initial premium will be $17/1.89 = 8.99$. The insured will then pay 8.99 in the first year, 17.99 in the second year and 6.75 in the third year.

5.3 Types of life insurance

Life insurance policies are traditionally classified into different types. *Term insurance* provides death benefits for a fixed number of years (similar to the temporary annuity). After the expiration of the term, coverage ceases and there are no more benefits. Example 5.1 involved such a policy, with a term of 3 years. *Whole life* insurance provides death benefits for life, so some payout on the policy is certain to occur. In our mathematical model we will not need to distinguish between the two types. We will assume that all policies will continue to age ω. For term insurance running for n years (which we refer to as n-year term) we simply will have $b_k = 0$ for $k \geq n$. Nonetheless, in the next chapter we will see that there are differences in the nature of these two types and that whole life insurance has a *savings* component as well as an *insurance* component. Another common type of contract is *endowment insurance*, which we will discuss in detail in the next section. A more modern development, known as *universal life* will be described in Chapter 13.

5.4 Combined insurance–annuity benefits

It is possible to combine both life insurance benefits and life annuity benefits in the same contract. One of the most popular types of such a policy is known as *endowment insurance*. This provides for a payment at some future time n if (x) is then alive, and in addition a death benefit if (x) dies before time n. Such a policy is known as *n-year endowment insurance* or *endowment insurance* at age $x+n$, since it combines a *pure endowment* (as defined in Section 4.2 with insurance. It is usually marketed as a savings plan, whereby the policyholder, by paying premiums each year, will accumulate a certain sum at time n. In addition, the policyholder is protected with life insurance if he/she dies before accumulating the desired amount.

It should be noted that a whole life policy is similar to endowment insurance since the policyholder is guaranteed some death benefit. Mathematically it can be viewed as endowment insurance at age ω, and in practice it actually is interpreted in this way. Since it is to the advantage of the policyholder, the insurer will assume that everybody dies by age ω, unlike the case of a life annuity, and pay the death benefit at age ω to all survivors.

To calculate the single premium for endowment insurances, we simply view it as two separate contracts and add the premiums.

Example 5.3 An insurance policy on (x) provides 1 unit payable at time n if (x) is then alive, plus 1 unit payable at the end of the year of death if (x) dies before time n. Level annual premiums of P are payable for n years. Find a formula for P.

Solution. The single premium for the death benefit is $A_x(1_n)$. The single premium for the pure endowment is $y_x(n)$. The premium pattern vector ρ is (1_n). The total single premium for the contract is then $A_x(1_n) + y_x(n)$ and

$$P = \frac{A_x(1_n) + y_x(n)}{\ddot{a}_x(1_n)}.$$

Example 5.3 is a very common type of endowment insurance with a level death benefit equal to the pure endowment amount, and level premiums payable for the full term. This is not essential, however, and many other combinations are possible.

Example 5.4 Consider a 20-year endowment insurance with the death benefit equal to 1 unit for the first 10 years, and 2 units for the second 10 years. The amount of the pure endowment is 3 units. Level annual premiums of P are payable for 15 years. Find a formula for P.

Solution. Calculating as in the previous example,

$$P = \frac{A_x(1_{10}, 2_{10}) + 3y_x(20)}{\ddot{a}_x(1_{15})}.$$

Example 5.5 (Figure 5.2). Suppose that $q_{60} = 0.2, q_{61} = 0.4$ and $i = 100\%$. A 2-year policy provides for benefits at the end of the year of death of 80 for death in the first year, and 75 for death in the second year. In addition, there is a pure endowment of 70 paid at age 62 if the policyholder is then alive. This is purchased by two-level annual premiums. Calculate the premium.

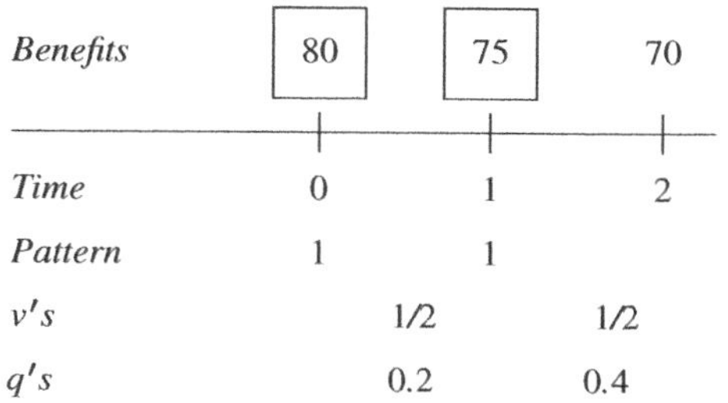

Benefits	80		75		70
Time	0		1		2
Pattern	1		1		
v's		1/2		1/2	
q's		0.2		0.4	

Figure 5.2 Example 5.5

Solution. The present value of death benefits is

$$A_{60}(80,75) = \left(80 \times \frac{1}{2} \times 0.2\right) + \left(75 \times \frac{1}{4} \times 0.8 \times 0.4\right) = 14.$$

The present value of the pure endowment is

$$\ddot{a}_{60}(0,0,70) = 70 \times \frac{1}{4} \times 0.8 \times 0.6 = 8.4.$$

The present value for the premium pattern vector is

$$\ddot{a}_{60}(\rho) = \ddot{a}_{60}(1,1) = 1 + \frac{1}{2} \times 0.8 = 1.4$$

By (5.3)

$$\pi_0 = \pi_1 = \frac{14 + 8.4}{1.4} = 16.$$

Another common type of combined policy is a deferred annuity that provides for death benefits during the deferred period. This is somewhat similar in nature to endowment insurance. The difference is that the accumulated savings are paid out as an annuity rather than as a single payment.

Example 5.6 (Figure 5.3). A contract on (40) provides for annuity benefits of 1 per year for life, beginning at age 65. If (40) dies before age 65, a death benefit of 10 will be paid at the end of the year of death. Level annual premiums of P are payable for 25 years. Find a formula for P.

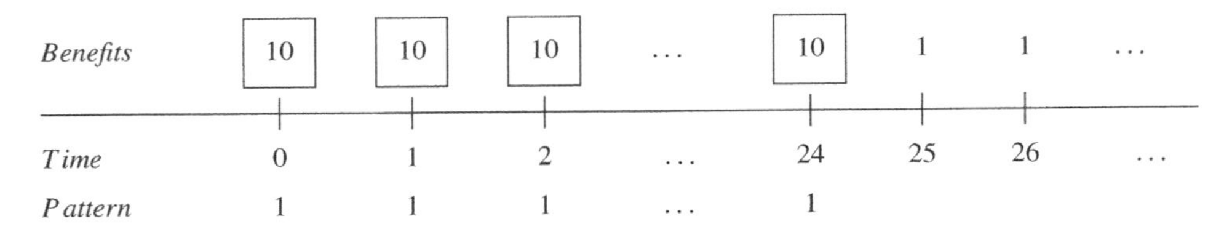

Figure 5.3 Example 5.6

Solution. We calculate this, as in the examples above, by adding the single premiums for the death benefits and the annuity benefits and dividing by the annuity for the premium pattern vector. The result is

$$P = \frac{10A_{40}(1_{25}) + \ddot{a}_{40}(0_{25}, 1_{\infty})}{\ddot{a}_{40}(1_{25})}.$$

The second term in the numerator can be written alternatively as $v(25)_{25}p_{40}\ddot{a}_{[40]+25}$.

An interesting variation on the above involves a popular marketing device, which is to issue policies with a return-of-premium feature.

Example 5.7 (Figure 5.4). A contract on (40) provides for annuity payments of 1 per year for life beginning at age 65. Should (40) die before age 65, there will be a return of all premiums (without interest) paid prior to death. Level annual premiums of P are paid for 25 years. Find a formula for P.

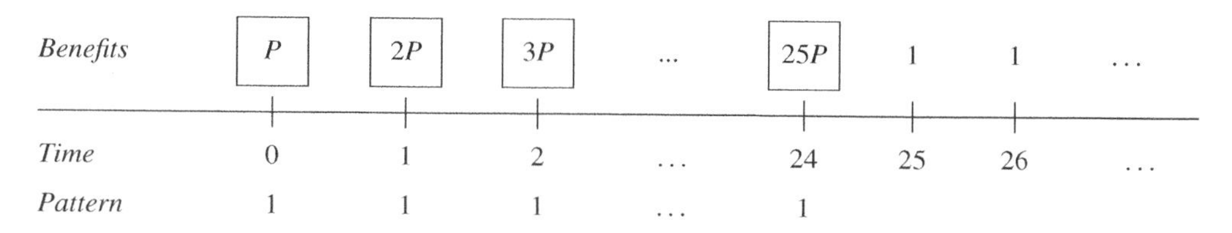

Figure 5.4 Example 5.7

Solution. This is basically the same type of problem as above, except that the benefits depend on the unknown P. We cannot obtain P directly, but we can set up an equation to solve for it. Suppose (x) dies between time k and $k + 1$, where k is between 0 and 24 inclusive. The insured will have paid $k + 1$ premiums of P and these will be returned at time $k + 1$. The death benefit vector is then

$$\mathbf{b} = (P, 2P, 3P, \ldots, 25P) = P\mathbf{j},$$

where $\mathbf{j} = (1, 2, \ldots, 25)$. Equating present values of the premiums and benefits,

$$P\ddot{a}_{40}(1_{25}) = PA_{40}(\mathbf{j}) + \ddot{a}_{40}(0_{25}, 1_{\infty}),$$

from which

$$P = \frac{\ddot{a}_{40}(0_{25}, 1_{\infty})}{\ddot{a}_{40}(1_{25}) - A_{40}(\mathbf{j})}. \tag{5.4}$$

Example 5.8 Consider the same policy as above, except that the premiums are returned *with interest* as determined by the function v. Find a formula for P.

Solution. One method is to simply use (5.4), where the vector $\mathbf{j}$ is changed from $j_k = k + 1$ to

$$j_k = \mathrm{Val}_{k+1}(1_{k+1}; v).$$

Another approach, which produces the solution in a much easier form for calculation is found through the following reasoning. In the first 25 years the accumulation is according to the discount function v rather than y_{25}. To see this clearly, consider the room-box description of Chapter 4. Upon death of a participant before age 65, the person's box is opened up, but the amount paid back as a death benefit is exactly the same as the amount that was there in the first place. The other participants will receive no survivorship earnings. Therefore, each survivor will have accumulated at age 65 the amount $P\,\mathrm{Val}_{25}(1_{25};v)$. This amount must be sufficient to provide an annuity of 1 unit per life beginning at age 65. Thus

$$P = \frac{\ddot{a}_{[40]+25}}{\mathrm{Val}_{25}(1_{25};v)}.$$

We will present an alternative solution to this problem in Section 6.7.

A generalization of the above is concerned with partial refunds.

Example 5.9 Now consider Example 5.8, except that the payment upon death is k times the accumulated premiums with interest, where $0 \le k \le 1$.

For each premium P, the portion kP accumulates at interest and $(1-k)P$ accumulates with interest and survivorship. Equating values at time 25,

$$k\,P\,\mathrm{Val}_{25}(1_{25}, v) + (1-k)\,P\,\mathrm{Val}_{25}(1_{25}; y_{40}) = \ddot{a}_{[40]+25}$$

and we can solve to get

$$P = \frac{\ddot{a}_{[40]+25}}{k\,\mathrm{Val}_{25}(1_{25};v) + (1-k)\ \mathrm{Val}_{25}(1_{25};y_{40})}.$$

For $k=1$ this reduces to the solution to Example 5.8 and for $k=0$, it is a special case of Example 4.5.

Remark For those preferring an alternative to the square bracket notation, there are various other ways in which to write the numerator of the last two examples. For example, $\mathrm{Val}_{25}({}^{25}1_{\infty}; y_{40})$ or $\ddot{a}_{40}(0_{25}, 1_{\infty})/y_{40}(25)$. Of course with constant interest, the simplest choice is just $\ddot{a}_{65}$.

The following variation is suggested by provisions in some pension plans where an employee must remain in the plan for a minimum period in order to get credit for premiums paid by the employer. A more realistic version will be given in Section 13.3.

Example 5.10 Redo Example 5.8 only with the provision that premiums are returned with interest upon death only if (40) lives to age 45. Nothing is returned if death occurs in the first 5 years.

Solution. There does not appear to be any convenient way to apply the simplifying approach of the last two examples. We can however use (5.4) with the vector $\mathbf{j}$ given by

$$j_k = \begin{cases} 0 & \text{if } 0 \le k < 5, \\ \mathrm{Val}_{k+1}(1_{k+1}; v) & \text{if } 5 \le k < 25. \end{cases}$$

Indeed, (5.4) is a flexible formula which can be adapted to a variety of premium return provisions, including cases where the interest rate on the premiums may differ from that used in the discount function.

We now present an example involving a more complex calculation.

Example 5.11 An annuity on (x), purchased by a single premium S, provides for 1 per year for life, beginning at time 1. If (x) dies before a total income of S has been paid out, the difference between S and the total income received will be refunded at the end of the year of death, so that the policyholder will at least receive total income equal to the single premium. (e.g., if S is 20 and (x) dies between time 5 and time 6, a death benefit of 15 would be paid at time 6.) Describe a procedure for calculating S.

Solution. The premium S must satisfy

$$S = \ddot{a}_x(0, 1_\infty) + A_x(\mathbf{b}), \qquad \text{where } b_k = \max\{S - k, 0\}.$$

Since $\mathbf{b}$ depends on S, there is no direct way to solve this and an iterative numerical procedure must be employed. One guesses at an initial value of S (it will be close to but greater than $\ddot{a}_x(0, 1_\infty)$), and then continues to adjust the value of S until the right hand side above also equals this value. This can be carried out automatically in Excel with the *goal-seek function.* Exercise 5.20 gives some particular examples.

The contract in Example 5.9 is usually termed a *cash refund annuity* and it is a variation on the instalment refund idea introduced in Exercise 4.15. In practice, both of these are somewhat more complicated than described since the refund is based on the gross premium rather than the net.

5.5 Insurances viewed as annuities

A comparison of (4.1) and the third expression in (5.1) shows that

$$A_x(\mathbf{b}) = \ddot{a}_x(\mathbf{c}), \quad \text{where } c_k = v(k, k+1)b_k q_{x+k}. \tag{5.5}‡$$

Let $\mathbf{w}_x$ denote the vector with entries $(w_x)_k = v(k, k+1)q_{x+k}$. We can then write (5.5) in compact form as

$$A_x(\mathbf{b}) = \ddot{a}_x(\mathbf{w}_x * \mathbf{b}). \tag{5.6}$$

We can verify this intuitively as follows. Suppose (x) wishes to pay a single premium for a life insurance policy running over several years, based on the death benefit vector $\mathbf{b}$. The insurer refuses, claiming that it only sells life insurance policies for 1 year at a time. It does, however, sell life annuities. The person therefore purchases a life annuity with benefits of $v(k, k+1)b_k q_{x+k}$ at time k if she is alive. She does not keep this annuity payment, but immediately returns it to the insurer to purchase a 1-year life insurance policy paying a death benefit at time $k+1$ if death occurs in the next year. This single premium will purchase a death benefit of exactly b_k. This follows from our previous discussion, but to re-emphasize the point, note that this premium will accumulate to $b_k d_{x+k}/\ell_{x+k}$ at the end of the year, and assuming l_x people engage in this scheme, this amount collected from each of ℓ_{x+k} survivors at time k will be enough to provide b_k to each of the d_{x+k} people who die during the year.

This illustration is of course fanciful, as all insurers sell policies for periods of more than 1 year. (Moreover, as a practical matter it overlooks the fact that 1-year premiums could change over time, as well as the fact that annuity and insurance premiums are based on different tables.) It does, however, provide a useful point of view. Each policyholder can look upon a life insurance policy as a life annuity, providing what is essentially that person's share of the death benefits. One could suppose that these annuity payments are then collected in a separate fund and used to pay all the benefits to those who die during the year. This is often a valuable way of looking at the situation, since all policies can be thought of as life annuities if we wish. The calculation of premiums and other quantities can then be reduced to the general principles outlined in Chapter 2.

This viewpoint allows us to immediately adapt all results obtained for life annuities to the insurance setting. For example, the *splitting identity* for life annuities (4.8) takes the following form for insurances:

$$A_x(\mathbf{c}) = A_x({}_k\mathbf{c}) + y_x(k)A_{[x]+k}(\mathbf{c} \circ k), \tag{5.7}$$

where the bracket on x indicates, as with annuities, that the discount function used is $v \circ k$.

We can also maintain the room-box visualization for life insurance policies, which we introduced for annuities in the previous chapter, and which will provide a useful guide in the next chapter.

The reader is cautioned that while (5.6) is useful conceptually and also for spreadsheet calculation (as illustrated in Section 5.9), it is not always the best for hand calculation. For such purposes, the procedure used in Example 5.1 is usually less subject to arithmetical errors.

5.6 Summary of formulas

In this section we summarize the procedure, developed in the last two chapters, for calculating an annual premium on a general life insurance–annuity contract. We first identify four pertinent vectors: $\mathbf{b}$, the death benefit vector; $\mathbf{c}$, the life annuity benefit vector; $\mathbf{u} = (0_g, \mathbf{w})$, the vector of payments that are guaranteed provided (x) lives to age $x + g$; and ρ, the premium pattern vector. In many cases only one or two of the first three vectors will be applicable, and the others will be set equal to the zero vector. We then calculate the initial premium from the equation

$$\pi_0 \ddot{a}_x(\rho) = A_x(\mathbf{b}) + \ddot{a}_x(\mathbf{c}) + v(g)\,{}_g p_x \ddot{a}(\mathbf{w}; v).$$

This allows for cases where the vectors on the right hand side can themselves depend on π_0.

5.7 A general insurance–annuity identity

5.7.1 The general identity

There is another useful relationship between insurances and annuities.

For the vector $\mathbf{w}_x$ introduced in Section 5.5, we can write

$$(w_x)_k = v(k, k+1) - v(k, k+1)p_{x+k} = 1 - v(k, k+1)p_{x+k} - d_k,$$

and it follows that

$$\mathbf{w_x} = \nabla 1_\infty - \mathbf{d}$$

in the notation of (2.14) with respect to the discount function y_x. From (5.6)

$$A_x(b) = \ddot{a}_x(\mathbf{w}_x * \mathbf{b}) = \ddot{a}_x(\nabla 1_\infty * \mathbf{b}) - \ddot{a}_x(\mathbf{d} * \mathbf{b})$$

and from (2.15) we obtain our main identity

$$A_x(\mathbf{b}) = \ddot{a}_x(\Delta\mathbf{b}) - \ddot{a}_x(\mathbf{d} * \mathbf{b}). \tag{5.8}$$

5.7.2 The endowment identity

We will use (5.8) to derive a well-known actuarial formula, which we call the *endowment identity*. Assume constant interest. The constant discount rate d then factors out as a constant multiple and $\ddot{a}_x(\mathbf{d} * \mathbf{b}) = d\ddot{a}_x(\mathbf{b})$. Suppose $\mathbf{b} = (1_n)$. Then $\Delta\mathbf{b} = (1, 0, 0, \dots, -1)$ where the -1 is in position indexed with n (i.e., in the $(n+1)$th entry since we start with 0). Equation (5.8) says that

$$A_x(1_n) = 1 - y_x(n) - d\ddot{a}_x(1_n).$$

Let $A_{x:\overline{n|}}$ be the net single premium for an n-year, 1-unit endowment insurance. That is, 1 is paid either at time n, or at the end of the year of death if that occurs before time n. (This is the standard symbol for such a premium.) Adding $y_x(n)$ to both sides of this equation, we get the endowment identity,

$$A_{x:\overline{n|}} = 1 - d\ddot{a}_x(1_n). \tag{5.9}$$

This identity is analogous to (2.16) and its derivation as given at the end of Section 2.9. For an interpretation, suppose I lend you 1 unit now, to be repaid in full at the end of n years, or at the end of the year of your death if this occurs before n years. You must also pay interest at the beginning of each year until the loan is paid. The present value of the loan must be equal to the present value of the principal repayments, plus the present value of the interest. The latter is just a temporary life annuity paying d units yearly for n years, beginning at time zero. The present value of the loan is just 1 and the present value of the principal repayments is just $A_{x:\overline{n|}}$, the net single premium for the endowment insurance. We obtain the equation $1 = A_{x:\overline{n|}} + d\ddot{a}_x(1_n)$, which gives (5.9).

If P denotes the level annual premium payable for n years for the 1-unit, n-year endowment insurance, we can also express P in term of annuities by dividing by $\ddot{a}(1_n)$ in (5.9) to obtain

$$P = \frac{1}{\ddot{a}_x(1_n)} - d. \tag{5.10}$$

This identity can be interpreted as follows. Suppose you invest 1 unit. This will provide you with interest earnings of d at the beginning of each year, until such time as you wish to terminate the investment and take back your principal of 1. An alternate scheme is to use the 1 unit to purchase an n-year life annuity, paying $1/\ddot{a}_x(1_n)$ at the beginning of each year for n years, and to use part of these proceeds to purchase a 1-unit, n-year endowment insurance, carrying level annual premiums of P for n years. The insurance will pay you back your principal at the end of n years, or at death if earlier. Before recovering your principal you will have net annual earnings of $1/\ddot{a}_x(1_n) - P$, and this must equal the income of d that you would get from the first alternative. This yields the given identity.

Remark This back-to-back annuity–insurance combination has recently become popular as an investment vehicle. At first glance, it appears that in practice it will produce a lower return than the straight investment, since one must pay expenses on both policies and, in addition, the different life tables used for annuity and insurances will work to the purchaser's disadvantage. (Our simplified model assumes no expenses and that the life table is the same in all cases.) However, the fact that the proceeds of the insurance and annuity contracts receive favourable tax treatment in many jurisdictions often means that the net after-tax return can actually be higher than that of the straight investment.

The above formulas are of course true for a 1-unit whole life insurance, which is just endowment insurance at age ω. We have

$$A_x = 1 - d\ddot{a}_x, \qquad P_x = \frac{1}{\ddot{a}_x} - d, \tag{5.11}$$

where P_x is the net level annual premium payable for life for a 1-unit whole life contract on (x).

5.8 Standard notation and terminology

5.8.1 Single-premium notation

The standard symbol for an insurance single premium is A, as we have given it. As with annuities, this is embellished with superscripts and subscripts to handle the common types of death benefit vectors. For example:

- A_x denotes $A_x(1_\infty)$ as we have already indicated
- $A^1_{x:\overline{n|}}$ denotes $A_x(1_n)$, the net single premium for a 1-unit, n-year term insurance. The superscript 1 above the x signifies that (x) must die before the expiration of the n-year period in order to collect.
- $A_{x:\overline{n|}}$ denotes $A_x(1_n) + y_x(n)$, the net single premium for a 1-unit n-year endowment insurance, as we have indicated above. The subscript $x : \overline{n|}$ signifies that the death benefit is paid upon the first 'failure' of (x) or the n-year period. The life (x) fails upon death and the n-year period fails at the end of n years. Recall that the same subscript is used in the temporary annuity standard symbol to signify that benefits are paid as long as both the life (x) and the n-year period are 'surviving'.

- $A_{x:\overline{n|}}^{\;\;1}$ is another symbol for ${}_nE_x$ (denoted by $y_x(n)$ in our notation). This is the net single premium for a 1-unit, n-year pure endowment. The superscript 1 is now is above the $\overline{n|}$ to signify that the n-year period must fail before (x) does in order for the contract to pay.

- ${}_{k|n}A_x$ denotes $A_x(0_k, 1_n)$. This is *deferred* insurance. A level death benefit of 1 unit begins after k years and continues for n years. Such a policy would not normally be sold by itself but may be combined with other policies.

- ${}_{k|}A_x$ stands for ${}_{k|\omega-x}A_x$ in accordance with the usual practice of omitting duration symbols when the contract continues for life.

- $DA_{x:\overline{n|}}^{1}$ denotes $A_x(n, n-1, \ldots, 1)$. The D stands for decreasing.

- $IA_{x:\overline{n|}}^{1}$ denotes $A_x(1, 2, \ldots, n)$. The I stands for increasing.

- DA_x and IA_x are respectively the above two symbols with $n = \omega - x$, in keeping with the general notational principle discussed in Section 4.7.

5.8.2 Annual-premium notation

The standard symbol for a level net annual premium is P. This is followed by the single-premium symbol. For those single premiums that begin with a capital A, the A is omitted. The premium payment duration t appears before the P on the lower left. If omitted, it means that premiums are paid for the natural duration of the contract. Examples follow:

- P_x is the annual premium, payable for life, for a 1-unit whole life policy on (x).

- ${}_tP_x$ is the annual premium payable for t years for a 1-unit whole life policy on (x).

- $P_{x:\overline{n|}}^{1}$ is the annual premium payable for n years for a 1-unit, n-year term policy on (x).

- ${}_tP_{x:\overline{n|}}$ is the annual premium payable for t years for a 1-unit n-year endowment insurance on (x).

- $P\left(IA_{x:\overline{n|}}^{1}\right)$ is the annual premium payable for n years for an n-year increasing term policy on (x). (In this case we insert the full single-premium symbol since it does not begin with an A.)

- $P({}_{n|}\ddot{a}_x)$ is the level annual premium payable for n years, for a deferred annuity providing income of 1 unit per year beginning at age $x+n$. Note here that the missing premium payment duration symbol is taken as n, the natural premium payment duration. Although the contract continues for life, it is not natural to continue paying premiums when the annuity payments begin.

All of these annual premium symbols are evaluated by taking the corresponding net single premium and dividing by $\ddot{a}_{x:\overline{t|}}$, where t is the premium paying duration. For example,

$${}_{10}P_{50:\overline{20|}} = \frac{A_{50:\overline{20|}}}{\ddot{a}_{50:\overline{10|}}}.$$

5.8.3 Identities

The same types of identity that we discussed in Section 4.7 arise with insurances as well. For example,

$$A_{x:\overline{n}|} = A^{1}_{x:\overline{k}|} + v(k)_{k}p_{x}A_{\{x\}+k:\,\overline{n-k}|}$$

The derivation will be left to the reader.

5.9 Spreadsheet applications

We modify the Chapter 4 Spreadsheet to handle death benefits. The death benefit vector is entered into Column G. The vector $\mathbf{w}_x$ is calculated in column L by putting the following formula in L10 and copying down

$$= G10 * (1 + B10)\hat{}(-1) * C10$$

In G8 we then put the same formula as in F8 except with L replacing F, and this will return $A_x(\mathbf{b})$.

The entry in I6 is changed to =(F8+G8)/H8.

The premium vector appears in Column I.

For a check, the formula in H8 can be copied to I8 and this cell should return a total of F8 and G8.

For a sample problem, compute the initial premium for a 1000-unit, 30-year endowment policy on (40) with premiums payable for 20 years, and the premium to double after 10 years. The interest rate is a constant 6%. The answer is 12.68.

Exercises

Type A exercises

5.1 Given that $\ell_{70} = 1000$, $\ell_{71} = 960$, $\ell_{72} = 912$, and that interest rates are a constant 10%, calculate $A_{70}(1_2)$.

5.2 A 3-year endowment insurance policy on (60) provides for benefits paid at the end of the year of death of: 500 if death occurs in the first year (i.e., between time 0 and time 1); 800 if death occurs in the second year; and 1000 if death occurs in the third year. In addition, there is a pure endowment of 1000 payable at age 63 if (60) is then alive. This is purchased by three annual premiums beginning at age 60. The second premium is double the initial premium and the third premium is three times the initial premium. You are given that that $q_{60} = 0.1, q_{61} = 0.2$ and $q_{62} = 0.25$. The interest rate is 25% for the first 2 years and 20% after that. Find the initial premium.

5.3 A 2-year term insurance policy on (60) provides for a death benefit of 100 payable at the end of the year of death. This is purchased by a single premium. If (60) lives to age

62, the single premium is returned without interest. Given that $q_{60} = 0.1, q_{61} = 0.15$, and interest is a constant 10%, find the single premium.

5.4 You are given $q_x = 0.053$, $q_{x+1} = 0.054$, $q_{x+2} = 0.055$, $i_0 = 0.06, i_1 = 0.08, i_2 = 0.10$. If $\mathbf{b} = (1, 2, 3)$, find the vector $\mathbf{c} = \mathbf{b} * \mathbf{w}_x$.

Type B exercises

5.5 A deferred life annuity on (40) provides for a yearly income of 1000 beginning at age 65 and continuing for life. It is to be purchased by a single premium of S payable at age 40. If death occurs during the deferred period (i.e., during the first 25 years), then the single premium is refunded without interest at the end of the year of death.

(a) Give a formula for S using the symbols $\ddot{a}$ and A.

(b) You are now given the following information. The same annuity contract without the premium-refund feature (i.e., nothing is paid during the deferred period) can be purchased for a single premium of 2000. In addition, a contract that provides the same annuity benefits, plus a level death benefit of 2000, payable at the end of the year of death, for death during the deferred period, can be purchased for a single premium of 2200. Calculate an exact numerical value for S.

5.6 A certain electrical appliance is sold with a 5-year guarantee. This provides that the full purchase price is refunded if the product fails within 2 years, and half of the purchase price is refunded if the product fails in the following 3 years. A study shows that out of a typical batch of 100 items, there will be 2 failures in the first year, 3 failures in the second year and 4 failures per year after that. Assuming that interest is a constant 5% and that reimbursement is made at the end of the year of failure, what is the cost of this guarantee to the manufacturer, as a percentage of the purchase price?

5.7 What is $A_x(1_n)$ if the constant interest rate $i = 0$? Give both a formal derivation, and a proof by general reasoning.

5.8 Suppose that interest is constant and q_y is a constant q for all y. Find an expression for A_x in terms of q and y.

5.9 An actuary calculates a single premium for a certain life insurance policy on (40), and then discovers there were two errors made. In the first place, the life table used showed a value of q_{40} that was only one-half of what the correct figure was. Second, the first-year death benefit was taken as 20 when it should have been 10. Will the correct premium be the same as, lower, or higher than the one calculated?

5.10 For a certain insurance contract, on (50), the death benefit for the first year of the contract is 1100, payable at the end of the year of death. The single premium for the whole contract is 600. This is based on an interest rate of 10% for the first year and a mortality table with $q_{50} = 0.20$. If the value of q_{50} is changed to 0.25, while all other value of q_x are unchanged, what is the new single premium?

5.11 There is a constant interest rate of 20%, and

$$A_{50} = 0.300, \qquad v^{10}\,{}_{10}p_{50} = 0.10, \qquad A_{61} = 0.400, \qquad q_{60} = 0.20.$$

Suppose that q_{60} is changed to 0.23, while all other values of q_x remain unchanged. What is the new value of A_{50}?

5.12 The cash flow vector $\mathbf{j} = (1, 2, 3, \ldots, 10)$. You are given that interest is constant and that

$$\ddot{a}_x(\mathbf{j}) = 30, \qquad A_x(1_{10}) = 0.10, \qquad \ddot{a}_x(1_{10}) = 7, \qquad v^{10}{}_{10}p_x = 0.48.$$

Find $A_x(\mathbf{j})$.

5.13 Consider a whole life policy on (x) with a level death benefit of 1. Suppose that, for some age $y > x$, the value of q_y is increased while all other values of q remain the same.

(a) Show that A_x is increased.

(b) Show that P_x is increased, where P_x is the level annual premium payable for life, for this policy.

(c) Show by example that the above statements are not necessarily true for a whole life policy with a non-constant death benefit.

5.14 (a) Suppose that, for all x, $q_{x+1} \geq q_x$. Show that $A_{x+1} \geq A_x$.

(b) Does the above remain true if we remove the monotone condition on q_x?

5.15 Show that for the vector $\mathbf{c}$ given by formula (5.5), and $k = 0, 1, 2 \ldots$, we have

(a) $\ddot{a}_x({}_k\mathbf{c}) = A_x({}_k\mathbf{b})$;

(b) $\ddot{a}_{[x]+k}(\mathbf{c}\circ k) = A_{[x]+k}(\mathbf{b}\circ k)$.

5.16 A deferred annuity on (40) provides for an income of 1000 per year for life beginning at age 60. If (40) lives to age 60, the first 10 annuity payments are guaranteed regardless of whether (40) is alive or not. Level annual premiums of P are payable for 10 years, beginning at age 40. If (40) dies before the annuity begins, all premiums paid prior to death are returned at the end of the year of death, and no annuity payments are made. Find a formula for P, assuming: (a) premiums are returned without interest; (b) premiums are returned with interest.

5.17 One could generalize our definition of an insurance contract by stipulating that for death at time t, the death benefit is paid at time $\tau(t)$ which is some function of t. (e.g., when benefits are payable at the end of the year of death, $\tau(t) = [t] + 1$, where $[\cdot]$ is the greatest integer function.) Show that a pure endowment contract can be considered as an insurance, in this sense.

5.18 An annuity on (x) provides 1 per year for life beginning at time 0, with the further provision that, n additional payments will be made after (x) dies, beginning at the end of the year of death. Interest is constant. Show that the net single premium is $\ddot{a}(1_n; v) + v^n \ddot{a}_x$. Derive this formula in two ways. (a) By using (5.11). (b) By using (5.2).

5.19 Redo Example 5.8 only now assuming that for death during the first 5 years, one-half of all premiums paid prior to death are returned at the end of the year of death.

5.20 A contract on (70) provides for a payment of 1000 at the end of 3 years if (70) is then alive, and is to be paid for by three-level annual premiums. For death in the first year, nothing is paid. For death in the 2nd or 3rd years, all premiums paid are returned with interest at the end of the year of death. You are given that $q_{70} = 0.1, q_{71} = 0.2, q_{73} = 0.3$ and $i = 25\%$. Find the annual premium.

5.21 Provide an algebraic proof that the two methods given for the solution of Example 5.8 produce the same answer. (Hint: Change the order of summation in a double sum.)

5.22 Provide an algebraic proof and give an intuitive explanation of the following identity. For any vector $\mathbf{c} = (c_0, c_1, \dots, c_n)$

$$\ddot{a}_x(\mathbf{c}) = v(n)_n p_x \mathrm{Val}_n(\mathbf{c}; v) + A_x(\mathbf{j})$$

where $j_k = \mathrm{Val}_{k+1}({}_k\mathbf{c}; v),\ k = 0, 1, \dots, n-1$ and ${}_k\mathbf{c}$ is as defined before Equation 2.18.

Spreadsheet exercises

The following exercises are to be done using the sample life table of Section 3.7.

5.23 A contract on (40) provides for death benefits if death occurs in the next 25 years. The amount of the death benefit is 50 000 for the first 10 years and 100 000 for the next 15 years. If (40) lives to age 65, he/she will receive a life annuity of 10 000 per year for life beginning at age 65. Premiums are payable for 15 years beginning at age 40. The premium doubles after 5 years. Interest rates are 5% for the first 20 years and 6% thereafter. Find the initial premium.

5.24 Interest rates are a constant 6%. A contract on (50) provides for a payment of 10 000 at age 70 if then alive. Level premiums of P are paid only at even-numbered times, that is, at age 50, 52, 54, ... If (50) dies before age 70 there is a return at the end of the year of death of all premiums paid prior to death. Find P.

5.25 Assume a constant interest rate of 4%. Find the premium for the annuity in Example 5.9 if: (a) $x = 40$;(b) $x = 70$.

6

Insurance and annuity reserves

6.1 Introduction to reserves

Given an insurance or annuity contract and a duration k, the *reserve* at time k is defined exactly as in Definition 2.7. It is the amount that the insurer needs at time k in order to ensure that obligations under the contract can be met. Calculating reserves for each policy is an important responsibility of the actuary, known as *valuation*. The insurer wants to be confident that funds on hand, together with future premiums and investment earnings, are sufficient to pay the promised future benefits. It is important to thoroughly master the concept of insurance and annuity reserves in order to properly understand and analyze the nature of these contracts.

Throughout this chapter we will deal with the following model. As usual we start with a fixed investment discount function v and a life table. We have a contract issued on (x) with death benefit vector $\mathbf{b}$, annuity benefit vector $\mathbf{c}$ and premium vector π. (For simplicity we will omit the possibility of guaranteed payments in our discussion, but this feature can easily be incorporated if desired.) Recall from Section 5.5 that we can view the death benefits as a vector of annuity benefits $\mathbf{b} * \mathbf{w}_x$, where $(\mathbf{w}_x)_k = v(k, k+1)q_{x+k}$. We can then form the *net cash flow* vector

$$\mathbf{f} = \boldsymbol{\pi} - \mathbf{b} * \mathbf{w}_x - \mathbf{c}, \tag{6.1}$$

which indeed represents the net cash flow on the contract from the insurer's viewpoint. The insurer will collect premiums of π and pay out death benefits in the form of $\mathbf{b} * \mathbf{w}_x$ and annuity benefits of $\mathbf{c}$. The reserve at time k on the contract is the reserve for the vector $\mathbf{f}$ with respect to the interest and survivorship discount function, as given in Definition 2.6. Denoting the reserve by ${}_kV$, we have

$${}_kV = {}_kV(\mathbf{f}; y_x) = -\mathrm{Val}_k\left({}^k\mathbf{f}; y_x\right). \qquad (6.2)‡$$

Fundamentals of Actuarial Mathematics, Third Edition. S. David Promislow.

Companion website: http://www.wiley.com/go/actuarial

It is often useful to write this in an alternate way that keeps benefits and premiums separate. That is, we note that the reserve at time k is the *value at time k of future benefits less the value at time k of future premiums*. This can be written in terms of the standard A and $\ddot{a}$ symbols as

$$ {}_kV = \frac{1}{y_x(k)}\left[A_x({}^k\mathbf{b}) + \ddot{a}_x({}^k\mathbf{c}) - \ddot{a}_x({}^k\boldsymbol{\pi})\right] \tag{6.3}$$

or, equivalently (and the way in which it usually appears in the literature) as

$$ {}_kV = A_{[x]+k}(\mathbf{b}\circ k) + \ddot{a}_{[x]+k}(\mathbf{c}\circ k) - \ddot{a}_{[x]+k}(\boldsymbol{\pi}\circ k). \tag{6.4}$$

Here we are using (4.8) and the corresponding statement for A.

Under the assumption that premiums are actuarially equivalent to benefits, we can also calculate the reserve retrospectively as

$$ B_k(\mathbf{f}; y_x) = \frac{1}{y_x(k)}\left[\ddot{a}_x({}_k\boldsymbol{\pi}) - \ddot{a}_x({}_k\mathbf{c}) - A_x({}_k\mathbf{b})\right]. \tag{6.5}$$

Under this formulation the reserve is the value at time k of the past premiums less the value at time k of past benefits. The reader is cautioned that the reserve cannot be so calculated when premiums and benefits are not actuarially equivalent.

Example 6.1 For the policy of Example 5.5, find the reserves at time 1 and time 2. See Figure 6.1. As in Chapter 2 we use an arrow to mark the point at which values are computed.

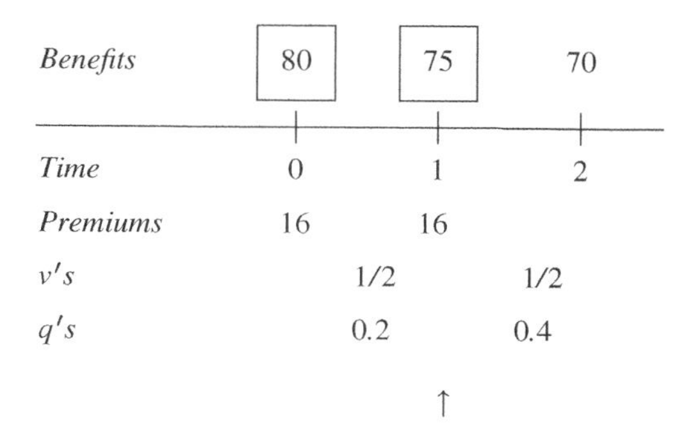

Figure 6.1 Example 6.1

Solution.

Value at time 1 of future death benefits = 75(1/2)(0.4) = 15.
Value at time 1 of future annuity benefits = 70(1/2)(0.6) = 21.
Value at time 1 of future premiums = 16.

$$ {}_1V = 15 + 21 - 16 = 20. $$

The above approach is recommended for hand calculation. All reserve calculations can be handled in exactly the same way, although there will generally be more than one summand in each of the three items.

Additional information can be obtained by first computing the net cash flow vector, which also provides the best procedure for spreadsheet calculation. To illustrate, the vector $\mathbf{w}_{60} = (0.1, 0.2, -)$. (We don't have information to compute the third entry but it is irrelevant, since it is multiplied by 0.) Then, $\mathbf{b} * \mathbf{w}_{60} = (8, 15, 0)$, $\mathbf{c} = (0, 0, 70)$, $\boldsymbol{\pi} = (16, 16, 0)$, so that the net cash flow vector $\mathbf{f}$ is $(8, 1, -70)$. For present purposes, we can forget about the particular death benefits and premiums. From the insurer's viewpoint the contract can be viewed as simply one of collecting 8 at time 0, 1 at time 1, and paying back 70 at time 2. From (6.2), ${}_1V$ is just the negative of the value at time 1, with respect to interest and survivorship, of the payments at times 1 and 2. This equals $-1 + 70(0.5)(0.6) = 20$, as above.

Similarly, ${}_2V$ is just the value at time 2 of the payment at time 2 which is 70. In general, for a contract running for n years, the reserve at time n is just the payment due at time n to the survivors. (Recall that from the convention introduced in Chapter 1, reserves are calculated before this payment.) For n-year term insurance, when there is nothing payable to survivors at the end, the nth year reserve will equal to 0.

Since we have used net premiums, we can calculate balances as a check.

$$B_1(\mathbf{f}) = \frac{8}{y_{60}(1)} = \frac{8}{0.4} = 20,$$

$$B_2(\mathbf{f}) = \frac{1}{y_{60}(2)}\ddot{a}_{60}({}_2\mathbf{f}) = \frac{1}{0.12}[8 + 0.4] = 70,$$

which agree with our previous calculations.

The following example exhibits a point of interest.

Example 6.2 For a 4-year endowment insurance on (60), $b_2 = 100, b_3 = 200$ and there is a pure endowment of 200 paid at age 64 if the insured is then alive. You are given that $\pi_2 = 10, \pi_3 = 20$, $q_{62} = 0.1$. The interest rate after 2 years is 25%. Find ${}_2V$ and ${}_3V$. See Figure 6.2.

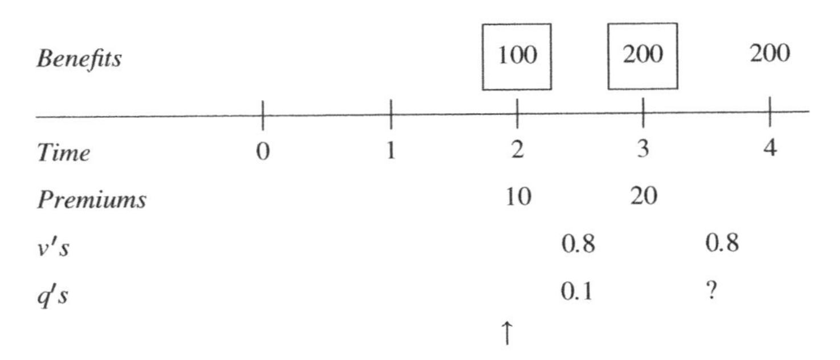

Figure 6.2 Example 6.2

Solution.

Value at time 2 of death benefits $= 100(0.8)(0.1) + 200(0.64)(0.9)q_{63}$.

Value at time 2 of annuity benefits $= 200(0.64)(0.9)p_{63}$.

Value at time 2 of premiums $= 10 + 20(0.8)(0.9) = 24.40$.

We don't know q_{63}, but it is not needed. Since $q_{63} + p_{63} = 1$, the two unspecified terms sum to 200(0.64)(0.9) = 115.20 and

$$_2V = 80 + 115.20 - 24.40 = 170.80.$$

Similarly, $_3V = 200(0.8) - 20 = 140$.

This example shows that for an endowment insurance (with benefits paid at the end of the year of death) where the pure endowment is the same amount as the final death benefit, we do not need to know the mortality rate for the final year. This is in fact evident, since the insured gets that amount whether they live or not.

Remark There is a subtle point involved with Equation (6.4), which is important to note for the actual calulation of reserves. We have already introduced the idea in Section 2.10.3. In our model we choose an investment discount function and a mortality table, and these remain fixed for the duration of a contract. In practice, when one reaches time k and actually wants to compute $_kV$ these assumptions may well have changed. The reserve computed at time k will be

$$A'_{x+k}(b \circ k) + \ddot{a}'_{x+k}(c \circ k) - \ddot{a}'_{x+k}(\pi \circ k)$$

where the primes indicate quantities that are calculated with the new interest and survivorship function y'_{x+k} as computed under what could be changed conditions at time k. Of course when there is no change in assumptions, y'_{x+k} will just equal $y_x \circ k$ and formula (4.6) shows that both formulas are the same. The point is then, that in practice, one does not really know what $_kV$ will be before time k. In this more realistic setting, we can view the original formula (6.4) as the best estimate one could make of $_kV$, if asked to compute it at time 0.

6.2 The general pattern of reserves

Are insurance reserves generally positive or negative? Paradoxically, we will motivate the answer to this question by providing an example where they are neither.

Example 6.3 An insurance policy provides a death benefit of 1 paid at the end of the year of death. Level annual premiums are payable for life. The interest rate is constant and the value of q_x is a constant q for all x. Find the reserves and give an explanation for the answer.

Solution. Let $p = 1 - q$ be the constant value of p_x. Then $_kp_x = p^k$, which is never 0. There is a positive probability of living to any age, so we have an example in which $\omega = \infty$. Our vectors will be of infinite length.

Let v denote the constant value of $v(k, k+1)$. Then $\mathbf{w}_x$ is a vector with a constant entry of vq and the premium pattern vector ρ is a vector with a constant entry of 1. The net level premium π is clearly vq, since that will make the vectors $\mathbf{w}_x * \mathbf{b}$ and π not only actuarially equivalent but actually equal to each other. The net cash flow vector $\mathbf{f}$ is then not just a zero-value vector but *actually equal to* the zero vector. All reserves will be 0.

What is happening here is that the premium of vq, collected each year, will accumulate to q at the end of the year, and this will be exactly sufficient to pay the death benefits due at that time. There will be nothing left over, so balances, and therefore reserves, equal zero.

The situation above is typical for many forms of insurance, such as automobile or property coverage, and insurers of such risks have little in the way of reserves. The given scenario is, however, not realistic for life insurance. The values of q_x are not constant but increase with age. If one paid for the insurance 1 year at a time, the yearly premium per unit of vq_x would rapidly increase and eventually become prohibitively high. As we noted in Chapter 5, the typical life insurance policies will level the premiums out. In most cases, policyholders are paying more in premiums than they need to in the early years, but not enough in the later years. At any point of time after time zero, future premiums will not be sufficient to cover the remaining benefits. The excess collected in the early years is used to cover this deficit. It is expected therefore that reserves are usually positive. This has important implications for the life insurance industry. It means that investing becomes a major activity, as life insurance companies tend to accumulate large amounts of assets. Some critics, with little understanding of insurance, look at these holdings of real estate, stocks and bonds, and claim that they represent unfair profits made at the expense of the policyholder. The truth is, however, that a large portion of these assets represent reserves, which in effect belong to the policyholders, as they will be used to pay the future benefits.

Negative reserves can arise on policies where the cost of the insurance benefits is decreasing each year. An example is a policy with a rapidly decreasing benefit amount, where despite the increase in q_x, the quantity $b_k v(k, k+1)q_{x+k}$ decreases. Some examples appear in the exercises. Similarly, negative reserves can arise in the case where the premiums increase rapidly rather than remaining level. Insurers try to avoid such a situation if possible. A negative reserve means, viewing things prospectively, that the policyholder owes money to the insurer, which will be provided by future premiums, or equivalently, looking at things retrospectively, that the policyholder has received coverage but not yet paid for it. The problem is that the policyholder may stop paying premiums on the policy, leaving an unpaid debt.

6.3 Recursion

In this section, we develop some important recursion formulas. It is convenient to make a slight alteration in notation. We will incorporate the annuity payments with the premiums and let $\boldsymbol{\pi}$ denote $\boldsymbol{\pi} - \mathbf{c}$. In other words, we think of annuity benefits as just negative premiums, which is really what they are, since the policyholder is receiving rather than paying these amounts. Our net cash flow vector $\mathbf{f}$ has entries $f_k = \pi_k - b_k v(k, k+1)q_{x+k}$, and from our basic recursion formula (2.26),

$$_{k+1}V = \left({}_kV + \pi_k - b_k v(k, k+1)q_{x+k}\right)y_x(k+1, k). \tag{6.6}$$

Since $y_x(k+1, k) = (1 + i_k)/p_{x+k}$ this is sometime written as

$$_{k+1}V = ({}_kV + \pi_k)\frac{1 + i_k}{p_{x+k}} - b_k\frac{q_{x+k}}{p_{x+k}}. \tag{6.7}$$

The recursion is started with the initial value $_0V = -\ddot{a}_x(\mathbf{f})$ which will be 0 under our standard assumption of net premiums.

It is instructive to note that the last term of q_{x+k}/p_{x+k} is equal to d_{x+k}/ℓ_{x+k+1}. From this, we see easily that it is the amount, per unit of death benefit, that each survivor must pay at the *end of the year* to provide the benefits paid to those who died during the year.

Formula (6.7) takes a retrospective viewpoint, and says that the reserve at time $k+1$ is obtained from that at time k, by adding the premium, accumulating at interest and survivorship for 1 year, and then subtracting enough to pay the death benefits. It is known as the *Fackler reserve accumulation formula*, named after one of the early North American actuaries, David Parks Fackler. In pre-computer days it was a popular method for calculating reserves. It is used infrequently for calculating purposes now, but is still useful for illustrating how the life insurance reserve changes from one period to the next.

Remark The quantity ${}_kV + \pi_k$ in the above formula is often called the *initial reserve* at time k as it represents the reserve at the beginning of the year, after the payment of the premium. In contrast, ${}_kV$ is sometimes referred to as the *terminal reserve* at time k, reflecting the fact that it the reserve at the end of the year, prior to the payment of the premium for the following year.

Alternate versions of this formula provide instructive information. We first give an important definition.

Definition 6.1 The quantity $b_k - {}_{k+1}V$ is known as *the net amount at risk* for the $(k+1)$th year and will be denoted by η_k (the subscript is chosen to correspond to b_k). (There are various other names in the literature, such as *death strain at risk.*)

Now, multiplying (6.7) by $p_{x+k} = 1 - q_{x+k}$ and rearranging, we get

$$ {}_{k+1}V = ({}_kV + \pi_k)(1 + i_k) - q_{x+k}\eta_k. \tag{6.8} $$

Formula (6.8) reflects the fact that we can also view the accumulation of funds on an insurance policy as an interest-only investment, rather than as an interest and survivorship investment. From this point of view, the policyholder keeps the reserve when he/she dies (the amount accumulated in her box), but then the insurer only needs to make up the difference as a death benefit. The insurer is therefore at risk only for the difference between the death benefit and reserve, which is the source of the name.

Readers who looked at Section 2.11 will note that this is a special case of the change of discount function that we investigated there. In this case y_x and b_k are replaced by v and η_k, respectively.

6.4 Detailed analysis of an insurance or annuity contract

In this section we use (6.8) to provide a detailed discussion of the workings of an insurance policy.

6.4.1 Gains and losses

In practice, interest and mortality rates will not conform exactly to those provided by our model. In a particular year, the insurer may earn more interest than predicted by the given discount function, which will result in gains. For an insurance contract, there may be fewer deaths than predicted by the given life table, also causing gains. If the insurer earns less interest than expected, and/or there are more deaths than expected, there will be losses. In any year, the actuary wants to analyze these gains or losses and see how much is due to investment earnings and how much is due to mortality.

Suppose we wish to measure the gain for a particular policyholder over the period running from time k to time $k+1$. At the beginning of the year, before premium payment the policyholder's box will contain the amount ${}_kV$. At the end of the year the insurer must make certain that the box has ${}_{k+1}V$ in order that future obligations can be met. Anything in excess of that amount can be considered as a gain, taken out and added to general surplus funds. On the other hand, if there is less than ${}_{k+1}V$, the insurer will have to make up the deficit from general surplus funds and there will be a loss. We will derive some general formulas. Suppose the actual interest rate earned during this year was i_k^* rather than i_k, and the actual rate of mortality was q_{x+k}^* rather than q_{x+k}. Then, the actual amount accumulated at time $k+1$ will be the right hand side of (6.8) with starred i and q. If we subtract the reserve, we obtain the total gain G_k from that policy for that year as

$$G_k = ({}_kV + \pi_k)\left(1 + i_k^*\right) - q_{x+k}^*\eta_k - {}_{k+1}V. \tag{6.9}$$

If we substitute for ${}_{k+1}V$ with the left hand side of (6.8) we can write this as

$$G_k = \left(q_{x+k} - q_{x+k}^*\right)\eta_k + \left(i_k^* - i_k\right)({}_kV + \pi_k). \tag{6.10}$$

which gives us a decomposition of the gain by source. The first term gives the gain due to mortality, and the second term gives the gain due to interest. That is, the mortality gain is the difference between the expected and actual mortality rates times the *net amount at risk*. The interest gain is the difference between actual and expected interest rates times the amount of funds at the beginning of the year, after payment of the premium.

Example 6.4 Refer back to the policy of Example 6.1. Suppose that in the first year of the policy, the interest earned was 50% instead of the predicted 100%, and the actual rate of mortality was 0.1 instead of the predicted 0.2. Find the total gain for the year, split into the portion due to interest and the portion due to mortality.

Solution. Substituting directly from (6.9), the mortality gain is

$$(0.2 - 0.1)(80 - 20) = 6,$$

while the interest gain is

$$(0.5 - 1)(0 + 16) = -8.$$

There is a total gain of −2, or in other words a loss of 2, for each policy.

We will verify this by working out an example in the aggregate. Suppose that 10 people age 60 buy this policy at a certain time. The insurer collects 160, and this accumulates at 50%

interest to 240 at the end of the year. Out of this, the insurer will pay a death benefit of 80 for the one death that occurred, leaving a total of 160. They have to put aside a total of 180, which is the reserve of 20 for each of the 9 survivors. There is a aggregate shortfall of 20, or 2 from each policy, which has to be drawn from surplus. (Note that the gain given by (6.8) is for each policy in force at the *beginning* of the year, not at the end.)

Formula (6.9) assumes that the premium paid is the net premium, calculated as in Chapter 5. In practice, the premiums actually charged on a policy will normally be different from the *valuation premiums* which are net premiums determined from the interest and mortality assumptions used to compute reserves. (See Section 6.5 for more detail on this.) This necessitates an adjustment in our analysis. Suppose the premium paid at time k is actually π_k^* rather than valuation premium π_k. The first term on the right of (6.8) is then

$$\left({}_kV + \pi_k^*\right)\left(1 + i_k^*\right) = \left({}_kV + \pi_k\right)\left(1 + i_k^*\right) + \left(\pi_k^* - \pi_k\right)\left(1 + i_k^*\right),$$

which leads to an extra source of gain or loss. We now have

$$G_k = \left(q_{x+k} - q_{x+h}^*\right)\eta_k + \left(i_k^* - i_k\right)\left({}_kV + \pi_k\right) + \left(\pi_k^* - \pi_k\right)\left(1 + i_k^*\right). \tag{6.11}$$

The third term represents the gain or loss arising from premiums that differ from the valuation premium. (We will elaborate on this point in Section 12.4.)

One application of the formulas in this section is to *dividend* calculation. Insurers frequently issue what are termed *participating policies*, in which gains resulting from favourable investment and mortality experience are returned to the policyholder in the form of dividends. We will not go into further detail on this topic, but note that formula (6.10) is a basic tool in computing these amounts.

Another use of gain and loss analysis is to determine how changes in the basis assumptions will affect the reserves. As an example, suppose that the reserve interest rate decreases. It is clear that valuation premiums will increase. In order to provide the same benefits with decreased earnings from investments, more must be collected from the policyholder. On the other hand, it is not clear how this will affect reserves, since at any time, both the present value of the future benefits and the present value of the future premiums will increase. Indeed, the effect will depend on the nature of the policy. In the usual case however with a level premium π and reserves that increase with time, the lowering of the interest rate will increase reserves. To see this, suppose that interest rates are constant and there is a change of rate from i to $i^* < i$. This will cause an interest loss of $({}_kV + \pi)(i - i^*)$ for the year running from time k to $k + 1$. Since the reserves are increasing with time, the losses will also be increasing with time. There will be of course a new level valuation premium of $\pi + \Delta$ to cover these losses. Due to the increasing losses, it must be that Δ is greater than the losses in early years and less in later years. In the early years, that portion of Δ not used to cover the loss, will be set aside as an extra reserve, to cover the greater losses to come in the future, and this will cause reserves to increase.

6.4.2 The risk–savings decomposition

We now look at a useful decomposition of the policy into a risk portion and a savings portion. Multiply Equation (6.8) by $v(k, k + 1)$ and rearrange to obtain

$$\pi_k = v(k, k + 1)q_{x+k}\eta_k + [v(k, k + 1)_{k+1}V -_k V].$$

This formula decomposes each premium into two parts. The first term is known as the *risk portion* of the premium, as it is that amount needed to buy insurance for 1 year for the net amount at risk. The remainder provides for the difference in reserves and is known as the *savings portion* of the premium.

Example 6.5 Find the decomposition for the premiums of Example 6.1.

Solution. In the first year the net amount at risk is $80 - 20 = 60$. The risk portion of the premium is then $\frac{1}{2} \times 0.2 \times 60 = 6$. The savings portion is $\frac{1}{2} \times 20 - 0 = 10$. (As a check, the two portions add up to the total premium.) In the second year the net amount at risk is $75 - 70 = 5$. The risk portion of the premium is $\frac{1}{2} \times 0.4 \times 5 = 1$ and the savings portion is $\frac{1}{2} \times 70 - 20 = 15$.

This decomposition shows that any policy can be viewed as being composed of two separate policies, the pure insurance part and the savings part.

For the pure insurance part, the premium is the risk premium and the death benefit paid is the net amount at risk. This part of the policy has zero reserves, as the risk premium is just sufficient to purchase coverage for the net amount at risk for 1 year. In the above example, the policyholder pays 6 in the first year, which is exactly enough to purchase the coverage for the net amount at risk of 60. He/she then pays 1 in the second year, which is exactly enough to purchase coverage for the net amount at risk of 5.

The savings part of the policy operates just like a bank account, with amounts accumulating at interest only. In the example above, the savings portion of 10 from the first premium accumulates to 40 at time 2 and the saving portion of 15 from the second premium accumulates to 30 at time 2. The total savings of 70 are then paid out as the pure endowment to all survivors at time 2. This is typical of endowment policies, including whole life, which operates as an endowment at age ω, as we have indicated. The accumulated amounts from the savings portion of the premium increase steadily to the pure endowment amount.

It is instructive to compare this with term insurance.

Example 6.6 Redo Example 6.1 for the corresponding 2-year term policy without the endowment. That is, $\mathbf{b}$ still equals (80, 75, 0), but $\mathbf{c} = \mathbf{0}$.

Solution. In this case the premium π will equal $(8 + 15(0.4))/1.4 = 10$, and we have $\mathbf{f} = (2, -5, 0)$. Then ${}_1V = 5$ and we know from the discussion following Example 6.1 that ${}_2V = 0$. In the first year, the net amount at risk is 75, so the risk portion of the premium is $\frac{1}{2} \times 0.2 \times 75 = 7.5$, and the savings portion is 2.5. In the second year the net amount at risk is 75, so the risk portion of the premium is $\frac{1}{2} \times 0.4 \times 75 = 15$ and the savings portion is -5. This shows the typical savings pattern on term policies. Modest savings are built up in early years, but these must be drawn on in later years when the premium is insufficient to pay for the insurance, resulting in a negative savings portion. As a check, the 2.5 deposited into the savings fund at time 0 will increase to 5 at time 1, which will then be withdrawn to make up for the deficit in the premium payable at that time.

Exactly the same analysis as above holds for life annuity as well as life insurance contracts. In this case, the death benefits are all of zero amount, so that the net amount at risk will be negative. This is natural enough and reflects the facts that extra deaths in the case of annuities result in gains. On single-premium annuities, we have $\pi_k = -c_k < 0$, for $k > 0$. Examples appear in the exercises.

6.5 Bases for reserves

Choosing the appropriate discount function and life table for the purpose of reserve calculation is another complex subject that we will only discuss briefly. As mentioned, the insurer must ensure that it has sufficient assets to cover their reserves, for if not, it will be in danger of being unable to meet future obligations. It is usually thought that reserve calculations should use conservative assumptions, so there is a built-in safety margin should experience prove to be adverse. In other words, the discount function would use somewhat lower interest rates than actually expected and the mortality table would show more deaths than actually expected (or fewer in the case of annuity contracts). In several jurisdictions, the bases used for reserves are specified by insurance regulatory bodies, whose main goal is to ensure protection for the policyholders. The following example illustrates the resulting effect on profitability.

Example 6.7 For the policy in Example 6.1, the company is required by legislation to compute reserves using a 50% interest rate rather than 100%. It actually does achieve the estimated 100% return on its investments, and mortality follows the predicted rates exactly. It still charges the premium of 16 based on the realistic interest rate of 100%. Analyze the effect of the conservative interest rate assumption on the company's gains and losses.

Solution. Redoing the calculations with an interest rate of 50% in place of 100% leads to a valuation premium of 544/23 and a first year reserve of 560/23, as the reader can verify. We will do an aggregate analysis. Suppose the insurer sells 2300 policies at age 60. It will collect total premiums of 36 800, which will accumulate with interest to 73 600 at the end of the year. Out of this it will pay 460 people a death benefit of 80 units, for a total death benefit payment of 36 800. This leaves 36 800. It now must set up a total reserve of 44 800, consisting of 560/23 for each of the 1840 survivors. Therefore the loss shown for the first year of the policy is 8000. Looking at formula 6.11, the large loss in the third term, more than offsets the gain from the second term.

In the second year, it starts with a reserve of 44 800. It collects another 16-unit premium from each of the 1840 survivors for a total of 29 440. This leaves a total of 74 240, which accumulates with interest to 148 480. Out of this it must pay a death benefit of 75 to each of the $736 = 1840 \times 0.4$ people remaining 1104 people who survived. The total benefit payments are 132 480, which leaves a gain for the year of 16 000.

For this group of policies, the insurer will show a loss of 8000 for the first year, and a gain of 16 000 for the second year. If the insurer had used the realistic interest rate of 100% interest, there would have been no gains or losses. In effect the more conservative reserve requires the insurer to borrow 8000 from surplus at the end of the first year, and then repay it with the 16 000 at the end of the second year, which is what the amount should be in view of the 100% interest rate.

This example shows that for our present model, with a single discount function, reserve assumptions do not affect the ultimate profitability of the insurer, since this depends solely on what actually happens. It can, however, change the incidence of this profit from year to year. This means that reserve assumptions can have an effect on profitabiltly when there are different interest rates involved, and we discuss this further in Section 12.4.3.

6.6 Nonforfeiture values

What should happen to policyholders who stop paying premiums before the term stated in the contract? This is known technically as *withdrawal* or *lapse*, or *surrender*. In consideration for the premiums they have already paid, they should be entitled to some reduced benefits under the policy. These are known as *nonforfeiture benefits* since they are benefits that were not forfeited by the cessation of premiums. In fact, in our simplified model, they should be entitled to take the reserve on their policy at any time they wish. Looking at this retrospectively, the reserve is the excess of the accumulated amount of their premiums over the accumulated cost of the insurance protection that they have received. It will be the amount that they have in their 'box'. In practice, insurers pay an amount that is somewhat less than the reserve for several reasons. This is a complex topic that we will only comment on briefly here. One reason is the high incidence of expenses in the early years of the policy (we discuss this more fully in Chapter 12). While these are accounted for by adding an amount to the premiums, the total amount of the initial expenses may not be recovered at the time of withdrawal.

Another factor is the phenomenon known as *anti-selection*. This is a well-established concept in insurance which is simply a recognition of the fact that policyholders will make choices according to their own self-interest, acting on knowledge that they have, but that the insurer may not have. (The prefix 'anti' refers to the fact that it is the policyholder doing the selecting against the insurer.) On a life insurance contract, the option to withdraw is less likely to be exercised by an unhealthy policyholder than a healthy one. After all, if someone is told they will die within a few months from a terminal disease, they would be foolish to give up the policy. Consequently, the group that does *not* withdraw can be expected to experience higher mortality than normal. There is therefore an anti-selection expense to withdrawal, in the form of these higher mortality rates of the remaining policyholders. The principle followed here is that this expense should be borne equitably by all the policyholders, not just by those who remain. This is done by adjusting the amounts paid out in the case of withdrawal.

The cash amount that will be paid to the withdrawing policyholder on a life insurance contract is known as the *cash surrender value* and is usually guaranteed at the time of issue for all durations. Normally, the policyholder is given the option of taking the nonforfeiture benefits in the form of a reduced level of insurance rather than in cash. The reduction can take the form of either a reduced amount of benefits, or a reduced term for the same benefits.

Life annuities also present an obvious possibility for anti-selection. Unhealthy annuitants would find it worthwhile to end the contract, take their reserve, leaving a group who could be expected to live longer than expected, and mortality losses would result. For this reason, nonforfeiture benefits would not be offered on single-premium life annuity contracts, once the benefits commence. Depending on the contract, they might be present during a deferred period.

6.7 Policies involving a return of the reserve

At the time of death, the policyholder (or, more accurately, the estate of the policyholder) receives the death benefit. Had he/she decided to lapse the policy an instant before death, he/she would have received an amount close to the reserve. Some people, who do not have a complete understanding of life insurance, have raised the complaint that the company is confiscating the person's reserve, since it is only returning the death benefit, and not the reserve, at death. The answer to this is that one must adopt one of two points of view. One can view the insurance policy as an interest and survivorship investment. In that case, it is true that the reserve is taken and spread among the survivors, but this is completely fair, as discussed in Chapter 4. Alternatively, one can view the policy as an interest-only investment. In this case the reserve is indeed available at death. But now, one must view the death benefit as the net amount as risk, rather than the originally stated amount.

It is possible to design a policy where a fixed amount, plus the reserve, is paid at death. The policyholder might then be told that the reserve is being paid in addition to the death benefit. This is of course just playing with words. What you really have is a policy with the pattern of death benefits worked out so that when you subtract the reserve you get some prescribed amount. Consider the following example.

Example 6.8 A policy on (x) provides for a payment, at the end of the year of death, of 1 plus the reserve, should death occur within n years. Level annual premiums of P are payable for n years. Find a formula for P.

Solution. A direct solution of this problem by the method outlined in Chapter 5 will cause difficulties, since P depends on the reserves, but the reserves in turn depend on P. While it may be possible to solve the resulting equations for small values of n, it is much better to take the interest-only view for the accumulation of money. That is, we use the discount function v in place of y_x and the net amount at risk η_k in place of b_k. Expressing the insurance as an annuity, we wish to solve

$$P\ddot{a}(1_n; v) = \ddot{a}(\eta * \mathbf{w}_x; v)$$

where η is the vector $(\eta_0, \eta_1, \ldots, \eta_{n-1})$. While this could be done on any policy, it would normally be completely impractical, since we would not know the net amounts at risk in advance. In this case, however, it works perfectly. We are given that $b_k = 1 + {}_{k+1}V$, so that $\eta_k = 1$ for all k. This leads to

$$P\ddot{a}(1_n; v) = \sum_{k=0}^{n-1} v(k+1)q_{x+k},$$

and we easily solve for P.

For another application of this idea, we can give a more formal solution to Example 5.8. We calculate the balance at time 25 and equate it to the reserve at time 25, which we know is $\ddot{a}_{\{40\}+25}$. To calculate the balance, we use the discount function v and death benefits of η_k. In this case, the actual death benefit is just equal to the reserve, so that $\eta_k = 0$ for all k. The balance is just the accumulated value of the premiums at interest, which is $Pv(25, 0)\ddot{a}(1_{25}; v)$, as we had before.

Note that reserve calculation by recursion is quite simple in this type of policy. One simply uses (6.8) where now η_k is just the face amount of the policy.

6.8 Premium difference and paid-up formulas

Consider any policy on (x) with death benefit vector $\mathbf{b}$ and with level annual premiums of P payable for h years, so that the premium vector $\pi = P(1_h)$. In this case, there are some other formulas that are useful for providing additional insight into the nature of balances and reserves. Throughout this section P is arbitrary and not necessarily the net premium.

6.8.1 Premium difference formulas

Fix a duration $k < h$. Let P_s be the level premium that *should* be charged for a policy with the same remaining benefits, if issued at time k to a person age $x + k$. That is, the value at time k of these new premiums should equal the value at time k of the death benefits after time k. Equating values at time k,

$$A_{[x]+k}(\mathbf{b} \circ k) = P_s \ddot{a}_{[x]+k}(1_{h-k}). \tag{6.12}$$

Since ${}_kV = A_{[x]+k}(\mathbf{b} \circ k) - P\ddot{a}_{[x]+k}(1_{h-k})$, we substitute in (6.10) to get

$${}_kV = (P_s - P)\ddot{a}_{[x]+k}(1_{h-k}). \tag{6.13}$$

Formula (6.11) is known as the *premium difference formula* for reserves. The quantity $P_s - P$ is the difference between what should be charged and what is actually charged after time k. The value at time k of this yearly deficit over the remaining premium payment period gives the reserve.

We can also obtain a retrospective premium difference formula. Let P_c be the premium that *could* have been charged to provide the benefits that were provided up to time k. That is,

$$A_x({}_k\mathbf{b}) = P_c \ddot{a}_x(1_k). \tag{6.14}$$

Since $B_k = [P\ddot{a}_x(1_k) - A_x({}_k\mathbf{b})]/y_x(k)$, we substitute in (6.14) to get

$$B_k = \frac{(P - P_c)\ddot{a}_x(1_k)}{y_x(k)}. \tag{6.15}$$

The quantity $P - P_c$ is the difference between what was actually charged and what could have been charged up to time k. The accumulated amount of this yearly excess gives the balance. When P is a net premium $B_k = {}_kV$ and (6.15) gives another formula for the reserve.

6.8.2 Paid-up formulas

Substituting for the annuity rather the insurance in (6.12) gives

$${}_kV = \left(1 - \frac{P}{P_s}\right) A_{\{x\}+k}(\mathbf{b} \circ k). \tag{6.16}$$

Similarly, from (6.15),

$$B_k = \left(\frac{P}{P_c} - 1\right)\frac{A_x({}_k\mathbf{b})}{y_x(k)}. \tag{6.17}$$

To interpret formula (6.16), note that the fraction of the future benefit that is purchased by the actual premiums is P/P_s. For example, if $P_s = 36$ and $P = 24$ then the policyholder is only paying two-thirds of what they should be for the future benefits. The difference of $(1 - P/P_s)$ must be that portion of the future benefits that has already been provided by the excess past premiums, so that multiplying this ratio by the present value of future benefits gives the reserve. In formula (6.17) the ratio $P/P_c - 1$ is that portion of the past benefits that were purchased but not needed. For example, if $P = 24$ and $P_c = 16$, then the policyholder has paid 1.5 times what they could have paid to provide those benefits. Multiplying this ratio by the value of these past benefits gives the balance.

Formula (6.16) is known as the *paid-up* formula. Suppose a policyholder lapses at time k and is given nonforfeiture benefits equal in value to the reserve. If the individual elects to take paid-up insurance for a reduced amount, this formula shows that the appropriate fraction is $(1 - P/P_s)$.

6.8.3 Level endowment reserves

We can use the premium difference formula to derive a very simple expression for reserves on level endowment insurance where a net level premium is paid for the full period and interest is constant. At duration k of an n-year contract, we know from (5.10) that

$$P = \frac{1}{\ddot{a}_x(1_n)} - d, \qquad P_s = \frac{1}{\ddot{a}_{x+k}(1_{n-k})} - d,$$

and, substituting in (6.13),

$${}_kV = 1 - \frac{\ddot{a}_{x+k}(1_{n-k})}{\ddot{a}_x(1_n)}, \tag{6.18}$$

reducing the calculation of reserves on such a policy to the calculation of annuity values.

6.9 Standard notation and terminology

The standard notation for reserves closely follows the notation for annual premiums as described in Section 5.8. The basic symbol for the reserve at time k is ${}_kV$ as we have adopted. This is embellished in exactly the same way as the symbol P was for annual premiums, with one exception. The premium payment period is moved to the upper left, from the lower left, since the latter is now used for the duration. If the upper left is empty, it signifies again that level premiums are paid for the natural duration of the contract. The following are some examples.

${}_kV_x$ is the reserve at time k for a 1-unit whole life policy on (x), with level annual premiums paid for life. The basic prospective formula (6.4) for this policy in standard symbols reads

$$ {}_kV_x = A_{x+k} - P_x\ddot{a}_{x+k}. $$

The corresponding retrospective formula is

$$ {}_kV_x = P_x\ddot{s}_{x:\overline{k}|} - A^{\,1}_{x:\overline{k}|}\frac{(1+i)^k}{{}_kp_x}. $$

To simplify retrospective reserve formulas, a symbol for the last term was introduced. Let

$$ {}_tk_x = A^{\,1}_{x:\overline{t}|}\frac{(1+i)^t}{{}_tp_x}. $$

This is often called the *accumulated cost of insurance* and denotes the single premium that each survivor would pay per unit of death benefit for t years, if this single premium were collected *at the end* of the k-year period, rather than at the beginning. (This would never be done in practice as it is not feasible to charge people at a time when they have no chance of collecting.)

${}_tV^{\,1}_{x:\overline{n}|}$ is the reserve at time t for a 1-unit, n-year term policy on (x) with level annual premiums payable for n years. For $t \le n$, the prospective formula for this quantity is

$$ {}_tV^{\,1}_{x:\overline{n}|} = A^{\,1}_{x+t:\overline{n-t}|} - P^{\,1}_{x:\overline{n}|}\,\ddot{a}_{x+t:\overline{n-t}|}, $$

while the retrospective formula is

$$ {}_tV^{\,1}_{x:\overline{n}|} = P^{\,1}_{x:\overline{n}|}\ddot{s}_{x:\overline{t}|} - {}_tk_x. $$

${}^h_tV_{x:\overline{n}|}$ is the reserve at time t for a 1-unit, n-year endowment insurance policy on (x), with level annual premiums payable for h years. For $t \le n$, this is given prospectively as

$$ {}^h_tV_{x:\overline{n}|} = \begin{cases} A_{x+t:\overline{n-t}|} - {}^hP_{x:\overline{n}|}\ddot{a}_{x+t:\overline{h-t}|} & \text{if } t < h, \\ A_{x+t:\overline{n-t}|} & \text{if } t \ge h, \end{cases} $$

or retrospectively as

$$ {}^h_tV_{x:\overline{n}|} = \begin{cases} {}^hP_{x:\overline{n}|}\ddot{s}_{x:\overline{t}|} - {}_tk_x & \text{if } t < h, \\ {}^hP_{x:\overline{n}|}\ddot{s}_{x:\overline{h}|}\frac{(1+i)^{t-h}}{{}_{t-h}p_x} - {}_tk_x & \text{if } t \ge h. \end{cases} $$

${}_kV({}_n|\ddot{a}_x)$ denotes the reserve at time k for a deferred annuity on x providing 1 unit for life beginning at age $x+n$ and with annual premiums payable for n years. For $k < n$, the retrospective formula is the easiest and is given by

$$ {}_tV({}_n|\ddot{a}_x) = P({}_n|\ddot{a}_x)\ddot{s}_{x:\overline{t}|}. $$

For $k \geq n$, the prospective formula is the easiest and is given by

$$_kV(_n|\ddot{a}_x) = \ddot{a}_{x+k}.$$

6.10 Spreadsheet applications

For reserve calculations we add two columns J and K to Chapter 5 spreadsheet. In cell J10, we enter

$$= I10 - F10 - L10$$

and copy down to get the net cash flow vector. Then in cell K10 we enter the formula

$$= -\text{SUMPRODUCT}(E10 : E\$129, J10 : J\$129)/E10,$$

and copy down.

This calculates $_kV$ in cell K 10+k, as the negative of the value of future net cash flows, divided by $y_x(k)$. (Division by 0 error terms will appear for large enough durations, but these occur after age ω and can be ignored If one prefers, the formula in K10 can be suitably modified with an IF statement to replace them with a blank.) As a test calculate $_{15}V$ for the test problem given in Section 5.9. The answer is 333.16.

We now have a final spreadsheet that will calculate premiums and reserves on all insurance and annuity contracts without guaranteed payments. Here is a complete summary.

INPUT FORMULAS

Column D: 1 in cell D10, Section 2.14 formula in cell D11. Copy down.

Column E: 1 in cell E10, Section 4.8 formula in cell E11. Copy down.

Column F: Section 4.8 formula in cell F8.

Column G: Section 5.9 formula in cell G8.

Column H: Copy cell F8 to cell H8.

Column I: Insert the formula =(F8 + G8)/H8 in I6. Copy Cell H8 to I8. Insert the formula =$I$6*H10 in I10 and copy down.

Column J: Copy down the formula for cell J10.

Column K: Copy down the formula for cell K10.

INPUT DATA FOR EACH PARTICULAR PROBLEM

Column B: Interest rates.

Cell C1: Age at issue.

Column F: Annuity benefit vector **c**.

Column G: Death benefit vector **b**.

Column H: Premium pattern vector ρ.

Column N: Life table. (For sample table, insert parameters in N3 and N4, Section 4.8 formula in N10. Copy down).

OUTPUT

$\ddot{a}_x(\mathbf{c})$ in F8.

$A_x(\mathbf{b})$ in G8.

Present value of all premiums in I8.

The premium vector π in column I.

Reserves in column K.

Exercises

Type A exercises

6.1 A 3-year endowment insurance on (60) provides for death benefits payable at the end of the year of death. The death benefit is 1000 for death in the first year, 2000 for death in the second year and 3000 for death in the third year. In addition, there is a pure endowment of 4000 paid at time 3 if the insured is then alive. This is purchased by three annual premiums, beginning at age 60. The first two premiums are equal and the third is double the amount of the initial premium. You are given $q_{60} = 0.10,\ q_{61} = 0.20,\ q_{62} = 0.25$. The interest rates are 25% for the first 2 years and 100% in the third year. Find the initial premium. Find ${}_kV$ for $k = 1, 2, 3$.

6.2 You are given $q_{60} = 0.20, q_{61} = 0.25$. Interest rates are 20% in the first year, 25% in the second year and 50% in the third year. A 3-year endowment insurance policy issued to (60) provides for death benefits, at the end of the year of death, of 1000 if death occurs in the first year and 2000 if death occurs in the third or second years. In addition, there is a pure endowment of 2000 paid at age 63 if the insured is then alive. Level annual premiums are payable for 3 years. Find the premium. Find ${}_kV$, for $k = 1, 2, 3$. Do this first by the basic prospective reserve formula, and check your answers by using the recursion formula (6.7).

6.3 For a 3-year term insurance policy on (x), level premiums are payable for 3 years. The following data are given.

k	q_{x+k}	i_k	b_k
0	0.10	0.20	50 000
1	0.15	0.25	20 000
2	0.20	0.30	15 000

(a) Find the vector $\mathbf{y_x}$.

(b) Find the net cash flow vector.

(c) Using your answers to (a) and (b), compute ${}_1V$ and ${}_2V$.

(d) Explain briefly why your answers to (c) are negative.

6.4 A 10-year endowment policy on (40) has death benefits of 900, payable at the end of the year of death, should this occur within 10 years, plus a pure endowment of 900 at age 50 if the insured is then alive. Level annual premiums of 20 are payable for

10 years. The interest rate is a constant 50% and $q_{48} = 0.25$. You are not given q_{49}. Find ${}_tV$ for $t = 8, 9, 10$.

6.5 An insurance policy has level premiums of 20 payable for the duration of the contract. If the reserve is calculated with a premium of 15, then ${}_0V = 100$. What is ${}_0V$ if the reserve is calculated with a premium of 16?

6.6 Refer to Exercise 6.2.

(a) Decompose each premium into the risk portion and savings portion.

(b) Suppose that during the second year of the policy the actual interest rate earned was 20% rather than 25% and the actual rate of mortality was 0.20 rather than 0.25. Find the per policy gain during this year from both interest and mortality.

6.7 Refer to Exercise 6.1. Decompose each of the three premiums into the risk portion and savings portion.

6.8 For a certain contract issued at age 60, ${}_5V = 200$, the premium payable at age 65 is 40, $q_{65} = 0.20$, the death benefit payable at age 66 for death between age 65 and 66 is 800, the interest rate for the sixth year of the contract (between age 65 and 66) is 20%.

(a) Find ${}_6V$.

(b) Decompose the premium payable at age 65 into the risk portion and the savings portion.

(c) Suppose that during the sixth year, the actual interest rate earned was 25% instead of 20% and the actual rate of mortality at age 65 was 0.15 instead of 0.20. Find the gain or loss during this year, from interest and from mortality, for each policy in existence at the beginning of the year.

Type B exercises

6.9 Refer again to Exercise 6.7. Explain briefly in words why the third premium has a negative risk portion.

6.10 A deferred life annuity on (60) provides for annuity benefits payable for 3 years, beginning at age 63, provided that (60) is alive. The first annuity payment is 1000, the second is 2000 and the third is 3000. Premiums are payable for 3 years, beginning at age 60. The second and third premiums are equal in amount and each *double* the amount of the initial premium. If (60) dies before age 63, there will be, at the end of the year of death, a return of all premiums paid prior to death, without interest. You are given that $q_{60} = 0.1$, $q_{61} = 0.2, q_{62} = 0.25, q_{63} = 0.3, q_{64} = 0.4$. The interest rates are 25% per year for the first 4 years, and 50% for the fifth year. Find (a) the initial premium, (b) ${}_1V$ and (c) ${}_4V$.

6.11 A life insurance policy on (x) has level death benefits of 1000. A life insurance on (y) has exactly the same premiums and reserves as the policy on (x) for the first 10 years, but different death benefits. Suppose that $q_{y+k} = 2q_{x+k}$ for $k = 0, 1, \ldots, 9$. If the common reserve ${}_8V = 300$, what is the death benefit on (y)'s policy for the year running from time 7 to time 8?

6.12 Use formula (6.8) to derive formula (4.9) with $k = 1$. (Note that for annuity contracts the death benefits are 0, and the premiums are the negative of the annuity benefits.)

6.13 You are given that $\ddot{a}_{81} = 3.2, q_{80} = 0.10, i = 0.20$.

(a) Find $\ddot{a}_{80}$.

(b) A single-premium whole life annuity on (80) provides for a constant benefit of 1 per year. Due to improvements in mortality, the actual mortality rate experienced during the first year of the contract is 0.07 rather than 0.1. On the other hand, the interest earned during that year was 22% rather than 20%. What is the total per contract gain for this year?

6.14 A 2-year term insurance policy on (50), with a constant death benefit of 1000, is purchased by two-level annual premiums. Policyholders who choose to lapse the policy at time 1 will receive ${}_1V$ as a cash value. Assume that mortality and interest follow the projected pattern, and that the premium charged is the net premium. Suppose, however, that a typical group of people age 51 who have purchased insurance at age 50 can be divided into two groups. Half of them are healthy and half are not. The non-healthy group can expect to have twice as many deaths over the following year as the healthy group. Suppose that $q_{51} = 0.03$. What is the loss on each remaining policy for the second year, assuming:

(a) All of the healthy policyholders at age 51 lapse the policy at that time, and none of the unhealthy ones do;

(b) One-half of the healthy policyholders, and one-quarter of the unhealthy policyholders, lapse the policy at age 51.

6.15 For a certain whole life policy on (x), given by a death benefit vector $\mathbf{b}$ and premium pattern vector ρ, reserves are calculated according to two mortality tables, with the 1-year probabilities of dying denoted by q_{x+k} and q^*_{x+k}, respectively. The same positive interest rates are used in both calculations, and in each case the premiums are the net premiums as determined by the particular table being used. Suppose the two 'curves' cross at one point. That is

$$q_{x+k} \geq q^*_{x+k}, \quad k = 0, 1, \dots, n,$$
$$q_{x+k} \leq q^*_{x+k}, \quad k = n, n+1, \dots, \omega - x - 1.$$

Assume that death benefits are nonincreasing. Show that the reserve at time n is higher for the starred rates (the steeper curve).

6.16 (a) Use (6.8) to derive the formula

$$A_{x+1}(1_{n-1}) - A_x(1_n) = iA_x(1_n) - q_x\big(1 - A_{x+1}(1_{n-1})\big).$$

(b) As an actuary for an insurance company, you receive an angry letter from a policyholder. The person, age 50, has just purchased a single-premium 3-year term insurance policy with a constant death benefit of 1. The complaint is that the person's friend, age 49, purchased a single-premium 4-year term insurance policy, with the same death benefits, for a lower single premium. The letter writer argues that both policies provide coverage for the ages 50–53, but that the friend gets an extra year of coverage, for the year running from age 49 to 50. Therefore, it is claimed, the friend's premium should be higher, not lower. What is your response?

6.17 A special 2-year term insurance policy on (70) is to be purchased by a single premium. Should death occur in the first year, the insured will receive, at the end of the year, 1000 plus the reserve at that time. If death occurs in the second year, the insured will receive at time 2 only a return of the single premium paid without interest. Suppose $q_{70} = 0.36$, $q_{71} = 0.40$, and the interest rate is 100%. Find (a) the single premium, (b) ${}_1V$.

6.18 A whole life insurance policy on (x) provides for death benefits, paid at the end of the year of death, of 1000 plus the reserve at that time. Level annual premiums of P are payable for life. Suppose that $q_x = 0.2$, $q_{x+1} = 0.2$ and $q_y = 0.4$ for *all* $y \geq x+2$. The interest rate is a constant 100%. Find (a) P (b) ${}_kV$, $k = 1, 2, \ldots$.

6.19 Two people, A and B, are both age x. A buys a 20-year endowment policy, with a constant death benefit of 1, and a pure endowment of 1 at time 20, paying level annual premiums for 20 years. B buys a whole life policy with a constant death benefit 1, paying level annual premiums for life, and in addition, each year, invests the difference between his/her premium and A's premium, in a savings account. At time 20, the reserve on B's policy, plus the amount in his savings account total 1. All premiums were calculated at a constant interest rate of i. B earned a constant rate of j on his investments. Is j greater than, less than or equal to i? Why?

6.20 A whole life policy with a level death benefit of 10 000 carries net annual premium of 100 payable for life. The rate of discount is a constant 0.04. At time n, the policyholder wishes to reduce their annual premium payment to 25, and is told they can do so, but the death benefit will be reduced to 7000. Assuming that the entire reserve is available as a nonforfeiture benefit, find ${}_nV$.

6.21 Suppose that a mortality table used to calculate reserves is altered by adding a positive constant to each value of q_x. Explain how the reserves will be affected, for whole life policies with a constant death benefit and level premiums.

Spreadsheet exercise

6.22 Suppose that mortality follows the sample life table of Section 3.7 and interest rates are 5% in the first 15 years, 6% for the next 15 years, and 7% thereafter. A contract issued at 40 provides for a life annuity beginning at age 65. The annual annuity payment is 1000 for the first 10 years and 2000 thereafter for life. If death occurs before annuity payments begin, there is a death benefit of 10 000 paid a the end of the year of death. Level annual premiums are payable for 15 years.

(a) Find the premium and all reserves.

(b) Suppose the interest rate in the first 15 years increases from 5% to 5.5%. Do reserves increase or decrease?

(c) If the interest rate in the first 15 years decreases to 4%, what happens to reserves?

7

Fractional durations

7.1 Introduction

Up to now, we have considered cash flows where the payments were at integer times. We now deal with the case where cash flows can occur at fractional durations. For example, if the basic time unit was a year, and payments were made monthly, these would be at times that are multiples of 1/12. In practice this is a common occurrence. Purchasers of life annuities often want the income to be paid monthly. Many purchasers of insurance policies wish to pay premiums monthly, or possibly quarterly or semiannually. One obvious method of handling this feature is to change the time unit. If we are dealing with annuities with monthly payments, we could just take our unit of time as 1 month, and payments would be at integer times. This option was not always feasible in pre-computer days. For cash flows discounted at interest only, calculations were done from tables showing yearly rates of interest, and elaborate formulas were developed to handle the fractional durations. This is no longer necessary, and the change-of-period approach is the modern way to handle the valuation of cash flows at compound interest.

For life annuities however it is still common to keep the time unit as a year. There are several reasons for this. In the first place, the year has been so ingrained as a measure of age that it seems hard to break away from this tradition. Another more important reason is that insurers often do not calculate premiums exactly for the different periods, but rely on approximate conversion formulas. For this purpose it is convenient to have some simple relationships established between the yearly annuity values and those with more frequent payments. Still another reason is that we also want to consider annuities with payments made *continuously*, which will be discussed in Chapter 8. To do this, it is convenient to first discuss the case in which payments are made m times per year, and then view continuous annuities as a limiting case as m approaches infinity.

We will follow the standard usage of referring to annuities with payments made at times which are multiples of $1/m$ as $m$thly annuities or annuities payable $m$thly.

Fundamentals of Actuarial Mathematics, Third Edition. S. David Promislow.

Companion website: http://www.wiley.com/go/actuarial

We will restrict attention to sequences of cash flows where the *periodic payments during each year are constant,* but can vary from year to year. If payments change during the year (e.g., they increase each month), then usually the only feasible approach is to change the period. The basic model is as follows. We are given a positive integer m, which is the number of payments made each year. We are given a cash flow vector, $\mathbf{c} = (c_0, c_1, \dots, c_N)$ where, as before, c_k is the *total* amount paid in year $k+1$, that is, between time k and $k+1$. Rather than being paid as a single cash flow however, there are payments of

$$\frac{c_k}{m} \text{ at time } k + \frac{j}{m}, \quad j = 0, 1, \dots, m-1. \tag{7.1}$$

Let y denote an arbitrary discount function. We denote the present value of the sequence of cash flows as described by $\ddot{a}^{(m)}(\mathbf{c}; y)$. The task at hand is to relate this quantity to the case where $m = 1$, namely $\ddot{a}(\mathbf{c}; y)$. Our first observation is that it is sufficient to find a formula for $\ddot{a}^{(m)}(\mathbf{c}; y)$ in the case that $\mathbf{c} = \mathbf{e}^0$, a single payment of 1 at time zero. This follows from the replacement principle. Let u_k be the value at time k of a sequence of payments of $1/m$ at time $k + (j/m), j = 0, 1, \dots, m-1$. We can write

$$u_k = \ddot{a}^{(m)}(\mathbf{e}^0; y \circ k).$$

The replacement principle tells us that we can replace all the payments between time k and $k+1$ by a single payment of $c_k u_k$ at time k. It follows that

$$\ddot{a}^{(m)}(\mathbf{c}; y) = \ddot{a}(\mathbf{c} * \mathbf{u}; y). \tag{7.2}$$

where $\mathbf{u} = (u_0, u_1, \dots, u_k, \dots)$.

The key to evaluating fractional annuities is to calculate the quantities u_k. We will discuss particular cases in the following two sections.

7.2 Cash flows discounted with interest only

Suppose that we are given yearly interest rates, and we want to consider a change of period to $1/m$ of a year. We use primed symbols to denote the quantities applicable to the new period, that is, v', i' and d' in place of v, i and d, respectively. We will start as usual by supposing that we are given the interest rates i_k for each nonnegative integer k. To determine the rates applicable to the new periods, the standard assumption made is that interest rates are constant over each year. To be precise, one postulates is that for all nonnegative integers k, (2.9) holds whenever $s, t, s+h, t+h$ all lie in the interval $[k, k+1)$.

Let i'_k denote the constant value of $v(s + 1/m, s) - 1$ and let d'_k denote the constant value of $1 - v(s, s + 1/m)$ for $k \le s < s + 1/m < k + 1$. An investment of 1 at time 0 will accumulate to $1 + i'_k$ at time $1/m$ and then to $(1 + i'_k)^2$ at time $2/m$, $(1 + i'_k)^3$ at time $3/m$ and so on. Arguing in the same way for the discount rate gives

$$\left(1 + i'_k\right)^m = 1 + i_k, \tag{7.3‡}$$

$$\left(1 - d'_k\right)^m = 1 - d_k, \tag{7.4‡}$$

since in (7.3) each side of the equation is the value at time $k+1$ of 1 unit invested at time k, and in (7.4) each side of the equation is the value at time k of 1 unit paid at time $k+1$.

In standard actuarial notation it is common to 'annualize' these rates by multiplying by m. That is, one defines

$$i_k^{(m)} = mi'_k, \qquad d_k^{(m)} = md'_k. \tag{7.5}$$

The quantities $i_k^{(m)}$ and $d_k^{(m)}$ are often known as *nominal* rates, to reflect the fact that they are rates in 'name only', and not really applicable to any period. They are sometimes referred to as the yearly interest (respectively discount) rates *convertible* or *compounded* $m$thly. The actual *rate* applicable to each $m$thly period in the year k to $k+1$ is found by dividing these nominal rates by m.

In the case of constant interest the subscript k is not needed and we simply use the symbols $i^{(m)}$ and $d^{(m)}$ for the nominal rates.

Example 7.1 Suppose that $i^{(m)} = 0.12$. Find i, the yearly rate of interest, if (a) $m = 2$, (b) $m = 12$.

Solution.

(a) We must divide by 2, to deduce that the rate per half-year period is 0.06. From (7.3), $(1+i) = 1.06^2 = 1.1236$ so that the corresponding yearly interest rate is 12.36%.

(b) Similarly, the corresponding yearly interest rate $= (1+0.01)^{12} - 1 = 12.68\%$.

It is important to distinguish between the actual yearly rates and nominal rates for different periods, when comparing quoted figures. For example, by custom, the rates on Canadian mortgages are quoted as nominal rates with $m = 2$, while for US mortgages, $m = 12$. These provisions would be evident in the actual written contract, but are not always made clear to a prospective borrower. As the above example indicates, the actual annual rates paid on Canadian or US mortgages would be higher than the quoted nominal rates.

We now deduce the present value of fractional annuities for the investment discount function. To simplify the notation, we first assume constant interest. A 1-year, 1-unit annuity payable $m$thly is simply an annuity paying $1/m$ units for m periods, discounted with the primed rates. From formulas (2.16) and (7.4) we can write

$$u_k = \ddot{a}^{(m)}(\mathbf{e}^0) = \frac{1}{m}\left(\frac{1-(1-d')^m}{d'}\right) = \frac{d}{d^{(m)}} \tag{7.6}$$

for all k, where the last quantity is taken as 1 if $d = 0$. From (7.2),

$$\ddot{a}^{(m)}(\mathbf{c}; v) = \frac{d}{d^{(m)}}\ddot{a}(\mathbf{c}; v).$$

For the general case of non-constant interest we have $u_k = d_k/d_k^{(m)}$, which is taken equal to 1 if $d^{(k)} = 0$.

Note that in the case of positive cash flows (and assuming the usual case of positive interest rates) the present value with payments paid $m$thly is less than the corresponding yearly present value. This can be shown mathematically by noting that d_k is less than $d_k^{(m)}$. For example, if $d_k = 0.36$ then, from (7.4), $d_k' = 1 - 0.64^{1/2} = 0.20$ and $d_k^{(2)} = 0.40$. (The reader is invited to provide a general proof.) This conclusion is also evident from a comparison of payments. In the yearly case, a full payment of c_0 would be made at time 0. In the $m$thly case, the annuitant receives c_0/m each period of length $1/m$, and will not have collected the full c_0 amount until one mth of a year before the year end. This is true for each year of the annuity. There is therefore a loss of interest, which is reflected in the lower present value for the $m$thly case.

7.3 Life annuities paid *m*thly

We now consider the case where we have the interest and survivorship discount function y_x. We will usually write $\ddot{a}_x^{(m)}(\mathbf{c})$ for $\ddot{a}^{(m)}(\mathbf{c}; y_x)$ as we did when $m = 1$.

As we are normally given mortality data only for integer ages, we are faced with the necessity of making some assumption in order to extrapolate for the data that we do not have. Many approaches are possible. In the next section we describe the most commonly used method. Other possibilities that have been proposed are described in Exercises 7.6, 7.13 and 7.14, as well as later in Chapter 8.

7.3.1 Uniform distribution of deaths

This preferred method is known as the assumption of a *uniform distribution of deaths over each year of age*, abbreviated as UDD. It can be viewed as just linear interpolation for ℓ_y. Suppose we assume that ℓ_y, rather than being defined just for integer values, is defined for all nonnegative y. The UDD assumption is that for any integer x and $0 < t < 1$,

$$\ell_{x+t} = (1 - t)\ell_x + t\ell_{x+1}.$$

The name comes from the fact that given $0 < t < t + h < 1$,

$$\ell_{x+t} - \ell_{x+t+h} = h(\ell_x - \ell_{x+1}) = hd_x.$$

In other words, during any interval of h years which lies between two integral ages, the number of deaths occurring in that period is just h times the total number of deaths for the yearly interval. We can therefore say that the deaths are spread uniformly over the year.

Example 7.2 Suppose $\ell_{60} = 1000, \ell_{61} = 940$. Assuming UDD, what is the probability that a person age $60\frac{1}{3}$ will die before reaching age $60\frac{1}{2}$?

Solution. The number of deaths in the $\frac{1}{6}$-year period between $60\frac{1}{3}$ and $60\frac{1}{2}$ equals 10, one-sixth of the total deaths in the year. Out of 1000 lives at age 60, there will 20 who die before reaching age $60\frac{1}{3}$ (one-third of the yearly deaths), leaving 980 alive at age $60\frac{1}{3}$. The probability is therefore 10/980.

An equivalent form of the UDD assumption is

$$ {}_tq_x = tq_x, \quad \text{for any integer } x \text{ and } 0 < t < 1, \tag{7.7}$$

which follows immediately by writing ${}_tq_x$ as $(\ell_x - \ell_{x+t})/\ell_x$.

Assuming UDD, we can easily calculate ${}_tp_y$ for all t and y (not necessarily an integer age) by the formula

$$ {}_tp_{x+h} = \frac{{}_{t+h}p_x}{{}_hp_x},$$

which follows from the multiplication rule. This points out an inconsistency in the UDD assumption. Suppose that the life table follows the expected pattern of increasing mortality by age. That is, for all integers $x \le y$ we have $q_x \le q_y$. Then it can be shown that under UDD, ${}_tq_x \le_t q_y$ for integers $x \le y$ and all $t \ge 0$, as we would expect. However, this inequality need not hold if x and y are not integers. See Exercise 7.15.

7.3.2 Present value formulas

We now derive formulas for fractional duration life annuities by making the UDD assumption for mortality and the normal assumption for interest that we used in Section 7.2. This is the standard procedure, but the different treatment of the two factors leads to some complications in the formulas. We first assume constant interest. For our purpose we require some new interest quantities. Let

$$\begin{aligned}
\beta(m) &= \frac{1}{vm^2}[v^{1/m} + 2v^{2/m} + \cdots + (m-1)v^{(m-1)/m}],\\
\alpha(m) &= \frac{d}{d^{(m)}} + d\beta(m).
\end{aligned}$$

We will later derive a simplified expression for $\beta(m)$.

We have

$$\ddot{a}_x^{(m)}(\mathbf{e}^0) = \frac{1}{m}[1 + v^{1/m}{}_{1/m}p_x + v^{2/m}{}_{2/m}p_x + \cdots + v^{(m-1)/m}{}_{m-1/m}p_x].$$

From (7.7), ${}_{j/m}p_x = 1 - (j/m)q_x$, so we can write

$$\begin{aligned}
u_0 = \ddot{a}_x^{(m)}(\mathbf{e}^0) &= \frac{1}{m}[1 + v^{1/m} + v^{2/m} + \cdots + v^{(m-1)/m}]\\
&\quad - \frac{q_x}{m^2}[v^{1/m} + 2v^{2/m} + \cdots + (m-1)v^{(m-1)/m}]\\
&= \ddot{a}^{(m)}(\mathbf{e}^0) - \beta(m)vq_x = \frac{d}{d^{(m)}} - \beta(m)vq_x,
\end{aligned}$$

where we refer back to (7.6). Similarly, for any positive integer k,

$$u_k = \frac{d}{d^{(m)}} - \beta(m)vq_{x+k}.$$

From (7.2) and (5.6) can now write

$$\ddot{a}_x^{(m)}(\mathbf{c}) = \frac{d}{d^{(m)}}\ddot{a}_x(\mathbf{c}) - \beta(m)A_x(\mathbf{c}). \tag{7.8}$$

Somewhat surprisingly, an insurance term has come into an annuity value formula, but it is easily explained. Apart from interest loss, a reason for the $m$thly annuity to be worth less than the yearly annuity is the loss of annuity income in the year of death. The annuitant receiving benefits $m$thly will only receive payments up to the time of death in their final year. For example, if they die shortly after the beginning of the year they will only get $1/m$ of the total yearly income. Of course, they could receive the total income by living until the last mth of the year. On average, annuitants should lose slightly less than half of the income in the year of death, which is reflected by the second term in (7.8). Under UDD the actual proportion is $\beta(m)$. This is close to $(m-1)/2m$ for small interest rates, as shown in Section 7.5 below.

To derive an alternate expression to (7.8) that uses only annuity present values we apply the general insurance-annuity identity (5.8) to the second term in (7.8) and obtain

$$\ddot{a}_x^{(m)}(\mathbf{c}) = \alpha(m)\ddot{a}_x(\mathbf{c}) - \beta(m)\ddot{a}_x(\Delta\mathbf{c}). \tag{7.9‡}$$

In the general case where the interest rate is constant over each year, but can vary from year to year, $\alpha(m)$ and $\beta(m)$ will be vectors with the entry corresponding to index k, equal to the corresponding value for the interest rate i_k. The general form of (7.9) will be

$$\ddot{a}_x^{(m)}(\mathbf{c}) = \ddot{a}_x(\mathbf{c} * \alpha(m)) - \ddot{a}_x(\Delta(\mathbf{c} * \beta(m))).$$

For the particular case of the level payment annuity with constant interest we have

$$\ddot{a}_x^{(m)}(1_n) = \alpha(m)\ddot{a}_x(1_n) - \beta(m)(1 - v(n)_n p_x)$$

Example 7.3 A woman age 60, earns a salary of 60 000 per year, payable monthly, and would normally expect to receive a raise of 2000 per year until retirement at age 65. She is injured and unable to work again. She plans to sue the parties responsible for her injury, for an amount equal to the present value of her lost salary. Find a formula to calculate this present value. Assume a constant interest rate.

Solution. We will assume the accident occurred just before her 60th birthday, and that the first raise is due 1 year later. The present value is

$$1000\left[\ddot{a}_{60}^{(12)}(\mathbf{c})\right] = 1000[\alpha(12)\ddot{a}_{60}(\mathbf{c}) - \beta(12)\ddot{a}_{60}(\Delta\mathbf{c})],$$

where $\mathbf{c} = (60, 62, 64, 66, 68)$, so that $\Delta\mathbf{c} = (60, 2, 2, 2, 2, -68)$.

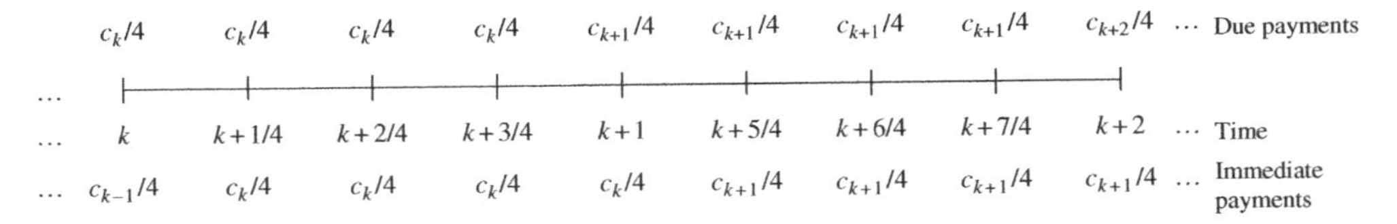

Figure 7.1 Due and immediate quarterly annuity payments

7.4 Immediate annuities

We now discuss an important variation. Refer back to (7.1). Suppose that instead of this scheme we assume payments of

$$\frac{c_k}{m} \text{ at time } k + \frac{j}{m}, \quad j = 1, 2, \ldots, m. \tag{7.10}$$

We can view this as having payments made at the *end* rather than the *beginning* of each period of length $1/m$. Annuities satisfying (7.10) are often referred to as *immediate*, while those satisfying (7.1) are referred to as *due*. The present value of the annuity with benefit vector given by (7.10) and discount function y is denoted by $a^{(m)}(\mathbf{c}; y)$. That is, the two dots are removed for immediate annuities.

The reader may wish to refer back to Section 2.14.1 where we commented on the somewhat unusual terminology. We also pointed out that the distinction between the two types is not needed in the annual case. However, when we postulate a constant rate of payment over the year, we do need to make the distinction in the $m$thly case. See Figure 7.1 for a comparison of the payments on a due and immediate quarterly annuity.

To go from the due annuity to the immediate annuity, we must do the following. At time zero, subtract a payment of c_0/m. At time 1, subtract a payment of c_1/m and add a payment of c_0/m. At time 2, subtract a payment of c_2/m and add a payment of c_1/m, etc. We then have, for *any* discount function y,

$$a^{(m)}(\mathbf{c}; y) = \ddot{a}^{(m)}(\mathbf{c}; y) - \frac{1}{m}\ddot{a}(\Delta\mathbf{c}; y). \tag{7.11}$$

Example 7.4 Find a formula for the present value of a life annuity on (40) consisting of payments of \$500 at the end of each month for 10 years. Assume a constant interest rate.

Solution. In this case,

$$\mathbf{c} = 6000(1_{10}), \qquad \Delta\mathbf{c} = 6000(1, 0_9, -1), \qquad \ddot{a}_{40}(\Delta\mathbf{c}) = 6000[1 - y_{40}(10)].$$

The present value is

$$\begin{aligned} 6000\left[a_{40}^{(12)}(1_{10})\right] &= 6000\left[\ddot{a}_{40}^{(12)}(1_{10}) - \frac{1}{12}(1 - y_{40}(10))\right] \\ &= 6000\left[\alpha(12)\ddot{a}_{40}(1_{10}) - \left(\beta(12) + \frac{1}{12}\right)(1 - y_{40}(10))\right]. \end{aligned}$$

7.5 Approximation and computation

In the previous section we developed some basic formulas. We now concentrate on the problem of obtaining numerical values.

For this purpose it is convenient to develop a closed-form formula for $\beta(m)$ in terms of standard interest symbols $i^{(m)}$ and $d^{(m)}$. Assume for the moment constant interest. From the definition of $\beta(m)$ we can write, using the discount function $v(k) = (1 + i')^{-k}$,

$$\beta(m) = \frac{1}{m^2}\mathrm{Val}_{m-1}(\mathbf{j}),$$

where $\mathbf{j} = (1, 2, \dots, m-1)$. From formulas (2.16) and (2.17),

$$\mathrm{Val}_{m-1}(\mathbf{j}) = (1 + i')^{(m-1)}\ddot{a}(\mathbf{j}) = \frac{[(1 + i')^{m-1} - 1]/d' - (m-1)}{d'}.$$

We can simplify the numerator since

$$\frac{(1 + i')^{m-1} - 1}{d'} = \frac{(1 + i')^m - (1 + i')}{i'} = \frac{i}{i'} - 1.$$

Substituting and noting that $mi' = i^{(m)}, md' = d^{(m)}$, we obtain the formula

$$\beta(m) = \frac{i - i^{(m)}}{i^{(m)}d^{(m)}}. \tag{7.12}$$

It then follows that

$$\alpha(m) = \frac{d}{d^{(m)}} + d\beta(m) = \frac{id}{i^{(m)}d^{(m)}}.$$

For varying interest we get the same relationships holding on a year-by-year basis. We need merely insert subscripts of k on all quantities.

Exact calculations of $\alpha(m)$ and $\beta(m)$ are not always done in practice. It is common to use an approximation that assumes a constant interest rate of 0. This implies that

$$\beta(m) = \frac{1 + 2 + \cdots + m - 1}{m^2} = \frac{m-1}{2m}.$$

Moreover $d/d^{(m)} = 1$ and $\alpha(m) = 1$, so we can write the approximate formula

$$\ddot{a}_x^{(m)}(\mathbf{c}) \doteq \ddot{a}_x(\mathbf{c}) - \frac{m-1}{2m}\ddot{a}_x(\Delta\mathbf{c}). \tag{7.13}$$

In much of the actuarial literature this approximation is not found by setting the interest rate equal to 0 as we have done, but rather by using linear interpolation for the discount function y_x. That is, it is assumed that for all nonnegative integers k and $0 < s < 1$,

$$y_{x+k}(s) = (1 - s) + sy_{x+k}(1).$$

Under this assumption

$$\begin{aligned}\ddot{a}_x^{(m)}(\mathbf{e}^0) &= \frac{1}{m}\left[1 + y_x\left(\frac{1}{m}\right) + y_x\left(\frac{2}{m}\right) + \cdots + y_x\left(\frac{m-1}{m}\right)\right]\\ &= \frac{m+1}{2m} + \frac{m-1}{2m}vp_x\\ &= 1 - \frac{m-1}{2m}[1 - vp_x],\end{aligned}$$

which is the same as $\ddot{a}_x^{(m)}(\mathbf{e}^0)$ calculated at 0 interest. By (7.2) both assumptions lead to the same approximation. This reason why this occurs is that the linearity of y_x, in addition to the UDD assumption, necessarily implies that the interest rate must be 0. The advantage of stating the latter condition from the outset is that it makes it clear that the approximation should be reasonable with relatively low interest rates. However, if rates are too large relative to mortality rates, this approximation can lead to inconsistencies.

Example 7.5 Suppose that $i = 0.21$, and $q_x = 0.004$. Calculate $\ddot{a}^2(\mathbf{e}^0; v)$ and $\ddot{a}_x^{(2)}(\mathbf{e}^0)$, using (7.13) for the latter.

Solution.

$$\ddot{a}^{(2)}(\mathbf{e}^0; v) = \frac{1}{2}[1 + 1.10^{-1}] = 0.955.$$

Using (7.13),

$$\ddot{a}_x^{(2)}(\mathbf{e}^0) = 1 - \frac{1}{4}\left[1 - \frac{0.996}{1.21}\right] = 0.956.$$

It is, however, inconsistent for an annuity in which payments are contingent upon survival to cost more than an annuity with identical payments that are certain to be made.

Equation 7.13 is part of a series of other possible approximation methods known as *Woolhouse's formulas*. This one is the two-term formula. The next in the series, the three-term formula, is not used often but appears to give very good results in certain cases. After developing the appropriate tools in the next chapter, we will investigate further the two-term formula inconsistency and derive the three-term formula. See Section 8.11.

*7.6 Fractional period premiums and reserves

Annual premium and reserve formulas remain basically unchanged when premiums are paid $m$thly. The only difference is that we replace $\ddot{a}_x$ with $\ddot{a}_x^{(m)}$ throughout. The purpose of this section is to derive a few useful identities.

Consider all policies with a particular issue age (x) and with level premiums payable for h years. The simplest type of such a policy will be a 1-unit, h-year term policy with annual premiums. Let P denote the net annual premium and ${}_kV$ denote the reserve at time k for this term policy. Given any other such policy, let $\tilde{P}$ denote the net annual premium if paid yearly, and let $\tilde{P}^{(m)}$ denote the net annual premium for this policy if paid $m$thly (i.e., $\tilde{P}^{(m)}/m$ is paid at the beginning of each mth of the year). Let ${}_k\tilde{V}$ denote the reserve at time k on this policy if premiums are paid annually and let ${}_k\tilde{V}^{(m)}$ denote the reserve at time k if premiums are paid $m$thly. For simplicity, we will assume constant interest.

We have

$$\tilde{P}^{(m)}\ddot{a}_x^{(m)}(1_h) = \tilde{P}\ddot{a}_x(1_h),$$

as both sides are equal to the present value of the benefits, so that

$$\frac{\tilde{P}}{\tilde{P}^{(m)}} = \frac{\ddot{a}_x^{(m)}(1_h)}{\ddot{a}_x(1_h)} = \frac{d}{d^{(m)}} - \beta(m)\frac{A_x(1_h)}{\ddot{a}_x(1_h)},$$

using (7.8) for the last equality, and we can write

$$\frac{\tilde{P}}{\tilde{P}^{(m)}} = \frac{d}{d^{(m)}} - \beta(m)P. \tag{7.14}$$

Since the value of insurance benefits is independent of the premium frequency, we have, for $k < h$,

$$\begin{aligned}{}_k\tilde{V}^{(m)} - {}_k\tilde{V} &= \tilde{P}\ddot{a}_{x+k}(1_{h-k}) - \tilde{P}^{(m)}\ddot{a}_{x+k}^{(m)}(1_{h-k}) \\ &= \tilde{P}^{(m)}\left[\frac{\tilde{P}}{\tilde{P}^{(m)}}\ddot{a}_{x+k}(1_{h-k}) - \ddot{a}_{x+k}^{(m)}(1_{h-k})\right].\end{aligned}$$

Now substitute from (7.8) for $\ddot{a}_{x+k}^{(m)}(1_h)$ and from (7.14) for $\tilde{P}/\tilde{P}^{(m)}$. The term $(d/d^{(m)})$ $\ddot{a}_{x+k}(1_{h-k})$ cancels out and we are left with

$${}_k\tilde{V}^{(m)} - {}_k\tilde{V} = \beta(m)\tilde{P}^{(m)}{}_kV. \tag{7.15}$$

The term on the right represents the additional reserve that must be held at time k in order to provide for the premiums that will not be collected in the year of death.

The significance of (7.14) and (7.15) is as follows. We consider a collection of policies with level premiums, common issue age and premium payment period. The benefits can be anything at all. We know the premium and reserve for the particular case of the level term policy. Then, using only those two quantities, we can easily adjust premiums and reserves from annual mode to $m$thly mode, for all other policies in the collection.

7.7 Reserves at fractional durations

Insurance companies do a complete calculation of reserves on December 31 (or whenever their particular fiscal year ends). However, policyholders are not always accommodating enough to purchase their policy on January 1. This means that most of the policy durations at the calendar year end will be fractional. In this section we derive formulas for calculating reserves at time $k + s$ where k is an integer and $0 < s < 1$. Consider the general annual premium policy with death benefit vector **b** and premium vector $\boldsymbol{\pi}$. Suppose first that the death benefit for those dying between time k and $k+s$ is paid at time $k+s$. The reserve at time $k+s$ would then be, by a calculation analogous to that used in deriving (6.7),

$$({}_kV + \pi_k)y_x(k+s,k) - b_k \frac{{}_sq_{x+k}}{{}_sp_{x+k}}.$$

That is, thinking retrospectively, the balance at time $k+s$ is obtained by taking the balance at time k, adding the premium, accumulating up to time $k+s$ at interest and survivorship, and then subtracting from each survivor enough to pay the death benefits for those dying between time k and time $k+s$.

We must, however, adjust this to reflect the fact that the death benefit will not be paid until the end of the year, and to do so we multiply the second term by $v(k+s,k+1)$. Note now that

$$\frac{v(k+s,k+1)}{{}_sp_{x+k}} = \frac{y_x(k+s,k+1)}{({}_{1-s}p_{x+k+s})({}_sp_{x+k})} = \frac{y_x(k+s,k+1)}{p_{x+k}}.$$

From the UDD assumption, ${}_sq_{x+k} = s(q_{x+k})$ and substituting these last two identities gives the desired reserve formula as

$$\begin{aligned} {}_{k+s}V &= ({}_kV + \pi_k)y_x(k+s,k) - sb_k \frac{q_{x+k}}{p_{x+k}} y_x(k+s,k+1) \\ &= (1-s)({}_kV + \pi_k)y_x(k+s,k) + s\left[({}_kV + \pi_k)y_x(k+1,k) - b_k \frac{q_{x+k}}{p_{x+k}}\right] y_x(k+s,k+1). \end{aligned}$$

Finally, substituting from (6.7) yields

$${}_{k+s}V = (1-s)[({}_kV + \pi_k)y_x(k+s,k)] + s[{}_{k+1}V\ y_x(k+s,k+1)].$$

In other words, under UDD, the reserve at some intermediate point in the year is obtained by linearly interpolating between the initial reserve, accumulated to that point with interest and survivorship, and the end-of-year terminal reserve, discounted to that point with interest and survivorship. Care must be taken not to use the terminal instead of the initial reserve at time k, as that would involve interpolating over a discontinuity.

A common simplification is to assume that interest and mortality rates are sufficiently small so that we can take y_x as identically equal to 1. This results in the approximate formula

$$\begin{aligned} {}_{k+s}V &= (1-s)({}_kV + \pi_k) + s({}_{k+1}V) \\ &= (1-s)_kV + s({}_{k+1}V) + (1-s)\pi_k. \end{aligned}$$

The third term above is known as the *unearned premium*. It is simply the total premium paid at the beginning of the year multiplied by the fraction of the year remaining.

When premiums are paid m times a year rather than annually, the same type of approximation as above is normally used. That is, we calculate the reserve at time $k+s$ by linearly interpolating between the terminal reserves at time k and $k+1$ and add the unearned premium.

To be precise, suppose that $s = h/m + r$ where h is an integer and $0 < r < 1/m$ and that the annual premium π_k is paid m times per year. Then a commonly used approximation is

$$_{k+s}V = (1-s)_kV + s(_{k+1}V) + \left(\frac{1}{m} - r\right)\pi_k.$$

Example 7.6 For a certain policy $_3V = 108$, $_4V = 180$ and for the fourth year of the policy the premium is 60, paid quarterly (i.e.,15 is paid at time 3, $3\frac{1}{4}, 3\frac{1}{2}$ and $3\frac{3}{4}$. Find the reserve at time 3 years and 7 months.

Solution. We have $s = 7/12 = 2/4 + 1/12$. The unearned premium is $\left(\frac{1}{4} - \frac{1}{12}\right) 60 = 10$. The above formula gives

$$_{3+7/12}V = \frac{5}{12} \times 108 + \frac{7}{12} \times 180 + 10 = 160.$$

An alternate way of calculating the unearned premium is to note that the reserve is taken at a time one-third through the quarter-year period, so the unearned premium is two-thirds of the premium payable at the beginning of the quarter, which is $\frac{2}{3} \times 15 = 10$.

7.8 Standard notation and terminology

Most of the standard notation for fractional payments has been discussed already. The one remaining symbol is $A^{(m)}$ which is used to denote net single premiums for an insurance in which the death payments are paid at the end of the mth of the year in which death occurs. For example if $m = 12$, the death benefit would be made at the end of the month of death.

The assumption of Exercise 7.14 below is known as the *Balducci hypothesis*, named for G. Balducci, an Italian actuary. It arose in conjunction with mortality studies designed to produce life tables.

Exercises

Assume UDD unless otherwise indicated.

Type A exercises

7.1 You are given that $\ddot{a}_{60} = 12$, $_{10}p_{50} = 0.8$, and interest is constant at 6%.

(a) Find the net single premium for an annuity on (50) with level benefit payments of 1000 per month, beginning at age 50 and continuing for life, with the provision that the first 120 monthly payments are guaranteed, regardless of whether (50) is alive or not.

(b) Redo (a), assuming now that the first payment is made at the end of 1 month.

7.2 You are given that $q_{70} = 0.2, q_{71} = 0.25, q_{72} = 0.30$. Interest rates are 20% for the first 2 years and 30% thereafter. A 3-year life annuity on (70) provides for benefits of 1000 in the first year, 2000 in the second year and 3000 in the third year, provided that (70) is then alive. Suppose that payments are to be made quarterly. Find the present value if: (a) the first payment is to be made at age 70; (b) the first payment is to be made at age $70\frac{1}{4}$.

7.3 An annuity on (50) provides for yearly payments for 30 years. The amount of the income is 10 000 for the first 10 years, 8000 for the next 10 years and 5000 for the last 10 years. The net single premium for this annuity is 100 000. You are given that the interest rate is a constant 5% and that ${}_{10}p_{50} = 0.9,\ {}_{20}p_{50} = 0.8,\ {}_{30}p_{50} = 0.7$. Find the premium if the same income is to be paid monthly instead of annually, assuming payments are made (a) at the beginning of each month, (b) at the end of each month.

7.4 You are given the following figures from a life table. $\ell_{60} = 1000, \ell_{61} = 700, \ell_{62} = 500$. Find the probability that a person now age $60\frac{1}{3}$ will die between the ages of $60\frac{1}{2}$ and $61\frac{3}{4}$.

7.5 Given $q_{70} = 0.2,\ q_{71} = 0.3,\ q_{72} = 0.4$, find the probability that a person age $70\frac{1}{2}$ will live to age $72\frac{1}{4}$.

7.6 Instead of assuming UDD, suppose that, for an integer x and $0 < t < 1$, we assume that ${}_tp_x = p_x^t$. Given $l_{60} = 100\,000,\ l_{61} = 81\,000,\ l_{62} = 41\,472$, find the probability that $(60\frac{1}{2})$ will die between the ages of $61\frac{1}{3}$ and $61\frac{2}{3}$.

7.7 A policy is issued on March 20, 1990. Given that ${}_{10}V = 2000,\ {}_{11}V = 3000$ and that the premium is a level 105 per month, find the reserve on December 31, 2000, using the standard approximation. Assume that each month is 30 days.

7.8 For a policy on (x), ${}_5V = 220,\ {}_6V = 80,\ b_5 = 1000,\ i_5 = 0.10,\ q_{x+5} = 0.2$. Find ${}_{5+1/4}V$.

Type B exercises

7.9 Let r be the present value of a life annuity on (x) providing 1 at the beginning of each period of length $1/m$, with payments guaranteed for n years. Let s be the present value of the same annuity but with the payments at the *end* of each period. If $v(n) = 0.9$ and ${}_np_x = 0.6$, find $r - s$. You are not given the value of m.

7.10 Show that under a constant interest rate of i, $a^{(m)}(1_n) = (i/i^{(m)})a(1_n)$.

7.11 You want to evaluate $\ddot{a}_x^{(4)}(\mathbf{k})$ at a constant interest rate of 20%, where $\mathbf{k} = (10, 9, 8, \ldots, 1)$. Assuming that UDD holds, what error is made if you use the zero-interest approximation? You are given that $\ddot{a}_x(\mathbf{k}) = 35$ and $\ddot{a}_x(1_{11}) = 6$.

*7.12 A 20-year term insurance policy on (50) has a constant death benefit of 1000, and net level annual premiums of 10, payable for 20 years. The reserve at time 15 is 60. Another policy on (50) has the same premium pattern but different death benefits. On this latter policy, the annual premium is 100, and ${}_{15}V = 500$. The interest rate is a constant 6%. What is the value of the annual premium and the reserve at time 15 on this latter policy, if the premiums are payable monthly instead of annually?

7.13 Suppose we assume that for all integers x and $0 < t < 1$,

$$\ell_{x+t} = \ell_x^{1-t} \ell_{x+1}^t.$$

(a) Show that this is equivalent to the assumption of Exercise 7.6.

(b) Show that if (2.9) holds for v, then it also holds for the discount function y_x whenever $s, t, s+h, t+h$ are all in the interval $[k, k+1)$ for some nonnegative integer k. (See Exercise 8.21 for more on this assumption.)

7.14 Suppose we assume that for all integers x and $0 < t < 1$,

$$\ell_{x+t}^{-1} = (1-t)\ell_x^{-1} + t\ell_{x+1}^{-1}.$$

Show that for all integers x and $0 < t < 1$, ${}_{1-t}q_{x+t} = (1-t)q_x$.

7.15 Assume UDD. Suppose that, a life table is such that $q_x < q_{x+1}$ for all nonnegative integers x.

(a) Show that for any $t > 0$ and integers $x < y$ we have ${}_tq_x \leq {}_tq_y$.

(b) Suppose that $q_{80} = 0.4$, $q_{81} = 0.45$. Show that ${}_{0.5}q_{80.5} >_{0.5} q_{81}$.

(c) Show that under the assumption of Exercise 7.6, it is true that for all $t > 0$ and positive numbers x, y, we have ${}_tq_x \leq {}_tq_y$.

Spreadsheet exercise

7.16 Modify the Chapter 4 spreadsheet so that by entering the frequency m, the spreadsheet will calculate both due and immediate annuities payable m times per year.

8

Continuous payments

8.1 Introduction to continuous annuities

In this section we consider annuities where payments are made continuously. Naturally, this is not physically possible, but we can picture these as a limiting case of *m*thly annuities as m goes to infinity. Suppose, for example, that you are to receive a total of 36 500 units each year. This could be done by paying 100 units per day, or 4 1/6 units per hour, or 0.001157 units per second and so on. If you can imagine payments coming in every nanosecond, you may get some feeling for what a continuous annuity would be like.

This may seem as a somewhat artificial concept, but there are many uses for continuous annuities. They can be used to approximate *m*thly annuity values for large values of m. Moreover, we will show that insurance contracts with the realistic provision of benefit payments at the moment of death, can be viewed as continuous annuities. For insurance contracts purchased by continuous premium payments we can derive some interesting mathematical relationships that are analogues of those that appeared in Chapters 5 and 6.

In the continuous case, we cannot specify an actual payment at any point of time. As shown by the figures above, this approaches zero as the frequency of payment increases. Instead we must speak of the *periodic rate* of payment. Consider a monthly annuity. If the payment in 1 month is 100, we could describe this by saying that the annual *rate* of payment for that month is 1200. This would mean that if the payments remained at the same monthly level for a year, then the total payment for that year would be 1200. Each of the annuities described above would be paid at the annual rate of 36 500. Of course in an annuity paid m times per year, the actual payment and therefore the periodic rate of payment could change every *m*th period. In our continuous annuity, it could change from moment to moment. (Picture a person standing under a chute, receiving money that is flowing down continuously. The speed at which it comes down could change at each instant.)

In place of a cash flow vector we now have a cash flow function c defined on $[0, N]$, where $c(t)$ is the periodic rate of payment at time t.

Fundamentals of Actuarial Mathematics, Third Edition. S. David Promislow.

Companion website: http://www.wiley.com/go/actuarial

We will assume that all our cash flow functions are *piecewise continuous*. That is, there is a partition, $0 = a_1 < a_2 < \cdots < a_n = N$, such that c is continuous on the intervals (a_{i-1}, a_i), $i = 1, 2, \ldots, n-1$.

Suppose we have an arbitrary discount function y and a continuous annuity with cash flow function c. The present value of the annuity will be denoted by $\overline{a}(c; y)$ or just $\overline{a}(c)$. (The 'bar' is a standard actuarial notational device to denote continuous as opposed to discrete quantities.) Consider an approximating annuity with payments made $m$thly. That is, payments are made at time j/m where j is an integer between 0 and $mN - 1$. We assume that N is an integer and the annual *rate* of payment at time j/m is $c(j/m)$. The actual payment at time j/m is therefore $(1/m)c(j/m)$, so the present value of this $m$thly annuity is

$$\sum_{j=0}^{mN-1} \frac{c(j/m)}{m} y(j/m). \tag{8.1}$$

Taking the limit as m goes to ∞ gives one of the key formulas of this chapter,

$$\overline{a}(c; y) = \int_0^N c(t)y(t)\mathrm{d}t, \qquad (8.2)‡$$

since the $m$thly present values are Riemann sums for this integral.

8.2 The force of discount

Throughout the book we use 'log' to refer to the natural logarithm (often denoted by ln or $\log_e$.)

The evaluation of continuous streams of cash flows leads naturally to a new function, which we will spend some time investigating before returning to the main topic. It can be written in several equivalent forms.

Definition 8.1 Given an arbitrary discount function y, define another function, $\delta_y(t)$, which we will call the *force of discount associated with* y, by

$$\begin{aligned}\delta_y(t) &= \lim_{h\to 0} \frac{y(t+h,t)-1}{h} = \lim_{h\to 0} \frac{y(t+h,0)-y(t,0)}{hy(t,0)} = \frac{\frac{\mathrm{d}}{\mathrm{d}t}y(t,0)}{y(t,0)} \\ &= \frac{\mathrm{d}}{\mathrm{d}t}\log y(t,0) = -\frac{\mathrm{d}}{\mathrm{d}t}\log y(t) = -\frac{\frac{\mathrm{d}}{\mathrm{d}t}y(t)}{y(t)}.\end{aligned} \tag{8.3}$$

(To derive the second last equality, recall that $y(t) = y(0,t) = y(t,0)^{-1}$.)

From the first equality in (8.3) we see that δ_y represents a *relative* rate of growth. Each unit invested at time t will accumulate to $y(t+h,t)$ units over the next h time periods, so the numerator gives us the relative growth rate over this time interval. We divide by h to get the relative rate of growth per *period*, and take limits to get the instantaneous periodic relative rate of growth at time t.

We are often given δ_y and wish to recover y.

Proposition 8.1

$$y(t) = \mathrm{e}^{-\int_0^t \delta_y(r)\mathrm{d}r}.$$

Proof. From the second last expression in (8.3), we integrate to get

$$\int_0^t \delta_y(r)\mathrm{d}r = -\int_0^t \frac{\mathrm{d}}{\mathrm{d}r}[\log y(r)]\mathrm{d}r = -\log y(t) + \log y(0).$$

We use the fundamental theorem of calculus for the second equality. Note that $\log y(0) = \log 1 = 0$, multiply by -1, and take exponentials to complete the proof. □

An immediate consequence of Proposition 8.1 is that for $s \leq t$,

$$y(s, t) = y(t)/y(s) = \mathrm{e}^{-\int_s^t \delta_y(r)\mathrm{d}r} = \mathrm{e}^{-\int_0^{t-s} \delta_y(s+r)\mathrm{d}r}. \tag{8.4}$$

Remark This traditional use of the force of discount is for the investment discount function v. In this case $\delta_v(t)$ will be denoted by just $\delta(t)$. The standard actuarial notation is δ_t, but we find it more convenient to use the bracket in place of a subscript to parallel our treatment for the symbol v. This quantity is commonly referred to as the *force of interest.* The terminology is justified since for an instantaneous period there is no difference between interest and discount. (This is shown precisely by the equality of the fourth and last expressions in (8.3)).

8.3 The constant interest case

Consider the constant interest discount function given by $v(t) = v^t$. Then $\delta(t)$ is also constant and denoted just by δ. Since $\log(v^t) = t\log(v)$, the second last expression in (8.3) shows that

$$\delta = -\log(v) = \log(1+i), \tag{8.5}$$

and Proposition 8.1 gives

$$v(t) = \mathrm{e}^{-\delta t}, \qquad (1+i)^t = \mathrm{e}^{\delta t}. \tag{8.6}$$

The second expression reflects the fact that under constant interest, invested capital grows exponentially and δ is just the usual *rate of exponential growth.*

Additional insight may be gained by comparing δ with the fractional interest quantities of Chapter 7. We have that

$$\delta = \lim_{m\to\infty} i^{(m)} = \lim_{m\to\infty} d^{(m)}.$$

This can be seen by writing

$$i^{(m)} = \frac{(1+i)^{1/m} - 1}{1/m}, \qquad d^{(m)} = \frac{1 - v^{1/m}}{1/m},$$

so that

$$\lim_{m\to\infty} i^{(m)} = \frac{\mathrm{d}}{\mathrm{d}t}(1+i)^t|_{t=0} = (1+i)^t \log(1+i)|_{t=0} = \delta.$$

The calculation for $d^{(m)}$ is similar.

The quantity δ is often thought of as an instantaneous rate of interest, but it is important to note that just like the rates $i^{(m)}$ and $d^{(m)}$, its value depends on the underlying period. The transformation for change of period though is much easier than for the interest rates, as it just involves a proportional change. For example, if the annual force of interest is 0.06, then the force of interest for a half year period will be simply 0.03.

We now return to continuous annuities, using a constant interest rate, and show that in certain cases we get expressions analogous to discrete annuities, but with δ replacing d. We will derive the continuous versions of (2.16) and (2.17). Define a function 1_n by

$$1_n(t) = \begin{cases} 1, & \text{if } t \le n, \\ 0, & \text{if } t > n. \end{cases}$$

This is the same notation as for the analogous vector, but the meaning should be clear from the context. We also define the function $I_n(t)$ by

$$I_n(t) = \begin{cases} t, & \text{if } t \le n, \\ 0, & \text{if } t > n. \end{cases}$$

From (8.2) and (8.6),

$$\overline{a}(1_n) = \int_0^n \mathrm{e}^{-\delta t}\mathrm{d}t = \frac{1-\mathrm{e}^{-\delta n}}{\delta}. \tag{8.7}$$

Integrating by parts,

$$\overline{a}(I_n) = \int_0^n t\mathrm{e}^{-\delta t}\mathrm{d}t = -t\frac{\mathrm{e}^{-\delta t}}{\delta}\bigg|_0^n + \frac{1}{\delta}\int_0^n \mathrm{e}^{-\delta t}\mathrm{d}t = \frac{\overline{a}(1_n) - n\mathrm{e}^{-\delta n}}{\delta}. \tag{8.8}$$

8.4 Continuous life annuities

8.4.1 Basic definition

The present value of an annuity issued to (x) that provides benefits paid continuously at the annual rate of $c(t)$ at time t, provided (x) is alive, will be denoted by $\overline{a}_x(c)$. Formula (8.2) gives the exact formula

$$\overline{a}_x(c) = \int_0^{\omega - x} c(t)v(t)\,{}_tp_x\mathrm{d}t. \tag{8.9‡}$$

Note that this formula follows essentially the same pattern that we observed after formula (5.1). Instead of summing we integrate, and for each t the integrand is the product of three factors: the interest discount factor; the 'amount paid ' at time t, (which we can intuitively think of as $c(t)dt$); and the probability that the payment is made.

8.4.2 Evaluation

We will generally only have available the values of ℓ_x for integer values of (x), and must interpolate for the remaining values in order to evaluate the integral. As in Chapter 7, we will assume UDD, which will allow us to express continuous annuities in terms of annual annuities. The procedure closely parallels that for $m$thly annuities and we could simply write down the formulas as limiting cases of those in Chapter 7. It is instructive however to proceed a new.

For simplicity of notation, we will at first assume constant interest. The initial step is to define the continuous analogues of the functions β and α introduced in Chapter 7. Let

$$\beta(\infty) = \mathrm{e}^{\delta}\overline{a}(I_1) = \frac{\mathrm{e}^{\delta} - 1 - \delta}{\delta^2} = \frac{i - \delta}{\delta^2},$$

invoking (8.7) and (8.8). Let

$$\alpha(\infty) = \overline{a}(1_1) + d\beta(\infty) = \frac{di}{\delta^2}.$$

It is clear that $\beta(\infty)$ and $\alpha(\infty)$ are the respective limits of $\beta(m)$ and $\alpha(m)$ as m approaches ∞.

We now use the same procedure as we used in Section 7.1 and replace all the payments in any 1 year by a single payment, equal to the value of these payments at the beginning of the year. This replacement principle holds in the continuous case as well. For a formal verification use the standard integration result

$$\int_0^b = \int_0^a + \int_a^b \quad \text{for } a < b.$$

Analogously to (7.2) this leads to

$$\overline{a}_x(c) = \ddot{a}_x(z), \tag{8.10}$$

where

$$z_k = \int_0^1 c(k+s)v^s{}_sp_{x+k}\mathrm{d}s = \int_0^1 c(k+s)v^s(1 - s \cdot q_{x+k})\mathrm{d}s. \tag{8.11}$$

This is easily evaluated when the rate of payment is constant over each year. For a nonnegative integer k, let c_k denote the constant value of $c(k+s), 0 \leq s < 1$. We can factor c_k outside of the integral sign in (8.11), which gives

$$z_k = c_k[\overline{a}(1_1) - q_{x+k}\overline{a}(I_1)] = c_k\left[\frac{d}{\delta} - \beta(\infty)vq_{x+k}\right].$$

Then, from (8.10),

$$\overline{a}_x(c) = \frac{d}{\delta}\ddot{a}_x(\mathbf{c}) - \beta(\infty)A_x(\mathbf{c}),$$

where $\mathbf{c} = (c_0, c_1, \ldots)$. Finally, from Theorem 5.1,

$$\overline{a}_x(c) = \alpha(\infty)\ddot{a}_x(\mathbf{c}) - \beta(\infty)\ddot{a}_x(\Delta\mathbf{c}). \tag{8.12}$$

As promised, this reduces the calculation to finding the present value of yearly annuities.

As in the $m$thly case, it is usual in practice to approximate this by calculating the adjusting factors at zero interest. We then have $\alpha(\infty) = 1, \beta(\infty) = \frac{1}{2}$, and the approximating formula is

$$\overline{a}_x(c) \doteq \ddot{a}_x(\mathbf{c}) - \frac{1}{2}\ddot{a}_x(\Delta\mathbf{c}).$$

In the case where $\mathbf{c} = 1_\infty$ so that $\Delta\mathbf{c} = (1, 0, 0, \ldots)$, we get the familiar equation

$$\overline{a}_x \doteq \ddot{a}_x - \frac{1}{2}. \tag{8.13}$$

This concludes the development of formulas for the constant interest case. When rates are only constant over each year, but can vary from year to year, we must replace the constants $d/\delta, \alpha(\infty)$ and $\beta(\infty)$ by vectors, analogously to the treatment in the discrete case.

8.4.3 Life expectancy revisited

Comparing the formula for life expectancy (3.8) with formula (4.1), we see that e_x is simply the present value of a life annuity of 1 per year, beginning at time 1, at a constant interest rate of 0. This make sense intuitively, since at zero rate of interest, the present value of the annuity will simply be the total amount paid, which in turn is just the whole number of years lived, which on average is just e_x. We can now write down a more rigorous formula for the *complete* life expectancy, as the present value of a life annuity paying 1 per year for life continuously, at zero interest. This gives

$$\mathring{e}_x = \int_0^\infty {}_tp_x \mathrm{d}t.$$

Assuming UDD, we apply the formulas given above. With $\mathbf{c} = 1_{\omega-x}$, and an interest rate of zero,

$$\ddot{a}_x = 1 + e_x, \qquad \ddot{a}_x(\Delta\mathbf{c}) = 1,$$

and from (8.13),

$$\mathring{e}_x = 1 + e_x - \frac{1}{2} = e_x + \frac{1}{2},$$

recovering the approximation we obtained intuitively in Chapter 3.

Similarly, the complete n-year temporary life expectancy is given by $\int_0^n {}_tp_x \mathrm{d}t$, which under UDD, is equal to $\sum_{k=1}^n {}_kp_x + \frac{1}{2}{}_nq_x$ as we obtained previously.

Here is a interesting use for life expectancy. Suppose you are given $\mathring{e}_x$ and the investment discount function v, but not the actual life table, and you want to approximate $\overline{a}_x$. One natural way to do it is simply take $\overline{a}(1_{\mathring{e}_x}; v)$ for this approximation. Indeed, some might even think that this (or an appropriate modification in the discrete case) should be exact, reasoning that the cost of a life annuity on a person should be the cost of providing the individual with income up to their precise life expectancy. The approximation should be close, but in the normal case of positive interest rates, this method always overstates the true value. A mathematical proof is outlined in Exercise 22.15 but we can easily explain this fact by general reasoning. Suppose an insurer sells 1-unit continuous life annuities to a group of people age x and charges them each the price of a fixed period annuity with the period equal to their life expectancy. The insurer can expect to gain money on those who die before living their life expectancy, but lose money on those who live longer, and these gains and losses should cancel out. However, since the gains come earlier, they will produce extra returns due to interest earnings and the insurer will end up with more than they need to provide the benefits. The break even premium will then be somewhat less than the premium for the fixed period annuity.

8.5 The force of mortality

Before proceeding with insurances, we introduce a new quantity that is analogous to the force of discount.

We first motivate this intuitively. What is the probability that (x) will die in the next instant of time? We could approximate this by looking at ${}_hq_x$, but if we assume continuity and take the limit as h approaches 0, we would get 0. The question is answerable but the answer of 0 is uninteresting and gives us no information about the mortality of (x) at that point. Instead, we can compute an annual *rate* of mortality at age (x), by dividing by h before taking the limit. This leads to the following definition.

Definition 8.2 The *force of mortality* at age (x) is the quantity

$$\mu(x) = \lim_{h\to 0} \frac{{}_hq_x}{h} = \lim_{h\to 0} \frac{\ell_x - \ell_{x+h}}{h\ell_x} = -\frac{\frac{\mathrm{d}}{\mathrm{d}x}\ell_x}{\ell_x} = -\frac{\mathrm{d}}{\mathrm{d}x}\log \ell_x. \tag{8.14}$$

The expression after the third equality is useful for obtaining additional insight. We have a group of lives declining over time due to mortality. The quantity $\mu(x)$ gives us the *relative* rate of decline in this group at age x.

In many cases we are looking at a fixed age x and we want the variable to be the time t. We will accordingly define

$$\mu_x(t) = \mu(x+t).$$

We can view $\mu_x(t)$ as the force of mortality at time t for an individual age x at time 0. From the fourth expression in (8.14),

$$\mu_x(t) = -\frac{\frac{\mathrm{d}}{\mathrm{d}t}{}_tp_x}{{}_tp_x}, \tag{8.15}$$

and we imitate the proof of Proposition 8.1 to derive

$${}_tp_x = \mathrm{e}^{-\int_0^t \mu_x(r)\mathrm{d}r}. \tag{8.16‡}$$

Example 8.1 Suppose that the force of mortality is given by

$$\mu(x) = \frac{1}{\omega - x}, \qquad x < \omega.$$

Find an expression for ${}_tp_x$.

Solution. For $t < \omega - x$,

$$-\int_0^t \frac{1}{\omega - x - r} dr = \log(\omega - x - t) - \log(\omega - x) = \log\left(\frac{\omega - x - t}{\omega - x}\right).$$

We then substitute in (8.16) to get

$${}_tp_x = \left(\frac{\omega - x - t}{\omega - x}\right) = 1 - \frac{t}{\omega - x}.$$

The following quantity is often useful.

Definition 8.3 Let $\lambda_x(t)$ denote the force of discount for the interest and survivorship function at age x

Note that

$$\lambda_x(t) = -\frac{\mathrm{d}}{\mathrm{d}t}\log y_x(t) = -\frac{\mathrm{d}}{\mathrm{d}t}\log v(t) - \frac{\mathrm{d}}{\mathrm{d}t}\log {}_tp_x = \delta(t) + \mu_x(t). \tag{8.17}$$

In other words, the force of discount for interest and survivorship is just the sum of the forces for these two components. We can then view the function μ_x as the particular case of δ_{y_x} when interest rates are zero.

8.6 Insurances payable at the moment of death

8.6.1 Basic definitions

Unlike annuities, the continuous version is the realistic model for insurance policies. It captures the idea that, in practice, claims are paid at the moment of death rather than the end of the year of death. We can also allow for the death benefit to vary continuously with time. In place of our death benefit vector, we have a *death benefit function*. This is a piecewise continuous function b defined on $[0, \omega - x)$ where $b(t)$ is the payment made at time t should death occur at that time. The net single premium for such a policy on (x) will be denoted by $\overline{A}_x(b)$. To calculate this, we will consider an approximating policy. Let m be a positive integer. If death occurs between time j/m and $(j+1)/m$, where $j = 0, 1, \ldots, m(\omega - x) - 1$, our approximating policy will pay $b(j/m)$ at time $(j+1)/m$. Up to now we have looked at the case of $m = 1$. If $m = 365$, for example, it would mean that death benefits were paid at midnight on the day of death, clearly getting close to what we want. If we let $A_x^{(m)}(b)$ denote the net single premium for this approximating policy, we will have

$$\overline{A}_x(b) = \lim_{m\to\infty} A^{(m)}(b).$$

Reasoning as in the middle expression of formula (5.1), we have

$$A_x^{(m)}(b) = b(0)v(1/m)\left(1 - {}_{1/m}p_x\right) + b(1/m)v(2/m)\left({}_{1/m}p_x - {}_{2/m}p_x\right) + \cdots .$$

Taking limits as m goes to ∞,

$$\overline{A}_x(b) = -\int_0^{\omega-x} b(t)v(t)\mathrm{d}_t p_x = -\int_0^{\omega-x} b(t)v(t)\frac{\mathrm{d}}{\mathrm{d}t}{}_t p_x \mathrm{d}t,$$

and from (8.15) we obtain our final formula

$$\overline{A}_x(b) = \int_0^{\omega-x} b(t)v(t){}_t p_x \mu_x(t)\mathrm{d}t. \tag{8.18‡}$$

Once again we have the familiar three-factor product; the amount paid; the interest factor; and the probability that payment is made. We can write the latter as being itself the product of two terms: ${}_t p_x$, the probability of living to time t; and $\mu_x(t)\mathrm{d}t$, which we can intuitively think of as the probability of dying at exact moment t.

As well, this formula illustrates our statement above that insurances payable at the moment of death can be viewed as continuous annuities. Analogously to (5.5),

$$\overline{A}_x(b) = \overline{a}_x(c), \quad \text{where } c(t) = b(t)\mu_x(t). \tag{8.19‡}$$

8.6.2 Evaluation

As with continuous annuities, we normally will have only the life table available and will use UDD in order to evaluate the integral in (8.18). We again will simplify the notation by first using constant interest. The most useful consequence of UDD for this purpose is

$${}_t p_y \mu_y(t) = -\frac{\mathrm{d}}{\mathrm{d}t}{}_t p_y = -\frac{\mathrm{d}}{\mathrm{d}t}(1 - tq_y) = q_y. \tag{8.20}$$

where y is an integer and $0 < t < 1$.

As in the annuity case, we have

$$\overline{A}_x(b) = \ddot{a}_x(\mathbf{z}), \tag{8.21}$$

where z_k is the value at time k of the benefits paid in the year k to $k + 1$. Substituting from (8.20)

$$z_k = \int_0^1 v^t b(k+t){}_t p_{x+k}\mu_{x+k}(t)\mathrm{d}t = vq_{x+k}\int_0^1 (1+i)^{1-t} b(k+t)\mathrm{d}t.$$

This leads to

$$\overline{A}_x(b) = A_x(\tilde{\mathbf{b}}), \quad \text{where } \tilde{b}_k = \int_0^1 (1+i)^{1-t} b(k+t)\mathrm{d}t, \tag{8.22}$$

thereby reducing the evaluation of moment of death insurances to insurances paid at the end of the year of death.

This is further simplified in the case where the benefit function b is constant over each year. If b_k is the constant value of $b(k+t)$ where k is an integer and $0 < t < 1$, then b_k factors out of the integral in (8.22) and

$$\tilde{b}_k = b_k \int_0^1 (1+i)^{(1-t)}\mathrm{d}t = b_k \frac{i}{\delta},$$

leading to the very simple formula

$$\overline{A}_x(b) = \frac{i}{\delta} A_x(\mathbf{b}), \tag{8.23‡}$$

where $\mathbf{b} = (b_0, b_1, \ldots)$.

Additional insight can be obtained by observing that

$$\frac{i}{\delta} = \frac{\mathrm{e}^{\delta} - 1}{\delta} \doteq 1 + \frac{\delta}{2} \doteq 1 + \frac{i}{2}.$$

(Here, $\doteq$ stands for 'approximately equal to'.) This reflects the fact that in the case where benefits are constant over the year, the difference between paying benefits at the end of the year of death and at the moment of death is that in the latter case the insurer will not earn interest from the time of death to the end of the year, and the premium must be higher to account for this. This time interval will vary with time of death, ranging from 0 for those dying at the end of the year, to 1 year for those dying at the beginning. On average it should be half a year.

In the case where interest is only constant over each year, we replace the right hand side above in (8.23) by $A_x(\mathbf{i}/\delta * \mathbf{b})$ where now $\mathbf{i}/\delta$ is a vector with entries of i_k/δ_k.

Example 8.2 An insurance contract on (x) provides for 2, paid at the moment of death if this occurs within n years, plus a pure endowment of 3 if x is alive at the end of n years. Find the present value given that interest rates are a constant 6%, $v^n{}_np_x = 0.4$ and $A_x(1_n) = 0.2$.

Solution. Care must be taken to apply the adjustment factor only to the term insurance, and not to the pure endowment, which is always paid at time n. At 6%, we have $i/\delta = 1.0297$ so the present value is equal to $2(0.2)(1.0297) + 3(0.4) = 1.612$.

When benefit payments are not constant over each year, we have to resort directly to (8.22) and try to evaluate the integral.

Example 8.3 An insurance policy on (x) provides for a death benefit of t paid at the moment of death, should this occur at time t. Assuming UDD, find a formula for the present value in terms of end-of-the-year-of-death insurances.

Solution. For an integer k and $0 < t < 1$, we have the death benefit $b(k+t) = k+t = (k+1) - (1-t)$. We can think of this as two policies, one paying $k+1$ for death between time k and time $k+1$, and the other paying $-(1-t)$ for death at time $k+t$, where $k = 0, 1, \ldots$. (Negative death benefits may be unrealistic in practice, but they are fine from a mathematical viewpoint.) From (8.23), the present value of the first policy is $(i/\delta)A_x(\mathbf{j})$, where $\mathbf{j} = (1, 2, 3 \ldots)$. From (8.22) the present value of the second policy is the same as that with a constant end of the year death benefit of

$$-\int_0^1 (1+i)^{1-t}(1-t)\mathrm{d}t = -\left(\frac{1+i}{\delta} - \frac{i}{\delta^2}\right) = -\frac{i}{\delta}\left(\frac{1}{d} - \frac{1}{\delta}\right)$$

after integration by parts. The total premium is therefore

$$\frac{i}{\delta}\left[A_x(\mathbf{j}) - \left(\frac{1}{d} - \frac{1}{\delta}\right)A_x\right].$$

It may appear more natural to have taken the split of the death benefit as k plus t. The split above was done to get an expression involving the death benefit vector $\mathbf{j}$, representing a linearly increasing benefit starting at 1 unit. This is often encountered in practice. The alternative method would result in a policy based on a vector with initial entry 0, which would not normally be found in a real-life policy.

In a more complicated case the integration could be difficult to carry out, necessitating the use of an approximate integration formula. This can arise in the case of death benefits in pension plans that may depend on salary and length of service and could vary continuously over each year. A frequently used, very simple rule, is simply to assume that the death benefit for the year from time k to $k+1$ is constant at the value which it takes at time $k + 1/2$.

8.7 Premiums and reserves

The realistic model for life insurance calls for death benefits to be paid at the moment of death and premiums to be paid periodically (yearly, monthly, quarterly, etc.). This means that we have to treat death benefits separately when computing premiums and reserves, so we cannot conveniently calculate the net cash flow vector $\mathbf{f}$ as we did in Section 6.1. However, this usually causes little difficulty as we can simply change $A_y(\mathbf{b})$ to $\overline{A}_y(b)$ in all premium and reserve formulas.

We also must define the shifted functions $b \circ k$ analogous to that for vectors. That is

$$(b \circ k)(t) = b(k+t), \quad 0 \le t \le \omega - x - k - t.$$

Example 8.4 Assume a constant interest rate. Consider a policy on (40) with death benefit function b (constant over each year). The total premium for the year k to $k+1$ is π_k, payable monthly. The interest rate is a constant 0.06. Write a formula for the reserve at time 10. Assume UDD.

Solution. We have

$$ {}_{10}V = [\overline{A}_{50}(b \circ 10) - \ddot{a}^{(12)}_{50}(\pi \circ 10)], $$

where $\pi = (\pi_0, \pi_1, \ldots)$. For $i = 0.06$,

$$ \frac{i}{\delta} = 1.0297, \qquad \alpha = 1.000\,28, \qquad \beta = 0.468\,10, $$

and we can write

$$ {}_{10}V = 1.0297A_{50}(\mathbf{b} \circ 10) - 1.000\,28\,\ddot{a}_{50}(\pi \circ 10) + 0.468\,10\,\ddot{a}_{50}\big(\Delta(\pi \circ 10)\big), $$

with $\mathbf{b} = (b_0, b_1 \ldots)$, where b_k is the constant value of $b(t)$ for $k \leq t < k+1$.

In order to avoid this dual treatment, it is sometimes convenient to consider a policy where premiums are payable continuously. This is unrealistic, but a reasonable approximation in the case of monthly premiums. It often results in simplifying the mathematics. We will look at cases where this is useful in the next two sections.

8.8 The general insurance–annuity identity in the continuous case

In this section we develop a connection between moment-of-death insurances and continuous annuities, which is analogous to that obtained in Section 5.7. We now assume that the associated benefit function b is differentiable except possibly at the endpoints of an interval. That is, we suppose there exist points r and s such that $b(t)$ is equal to 0 for $0 < t < r$ and $s < t < \omega - x$, and that the derivative $b'(t)$ exists for $r < t < s$.

Then

$$ \overline{A}_x(b) = \int_r^s b(t)v(t)\,{}_tp_x\mu_x(t)\mathrm{d}t = -\int_r^s b(t)v(t)\frac{\mathrm{d}}{\mathrm{d}t}{}_tp_x\,\mathrm{d}t. $$

Integration by parts gives

$$ \overline{A}_x(b) = b(t)v(t)\,{}_tp_x\big|_s^r + \int_r^s [b'(t)v(t) - b(t)v(t)\delta(t)]\,{}_tp_x\,\mathrm{d}t. $$

Simplifying,

$$ \overline{A}_x(b) = b(r)v(r)\,{}_rp_x - b(s)v(s)\,{}_sp_x + \overline{a}_x(b') - \overline{a}_x(\delta b). \tag{8.24} $$

Here δb is just the function defined by $\delta b(t) = \delta(t)b(t)$.

This is in fact a continuous analogue of (5.8). We need the two additional terms at the beginning to handle the possible jumps in the function b.

We can use this to derive a continuous version of the *endowment identity* of Section 5.7.2. Assume a constant force of interest of δ. For a 1-unit, n-year endowment insurance with benefits paid at the moment of death, $\overline{A}_{x:\overline{n}|}$ denotes the net single premium, and $\overline{P}_{x:\overline{n}|}$ denotes the rate of annual premium, payable continuously for n years. In this case the function b is constant on the interval $(0, n)$, so $b' = 0$. Taking $b = (1_n)$, the first two terms in (8.24) are $1 - v^n{}_np_x$, which leads to

$$\overline{A}_{x:\overline{n}|} = \overline{A}_x(1_n) + v(n){}_np_x = 1 - \delta\overline{a}_x(1_n),$$

and

$$\overline{P}_{x:\overline{n}|} = \frac{\overline{A}_{x:\overline{n}|}}{\overline{a}_x(1_n)} = \frac{1}{\overline{a}_x(1_n)} - \delta.$$

8.9 Differential equations for reserves

The goal of this section is to develop a differential equation for reserves, which is analogous to the recursion formulas of Sections 2.10.4 and 6.3.

Suppose we have a transaction with continuous cash flows defined by the cash flow function c defined on $[0, N]$, and a general discount function y. The *reserve* at time t, which we will denote by ${}_t\overline{V}(c; y)$, is defined exactly as in the discrete case, namely as the negative of the value at time t of all future cash flows. So

$${}_t\overline{V}(c; y) = -\int_0^{N-t} c(t + r)y(t, t + r)\mathrm{d}r.$$

It is convenient to make a change of variable from r to $s = r + t$ and write

$${}_t\overline{V}(c; y) = -\int_t^N c(s)y(t, s)\mathrm{d}s = -y(t, 0)\int_t^N c(s)y(s)\mathrm{d}s.$$

Differentiating by the product rule, using (8.3) to differentiate the first factor in the last expression, and the fundamental theorem of calculus to differentiate the second factor, gives

$$\frac{\mathrm{d}}{\mathrm{d}t}{}_t\overline{V}(c; y) = \delta_y(t){}_t\overline{V}(c; y) + c(t), \tag{8.25}$$

a continuous analogue of (2.26).

Consider now an insurance policy on (x) with death benefits of $b(t)$ paid at the moment of death, and premiums paid continuously at the rate of $\pi(t)$ at time t. Let $c(t) = \pi(t) - b(t)\mu_x(t)$, the annual rate of continuous cash flow. The reserve at time t on this policy is just $-{}_t\overline{V}(c; y_x)$, which we will denote by just ${}_t\overline{V}$. Substituting in (8.25), and using (8.17),

$$\frac{\mathrm{d}}{\mathrm{d}t}{}_t\overline{V} = {}_t\overline{V}(\mu_x(t) + \delta(t)) + \pi(t) - b(t)\mu_x(t),$$

or

$$\frac{\mathrm{d}}{\mathrm{d}t}{}_t\overline{V} = \delta(t)({}_t\overline{V}) + \pi(t) - \mu_x(t)\big(b(t) - {}_t\overline{V}\big).$$

These last two formulas are the continuous versions of (6.7) and (6.8), respectively. In the first one, we view accumulation at interest and survivorship. In the second, we view the accumulation at interest only, and take the death benefit as the net amount at risk, $b(t) - {}_t\overline{V}$. This result is often referred to as *Thiele's differential equation* named after the Danish actuary T. N. Thiele, who discovered it in the late 19th century.

In certain simple cases, for example, constant μ and δ, this equation can be solved analytically. See, for example, Exercise 8.21(b). In general however, the only feasible approach is a numerical one. It is of interest to see what the solution will look like when we employ *Euler's method*, one of the main procedures for numerical solutions of differential equations. We describe this first for a general first order equation of the form

$$f'(t) = g\big(t, f(t)\big)$$

given an initial value $f(t_0)$.

We pick a step size $h > 0$ and then simply replace the derivative with the difference quotient $f(t+h) - f(t)/h$ leading to a recursion

$$f(t+h) \doteq f(t) + hg\big(t, f(t)\big) \tag{8.26}$$

We then work backwards and forwards from t_0, starting with given value of $f(t_0)$. For example, from (8.26) we get an approximate value of $f(t_0 + h)$ and we then again apply (8.26) with that value to get an approximate value of $f(t_0 + 2h)$, and so on. Iterating the procedure we can compute approximate values of $f(t)$, where t differs from t_0 by an integral multiple of h. For sufficiently small h, this should give a reasonable approximations to the true values. Applying this to Thiele's equation we get

$${}_{t+h}V = {}_tV(1 + h\delta_t) + h\pi_t - h\mu_x(t)(b_t - {}_tV). \tag{8.27}$$

Note that what we get is very close to Equation (6.8) applied to time units of length h, where for the period running from time t to time $t + h$, we take the interest rate for this period as $h\delta_t$, the mortality rate for this period as $h\mu_x(t)$, and the premium paid for this period as $h\pi_t$. The only difference is that (8.27) does not apply the interest factor to the premium, assuming in effect that it is paid at the end of the period. For very short periods this will make little difference.

8.10 Some examples of exact calculation

As we indicated above, the normal approach to calculating values in the continuous case is to invoke UDD and reduce to the yearly case. Conceivably, one could have some nice analytic expressions for mortality functions and therefore be able to calculate values exactly. This is rarely done in practice. In the first place, it seems to be extremely difficult to find functions in closed form that provide an accurate picture of observed mortality. Second, even with such functions, it may be difficult to actually carry out the integration. For illustrative purposes, however, we will look at a few particularly simple cases. These are not meant to be typical of modern human mortality. In Chapter 14, we will introduce some more realistic mortality functions.

8.10.1 Constant force of mortality

Suppose that $\mu(x)$ is a constant μ for all x. This of course is unrealistic, since we would expect, as verified by observed data, that in general $\mu(x)$ increases with x. This is precisely what we mean by the *aging process*. There are exceptions. In the very early years there appears to be a decrease in $\mu(x)$, reflecting the effect of early childhood diseases. There is also a noted decrease in the early adult ages, around 25 or so. Many attribute this to the effect of a large number of automobile accident deaths for people in their early twenties. From age 30 onwards there is a rapid, exponential growth in $\mu(x)$, until around age 70, where $\mu(x)$ continues to increase but the exponential growth dampens out. There are some who predict that with medical and genetic advances, we may be approaching a situation where we can freely transplant or replace any genes or body parts that go awry. They postulate in essence that we will then have eliminated aging and achieved a constant force of mortality. This does not mean that people will no longer die, but only that the 90-year-old will be no more likely to die at any time than the 20-year-old. At the present time, however, we are not at all close to this state and it is still in the realm of science fiction.

In any event, we look at some of the mathematical consequences of this assumption. In the first place, note that ω is now ∞, since ${}_tp_x = \mathrm{e}^{-\mu t} > 0$, which means that there is at any time, a positive probability of continuing to survive.

In our examples we will also assume a constant force of interest δ. Our first observation is that with these assumptions,

$$\overline{a}_x = \int_0^\infty \mathrm{e}^{-\delta t}\mathrm{e}^{-\mu t}\mathrm{d}t = \int_0^\infty \mathrm{e}^{-(\mu+\delta)t}\mathrm{d}t = \frac{1}{\mu+\delta}.$$

This is easily verified intuitively. As we have seen, the force of discount for the interest and survivorship discount function under our assumptions is the constant $\mu + \delta$. Therefore, an investment of $1/(\mu + \delta)$ will provide a continuous return at the annual rate of $[1/(\mu + \delta)] \cdot [\mu + \delta] = 1$.

Taking $\delta = 0$, we see as well that

$$\mathring{e}_x = \frac{1}{\mu},$$

which again makes sense, since as μ increases, life expectancy should decrease.

Turning to insurance, the constant forces of interest and mortality lead to

$$\overline{A}_x = \int_0^\infty \mathrm{e}^{-\delta t}\mathrm{e}^{-\mu t}\mu\mathrm{d}t = \frac{\mu}{\mu+\delta}. \tag{8.28}$$

Note, as a check, that when $\delta = 0$ we get $\overline{A}_x = 1$. This is obvious, since the insured is sure to die sometime and receive 1 unit, which has a present value of 1 under a zero interest rate.

This formula has an interesting alternate derivation which avoids any integration. Suppose you have wealth A invested at a constant force of interest δ and each instant of time, you use the interest earnings of δA to buy an instant of term insurance, which will cost you μ per unit. So the amount of insurance purchased at each unit of time will be $(\delta/\mu)A$. At death you will have the principle of A plus the insurance benefit, so to make the total wealth at death equal to 1, we want $(\delta/\mu)A = 1 - A$. Solving for A gives precisely the above quantity.

For a 1-unit whole-life policy, with level premiums payable continuously for life, the annual rate of premium payment is given by

$$\overline{P} = \frac{\overline{A}_x}{\overline{a}_x} = \mu,$$

and the reserve on such a policy at any time t is given by

$$\overline{A}_{x+t} - \overline{P}\overline{a}_{x+t} = 0,$$

as we see from substituting from above. In fact, this is just the continuous analogue of Example 6.2. Since the cost of providing the insurance does not increase with time, the amount collected at any instant is exactly what is needed at that instant, and there is no accumulation of a reserve.

8.10.2 Demoivre's law

Another extremely simple mortality assumption states that for all ages x,

$${}_tp_x = 1 - \frac{t}{\omega - x}, \quad 0 \le t \le \omega - x.$$

We have already encountered this law in Example 8.1. It is known as *Demoivre's law*, named after Abraham Demoivre, an eighteenth-century mathematician, perhaps best known for his trigonometric identities. He is reputed to have used the above formula in making some life table calculations. We can look upon his reasoning as follows. We know that for fixed x, the function ${}_tp_x$ decreases from a value of 1 at $t = 0$ to a value of 0 at $t = \omega$. Equipped with no other information at all, Demoivre made the simplest possible assumption, namely that the graph of the function was a straight line. It is obviously far too simplistic to be representative of actual mortality, but we will nonetheless investigate its mathematical consequences.

We get a very simple expression for life expectancy:

$$\mathring{e}_x = \int_0^\infty \left(1 - \frac{t}{\omega - x}\right) \mathrm{d}t = \frac{1}{2}(\omega - x).$$

That is, at any age, one can expect to live one-half of the maximum remaining lifetime.

Similarly, taking an upper limit of n in the above integral, we calculate the n-year temporary life expectancy as $n - n^2/2(\omega - x) = n\left({}_{n/2}p_x\right)$. That is, we multiply the maximum possible future lifetime over the next n years by the probability of living to the middle of this period.

The force of mortality takes a particularly simple form.

$$\mu_x(t) = -\frac{\frac{\mathrm{d}}{\mathrm{d}t}\,{}_tp_x}{{}_tp_x} = \frac{1}{\omega - x - t}, \quad 0 \le t < \omega - x$$

which verifies Example 8.1, where we did the inverse calculation. This at least has the desirable feature of being increasing.

This leads to

$$ {}_tp_x\mu_x(t) = \frac{1}{\omega - x} = q_x, \quad 0 \leq t < \omega - x. $$

We noticed above that this formula held under UDD for an integer x and t in the interval (0,1). Under Demoivre's law, it holds for all x and t. As a consequence,

$$ \overline{A}_x = \frac{1}{\omega - x} \int_0^{\omega - x} v(t)\mathrm{d}t = \frac{1}{\omega - x} \overline{a}(1_{\omega - x}; v). \tag{8.29} $$

Once again, we can check that under zero interest, we get $\overline{A}_x = 1$. Similarly,

$$ \overline{A}_x(1_n) = \frac{1}{\omega - x} \overline{a}(1_n; v). $$

A variation of Demoivre's law, which we will call the *modified Demoivre's law*, is given by

$$ \mu_x(t) = \frac{\alpha}{\omega - x - t}, \quad 0 \leq t < \omega - x - t, $$

for some $\alpha > 0$. Our original form was with $\alpha = 1$. Using (8.16), this means that

$$ {}_tp_x = \left(1 - \frac{t}{\omega - x}\right)^{\alpha}, \quad 0 \leq t < \omega - x, $$

which leads to a simple life expectancy formula,

$$ \overset{\circ}{e}_x = \int_0^{\omega - x} \left(1 - \frac{t}{\omega - x}\right)^{\alpha} \mathrm{d}t = \frac{\omega - x}{\alpha + 1}. $$

8.10.3 An example of the splitting identity

The following is an example showing the use of the splitting identity in the continuous case.

Example 8.5 Suppose that the force of mortality is given by

$$ \mu(x) = \begin{cases} 0.02, & 30 \leq x \leq 50, \\ \frac{1}{100 - x}, & 50 \leq x < 100. \end{cases} $$

The force of interest δ is a constant 5 %. Calculate $\overline{A}_{30}$.

Solution. By the splitting identity,

$$ \overline{A}_{30} = \overline{A}_{30}(1_{20}) + v(20)_{20}p_{30}\overline{A}_{50}. $$

Proceeding as in the previous examples of this section,

$$\overline{A}_{30}(1_{20}) = \frac{\mu}{\mu+\delta}(1-e^{-20(\mu+\delta)}), \quad \overline{A}_{50} = \frac{1}{50}\left(\frac{1-e^{-50\delta}}{\delta}\right), \quad v(20)_{20}p_{30} = e^{-20(\mu+\delta)}.$$

Substituting for μ and δ, we get $\overline{A}_{30} = 0.3058$.

8.11 Further approximations from the life table

Suppose you want to approximate the force of mortality given only a life table. For ages that are not integers, the UDD assumption gives an easy answer. For an integer x and $0 < t < 1$, we obtain from (8.20) that

$$\mu(x+t) \doteq \frac{q_x}{1 - {}_tq_x}.$$

Suppose now we want to approximate $\mu(x)$ where x is an integer. We could take $t = 0$ in the above, to get an estimated value of $\mu(x)$ as just q_x, but we could also replace x by $x-1$ and take $t = 1$ to get an estimated value of $\mu(x)$ as $q_{x-1}/(1-q_{x-1})$. These are unlikely to be the same. It is easy to see where the problem lies. From UDD, the function ${}_tp_x$ is piecewise linear as a function of t and the derivative will not exist at integer values of t. We are in effect getting the different right and left hand forces of mortality at the integer points. To get a unique answer we need to consider an interval containing x as a interior point, such as $(x-1, x+1)$. One such approach is to note that

$$p_{x-1}p_x = {}_2p_{x-1} = e^{-\int_0^2 \mu_{x-1}(t)\mathrm{d}t}.$$

We now approximate the integral by one of the simplest methods for doing so, namely the *mid-point rule*. This simply says that to approximate an integral of a nonnegative function over an interval you take the value of the function at the midpoint of the interval and multiply by the length of the interval. (This approximation is exact when the function is linear.) With this method, the right hand side above is approximated by $e^{-2\mu(x)}$, and solving gives the formula

$$\mu(x) \doteq -\frac{\log p_{x-1} + \log p_x}{2}. \tag{8.30}$$

(See Exercise 8.24 for a check on the accuracy of this approximation.)

This can be furthered simplified by using the fact that for small values of r, $-\log(1-r)$ is very close to r, which leads to the approximation

$$\mu(x) \doteq \frac{q_{x-1} + q_{x+1}}{2}, \tag{8.31}$$

a average of two values of q at surrounding ages, which has a certain intuitive appeal.

We now return to developing the three-term Woolhouse formula introduced in 7.5, which offers an alternative method to the $\alpha(m) - \beta(m)$ method for fractional annuity computations.

For motivation we revisit Formula (7.13) which arose from the assumption that for all integers x, and $0 \le s \le 1$, the quantity $y_x(s)$ is of the form $1 - as$ where $a = 1 - vp_x$. This assumption, together with constant interest, implies that

$${}_sp_x = e^{\delta s}(1 - as),$$

so that

$$\frac{\mathrm{d}}{\mathrm{d}s}{}_sp_x = \mathrm{e}^{\delta s}[\delta(1 - as) - a],$$

and at $s = 0$ this derivative equals $\delta - a$. This will be positive when $\delta > a = 1 - vp_x$ which is equivalent after some manipulation to

$$p_x > \mathrm{e}^{\delta}(1 - \delta).$$

This explains the inconsistency observed in Example 7.5. If the above condition holds, and we assume the piecewise linearity of $y_x(s)$ as given above, the positivity of the derivative at $s = 0$ means that ${}_sp_x$ actually takes values greater than 1 for sufficiently small values of s! Writing this in terms of ℓ_x we can interpret it fancifully as meaning that people who have died are coming back to life in even greater numbers than were there before, causing the life annuity to be worth more than the one with fixed payments.

Suppose we instead make the next simplest assumption. Assume that y_x is a quadratic on $[0, 1]$ so that for some constants a, b,

$$y_x(s) = 1 - as - bs^2, \quad 0 \le s \le 1. \tag{8.32}$$

This is the only assumption we make now. We do not need constant interest.

Choose any integer m and invoke well-known formulas for the sum of consecutive integers (as we did in Chapter 7) as well as the formula for the sum of their squares. These result in

$$\sum_{i=0}^{m-1} \frac{i}{m} = \frac{m-1}{2}, \qquad \sum_{i=0}^{m-1} \left(\frac{i}{m}\right)^2 = \frac{m-1}{2} - \frac{m^2-1}{6m}.$$

We now follow the method introduced in Section 7.1 to compute the relevant vector $\mathbf{u}$. It follows from (8.32) that

$$u_0 = \frac{1}{m}\sum_{j=0}^{m-1} y_x(j/m) = 1 - (a+b)\left(\frac{m-1}{2m}\right) + 2b\left(\frac{m^2-1}{12m^2}\right).$$

Next, calculate the coefficients $a + b$ and $2b$. From (8.32)

$$a + b = 1 - y_x(1) = (\nabla\, 1_\infty)_0. \tag{8.33}$$

using the notation of (2.14) with respect to the discount function y_x.

We now consider the interest and survivorship force of discount $\lambda_x(s) = -y_x'(s)/y_x(s)$, and let λ_x denote the vector whose entry at index k is $\lambda_x(k)$. From (8.32)

$$y_x'(s) = -a - 2bs,$$

so that assuming differentiability at integer points,

$$2b = y_x'(0) - y_x'(1) = -\lambda_x(0) + y_x(1)(\lambda_x(1) = -(\nabla\lambda_x)_0. \tag{8.34}$$

From (8.33) and (8.34)

$$u_0 = 1 - \left(\frac{m-1}{2m}\right)(\nabla 1_\infty)_0 - \left(\frac{m^2-1}{12m^2}\right)(\nabla\lambda_x)_0.$$

Similarly we can verify that

$$u_k = 1 - \left(\frac{m-1}{2m}\right)(\nabla 1_\infty)_k - \left(\frac{m^2-1}{12m^2}\right)(\nabla\lambda_x)_k.$$

Then by invoking (7.2) and (2.15) we arrive at the three-term Woolhouse formula.

$$\ddot{a}_x^{(m)}(\mathbf{c}) \doteq \ddot{a}_x(\mathbf{c}) - \frac{m-1}{2m}\ddot{a}_x(\Delta\mathbf{c}) - \frac{m^2-1}{12m^2}\ddot{a}_x(\lambda_x * \Delta\mathbf{c}). \tag{8.35}$$

The particular form in the case of level payments is

$$\ddot{a}_x^{(m)}(1_n) \doteq \ddot{a}_x(1_n) - \frac{m-1}{2m}\left(1 - v(n)_n px\right) - \frac{m^2-1}{12m^2}\left(\lambda_x(0) - \lambda_x(1)v(n)_n p_x\right).$$

A direct calculation from (8.35) is not practical, as normally we will be computing from the life table and will not have available the exact values of $\mu(x)$ needed to compute λ_x. However we can simply use the approximations given by (8.30) or (8.31) in order to derive numerical results.

Formula (8.35) seems to avoid for the most part the inconsistency of a life annuity having a higher present value than the corresponding fixed interest period annuity, but this can still arise in extreme cases. See Exercise 8.27.

8.12 Standard actuarial notation and terminology

We first review some terminology. Policies where benefits are paid at the moment of death, and premiums are paid continuously are sometimes referred to as *fully continuous*. Policies where benefits are paid at the end of the year of death, and premiums are paid yearly, are sometimes refereed to as *fully discrete*. Neither of these are realistic. The normal life insurance contract involves benefits at the moment of death, and premium payable yearly (or possibly monthly, quarterly, etc). These are referred to as *semi-continuous*.

The single premium notation for standard insurance and annuity contracts is the same as in the yearly case, except, as we have already introduced, A is replaced by $\overline{A}$ and $\ddot{a}$ is replaced by $\overline{a}$. For annual premium contracts, the rules given in 5.8.2 apply. When premiums are payable continuously at a level rate, the same notation is used with $\overline{P}$ replacing P. So, for example,

- ${}_hP(\overline{A}_{x:n})$ denotes the net annual level premium, payable for h years, for a, 1-unit n-year endowment insurance on (x) with death benefits payable at the moment of death. (Note that we do not omit the single premium symbol, as it starts with $\overline{A}$ rather than A.)

- $\overline{P}(\overline{A}^{\,1}_{x:n})$ denotes the annual rate of premium payment, when premiums are paid continuously at a level rate for n years, for a 1-unit n-year term insurance on (x) with benefits payable at the moment of death.

Symbols for reserves follow the rules in Section 6.9, which say essentially that you simply take the premium symbol and replace P by V, or $\overline{P}$ by $\overline{V}$, and move the duration symbol to the top rather than bottom left. So, for example, the tth year reserve for the two above policies would be denoted by ${}^h_tV(\overline{A}_{x:n})$ and $\overline{V}(A^{\,1}_{x:n})$, respectively.

The assumption of Exercise 7.13 is sometimes referred to as the *constant force assumption* since it implies that the force of mortality is constant over *each year*, as shown by Exercise 8.21 below. It should not be confused with the assumption of a constant force over the entire span of life, as we discussed in Section 8.10.

Notes and references

The derivation of (8.18) uses a Riemann–Stieljes integral (see Rudin 1976, Theorem 6.17).

For an alternate derivation of the general Woolhouse formula (which follows the original derivation of W. Woolhouse in 1859) see Dickson *et al.* (2013), Appendix B.2. In Section 5.13 of that work, the author's take a known function for μ_x and show that the three-term formula gives remarkably close approximations to the true value. See also Exercise 8.26 for more confirmation of this.

Exercises

Type A exercises

8.1 A life annuity on (x) provides for benefits made continuously for 2 years, provided that (x) is alive. The annual rate of payment at time t is $c(t)$, defined by

$$c(t) = \begin{cases} t, & 0 < t < 1, \\ 1, & 1 < t < 2. \end{cases}$$

The interest rate is 0, $q_x = 0.3$ and $q_{x+1} = 0.4$. Find the present value, assuming UDD.

8.2 Given $q_x = 0.2$, $q_{x+1} = 0.25$, $i = 0.25$ and a vector $\mathbf{c} = (100, 300)$, calculate $\overline{a}_x(\mathbf{c})$, assuming UDD.

8.3 You are given that $q_{40} = 0.1, q_{41} = 0.2$, the rate of interest is a constant 25% and that UDD holds. Find the present value of a 2-year endowment insurance on (40), paying 1000 at the moment of death if this occurs before age 42, and 1000 at age 42 if (40) is alive at that time.

8.4 Suppose that $q_{50} = 0.2$. Assume UDD.

(a) Calculate $\mu_{50}(0.2)$.

(b) Given that the interest rate is 25%, find the single premium for a 1-year term insurance policy on (50) that pays 1 unit at the moment of death should this occur within 1 year.

8.5 Suppose that the force of mortality is given by

$$\mu_x = B(1.09)^x$$

and

(a) Find a value for B such the resulting mortality table is the illustrative table of Chapter 3.

(b) Using this value of B compare the true value of μ_{50} with the estimated value using formula (8.30).

8.6 A life annuity on (80) provides for benefits made continuously for 2 years, at the annual rate of $c(t)$ at time t, provided that (x) is alive. Suppose that $c(t) = t, 0 < t < 2$, and the interest rate is 0. Find the present value under each of the following assumptions.

(a) $q_{80} = 0.09, q_{81} = 0.12$, and UDD holds.

(b) The force of mortality is given by

$$\mu_{80}(t) = \frac{1}{10 - t}, \quad 0 < t < 10.$$

8.7 For a certain mortality basis, $q_{50} = 0.2$. Suppose that the force of mortality $\mu_{50}(t)$ is changed to a new force of mortality $\hat{\mu}$ given by

$$\hat{\mu}_{50}(t) = 2\mu_{50}(t) - 0.1, \quad 0 < t < \omega - 50.$$

Find the new value of q_{50}.

8.8 The force of mortality is a constant 0.04 and the force of interest is a constant 0.06. A term insurance policy has a level death benefit of 1 payable at the moment of death if this occurs within 40 years. Level annual premiums are payable continuously for 20 years at the annual rate of of π. Find (a) π, (b) ${}_{10}V$.

8.9 Suppose that the force of mortality for ages over 60 is given by

$$\mu_{60}(t) = \begin{cases} t, & 0 < t < 1, \\ 1, & 1 < t. \end{cases}$$

Find the probability that (60) will die within 2 years.

8.10 Suppose mortality follows Demoivre's law with $\omega = 110$ and that the interest rate is 0. A term insurance policy on (60) has a level death benefit of 1000 payable at the moment of death provided this occurs within 40 years. Net level premiums are payable continuously for 20 years at the annual rate of π. (a) Find π. (b) Find ${}_{10}V$ by both the prospective and retrospective methods.

8.11 Two actuaries, A and B, agree that the probability that a female age 60 will die within 10 years is 0.36.

(a) Actuary A decides that the force of mortality for male lives is 1.5 times the force of mortality for females lives, at all ages over 60. What is the probability that a male age 60 will die within 10 years, under A's assumption?

(b) Actuary B disagrees and decides that the force of mortality for males lives is obtained by adding a constant of 0.01 to the force of mortality for female lives, at all ages over 60. What is the probability that a male age 60 will die within 10 years under B's assumption?

8.12 An insurance policy provides for death benefits payable at the moment of death. The amount payable for death at time t is $e^{0.08t}$, for all $t > 0$. The force of mortality is a constant 0.06 and the force of interest is a constant 0.04. Net level premiums are payable at a constant rate for life. Find the rate of premium payment.

Type B exercises

8.13 Show that under UDD, for an integer x and $0 < t < 1$, $\mu_x(t) = q_x/(1 - tq_x)$.

8.14 The central rate of mortality at age x is defined by

$$m_x = \frac{d_x}{\int_0^1 \ell_{x+t} \mathrm{d}t}.$$

(a) Show that m_x can be expressed as a continuous weighted average of values of the force of mortality.

(b) Show that under UDD, $m_x = \mu_x\left(\frac{1}{2}\right)$.

8.15 A whole-life insurance, with net level premiums for life payable continuously at the annual rate of π, provides for death benefits of $e^{\gamma t}$ at the moment of death, for death at time t. Given that the force of mortality is a constant μ and the force of interest is a constant δ, such that $\mu + \delta > \gamma$, find π in terms of γ, μ and δ.

8.16 A 1-year deferred, 1-year life annuity on (x) provides for continuous payments from time 1 to time 2 provided (x) is alive. The rate of payment at time $1 + t$ is $e^{0.1t}$, $0 < t < 1$. This is purchased by a single premium. For death during the first year, the single premium is returned without interest at the moment of death. You are given that $q_x = 0.1$, $q_{x+1} = 0.2$, and that the force of interest δ is a constant 0.10. Assuming UDD, find the single premium.

8.17 An n-year term insurance policy on (30) has a constant death benefit of 1, payable at the moment of death. Premiums are payable continuously at a constant rate for n years. Mortality follows Demoivre's law with $\omega = 100$, and the interest rate is $i = 0$. Find the rate of premium payment as a function of n.

8.18 The force of interest is a constant δ. The force of mortality is a constant μ_1 for the first n years, and a constant μ_2 after n years. Derive expressions for $\overline{a}_x$ and $\overline{A}_x$ in terms of μ_1, μ_2, δ and n.

8.19 Suppose Demoivre's law holds and the force of interest is a constant δ. Find an expression for $\overline{a}_x(1_n)$ in two ways. First use formula (8.9) directly, and second apply (8.23) to the expression given for $\overline{A}_x(1_n)$ under the same assumptions. Verify that as δ approaches 0, you get the formula for the n-year temporary life expectancy.

8.20 A 'sawtooth' life insurance policy on (x) provides for a death benefit of t paid at death if this occurs at time $k + t$, where k is an integer and $0 < t < 1$. Assume UDD and constant interest. Show that the net single premium for this policy is equal to $\beta(\infty)A_x$.

8.21 Suppose that the force of interest is a constant δ and the force of mortality is a constant μ. Consider an n-year endowment insurance with net premiums payable continuously at the constant annual rate of π. Death benefits of b_t are payable at the moment of death for death at time $t < n$. There is a payment of 1 at time n if the insured is then alive.

(a) Suppose that $b_t = 1$ for all $t < n$. Use the reserve differential equation to show that

$$_tV = (\pi - \mu)\frac{e^{t(\mu+\delta)} - 1}{\delta + \mu}, \qquad \pi = \mu + \frac{\mu + \delta}{e^{n(\mu+\delta)} - 1}.$$

Give an intuitive explanation of these results. Verify that these agree with the formulas given at the end of Section 8.8. Show that the reserve formula is a continuous version of formula (6.15).

(b) Suppose now that $b_t = 1 + {}_tV$ for all $t < n$. Use Thiele's differential equation to derive formulas for ${}_tV$ and π.

8.22 Show that under the assumption of Exercise 7.13, the force of mortality is constant over each year. That is, $\mu_x(t) = \mu_x(s)$ where x is a nonnegative integer and s and t are between 0 and 1.

8.23 Give an alternate derivation of (8.12) by using the continuous endowment identity, as given at the end of Section 8.8, and (8.23). First prove it for $\mathbf{c} = \mathbf{e}^0$.

8.24 Suppose that modified Demoivre's law holds with $\alpha = 2$. If $\mu_x = 30$ what is $\mathring{e}_x$?

8.25 Suppose that the force of mortality is given by

$$\mu(x) = B(1.09)^x.$$

(a) Find a value for B such the resulting mortality table is the illustrative table of Chapter 3.

(b) Using this value of B compare the true value of $\mu(50)$ with the estimated value using formula (8.30).

8.26 Suppose that the force of mortality is a constant 0.2 and the force of interest is a constant 0.1. Compute $\ddot{a}_x^{(4)}(1_1)$ by four methods. (a) formula (7.9), (b) formula (7.13), (c) formula (8.35) and (d) exactly. Compare the results.

8.27 Find an example of some life table values and an interest rate, such that using (8.35) together with (8.30), yields an inconsistent result. Hint: Choose life table values so that the estimated value μ_{x+1} is much higher than the estimated values of μ_x.

8.28 A life is subject to a constant force of mortality of 0.15. The force of interest is a constant 0.10. An insurance contract on this life provides for a death benefit of $e^{.05t}$ for death at time t, with level premiums payable continuously for life.

(a) Find ${}_tV$.

(b) Write down Thiele's differential equation for the reserve, and verify that your answer to (a) satisfies this equation.

Spreadsheet exercise

8.29 Modify previous spreadsheets to handle the case of death benefits payable at the moment of death.

9

Select mortality

9.1 Introduction

In this chapter we will discuss a refinement to our basic model. In previous chapters we assumed that for a sufficiently homogeneous group – North American male nonsmokers, for example – mortality rates are a function of attained age only. This may be a reasonably accurate assumption for the totality of people in this group, but it need not apply for certain subsets. An insurer, after all, is not interested in the mortality of the general population, but only the subset of those who buy policies. Observed data show that for purchasers of life insurance, mortality depends not only on age, but on *duration since the policy was purchased.* To see why this should be true, let us compare two people both age 60, alike in all respects except that A just recently purchased an insurance policy, and B purchased a policy 1 year ago at age 59. During the next year, we would expect that A will have a higher probability of living to age 61 than B will. This arises from the fact that the insurer does not have to accept everyone who applies for a policy. Applicants can be requested to undergo medical exams, or to furnish information regarding their health, lifestyle and other factors, in order to verify that they are reasonable risks. Individual A has just been so certified. Individual B, on the other hand, was certified as a good risk 1 year ago, but this condition could have changed. B may have contracted a fatal disease, or taken up life-threatening habits. Moreover, if we consider a third individual, C, also age 60, similar to A and B, except that he/she was sold a policy at age 58, his/her chance of living over the next year can be expected to be even less that that of B. This person has had 2 years in which to deteriorate.

The standard actuarial method to handle this situation is to use the same notational device that we introduced in Section 4.3.2. We keep the convention that subscripts denote attained age, but put square brackets around the age that the policy was issued, in order to separate age and duration. Therefore, the symbol ${}_sq_{[x]+t}$ denotes the probability that a person now age $x+t$, who was sold insurance at age x, will die within the next s years. This is known

Fundamentals of Actuarial Mathematics, Third Edition. S. David Promislow.

Companion website: http://www.wiley.com/go/actuarial

as a *select* mortality rate, as it incorporates the effects of the insurer's ability to select. We define

$$ {}_s p_{[x]+t} = 1 - {}_s q_{[x]+t}. $$

As usual we omit left subscripts of 1 for q and p. Our discussion above showed that

$$ q_{[60]} \leq q_{[59]+1} \leq q_{[58]+2} \cdots, $$

and, in general,

$$ q_{[x]+t} \leq q_{[x-s]+t+s}, \tag{9.1} $$

for all nonnegative x and t and $0 \leq s \leq x$.

9.2 Select and ultimate tables

To fully model this phenomenon we would need a separate mortality table for each age x to cover policies issued at that age. Of course the tables would get shorter with increasing issue age. The age x table would consist of the $\omega - x$ entries $\{q_{[x]+t}, t = 0, 1, \ldots, \omega - x - 1\}$. Observations show, however, that the selection effects decrease with time and can be assumed to wear off after a certain duration, known as *the select period*. If the select period is r, we can expect that two individuals of the same age, who have been issued insurance r or more years ago, will exhibit the same mortality rates, even if the issue age is different. This allows for some simplification. We can revert to the symbol q_y to mean the probability that a person now age y who was issued insurance r or more years ago will die within a year. In other words, we postulate that equality will hold in (9.1) when $t \geq r$, so we can remove the square bracket and denote the common value by q_{x+t}. The rates q_y are known as *ultimate* rates, and the resulting life table is known as a *select and ultimate table*. A common choice for select period is 15 years, although some recent tables have used as much as 25. The simple example below has a select period of 2, but that is enough to illustrate the basic idea.

x	$q_{[x]}$	$q_{[x]+1}$	q_{x+2}	$x+2$
60	0.10	0.12	0.15	62
61	0.11	0.14	0.17	63
62	0.12	0.15	0.18	64

To find the mortality rates applicable to a policy issued at age x, we start at the left column, read across until we get to the ultimate column, and then read down. So, for example, the rates used for a policy issued at age 60 would be in order $q_{[60]}, q_{[60]+1}, q_{[60]+2} = q_{62}, q_{63}, q_{64}, \ldots,$ which in this example are 0.10, 0.12, 0.15, 0.17, 0.18,.... We can picture the flow of mortality rates as several streams, all running into the same river, which is the ultimate column. Therefore, we do not need a complete separate table for each age, but only $r+1$ columns.

Tables with a select period of 0 years, which is precisely what we have been discussing in previous chapters, are known as *aggregate tables*.

Some people prefer to construct a select and ultimate table of ℓ_x, although this is not really necessary, for as we have seen, one can always work directly with the values of q_x. The process is straightforward, although it requires a little care. One completes the ultimate column first, proceeding as in Chapter 3 to calculate values of ℓ_x starting with an arbitrary value for ℓ_r which is the first entry in this column. One then simply works backwards to complete the other entries, using the recursion

$$\ell_{[x]+k} = \frac{\ell_{[x]+k+1}}{1 - q_{[x]+k}}.$$

For example, in the above table, if $\ell_{63} = 5000$, then $\ell_{[61]+1} = 5000/0.86 = 5814$ and $\ell_{[61]} = 5814/0.89 = 6353$.

9.3 Changes in formulas

To change our previous formulas to incorporate select mortality is usually just a matter of inserting a square bracket around the issue age. We will look at some of these in detail.

Chapter 3

The multiplication rule (3.6) becomes

$$_{s+t}p_{[x]} = {_s}p_{[x]}\ {_t}p_{[x]+s}. \tag{9.2}$$

Chapter 4

We will denote the interest survivorship discount function for a policy issued at age x by $y_{[x]}$. From (9.2) we derive

$$y_{[x]}(k, k+n) = \frac{y_{[x]}(k+n)}{y_{[x]}(k)} = v(k, k+n)\,{_n}p_{[x]+k}. \tag{9.3}$$

Taking $n = 1$, the recursion formula (4.5) now becomes

$$y_{[x]}(k+1) = y_{[x]}(k)v(k, k+1)p_{[x]+k}.$$

It is important to note that with select mortality, (4.6) and (4.7) no longer hold. As mentioned, we need a separate table for each issue age, and there is no longer any necessary relation between the ages. Note, for example, that

$$y_{[x+k]}(n) = v(n)\,{_n}p_{[x+k]},$$

and even if $v(k, k+n) = v(n)$ this is not the same as the right hand side of (9.3) unless k is greater than or equal to the select period.

The subscript of $[x]+k$ on annuity symbols is used in the same way as introduced in Section 4.3.2, except that now, modification to the mortality as well as the interest is needed. That is

$$\ddot{a}_{[x]+k}(\mathbf{c}) = \ddot{a}(\mathbf{c}; y_{[x]} \circ k) = \sum_{j=0}^{\omega-x-k-1} c_j v(k, k+j)\,{}_jp_{[x]+k}.$$

Note that $\ddot{a}_{[x]+k}(\mathbf{c}) = \ddot{a}_{x+k}(\mathbf{c})$, if k is greater than the select period, and interest is constant. With non-constant interest however, these are not necessarily equal even if k is greater than the select period, as we observed in Chapter 4.

Chapter 5

In the third formula in (5.1), q_{x+k} becomes $q_{[x]+k}$.

The subscript $[x]+k$ is used for insurances in the same way to that noted above for annuities. That is

$$A_{[x]+k}(\mathbf{b}) = \sum_{j=0}^{\omega-x-k-1} b_j v(k, k+j+1)\,{}_jp_{[x]+k}\ q_{[x]+k+j}.$$

Chapter 6

No changes are necessary in this chapter except to note that the subscripts of $[x]+k$ in reserve formulas are modified as we noted for Chapters 4 and 5 above.

Chapter 8

We define

$$\mu_x(t) = \lim_{h\to 0} \frac{{}_hq_{[x]+t}}{h} = \lim_{h\to 0} \frac{\ell_{[x]+t} - \ell_{[x]+t+h}}{h\ell_{[x]+t}}.$$

(We are following standard notation here. The square bracket is not necessary since the x is already separated from the t.) The basic identity (8.16) can then be written

$${}_tp_{[x]} = \mathrm{e}^{-\int_0^t \mu_x(r)\mathrm{d}r}.$$

Of course, if the select period is 0 then $\mu_x(t)$ is just equal to $\mu(x+t)$ and agrees with the notation that we have already introduced in Chapter 8.

Spreadsheets

The interested reader may want to adapt the previous spreadsheets that have been introduced to incorporate a select and ultimate table. However, without doing this fully, one can still do calculations directly on the existing spreadsheet for each particular problem. For a policy issued at age (x) on a table with a select period of r years, one simply replaces the entries of q_{x+t} with $q_{[x]+t}$ for $t = 0, 1, \dots, r-1$.

9.4 Projections in annuity tables

There are other situations where a two-dimensional approach to mortality rates is called for. One such example is annuity tables. We need however a different formulation than that presented above. The select tables we have described would not normally be used for annuities. Purchasers of annuities are generally people who expect to live longer than the average or they would be unlikely to enter into such a contract. In any event the insurer is not worried about deteriorating mortality with duration. The problem here is exactly the opposite, namely, people may live longer than the tables predict, requiring that more must be paid out in benefits than the premiums allow for. This has proved to the case in recent years, as advances in medical and public health knowledge, better lifestyles and other factors have led to continually improving mortality, causing issuers of annuities as well as pension plans, to be confronted with what has been termed *longevity risk*. In addition to age, *year of birth* becomes an important variable when pricing annuities. A 60 year old born say in 1950 could be expected to have less chance of dying in the next year, than a 60 year old born in 1930, due to improvements in mortality that have come about over the 20 year period. In place of the rates q_x, we would like a two variable function $q_{x,n}$ where x denotes age and n denotes year of birth.

We will alter this notation somewhat in order to point out the parallel nature of this situation with our original discussion of select mortality. Suppose we have picked a base year, say the year a mortality study was completed. We will now let $q_{[x]+t}$ denote the probability of dying within 1 year, for a person now age $x + t$ who was age x in the base year. In other words we have just extended the select notation by allowing t to denote time elapsed since some particular event, which could be anything at all. In our original application, the event was observation for insurance purposes. For annuity tables, it is the year of the mortality study that produced the table. In this case, we have in place of (9.1)

$$q_{[x]+t} \geq q_{[x-s]+t+s},$$

since, letting b denote the base year, the term on the left refers to someone born in year $(b - x)$ while that on the right refers to some one born in a later year $(b - x + s)$.

There is a however a major problem in constructing a table to show such rates. Suppose for example that a mortality study of annuitants was completed in the year 2000. The result would be a table showing $q_{[x]}$, the one year mortality rates for people age x in the year 2000. Now suppose we want to know what $q_{[60]+1}$ is. This would be the rate for a person age 61 who was born in 1940, but in the year of the study no such person would ever have existed. The rate $q_{[60]+1}$ (as well as all rates with $t > 0$) could not be based purely by a statistical analysis of past mortality data, but would involve the prediction of future trends in mortality improvement. This is rather difficult task, and present actuarial usage involves rather simple tools. The usual method is to assign a yearly projection factor r_y for each age, which represents the reduction in mortality expected at that age for each year. That is, it is assumed that,

$$q_{[x]+t} = q_{[x+t]}(1 - r_{x+t})^t.$$

Following is an example taken from an actual table known as the Uninsured Pensioner (UP) 1994 Table with projection scale AA.

Example 9.1 The UP 1994 table shows male mortality rates for age 60–62 of 0.008576, 0.009633, 0.010911, respectively and projection factors of 0.016, 0.015, 0.015, respectively. A male is age 60 in the year 2010. Using this table find the probability that he will live to age 63. Compare this with the probability that a male who was age 60 in 1994 would live to age 63.

Solution. The required probability is

$$[(1-(0.008576)(0.984)^{16}][(1-0.009633)(0.985)^{17}][(1-0.010911)(0.985)^{18}] = 0.9778.$$

For the male age 60 in 1994, the probability would be

$$[(1-(0.008576)][(1-0.009633)(0.985)][(1-0.010911)(0.985)^{2}] = 0.9716.$$

There is no reason to believe that mortality improvement will stop, so the concept of a select and ultimate table does not appear in this context. Therefore it is not usual in the actuarial literature to view the table that we have described as a select table. Rather it is viewed as a collection of separate tables, one for each age x in the base year. Each of these separate tables is often referred to as a *generational* table, as it traces the mortality of people all born in the year $b-x$. In the construction above each generational table would appear as the row of the select table as described. Each column of this select table can be considered as a version of the original table projected to a future year, and is referred to as the *static table* for that year. It practice, insurers often simplify calculation by making use of the appropriate static table for setting annuity rates in a particular year, rather than using the more accurate generational tables, which would involve a separate table for each age.

9.5 Further remarks

There are many other situations in which the concept of select mortality applies. For another example involving rates decreasing with duration, suppose we are studying the effects of a certain treatment for cancer patients. This could depend both on age and the duration since treatment. In this case, if a person was treated a year ago, and is still alive, there is some indication that the treatment is working, and that such person is more likely to live than someone of the same age who has just received the treatment.

Exercises

9.1 A person now age 62 was sold insurance 1 year ago at the age of 61. Mortality is given by the table at the beginning of the chapter. Find the probability that this individual will die between the ages of 63 and 65.

9.2 A 4-year term insurance policy sold to a person age 61 provides for 1000 paid at the end of the year of death should this occur before age 65. Level annual premiums are payable for 4 years. The interest rate is a constant 5% and mortality is given by the table at the beginning of the chapter.

(a) Find the premium.

(b) Find ${}_2V$.

9.3 Explain in words what the following indicates:

$$({}_6p_{[30]+10})({}_8q_{[30]+16}).$$

9.4 For the table given in Section 9.2, construct the corresponding table of $\ell_{[x]+t}$, given that $\ell_{62} = 5000$.

9.5 An annuity mortality table, with projection factors for improvement, constructed in the year 2000, shows

$$q_{70} = 0.02, \quad q_{71} = 0.03 \quad q_{72} = 0.04, \quad r_{70} = 0.02, \quad r_{71} = 0.018, \quad r_{72} = 0.016.$$

Find the probability that a person age 70 in the year 2010 will die between the ages of 72 and 73.

Spreadsheet exercises

9.6 Consider a select and ultimate version of the sample table, with a 15-year select period, as follows. For $x \geq 50$, and $t = 0, 1, \ldots, 14$ we take

$$q_{[x]+t} = 1 - (1.00001)^{15-t} e^{-0.00005(1.09)^{x+t}}.$$

Interest rates are a constant 6%. For a 20-year endowment insurance on (50) with a level death benefit of 100 000 and a pure endowment of 100 000 at age 70, purchased by level net premium for 20 years, calculate (a) the premium, (b) ${}_{10}V$.

9.7 Using the same table as in Exercise 9.6, calculate (a) $e_{[60]}$, (b) $e_{[50]+10}$.

10

Multiple-life contracts

10.1 Introduction

In previous chapters we have studied insurance and annuity contracts sold to a single individual. There are many situations when such contracts are sold to a group of several people. We will concentrate on the case of two lives, the most common arrangement, which is sufficient to exhibit most of the complications.

The following are a few of the more usual situations. A married couple may wish to purchase an annuity that pays income while either of them is alive. They may also want an insurance policy that pays benefits on the second death, which in some jurisdictions enables them to pass on assets without paying estate taxes. Two business partners may desire a policy that pays a death benefit upon the first death of the two. A person may arrange for an annuity to be paid to a dependent, where payments begin only on the death of the supporting individual. We will discuss these and more in the following sections. For simplicity we assume aggregate mortality in this chapter. It is not too difficult to modify the results for the effects of selection.

10.2 The joint-life status

Consider a pair of lives, (x) and (y). Suppose that we consider the pair to be in a state of *survival* when *both* of them are alive. In other words, the pair *fails* on the first death. Such a pair is commonly known as a *joint-life status*. An annuity sold on a joint-life status would provide income as long as the status survived, that is, as long as both individuals were living. Annuity benefits would stop upon the first death. An insurance policy sold on a joint-life status would pay a death benefit upon the first death.

While it is possible to construct life tables in this case, it is awkward, and it is best to work directly with probabilities. The basic quantity we need is the probability that *both* of (x) and (y) will be alive in t years. We will denote this probability by ${}_tp_{xy}$.

Fundamentals of Actuarial Mathematics, Third Edition. S. David Promislow.

Companion website: http://www.wiley.com/go/actuarial

We will now derive an expression for ${}_tp_{xy}$ that can be calculated from the life table for single lives. The method discussed will be well known to the reader who is familiar with the concept of independence in probability theory (see Section A.2).

Suppose, for example, that ${}_tp_x = 0.9$, ${}_tp_y = 0.8$. Imagine that we start by observing 100 pairs of lives, where one is age x and the other is age y. Of these 100 pairs, we can expect that in 90 cases, the life age x will be alive in t years. Out of these 90 cases we can expect that 80% of the time the life age y will survive t years. This leaves on average 72 pairs of the 100 where both lives survive, so we should have ${}_tp_{xy} = 0.72$.

It is instructive to derive this in another way by looking at deaths rather than survivors. We expect 20 deaths altogether from the age y individuals. On average, two of these age y deaths should occur among the 10 cases where the age x individual dies. This leaves 18 age y deaths among the 90 cases of age x survivors, and therefore $90 - 18 = 72$ cases where both survive.

The particular numbers 0.9 and 0.8 do not matter and the same arguments show that in general we should have

$$ {}_tp_{xy} = {}_tp_x \, {}_tp_y. \tag{10.1} $$

We define

$$ {}_tq_{xy} = 1 - {}_tp_{xy}, $$

the probability that the joint-life status will fail within t years, which is the probability that at least one of (x) and (y) will die within t years.

The reader should carefully note that ${}_tq_{xy}$ is *not* equal to ${}_tq_x \, {}_tq_y$. The latter expression is the probability that *both* lives will die within t years, which is certainly less than the probability that at least one will die in this time interval. To express ${}_tq_{xy}$ in terms of single-life rates, we have ${}_tq_{xy} = 1 - {}_tp_x \, {}_tp_y = 1 - (1 - {}_tq_x)(1 - {}_tq_y)$, which gives

$$ {}_tq_{xy} = {}_tq_x + {}_tq_y - {}_tq_x \, {}_tq_y. \tag{10.2} $$

We now confess that although (10.1) and (10.2) are almost universally used, the argument we gave above is not quite accurate. It would be perfectly valid if the two lives were completely independent of each other. This is unlikely to be true, however, for two people who would choose to buy an insurance or annuity contract together, such as a married couple or two business partners. One can expect that their lives are intertwined to the extent that life-threatening occurrences for one would also affect the other.

For example, suppose that the two lives are business partners who frequently fly together. If one of them were to die in a plane crash, it is likely that the other one would as well. To illustrate the point, we take an extreme case. Return to the example given above, but suppose that the 100 pairs we start with consist in every case of two people who always fly together. We claimed above that there would be on average 2 age y deaths from the 10 cases where the individual aged x died. But if we expect that the lives are likely to die together, we can certainly expect more than this. For example, if all the age x deaths were from plane crashes, we would necessarily have 10 age y deaths as well. This is extreme, but suppose that we have even 3 instead of 2 age y deaths. This leaves 17 instead of 18 deaths from the 90 age x

survivors and we would have ${}_tp_{xy} = 0.73$. This shows that rather than (10.1) we would expect for the typical buyers of an insurance or annuity contract that

$$ {}_tp_{xy} > {}_tp_x \, {}_tp_y. \tag{10.3} $$

For married couples there is an additional factor that supports (10.3). Statistics bear out the fact that upon the death of one partner, the survivor, deprived of the care and companionship of the deceased, often tends to die earlier than otherwise expected. This has been referred to as the 'broken heart syndrome'.

For convenience, we will assume (10.1), or equivalently (10.2), unless otherwise stated. The reader should note, however, that most of the key formulas remain true without this assumption. The main use of (10.1) is to conveniently calculate numerical quantities, given the single-life data.

Generalizations of the joint-life status can be found in Chapters 17 and 19. On the question of dependence, see in particular Sections 17.5 and 17.6.

10.3 Joint-life annuities and insurances

Consider a joint-life annuity contract paying c_k each year provided that both (x) and (y) are alive. In order for a payment to be made at time k we need both lives to have survived to that time, and the probability of this is just ${}_kp_{xy}$. The present value of the benefits, denoted by $\ddot{a}_{xy}(\mathbf{c})$, is given by a formula exactly the same as (4.1), except ${}_kp_{xy}$ replaces ${}_kp_x$:

$$ \ddot{a}_{xy}(\mathbf{c}) = \sum_{k=0}^{N-1} c_k v(k) {}_kp_{xy}, \qquad (10.4)‡ $$

where $N = \min\{\omega - x, \omega - y\}$.

We use N in this way throughout the chapter. To verify that it is the appropriate upper bound, note for example that if $x = 40, y = 70$ and $\omega = 110$, the last possible payment would be at time 39. At time 40, (70) would not be living according to our model. As with single lives, we will use 1_∞ to denote a vector with entries of 1, running to duration $N - 1$, and by convention the omission of a benefit vector or function implies that it is 1_∞.

In the continuous case we have the formula

$$ \bar{a}_{xy}(c) = \int_0^N c(t)v(t){}_tp_{xy}\mathrm{d}t, \qquad (10.5)‡ $$

for the present value of an annuity, with payments made continuously at the annual rate of $c(t)$ at time t, provided both (x) and (y) are alive.

We can also consider a *joint-life insurance* policy that pays b_k at the end of the year of the first death of (x) and (y) when this occurs between time k and time $k + 1$. In other words, payment will be made at time $k + 1$ if both lives survive k years, but both do not survive $k + 1$ years. The probability of this is just

$$ {}_kp_{xy} - {}_{k+1}p_{xy}. $$

This probability can also be expressed as

$$ {}_kp_{xy}\, q_{x+k:y+k}, $$

since for such an event to happen, both lives must survive for k years and then one must die in the subsequent year. (A colon is used to separate the two ages in a joint-life status if needed for clarity.) This latter expression can also be derived from the joint-life version of the multiplication formula,

$$ {}_{k+1}p_{xy} = {}_kp_{xy}\, p_{x+k:y+k}, $$

since for both (x) and (y) to survive $k+1$ years, they must first both survive k years, and then, being age $x+k$ and $y+k$ respectively, they must both survive one more year.

The present value of the joint-life insurance, denoted by $A_{xy}(\mathbf{b})$, can be written, analogously to the second and third expressions in (5.1), as

$$ \begin{aligned} A_{xy}(\mathbf{b}) &= \sum_{k=0}^{N-1} b_k v(k+1)({}_kp_{xy} - {}_{k+1}p_{xy}) \\ &= \sum_{k=0}^{N-1} b_k v(k+1)\,{}_kp_{xy}\, q_{x+k:y+k}. \end{aligned} \qquad (10.6)‡ $$

Suppose that such a joint-life insurance were sold with annual premiums, based on the premium pattern vector ρ. Premiums will cease upon the first death, so the premium payments form a joint-life annuity. We determine the initial premium π_0 in exactly the same way as with a single-life insurance, namely,

$$ \pi_0 = \frac{A_{xy}(\mathbf{b})}{\ddot{a}_{xy}(\rho)}. \qquad (10.7)‡ $$

10.4 Last-survivor annuities and insurances

10.4.1 Basic results

Consider a pair of lives (x) and (y), which we consider to be in state of survival if *either* of them is alive. In other words, the pair fails upon the *second* death. This is known as a *last-survivor status*. The standard symbol for this is $\overline{xy}$, which distinguishes it from a joint-life status. The probability that this status will be in a state of survival at time t is denoted by ${}_tp_{\overline{xy}}$. To calculate this, we use (A.3), a basic fact from probability theory, already illustrated above in (10.2), which says that the probability that at least one of two events occurs is given by the sum of the probabilities of each occurring, less the probability that they both occur. This gives

$$ {}_tp_{\overline{xy}} = {}_tp_x + {}_tp_y - {}_tp_{xy}. \qquad (10.8) $$

We let ${}_tq_{\overline{xy}}$ denote the probability that $\overline{xy}$ will fail before time t. Clearly ${}_tq_{\overline{xy}} = 1 -_t p_{\overline{xy}}$, so that if our independence assumption (10.1) holds, we can substitute in (10.8) to obtain

$$ {}_tq_{\overline{xy}} = {}_tq_x \, {}_tq_y. \tag{10.9} $$

Alternatively, this can be derived directly by noting that in order for this status to fail within t years, we need to have both lives fail within t years.

The probability that (x) fails between time t and time $s + t$ is given either by ${}_tp_{\overline{xy}} -_{s+t} p_{\overline{xy}}$ or by ${}_{s+t}q_{\overline{xy}} -_t q_{\overline{xy}}$. Assuming independence, the latter expression simplifies to

$$ ({}_{s+t}q_x)({}_{s+t}q_y) - ({}_tq_x)({}_tq_y). $$

Formulas for annuities and insurances based on the last survivor status are easily obtained. As in the joint-life case we simply replace the subscript (x) by $(\overline{xy})$. So, for an annuity paying c_k at time k, provided that both (x) and (y) are alive, the present value is just

$$ \ddot{a}_{\overline{xy}}(\mathbf{c}) = \sum_{k=0}^{M-1} v(k)c_k \, {}_kp_{\overline{xy}} = \ddot{a}_x(\mathbf{c}) + \ddot{a}_y(\mathbf{c}) - \ddot{a}_{xy}(\mathbf{c}), \tag{10.10} $$

where the upper limit M now denotes the maximum of $\omega - x$ and $\omega - y$. The final expression results directly from (10.8).

The analogous formula holds for continuous annuities with $\bar{a}$ replacing $\ddot{a}$.

Similarly, the present value of an insurance contract paying b_k at the end of the year of the second death of (x) and (y), if this occurs between time k and $k + 1$ is given by

$$ A_{\overline{xy}}(\mathbf{b}) = \sum_{k=0}^{M-1} v(k+1)b_k({}_kp_{\overline{xy}} -_{k+1} p_{\overline{xy}}) = A_x(\mathbf{b}) + A_y(\mathbf{b}) - A_{xy}(\mathbf{b}), \tag{10.11} $$

where we again use (10.8), as well as the second formula in (5.1) to derive the final expression.

Formula (10.11) can be verified directly. Suppose, for example that (x) dies first, at a time between j and $j + 1$. Then the first term on the right hand side pays b_j and the third term pays $-b_j$, so a net amount of 0 is paid on the first death. When (y) dies later at a time between k and $k + 1$ only the second term pays and y will get the required death benefit of b_k. The same type of reasoning can be applied to verify (10.10) as well as many other types of multiple-life contracts.

These last survivor annuities and insurances are a special case of more general contracts that we will investigate in Sections 10.6 and 10.7.

10.4.2 Reserves on second-death insurances

Reserves on last survivor insurances present a particular problem. For a second-death insurance, the true reserve at time k depends on the state of the pair. There are three possible cases. (For simplicity in notation we assume constant interest. In the general case one uses the investment discount function $v \circ k$ in computing A and $\ddot{a}$.)

$$ {}_kV^{(0)} = A_{\overline{x+k:y+k}}(\mathbf{b} \circ k) - \ddot{a}_{\overline{x+k:y+k}}(\pi \circ k) \quad \text{if both lives survive } k \text{ years.} $$

$$ {}_kV^{(1)} = A_{x+k}(\mathbf{b} \circ k) - \ddot{a}_{x+k}(\pi \circ k) \quad \text{if (x) only survives } k \text{ years.} $$

$$ {}_kV^{(2)} = A_{y+k}(\mathbf{b} \circ k) - \ddot{a}_{y+k}(\pi \circ k) \quad \text{if (y) only survives } k \text{ years.} $$

The above formulas also assume that the premium pattern remains unchanged on the first death. If the contract called for reduced premiums in such an event, then suitable adjustments would be required in the second term of the formulas for ${}_kV^{(1)}$ and ${}_kV^{(2)}$.

It may seem puzzling at first that the reserve on this contract should make a sudden change on the first death, but a moment's reflection will verify that this is perfectly logical. The pairs for which the first death has already occurred present a greater liability and a higher reserve is needed. Taking a retrospective, point of view, what happens is that there is a transfer each period from the funds of the pairs with both individuals still living, to the funds of those where one has already died.

The problem is however that there is no reason for a pair of lives to report the first death to the insurer, unless so stipulated in the contract. This means that the insurer might well not be aware of the state of the various pairs. In this case an appropriate solution would be to hold a weighted average of the different reserves. That is, for each contract the reserve at time k would be

$$\frac{{}_kp_{xy}\ {}_kV^{(0)} + ({}_kp_x - {}_kp_{xy})\ {}_kV^{(1)} + ({}_kp_y - {}_kp_{xy})\ {}_kV^{(2)}}{{}_kp_{\overline{xy}}}.$$

If the actual mortality experience is close to that of the weights used above, then this method should produce aggregate reserves which are close to what they should be by the more exact method.

Note however that it would not be reasonable to base cash values on such a blended reserve figure, which will clearly be higher than ${}_kV^{(0)}$. This would create a definite anti-selection opportunity as those pairs for which both are still living, could cash in the policy and receive an amount which is more than that which they are entitled to.

10.5 Moment of death insurances

To handle moment of death insurances for two-life statuses, we need the joint-life analogue of the force of mortality,

Definition 10.1 The *force of failure* for the joint-life status (xy) at time t is the quantity

$$\mu_{xy}(t) = -\frac{\mathrm{d}/\mathrm{d}t\ {}_tp_{xy}}{{}_tp_{xy}} = -\frac{\mathrm{d}}{\mathrm{d}t}\log({}_tp_{xy}).$$

The independence assumption (10.1) lets us calculate this from the single life functions.

$$\mu_{xy}(t) = -\frac{\mathrm{d}}{\mathrm{d}t}(\log {}_tp_x + \log {}_tp_y) = \mu_x(t) + \mu_y(t). \qquad (10.12)$$

Consider an insurance policy that pays $b(t)$ at time t if the first death of (x) and (y) occurs at that time. The present value will be denoted by $\bar{A}_{xy}(b)$ and given, analogously to (8.18), as

$$\bar{A}_{xy}(b) = \int_0^N b(t)v(t){}_tp_{xy}\mu_{xy}(t)\mathrm{d}t. \qquad (10.13)‡$$

The present value of an insurance paying $b(t)$ at time t if the *second* death of (x) and (y) occurs at that time is denoted by $\bar{A}_{\overline{xy}}(b)$. The same reasoning as used after formula (10.11) establishes the analogous formula

$$\bar{A}_{\overline{xy}}(\mathbf{b}) = \bar{A}_x(\mathbf{b}) + \bar{A}_y(\mathbf{b}) - \bar{A}_{xy}(\mathbf{b}). \tag{10.14}$$

As with the single-life case, we are faced with the problem of evaluating $\bar{A}_{xy}$ from the life table. Assume that b takes the constant value of b_k over the year running from time k to time $k+1$. It is natural to ask if the same i/δ correction that we used for single-life insurances is also applicable to the joint-life case. Recall that this correction came from (8.20), so the question is whether the corresponding statement is true for joint-life statuses. That is, is it true that ${}_tp_{xy}\left(\mu_x(t) + \mu_y(t)\right) = q_{xy}$ for $0 < t < 1$? Assume UDD and therefore (8.20) for single lives. Then, for $0 < t < 1$,

$$\begin{aligned}
{}_tp_{xy}\left(\mu_x(t) + \mu_y(t)\right) &= {}_tp_y\, q_x + {}_tp_x\, q_y \\
&= (1 - tq_y)q_x + (1 - tq_x)q_y \\
&= q_{xy} + (1 - 2t)q_x q_y,
\end{aligned}$$

where we use (10.2) for the last equality. We see that the required condition does not hold exactly, due to the extra term of $(1 - 2t)q_x q_y$. Arguing as in Section 8.6.2, we deduce that

$$\bar{A}_{xy}(b) = A_{xy}\left(\frac{\mathbf{i}}{\delta} * \mathbf{b}\right) + R, \tag{10.15}$$

where $\mathbf{b} = (b_0, b_1, \ldots)$ and

$$R = \sum_{k=0}^{N-1} r(k)b(k)v(k+1)\,{}_kp_{xy}\, q_{x+k}q_{y+k}, \tag{10.16}$$

with $r(k) = \int_0^1 v(k+1, k+t)(1-2t)\mathrm{d}t$. Analysis shows that $r(k) > 0$ so that R is positive, but likely to be quite small and safely ignored in practice. Using the i/δ correction will slightly understate the value of the insurance, but it is a reasonable approximation. R is the present value of a contract that pays off only if (x) and (y) die in the same year, an event that will have small probability for most ages. Moreover, since $1 - 2t$ is positive for $0 \le t < \frac{1}{2}$ and negative for $\frac{1}{2} < t \le 1$, the cancellation in the integral will tend to make $r(k)$ small.

10.6 The general two-life annuity contract

There is a demand for two-life annuities that are more flexible than ones we described in previous sections. One popular arrangement is for benefits to continue while either person is alive, but for the amount to reduce when only one of the parties is alive. For example, a married couple may wish to provide for an annuity at retirement that will pay 60 000 yearly while both are alive, reducing to 40 000 yearly when only one is alive. This one-third reduction is a common one, reflecting the fact that while one person can live more cheaply than two, he/she will need more than half of the amount, since many expenses, such as housing, will not necessarily reduce. Many variations are possible. Benefits need not be symmetric and could vary according to the particular survivor. A common provision in pension plans is that the income stays level as long as the employee is alive, but will continue to the spouse at a reduced level upon the death of the employee. We present a single formula that covers all possibilities.

A general life annuity on the pair (x) and (y) can be described by three annuity benefit vectors: $\mathbf{f}$, where f_k is the amount paid at time k if *only* (x) is alive; $\mathbf{g}$, where g_k is the amount paid at time k if *only* (y) is alive; and $\mathbf{h}$, where h_k is the amount paid at time k if *both* (x) and (y) are alive. Let $\mathbf{j} = \mathbf{h} - \mathbf{f} - \mathbf{g}$. We can think of the contract as three separate annuities. One is a life annuity on (x) with benefit vector $\mathbf{f}$. Another is a life annuity on (y) with benefit vector $\mathbf{g}$. These two will provide for the required payments when only one of the pair is living. The third contract must adjust the payment when both are alive. If both are living at time k, the first two annuities will provide $f_k + g_k$, so the third must provide the difference, $j_k = h_k - f_k - g_k$. The present value of the complete contract is then the sum of the three separate present values, which is just

$$\ddot{a}_x(\mathbf{f}) + \ddot{a}_y(\mathbf{g}) + \ddot{a}_{xy}(\mathbf{j}). \qquad (10.17)\ddagger$$

This formula reduces the calculation of present values for the general two-life annuity to calculating those for single or joint-life annuities. (See Section 10.12 for spreadsheet methods to calculate the latter.)

Example 10.1 Verify that (10.17) reduces to (10.10), when the yearly benefit is a constant 1 unit.

Solution. We have $f_k = g_k = h_k = 1$ for all k, so $j_k = -1$ for all k and (10.17) gives

$$\ddot{a}_{\overline{xy}} = \ddot{a}_x + \ddot{a}_y - \ddot{a}_{xy}.$$

Example 10.2 An annuity on (x) and (y) provides yearly payments as long as either (x) or (y) are alive. Payments begin at 12, but reduce to 8 if (x) only is alive or to (6) if (y) only is alive. Find the present value

Solution. $f_k = 8$, $g_k = 6$, and $h_k = 12$, so $j_k = -2$ for all k. This gives a present value of

$$8\ddot{a}_x + 6\ddot{a}_y - 2\ddot{a}_{xy}.$$

Example 10.3 An annuity pays 1 unit each year, provided that at least one of (50) and (65) is living and is over age 70, but not if (50) is alive and under age 65. Find the present value.

Solution. Letting x refer to (50) and y to (65),

$$\mathbf{f} = (0_{20}, 1_{\infty}), \qquad \mathbf{g} = (0_5, 1_{\infty}), \qquad \mathbf{h} = (0_{15}, 1_{\infty}),$$

so $\mathbf{j} = -(0_5, 1_{10}, 0_5, 1_\infty)$. The present value is then

$$\ddot{a}_{50}(0_{20}, 1_\infty) + \ddot{a}_{65}(0_5, 1_\infty) - \ddot{a}_{50:65}(0_5, 1_{10}, 0_5, 1_\infty).$$

Example 10.4 Find a formula for the present value of an annuity that pays c_k at time k, provided that (y) is alive, but (x) is not alive.

Solution. $f_k = 0$, $g_k = c_k$, $h_k = 0$, so that $j_k = -c_k$, for all k. The present value is

$$\ddot{a}_y(\mathbf{c}) - \ddot{a}_{xy}(\mathbf{c}).$$

A contract of the type described in Example 10.4 is known as a *reversionary annuity*. It provides a life annuity on one life, which does not begin until another life has died. It can be used for a similar purpose as life insurance, except that the proceeds are paid out as a life annuity to another person, rather than as a lump sum. As well, reversionary annuities are often used to put a value on certain inheritances that stipulate that the income from a certain asset will first go to a certain individual and then, upon the death of that party, revert to another person.

Remark We comment briefly here on the effect of using assumption (10.1) in which ${}_tp_{xy}$ was approximated by a quantity that is normally too low, as we saw in (10.3). It follows that this approximation will give values for $\ddot{a}_{xy}(\mathbf{b})$ that are slightly too low (assuming the entries of $\mathbf{b}$ are nonnegative). Since the coefficient of this term is negative in the most common examples of two-life annuities, as illustrated above, the standard assumption of independence means that most two-life annuity premiums are somewhat higher than they would be if a more realistic model was used.

10.7 The general two-life insurance contract

The most common type of policy sold to two lives is the joint-life insurance described above, where payment is made on the first death. In some cases, however, people want policies that will pay on the second death. As we indicated in the introduction to this chapter, this can be used as a means of minimizing estate taxes. We will in fact consider a general type of contract where death benefits can be paid on both deaths. Assuming payment at the moment of death, such a policy will be described by two death benefit functions: $b(t)$, the amount paid at the time of the first death; and $d(t)$, the amount paid at the time of the second death. We can view this as two separate insurances, one on the status xy with benefit function b, and one on the status $\overline{xy}$ with benefit vector d. Substituting from (10.14), the present value of benefits is

$$\bar{A}_{xy}(b) + \bar{A}_{\overline{xy}}(d) = \bar{A}_x(d) + \bar{A}_y(d) + \bar{A}_{xy}(b - d). \tag{10.18}$$

This can be verified directly by the same type of reasoning as that used after formula (10.11).

10.8 Contingent insurances

In some insurance policies written on two lives, the benefits depend on the *order* of death. In the most common cases, a designated individual must die first or second in order to receive benefits. If this does not occur, then no benefits are paid. For this reason these are known as *contingent* insurances. We will first consider such contracts with benefits payable at the end of the year of death. It is assumed throughout this section that there is zero probability that two lives will die at exactly the same instant of time.

10.8.1 First-death contingent insurances

We first need to compute some new probabilities. Let q^{1}_{xy} denote the probability that in the next year (x) will die and that (y) will be alive at the time of the death of (x). The symbol of 1 above x indicates that (x) is the *first* to die. Note that (y) does not have to die within the 1-year period in order for the event in question to occur.

The first problem is to estimate this probability from the life table. We will derive the answer here intuitively. A more formal derivation appears in Chapter 17.

We note that q^{1}_{xy} can be approximated by another probability. Consider the following events: event A consists of (x) dying within the year, before the death of (y); event B consists of (x) dying within the year, and (y) surviving to the middle of the year. Then q^{1}_{xy} is the probability of A. We claim that this is close to the probability of event B. These events are of course not the same. If (x) and (y) both die in the first half of the year, with (x) dying first, then A occurs but not B. On the other hand, if (x) and (y) both die in the second half of the year with (y) dying first, then B occurs but not A. However, both these latter situations are relatively rare, involving both lives dying in the same year. In any event, as long as the probabilities of these two death situations are roughly the same, they will tend to cancel each other out, and we can assume that the probability of A is approximately the same as the probability of B.

The probability of B can be readily computed from the life table. The probability that (x) dies during the year is just q_x. Assuming UDD, the probability that (y) will be alive in the middle of the year is ${}_{1/2}p_y = 1 - q_y/2$. Making the usual independence assumption that we did in deriving (10.1), we deduce that the probability that both of these occurrences will happen is the product of probabilities. In other words, assuming UDD

$$q^{1}_{xy} \doteq q_x - \frac{1}{2}q_x q_y. \qquad (10.19)‡$$

As a check on this, let us compute another expression for q_{xy}, the probability that at least one of the two lives will die within the year. This can happen in two mutually exclusive ways. Either (x) dies during the year, with (y) alive at that time, or (y) dies during the year, with (x) alive at that time. By basic rules of probability theory we can sum the two probabilities to get the probability that either will happen. We must have that

$$q_{xy} = q^{1}_{xy} + q^{1}_{yx}. \qquad (10.20)$$

Adding the right hand side of (10.19) to that with (x) and (y) interchanged does indeed give q_{xy}, as shown by (10.2).

Consider now an insurance contract where a designated life must die before another in order to collect. This arises in the following situation. In any insurance policy, a particular individual, known as the *beneficiary*, is designated to receive the death benefit upon death of the insured. If the beneficiary dies before the insured, a new beneficiary is normally chosen. Suppose, however, that an insured has no other beneficiary in mind. To handle this, insurers are willing to offer a contract that only pays if a stated beneficiary is alive at the death of the insured.

Suppose that a policy sold on the lives (x) and (y) provides for benefits to be paid at the end of the year of the death of (x) provided that (y) is alive at the time of such death. If (y) dies before (x), nothing is paid. For a death benefit vector $\mathbf{b}$, the present value of the benefits for this contract will be designated by $A^{\,1}_{xy}(\mathbf{b})$. We calculate this by the standard formula for an insurance present value, as given in (5.1) and again in (10.6). We need only insert, for each k, the probability that the conditions for payment will occur in the year k to $k+1$. Focus on the second expression in the right hand side of (10.6) where we wrote the probability that the first death of (x) and (y) will occur between time k and time $k+1$ as ${}_kp_{xy}\, q_{x+k:y+k}$. In the present case, in order that the death benefit is paid at time $k+1$, we still require that both (x) and (y) are alive at time (k) but then we need that $(x+k)$ dies in the next year, and to be the first of the two lives to die. In place of $q_{x+k:y+k}$ we want $q^{\,1}_{x+k:y+k}$. This gives

$$A^{\,1}_{xy}(\mathbf{b}) = \sum_{k=0}^{N-1} b_k v(k+1)\, {}_kp_{xy} q^{\,1}_{x+k:y+k}. \tag{10.21‡}$$

As a check, we can use (10.20) to see that

$$A_{xy}(\mathbf{b}) = A^{\,1}_{xy}(\mathbf{b}) + A^{\,1}_{yx}(\mathbf{b}), \tag{10.22}$$

which must be true, since a joint-life insurance can be considered as two separate contingent insurances, each paying benefits if a particular life dies first.

10.8.2 Second-death contingent insurances

Suppose we have a policy where death benefits are paid at the end of the year of death only if (x) is the *second* of the two lives to die. Denote the present value of this by $A^{\,2}_{xy}(\mathbf{b})$. We calculate this as follows. Think of a regular single-life policy on (x) as two separate contracts. The first pays if (x) dies before some other life (y), and the second pays if (x) dies after (y). We must have that

$$A_x(\mathbf{b}) = A^{\,1}_{xy}(\mathbf{b}) + A^{\,2}_{xy}(\mathbf{b}),$$

so that

$$A^{\,2}_{xy}(\mathbf{b}) = A_x(\mathbf{b}) - A^{\,1}_{xy}(\mathbf{b}). \tag{10.23‡}$$

10.8.3 Moment-of-death contingent insurances

Moment-of-death contingent insurances can be calculated by a formula similar to (10.21) but with integrals replacing sums, and μ replacing q. If $\bar{A}^{\ 1}_{xy}(b)$ is the present value of a contract paying $b(t)$ at time t provided that (x) dies at time t and (y) is alive that time, then

$$\bar{A}^{\ 1}_{xy}(b) = \int_0^N b(t)v(t)_t p_{xy}\mu_x(t)\,\mathrm{d}t. \qquad (10.24)‡$$

We can intuitively verify this by noting that ${}_t p_{xy}\mu_x(t)\mathrm{d}t$ represents the probability that both (x) and (y) survive up to time t and then (x) dies at that time. Note also, from (10.12), that

$$\bar{A}_{xy}(b) = \bar{A}^{\ 1}_{xy}(b) + \bar{A}^{\ 1}_{yx}(b),$$

which is obviously required.

For second-death insurance, the same reasoning as in (10.23) shows that

$$\bar{A}^{\ 2}_{xy}(b) = \bar{A}_x(b) - \bar{A}^{\ 1}_{xy}(b). \qquad (10.25)$$

Example 10.5 Suppose (x) is subject to a constant force of mortality μ, (y) is subject to a constant force of mortality v, and the force of interest is a constant δ. Find $\bar{A}^{\ 1}_{x:y}$.

Solution. From (10.24),

$$\bar{A}^{\ 1}_{xy} = \int_0^\infty \mathrm{e}^{-\delta t}\mathrm{e}^{-(\mu+v)t}\mu\mathrm{d}t = \frac{\mu}{\mu + v + \delta}.$$

Normally, we must evaluate contingent insurances directly from the life table. In the case that benefits are constant over each year, the procedure is similar to (10.15) except that we get only one-half of the remainder term. This necessarily follows from (10.22) and the fact that R is symmetric in (x) and (y). That is, assuming UDD for each of the lives (x) and (y),

$$\bar{A}^{\ 1}_{xy}(\mathbf{b}) = A^{\ 1}_{xy}\left(\frac{\mathbf{i}}{\delta} * \mathbf{b}\right) + \frac{R}{2}, \qquad (10.26)$$

where R is as defined in (10.16).

10.8.4 General contingent probabilities

Let ${}_s q^{\ 1}_{xy}$ denote the probability that within s years (x) will die and (y) will be alive at the time of the death of (x). (So $q^{\ 1}_{xy}$, as we have defined it above is just this symbol with $s = 1$.) Similarly, let ${}_s q^{\ 2}_{xy}$ denote the probability that within s years (x) will die having been predeceased by (y).

Consider first the case where s is a positive integer n. We can evaluate this easily by adding up the probabilities of the required event occurring in each year to obtain

$$ {}_nq_{xy}^{1} = \sum_{k=0}^{n-1} {}_kp_{xy} q_{x+k:y+k}^{1}. \tag{10.27‡}$$

From (10.21) this is just $A_{xy}^{1}(1_n)$ at zero interest. For the general duration s we evaluate the probability by taking the corresponding continuous insurance formula (10.24) at zero interest:

$$ {}_sq_{xy}^{1} = \int_0^s {}_tp_{xy}\mu_x(t)\mathrm{d}t. \tag{10.28‡}$$

Note that we can take $s = \infty$ in the above to get the probability that (x) dies before (y). (There was no point in doing this in the single life case since ${}_\infty q_x$ is just equal to 1). Some basic identities are

$$ {}_\infty q_{xy}^{1} + {}_\infty q_{yx}^{1} = 1, \qquad {}_\infty q_{xy}^{1} + {}_\infty q_{xy}^{2} = 1. $$

10.9 Duration problems

Certain multiple-life problems involve a duration that runs from the time of death of an individual rather than from time zero. Normally, the best approach to finding present values is to refer to our basic formula where we sum or integrate the product of three factors. Complications can arise when computing the probability that payment will be made. In many cases a formula for this probability will vary with time, and we will need to break up our total time interval into appropriate subintervals, each with its own formula. Evaluation of the resulting integrals often involves the techniques of changing variables and reversing the order of integration or summation. We will illustrate with several examples involving continuous annuities or moment of death insurances. For simplicity, we assume throughout constant interest and independence of the two lives involved. In all cases we wish to find a formula for either a present value or a probability. Moreover we want the resulting formula to involve either single-life or joint-life statuses, or contingent probabilities and insurances.

Example 10.6 An insurance contract provides for 1 to be paid at the moment of the death of (y) provided that (y) dies at least n years after the death of (x). Find a formula for the present value.

Solution. We must consider two separate time periods. For a time t between 0 and n there is no chance of payment being made. For a time $t \geq n$, payment is made provided y dies at time t and x was not alive at time $t - n$. The present value is

$$ \int_n^\infty v^t {}_tp_y\mu_y(t)(1 - {}_{t-n}p_x)\mathrm{d}t. $$

To write this in terms of our standard symbols, we change the variable from t to $s = t - n$, which gives us an integral with lower limit 0. Then the integral becomes

$$\int_0^\infty v^{s+n}{}_{s+n}p_y\mu_y(s+n)(1 -_s p_x)\mathrm{d}s = v^n{}_np_y \int_0^\infty v^s{}_sp_{y+n}\ \mu_{y+n}(s)(1 -_s p_x)\mathrm{d}s.$$

In the second equality above, we use the multiplication rule as well as the fact that by definition, $\mu_z(r)$ depends only on the sum $z + r$. From the above formula we can write the solution in the required form as

$$v^n{}_np_y\left(\bar{A}_{y+n} - \bar{A}^{\,1}_{y+n:x}\right) = v^n{}_np_y\bar{A}^{\,2}_{y+n:x}.$$

We can deduce this intuitively if we do some fanciful reasoning. We tell (y) to age n years while we keep (x) at the same age. We then let the two lives, now age $y + n$ and x, 'race' to see who dies second. We must multiply the result by an interest factor of v^n to account for the delay in the race, and also by ${}_np_y$ which is the probability that y will survive to be able to participate in the race in the first place.

Example 10.7 What is the probability that (x) and (y) will die at least n years apart from each other?

Solution. The probability that (y) will die n or more years after the death of (x) is given by the solution to Example 10.6 with 0 interest. This is

$${}_np_y\left(1 -_\infty q^{\,1}_{y+n:x}\right) =_n p_y\ {}_\infty q^{\,1}_{x:y+n}.$$

Adding to this the probability that (x) dies n or more years after the death of (y) gives the solution of

$${}_np_y\ {}_\infty q^{\,1}_{x:y+n} +_n p_x\ {}_\infty q^{\,1}_{y:x+n}.$$

Example 10.8 A temporary life annuity on (y) provides for payments made continuously at the annual rate of 1 for n years, but with the payments beginning at the death of (x), rather than at time 0. Payments stop at the death of (y). Find a formula for the present value.

Solution. For $0 \leq t < n$, payment will be made provided (y) is alive and (x) is not. For $t \geq n$, the payment will be made provided (y) is alive, (x) is not alive but was alive n years before. The present value is

$$\int_0^n v^t{}_tp_y(1 -_t p_x)\mathrm{d}t + \int_n^\infty v^t{}_tp_y({}_{t-n}p_x -_t p_x)\mathrm{d}t = \bar{a}_y(1_n) - \bar{a}_{xy} + \int_n^\infty v^t{}_tp_y\ {}_{t-n}p_x\mathrm{d}t,$$

where we rearrange and combine terms to get the expression on the right. Using the change of variable technique of Example 10.6 on the final term yields the answer of

$$\bar{a}_y(1_n) - \bar{a}_{xy} + v^n {}_np_x \bar{a}_{x+n:y}.$$

Note that this is similar to the usual reversionary annuity formula except (using the same fanciful reasoning as in Example 10.6), after n years we only award payment to (y) if living jointly with an individual who was age x and was then allowed to age n years while (y) stayed the same.

Another contract of this type, which does not involve a duration directly, but is based on a similar theme, is a reversionary annuity where the benefits can depend on the time of death of the first person as well as the time they are made.

To illustrate the technique, we first revisit a continuous version of the standard reversionary annuity as given in Example 10.4. Consider a contract with continuous payments at the annual rate of $c(s)$ at time s, made if (y) is alive but (x) is not. As an alternate approach to that used before, we will view this as a contingent insurance, with a death benefit of $\bar{a}_{y+t}(c \circ t)$ paid at time t, the moment of death of (x) if this occurs before the death of (y). Note the time shifting necessary here. For example, if (x) dies at time 2, the rate of payment at a time 3 periods after the annuity begins is $c(5) = (c \circ 2)(3)$. We can write the present value as

$$\int_0^\infty v^t {}_tp_x \, {}_tp_y \mu_x(t) \bar{a}_{y+t}(c \circ t) \mathrm{d}t. \qquad (10.29)$$

Note now that

$$v^t {}_tp_y \bar{a}_{y+t}(c \circ t) = \int_t^\infty c(s) v^s {}_sp_y \mathrm{d}s,$$

since each side is the present value of a t-year deferred life annuity on (y) where the payment at time s for $s \geq t$ is $c(s)$. Substituting this into the integral gives us the present value as

$$\int_0^\infty \int_t^\infty c(s) v^s {}_sp_y \, {}_tp_x \mu_x(t) \, \mathrm{d}s \, \mathrm{d}t = \int_0^\infty \int_0^s c(s) v^s {}_sp_y \, {}_tp_x \mu_x(t) \mathrm{d}t \, \mathrm{d}s, \qquad (10.30)$$

by reversing the order of integration. Now,

$$\int_0^s {}_tp_x \mu_x(t) \mathrm{d}dt = \int_0^s -\frac{\mathrm{d}}{\mathrm{d}t} {}_tp_x \mathrm{d}t = 1 - {}_s p_x,$$

the probability that x dies before time s. From this we conclude that the present value of the contract is

$$\int_0^\infty c(s) v^s {}_sp_y (1 - {}_s p_x) \mathrm{d}s = \bar{a}_y(c) - \bar{a}_{xy}(c).$$

This appears as a very complicated way of arriving at the same form of answer as that obtained in a much simpler fashion in Section 10.6. However, one benefit of this insurance approach is that it works just as well with the alternative type of reversionary annuities we introduced above, and provides a convenient method for handling those. Suppose, for example that upon the death of (x), the periodic rate of payment to a surviving (y) is $c(r)$ at a time r as measured from the *start of the annuity*. The annuity would then begin with the periodic rate of $c(0)$. In (10.29) we then have c in place of $c \circ t$. For death at time t the rate of payment at a time $s > t$ would be $c(s-t)$, and that would appear in 10.30 in place of $c(s)$. This is will normally make for a rather complicated integration, but there are cases where it can be worked out, as in the following example.

Example 10.9 Suppose that the force of mortality for x is a constant μ, the force of mortality for y is a constant ν and the force of interest is a constant δ. Upon the death of (x) an annuity will be paid to (y), if surviving, where the rate of payment at a time r after the annuity payments begin is $e^{\gamma r}$, with $0 \le \gamma < \nu + \delta$. Find the present value.

Solution. From the right side of 10.30, substituting $c(s-t) = e^{\gamma s}e^{-\gamma t}$ for $c(s)$ as outlined, the integral involving the variable t will be

$$\int_0^s e^{-\gamma t}e^{-\mu t}\mu dt = \frac{\mu(1 - e^{-(\mu+\gamma)s})}{\mu + \gamma}.$$

Substituting and integrating with respect to s gives the present value as

$$\frac{\mu}{\mu+\gamma}\int_0^\infty e^{-\delta s}e^{-\nu s}e^{\gamma s}(1 - e^{-(\mu+\gamma)s})\,ds = \frac{\mu}{\mu+\gamma}\left[\frac{1}{\nu+\delta-\gamma} - \frac{1}{\nu+\delta+\mu}\right].$$

*10.10 Applications to annuity credit risk

Suppose that you are a prospective annuity purchaser (or an advisor to such), and have a list of various companies and the premiums they charge. The prices will vary, due to different interest and mortality assumptions, as well as a different treatment of many other factors that go into gross premium calculation. A natural tendency would be to select the one with the lowest premium, but that is not necessarily the best choice. This ignores *credit risk* which is the risk that the company may be unable to meet the promised payments due to financial difficulties. One may want to adjust the prices to reflect the different degrees of this risk among the various issuers. This could be an extensive task, and we will not go into details here. Our goal in this section is to indicate briefly how the theory of multiple-life contracts can be applied towards this endeavour.

A first step might be to compare for a fixed interest and mortality basis, the cost of a life annuity calculated in the usual way with one that provides for payments only if a chosen company is viable. For this purpose we just treat the company as another individual, which we'll denote by γ, which is a survival state if it is in position to meet its financial obligations. We need to model ${}_tp_\gamma$, which is the probability that the company is in a survival state at time t. This will produce the associated force of failure, $\mu_\gamma(t) = -d/dt \log {}_tp_\gamma$. We can then consider two-life contracts based on (x) and γ. For simplicity we confine our discussion to a life annuity on (x) that provides continuous payments at a level annual rate of K.

Taking into account the credit risk, the worth of the contract to the purchaser should be the joint-life annuity $K\bar{a}_{x\gamma} = K\int_0^\infty {}_tp_x\,{}_tp_\gamma\,dt$ rather than $K\bar{a}_x$. This assumes that no benefits are paid after failure of the company. In practice however, there is often a recovery rate r whereby the company, upon financial failure, will continue to pay an income of r for every 1 unit of promised annuity income. In this case the purchaser will get Kr if (x) only is surviving, and K if both x and γ are surviving so that our general two-life annuity formula gives the value to the purchaser as

$$K[r\bar{a}_x + (1-r)\bar{a}_{x\gamma}],$$

which of course reduces to $K\bar{a}_x$ when $r = 1$.

Another complication arises in cases where government regulatory bodies provide guarantees for annuities. There are many possible options concerning the amount of the guarantees and how they relate to the recovery rates of the failing company. We will consider the following particular example, which is encountered in some jurisdictions. Upon failure, the insurer continues to pay at their recovery rate. The guarantor makes up the difference, but the value, at interest and survivorship, at the time of failure, of all payments made by the guarantor is limited to an amount G. Suppose, for example that $K = 30,000, G = 100,000, r = 0.4$. In the event of default, the issuing company would pay 12,000 each year, and the government body would provide the annual shortfall of 18,000 but would only pay it only until the present value of 100,000 was exhausted, which should be somewhat over 6 years under normal interest and mortality bases. We will arrive at a formula to compute the present value of a K-unit, whole life annuity subject to these default arrangements. We assume that the mortality and interest bases are the same for the issuer and the guaranteeing party.

Let $s_0 = \inf\{s : \bar{a}_{x+s} \le G/(1-r)K\}$. Note that s_0 would equal 0 if $\bar{a}_x < G/(1-r)K$. For failure of the issuer at a time t before time s_0 the guarantee will not cover the full annuity. In that case the annuitant, who, in the absence of default, would have received total future payments with a value at time t of $K\bar{a}_{x+t}$, will instead receive payments with a reduced value at time t of $G + rK\bar{a}_{x+t}$. The deficit is $(1-r)K\bar{a}_{x+t} - G$. We must subtract from the usual present value, the present value of a a contingent insurance that provides this deficit on the failure of γ before time s_0, provided that (x) is then alive. Referring to equation (10.24) the credit risk adjusted present value of the annuity is

$$K\bar{a}_x - \int_0^{s_0} v^t\,{}_tp_x\,{}_tp_\gamma\,\mu_\gamma(t)[(1-r)K\bar{a}_{x+t} - G]\mathrm{d}t.$$

10.11 Standard notation and terminology

Particular examples of the multiple-life notation have already been encountered in the standard notation for term and endowment insurances and temporary annuities. In place of the life (y), we dealt with a period of n years, denoted by $\overline{n|}$, which 'survives' for exactly n years and then 'fails' at time n. So the standard symbol for the present value of a 1-unit, n-year term insurance premium, $A^1_{x:\overline{n|}}$ does indeed signify that the policy provides for 1 unit at the end of the year of death provided (x) dies before the 'failure' of $\overline{n|}$. Similarly, the standard symbol for the present value of a 1-unit, n-year endowment insurance $A_{x:\overline{n|}}$ indicates that the contract provides for 1 unit to be paid at the death of (x) or at time n if earlier, that is, on the first failure of x and $\overline{n|}$. The 1-unit, n-year, temporary annuity present value symbol $\ddot{a}_{x:\overline{n|}}$ indicates that the contract provides for payments to continue as long as both (x) is alive and $\overline{n|}$ is 'alive', meaning that the n-year period has not run out.

10.12 Spreadsheet applications

We can adapt the Chapter 6 spreadsheet to be applicable to joint-life annuities and insurances. In the first place we will make provision to insert two life tables, one for (x), and one for (y). This is common in practice as often one life will be male, and the other female, so different tables apply. Start by inserting a new column D, which we intend show the values of q_{y+t}. This will move all the existing columns over one to the right. In cell D1 we put the age y. Copy the formula from C10 to D10. It should be the same except with C1 replaced by D1 and a reference to column P. (the reference is column C will be automatically changed from column N to column O).

We then can insert a new table in Column P by putting the parameters in P3 and P4, coping the formula from 010 to P10 and copying down.

In column F, which now contains y_{xy}, we change the formula in F11 by appending $*(1 - D10)$ and copy down.

In Column M, which now contains $w_{xy} * b$, we change the formula in M10 by replacing C10 by (C10 + D10 − C10*D10) in order to multiply by q_{xy} rather that q_x.

As a test exercise suppose that our example table applies to females and that for the male table the mortality at each age up to 119 is 1.25 times the female rate. Interest rates are a constant 6%. An insurance on a male age 50 and a female age 45 provides 10,000 at the end of the year of the first death. Find the net level premium payable yearly while both lives are living. The answer is 208.75.

Notes and references

An extensive study of dependence in two-life annuity contracts can be found in Frees *et al.* (1996).

More advanced examples of multiple-life contracts, involving an arbitrary number of lives, can be found in Jordan (1967, Part II), or in Bowers *et al.* (1997, Chapter 18).

The result from real analysis that justifies the reversal of integration in formula (10.30) can be found in Royden (1988) Theorem 12.19 (iii).

Exercises

In all the exercises for this chapter, we assume that (10.1) holds.

Type A exercises

10.1 You are given the following information from a life table: $\ell_{80} = 100$, $\ell_{81} = 80$, $\ell_{82} = 40$, $\ell_{83} = 20$. You also have the vector $\mathbf{c} = (1000, 1200, 1500)$ and the vector $\mathbf{b} = (1000, 2000)$. The interest rate is a constant 25%. Find (a) $p_{80:81}$ and ${}_2p_{80:81}$, (b) $p_{\overline{80:81}}$ and ${}_2p_{\overline{80:81}}$, (c) $\ddot{a}_{80:81}(\mathbf{c})$, (d) $A_{80:81}(\mathbf{b})$, (e) $\ddot{a}_{\overline{80:81}}(\mathbf{c})$, (f) $A_{\overline{80:81}}(\mathbf{b})$.

10.2 Suppose that ${}_5p_x = 0.8,\ {}_6p_x = 0.7,\ {}_5p_y = 0.6,\ {}_6p_y = 0.4$. Find the probability that

both (at least one) (exactly one) (neither) of the two lives
live 5 years (die within the next 5 years) (die between time 5 and time 6).

There are 12 answers required here.

10.3 You are given that $p_{70} = 0.9,\ {}_2p_{70} = 0.8,\ {}_3p_{70} = 0.7$. The interest rate is 20% for the first year and 25% for the second year. Find the present value of the benefits on a 2-year term insurance policy sold to two lives (70) and (71) that provides benefits upon the first death, payable at the end of the year of death. The amount of the death benefit is 1000 if the first death occurs in the first year, and 2000 if it occurs in the second year.

10.4 An insurance policy pays 1 unit at the moment of the second death of (x) and (y). Level premiums are payable continuously as long as either of the two is alive. Given that (x) is subject to a constant force of mortality of 0.1, (y) is subject to a constant force of mortality of (0.3) and the force of interest is a constant 0.1, what is the annual rate of premium payment?

10.5 For two lives (x) and (y), the forces of mortality are given by $\mu_x(t) = 0.04$ for all t, and $\mu_y(t) = 1/(20 - t)$ for $0 \le t < 20$.

(a) Find the probability that the joint-life status (xy) will survive to time 10.

(b) Find the probability that the last-survivor status $\overline{xy}$ will fail before time 10.

(c) An insurance contract provides for a death benefit paid at the moment of the first death of (x) and (y). The amount of the benefit is $e^{0.1t}$ for death at time t. The force of interest is a constant 0.06. Find the present value of the benefits.

10.6 Suppose that Demoivre's law holds with $\omega = 100$. Consider two lives (80) and (60).

(a) What is the probability that the first death will occur between time 5 and time 10?

(b) What is the probability that the second death will occur between time 5 and time 10?

10.7 The following is a portion of a select and ultimate table with a select period of 2 years.

x	$q_{[x]}$	$q_{[x]+1}$	q_{x+2}	$x+2$
60	0.08	0.14	0.22	62
61	0.09	0.20	0.25	63
62	0.10	0.20	0.30	64

A and B are both age 62. A is a newly selected life, while B was first selected at age 61. What is the probability that the second death of A and B will occur between 2 and 3 years from now?

10.8 An insurance contract sold to (x) and (y) provides for a death benefit of 1 unit at the moment of the first death and 3 units at the moment of the second death. Annual premiums are payable while at least one of the two is living. The annual rate of premium payment reduces to three quarters of the initial rate upon the first death. You are given that (x) is subject to a constant force of mortality of 0.2, and (y) is subject to a constant force of mortality of 0.3. The force of interest is a constant 0.1. Find the initial annual rate of premium payment.

10.9 A contract on two lives (x) and (y) provides that if (y) dies first, (x) will receive a life annuity of 1 per year, starting at the end of the year of (y)'s death, while if (x) dies first, (y) will receive a life annuity of 2 per year, starting at the end of the year of (x)'s death. Level annual premiums of π are payable while both (x) and (y) are alive. Given that $\ddot{a}_x = 16, \ddot{a}_y = 13$ and $\ddot{a}_{xy} = 10$, find π.

10.10 You are given two independent lives (x) and (y) for which (x) is subject to a constant force of mortality of log(4/3), while (y) is subject to a force of mortality at time t of $1/(10-t)$ for $0 \leq t < 10$.

(a) What is the probability that the first death among these two will occur between time 2 and time 3?

(b) What is the probability that the second death among these two will occur between time 2 and time 3?

10.11 You are given that $q_{70} = 0.2,\ q_{71} = 0.25,\ q_{72} = 0.4, q_{73} = 0.6$. The interest rate is a constant 25%. The vector $\mathbf{b}$ is (1000, 2000, 3000). Compute (a) $A^1_{70:71}(\mathbf{b})$, (b) $A^1_{71:70}(\mathbf{b})$, (c) $A^2_{70:71}(\mathbf{b})$, (d) $A^2_{71:70}(\mathbf{b})$.

Type B exercises

10.12 An annuity contract issued to two lives, (40) and (50), provides for payments of 1 per year, to be made provided that either or both of the following conditions hold.

(i) (40) is alive and under age 60;

(ii) (50) is alive and over age 60.

Find a formula for the present value of the benefits using terms of the form $\ddot{a}_x(\mathbf{c})$ or $\ddot{a}_{xy}(\mathbf{c})$.

10.13 Repeat Exercise 10.12, but now with (ii) stating that (50) is alive and between age 60 and age 80.

10.14 An insurance policy sold to (x) and (y) provides for a payment of 2 units on the first death and 3 units on the second death. Level annual premiums are payable for as along as either individual is living, with the premiums reducing by one-third upon the first death. Find an expression for the initial premium in terms of $\bar{A}_x, \bar{A}_y, \bar{A}_{xy}, \ddot{a}_x, \ddot{a}_y$, and $\ddot{a}_{xy}$.

10.15 An annuity sold to (x) and (y) provides for annual payments for 20 years, provided at least one of the two is alive. The annual payment begins at 12 units. During the first

10 years, the payments remain at 12. However, during the second 10-year period, the payment reduces to 6 if (x) only is alive, or to 8 if (y) only is alive. Write the present value in terms of single- and joint-life annuities.

10.16 An annuity contract sold to (40) and (50) provides for yearly payments if either life is living and under age 70. The yearly payment is 6 if both are alive, 4 if (40) only is alive, and 5 if (50) only is alive. Find a formula for the present value of the benefits using terms of the form $\ddot{a}_x(\mathbf{c})$ and $\ddot{a}_{xy}(\mathbf{c})$.

10.17 For the policy of Exercise 10.8, calculate ${}_tV$.

10.18 Suppose that Demoivre's law holds. Show that (10.19) holds exactly for a general left subscript n in place of 1. That is,

$$ {}_nq^1_{xy} = {}_nq_x - \frac{1}{2}{}_nq_x\,{}_nq_y. $$

10.19 For the two lives of Exercise 10.10, find the probability that (x) will die before (y).

10.20 Given two lives (x) and (y) and times $s < t$, consider the following four events: A, the joint-status (xy) fails between time s and time t; B, the last-survivor status $(\overline{xy})$ fails between time s and time t; C, at least one of (x) and (y) fails between time s and time t; and D, both (x) and (y) fail between time s and time t. Rank the probabilities of these from largest to smallest to the extent that you are able, showing clearly those events that you cannot compare without further information.

10.21 Suppose that interest is a constant 25% and that $q_x = 0.2$. Assume UDD. Find the ratio of $\bar{A}_{xx}(1_1)$ to $(i/\delta)A_{xx}(1_1)$.

10.22 The probability ${}_nq^1_{xy}$ can be generalized to ${}_nq^1_{x_1:x_2:\ldots x_k}$, which is the probability that out of a group of k lives, $(x_1), (x_2), \ldots, (x_k)$, (x_1) will die within n years and be the first to die. Express the following in terms of first-death probabilities:

(a) the probability that (x) will die within n years and will be the *second* to die among a group of three lives (x), (y), and (z);

(b) the probability that (x) will die within n years and be the *third* to die among (x), (y), and (z).

10.23 Find a formula to calculate $\bar{A}^2_{yx}(\mathbf{b})$ in terms of

(a) $\bar{A}^1_{xy}(\mathbf{b}), \bar{A}_y(\mathbf{b}), \bar{A}_{xy}(\mathbf{b})$,

(b) $\bar{A}^2_{xy}(\mathbf{b})$, $\bar{A}_{\overline{xy}}(\mathbf{b})$.

10.24 Suppose that ${}_nq_x = 0.2$, and ${}_nq_y = 0.3$. What is ${}_nq^1_{xy} - {}_nq^2_{yx}$?

10.25 An insurance pays 1 at the death of (y) provided that (x) was alive n years before this death. Assume constant interest. Find a formula for the present value in two ways. First by setting up and evaluating the necessary integrals. Secondly, by using Example 10.6 to avoid any integration.

10.26 An annuity provides for continuous payments at the annual rate of 1, while both (x) and (y) are alive, and in addition continues the payments for n more years after the first death, as long as the survivor is living. Payments stop completely upon the second death. Find a formula for the present value. Verify that your answer gives the correct result in the case that $n = 0$.

10.27 In Example 10.9 take $\nu = \delta = 0$ and $\gamma < 0$.

(a) Evaluate the present value and give an intuitive explanation of the answer.

(b) Take the parameters as in part (a), but now consider the simpler type of reversionary annuity where the rate of payment at time r (measured as usual from 0) is $e^{\gamma r}$. Show that the present value is an increasing function of μ and explain why this is so.

10.28 In this problem all death benefits are payable at the moment of death, and all premiums are payable continuously at a level rate. A life insurance contract on (x) and (y) provides for death benefits as follows. If (x) dies first at time t, the amount paid is the reserve at time t on a 1-unit whole life insurance on (y) with premiums payable for life. If (y) dies first at time t, the amount paid is the reserve at time t for a 1-unit whole life insurance on (x) with premiums payable for life. Premiums are payable until the first death. Show that the rate of premium payment is $\bar{P}_x + \bar{P}_y - \bar{P}_{xy}$, where $\bar{P}_x$ denotes the rate of premium payment for a 1-unit whole life insurance on x with premiums payable for life, and $\bar{P}_y$ and $\bar{P}_{xy}$ denote similar quantities for y and xy, respectively. You should demonstrate this mathematically and also give a verbal explanation.

10.29 A contract on (x) and (y) provides the following death benefits, payable at the moment of death. If (x) dies first, 10 is paid upon the death of (x) and 30 is paid upon the death of (y). If (y) dies first, 20 is paid upon the death of (x) and 40 is paid upon the death of (y). Find the present value given that

$$\bar{A}_x = 0.30, \qquad \bar{A}_y = 0.45, \qquad \bar{A}_{xy} = 050, \qquad \bar{A}^{\,1}_{xy} = 0.20.$$

10.30 An annuity provides continuous payments at the annual rate of 1 to (y) while living, beginning n years after the death of (x). Assuming constant interest, find a formula for the present value.

Spreadsheet exercise

10.31 Suppose that our sample table applies to females and the mortality rate for males is 1.25 times that for females up to age 119. A is a male age 50 and B is a female age 40. The interest rate is a constant 5%.

(a) Find the level annual premium for an insurance paying 10 000 at the end of the year of first death of A and B with premiums payable while both are alive.

(b) Find the level annual premium for an insurance paying 10 000 at the end of the year of the death of A provided this occurs before the death of B. Premiums are payable while both are alive.

(c) Assuming UDD, find the probability that A will die before B.

11

Multiple-decrement theory

11.1 Introduction

Our discussion of insurance contracts up to now has been concerned with benefits payable upon the occurrence of death. We now investigate situations when an insured is at the same time subject to several different events that can have financial impact.

One example, which we mentioned earlier, is the event of *withdrawal* or *lapse*, whereby an insured life terminates the contract and receives a cash value. To properly model a life insurance contract, one must consider two causes of termination, death and withdrawal, operating simultaneously.

Sometimes the insurer must distinguish between different causes of death. For example, some policies have a feature that provides additional death benefits for accidental death as opposed to death from natural causes.

Some policies include disability benefits, providing income for people who can no longer work. For such, the insurer must consider the possibility of disability, as well as death and withdrawal.

For employees covered under a pension plan, there are at least four events of interest: disability; death; termination of employment; and retirement.

These are just a few of the possibilities, and many more can arise. In this chapter we analyze the general situation.

11.2 The basic model

We suppose that we have m different causes of *failure* operating simultaneously on a group of lives. These causes are often referred to as *decrements*, since they bring about a decrease in the number of lives under observation. The insurance policies introduced in Chapter 5 involved only one decrement, namely that of death. When there are several decrements we are concerned with what is called in the actuarial literature as *multiple-decrement theory*, or, in biostatistical contexts, the theory of *competing risks*.

Fundamentals of Actuarial Mathematics, Third Edition. S. David Promislow.

Companion website: http://www.wiley.com/go/actuarial

There is a generalization of these ideas to multi-state insurances and annuities, where individuals can transfer freely between one of several states. We deal with this in Chapter 19, which requires a basic knowledge of Markov chains, as given in Chapter 18.

11.2.1 The multiple-decrement table

It is convenient to make use of a generalized life table, known as a *multiple-decrement table*. We will number our causes from 1 to m and use a superscript (j) to refer to cause j. We will also use a superscript (τ), which the reader can interpret as meaning *total*. We begin with an arbitrary number $\ell_0^{(\tau)}$ of lives age 0. We let $\ell_x^{(\tau)}$ denote the number of individuals from this group who are still surviving at age x. That is, they have not succumbed to any of the m causes. We let $d_x^{(j)}$ denote the number of lives who will fail *first* from cause (j) between the ages of x and $x+1$. Let

$$d_x^{(\tau)} = \sum_{j=1}^{m} d_x^{(j)}, \tag{11.1}$$

which is the number of people who will fail from *some* cause between the ages of x and $x+1$. It follows that

$$\ell_{x+1}^{(\tau)} = \ell_x^{(\tau)} - d_x^{(\tau)}. \tag{11.2}$$

That is, the number of survivors at age $x+1$ is equal to the number of survivors at age x, less those who failed from some cause between the ages of x and $x+1$. The following is a portion of a sample table with two decrements:

x	$\ell_x^{(\tau)}$	$d_x^{(1)}$	$d_x^{(2)}$
0	1000	50	100
1	850	60	105
2	685	70	120

It is important to note the word 'first' used in the definition of $d_x^{(j)}$. The model assumes that any cause of failure results in the individual leaving the group, so they are no longer under observation. For example, if cause 1 denotes death and cause 2 withdrawal, a policyholder who withdraws at age $60\frac{1}{2}$ and then dies at age $60\frac{3}{4}$ would be included in $d_{60}^{(2)}$, but not in $d_{60}^{(1)}$. This is what we want in the common applications. In the case of withdrawal, the insurer would pay the policyholder their cash value, the policy would terminate, and the insurer would have no further interest in the time of death of this person. Therefore, whenever we refer to failing from a certain cause j in the multiple-decrement model, it is understood that this means that this occurs before failure from any other cause.

We defined a multiple-decrement table starting at age 0, but this could be some other age, depending upon the particular application. For example, the multiple-decrement table for an employee pension plan would begin at the first age the employees become eligible for the plan, perhaps age 25 or so.

Our model assumes that the multiple-decrement table, as given, can be used for all ages x. That is, it is an aggregate table that does not show selection effects. We could define for each x, as we did for the single-decrement case, a *select* multiple-decrement table to apply to people first observed at age x. This is important in practice since many causes of decrement, such as withdrawal, will certainly depend more on duration since the policy began than on attained age. However, we will not get into these details in this chapter and will work with the aggregate model. The more general case can be deduced from the material of Chapter 17.

11.2.2 Quantities calculated from the multiple-decrement table

We first define probabilities of failure. Let

$$q_x^{(j)} = \frac{d_x^{(j)}}{\ell_x^{(\tau)}}$$

be the probability that (x) will fail first from cause j within 1 year. In practice, one would start with these probabilities and construct the multiple-decrement table inductively, by calculating

$$d_x^{(j)} = \ell_x^{(\tau)} q_x^{(j)}$$

and then using (11.1) and (11.2) to complete the table. Let

$$_k p_x^{(\tau)} = \frac{\ell_{x+k}^{(\tau)}}{\ell_x^{(\tau)}} \tag{11.3}$$

be the probability that (x) will survive to age $x + k$ without succumbing to any cause.

Probabilities are calculated from the table with similar formulas as in the single-cause life table. One must only remember that the denominators are $\ell_x^{(\tau)}$. For example,

$$_n q_x^{(j)} = \frac{\sum_{k=0}^{n-1} d_{x+k}^{(j)}}{\ell_x^{(\tau)}}$$

is the probability that (x) will fail as a result of cause j within n years. Similarly,

$$\frac{d_{x+k}^{(j)}}{\ell_x^{(\tau)}} = {_k p_x^{(\tau)}} q_{x+k}^{(j)} \tag{11.4}$$

is the probability that (x) will fail from cause j in the time interval k to $k + 1$.

Another quantity of interest is

$$\ell_x^{(j)} = \sum_{k=0}^{\omega-x-1} d_{x+k}^{(j)}.$$

This represents the number of people in our group who will fail from cause j sometime after age x. We assume in our model that everyone will eventually fail from some cause (which is obvious if death is one of the causes) so we can write

$$\ell_x^{(\tau)} = \sum_{j=1}^{m} \ell_x^{(j)}.$$

Knowing the value of $\ell_x^{(j)}$ for all integral values of x and all j allows us to complete the table since

$$d_x^{(j)} = \ell_x^{(j)} - \ell_{x+1}^{(j)}.$$

Note that we can write

$$_nq_x^{(j)} = \frac{\ell_x^{(j)} - \ell_{x+n}^{(j)}}{\ell_x^{(\tau)}}. \tag{11.5}$$

Another symbol we can define is

$$_np_x^{(j)} = \frac{\ell_{x+n}^{(j)}}{\ell_x^{(\tau)}} \tag{11.6}$$

which is the probability that (x) will fail first from cause j after time n. Here, instead of thinking of the basic symbol $_np$ as denoting survival to time n, we think of the equivalent formulation as failing after time n.

11.3 Insurances

Consider an insurance policy on (x) that provides, for all relevant values k and j, a payment of $b_k^{(j)}$ at time $k+1$, provided that failure occurs from cause j in the year k to $k+1$. We can view this as m separate policies, where the jth policy pays benefits only for failure from cause j. The formulas are exactly the same as (5.1), but with the probabilities taken from (11.4). That is, the present value of the jth policy will be

$$\sum_{k=0}^{\infty} b_k^{(j)} v(k+1) \frac{d_{x+k}^{(j)}}{\ell_x^{(\tau)}} = \sum_{k=0}^{\infty} b_k^{(j)} v(k+1)\,{}_kp_x^{(\tau)} q_{x+k}^{(j)}, \tag{11.7‡}$$

and the total present value would be obtained by summing over all j.

To handle insurances payable at the moment of failure we will need the following:

Definition 11.1 For each cause j, let

$$\mu^{(j)}(x) = \lim_{h\to 0} \frac{{}_hq_x^{(j)}}{h} = \lim_{h\to 0} \frac{\ell_x^{(j)} - \ell_{x+h}^{(j)}}{h\ell_x^{(\tau)}} \tag{11.8}$$

The *force of decrement at time t*, from cause (j) for (x) is the quantity given by

$$\mu_x^{(j)}(t) = \mu^{(j)}(x+t). \tag{11.9}$$

(In the case of selected multiple-decrement tables we would need to define $\mu_x^{(j)}(t)$ separately for each age x.)

Analogously to (8.15) and noting that the quantities in (11.5) and (11.6) differ by a constant we can write

$$\mu_x^{(j)}(t) = \frac{\frac{-d}{dt}\,{}_tp_x^{(j)}}{{}_tp_x^{(\tau)}} = \frac{\frac{d}{dt}\,{}_tq_x^{(j)}}{{}_tp_x^{(\tau)}}. \tag{11.10}$$

Consider the continuous analogue of the insurance whose present value is given by (11.7). This is a contract that pays a death benefit of $b^{(j)}(t)$ at time t, should failure from cause j occur at that time. We denote the present value by $\bar{A}_x^{(j)}(\mathbf{b}^{(j)})$. It is given by

$$\bar{A}_x^{(j)}(\mathbf{b}^{(j)}) = \int_0^\infty b^{(j)}(t)v(t){}_tp_x^{(\tau)}\mu_x^{(j)}(t)\mathrm{d}t. \tag{11.11}‡$$

This reflects the fact that in order to collect at time t, (x) must survive *all* causes up to time t, and then succumb to cause j at time t. We discuss the evaluation of this integral in Section 11.5.

For policies which are purchased by annual premiums, payable until the first failure, we determine the initial premium as in the single decrement case, by simply dividing the present value by

$$\ddot{a}_x^{(\tau)}(\rho) = \sum_{k=0}^{\infty} v(k)\rho_k\,{}_kp_x^{(\tau)},$$

where ρ is the premium pattern vector.

11.4 Determining the model from the forces of decrement

The multiple-decrement model is often given from the outset by specifying the forces of decrement, and we must use these to calculate probabilities. Note that in the last expression in (11.8) the numerator involves $\ell^{(j)}$ while the denominator involves $\ell^{(\tau)}$. This means that we cannot apply a version of (8.16) directly. We can, however, define the total force

$$\mu_x^{(\tau)}(t) = \sum_{j=1}^{m} \mu_x^{(j)}(t) = \lim_{h\to 0} \frac{\ell_{x+t}^{(\tau)} - \ell_{x+t+h}^{(\tau)}}{h\ell_{x+t}^{(\tau)}}.$$

So $\mu_x^{(\tau)}$ is the same type of quantity as the single-life force of mortality introduced in Chapter 8, except based on the total decrement rather than just on failure by death. We can now apply (8.16) to deduce that

$${}_tp_x^{(\tau)} = \mathrm{e}^{-\int_0^t \mu_x^{(\tau)}(r)\mathrm{d}r}. \tag{11.12}‡$$

Arguing as we did in deriving (10.28), apply (11.11) with a constant interest rate of 0, and $b^{(j)}(t) = 1$ for $0 \le t < s$, and 0 elsewhere, to conclude that

$$ {}_sq_x^{(j)} = \int_0^s {}_tp_x^{(\tau)} \mu_x^{(j)}(t)\,\mathrm{d}t. \qquad (11.13)\ddagger $$

Example 11.1 In a model with two decrements, you are given that $\mu_x^{(1)}(t) = 0.02$ and $\mu_x^{(2)}(t) = 0.03$ for all $t \ge 0$. Find ${}_3q_x^{(1)}$.

Solution. $\mu_x^{(\tau)} = 0.05$, so that ${}_tp_x^{(\tau)} = \mathrm{e}^{-0.05t}$ and, from (11.13),

$$ {}_3q_x^{(1)} = \frac{2}{5}(1 - \mathrm{e}^{-0.15}). $$

11.5 The analogy with joint-life statuses

We have already encountered a multiple-decrement model in Chapter 10. The difference is that, rather than single lives, the basic objects were joint-life statuses consisting of a pair of lives (x) and (y). These pairs were subject to two distinct causes of failure, namely the death of (x) and the death of (y). The notation differs somewhat, but the reader should notice that, with this point of view, the contingent insurance present value given in (10.24) is a special case of that given in (11.11), and the probability in (10.28) is a special case of that in (11.13).

It follows that if we assume that the causes of decrement are acting independently, we can evaluate the integral (11.11) from the multiple-decrement table as in Chapter 10. That is, if the benefits are constant over each year, and the particular decrement can be assumed to be uniformly distributed over each year, the i/δ correction will result in a reasonable approximation.

11.6 A machine analogy

The difficult part of multiple-decrement theory deals with relationships between the different decrements. In order to better understand and motivate the ideas, we first look at what appears to be a different situation, but which we will eventually relate to our model above. To facilitate this, we will use notation for this new example that parallels that which we have already introduced.

Suppose we have a machine with two components, part 1 and part 2, which work completely independently of each other so that the condition of one part does not affect the operation of the other. In order for the machine to work, both parts must be working, so if either part fails, the machine will fail, even though the other part may be in perfect order. We assume also that both parts cannot fail simultaneously, so we can always identify which part caused the machine to fail. Suppose we want to compute probabilities of failure over some time period, say a year. We have four quantities of interest. For $j = 1, 2$, let $q'^{(j)}$ be the probability that part j will fail, and let $q^{(j)}$ be the probability that the machine will fail due to the failure of part j, meaning that part j failed during the year and was the first of the two parts to fail. What are the relationships between these? Since the failure of the machine due to the failure of part j obviously implies that part j failed, it is clear from elementary probability theory that

$$q'^{(j)} \geq q^{(j)}. \tag{11.14}$$

It is is also clear that, in general, the two quantities are not equal and we would expect the left hand side to be strictly greater than the right. Suppose, for example, that sometime during the year part 1 fails, causing the machine to fail, and sometime after that but before the end of the year, part 2 fails. Then the event of the failure of part 2 would have occurred, but *not* the event of part 2 causing the failure of the machine.

After introducing some additional notation, we will state some further relationships. Let $p'^{(j)} = 1 - q'^{(j)}$ denote the probability that part j will be working at the end of the year, $q^{(\tau)}$ denote the probability that the machine will fail within the year, and $p^{(\tau)} = 1 - q^{(\tau)}$ denote the probability that the machine is working at the end of the year. The machine can fail in one of two mutually exclusive ways, namely, failure of part 1 or failure of part 2. By elementary probability theory we have

$$q^{(\tau)} = q^{(1)} + q^{(2)}.$$

On the other hand, in order that the machine be working at the end of the period, both parts must be working. Since the parts work independently, we can multiply probabilities to calculate this (as we illustrated in Chapter 10), which gives

$$p^{(\tau)} = p'^{(1)}p'^{(2)}.$$

From the above two equations we get the fundamental identity

$$q^{(1)} + q^{(2)} = 1 - p'^{(1)}p'^{(2)}, \tag{11.15}$$

which is often written in the form

$$q^{(1)} + q^{(2)} = q'^{(1)} + q'^{(2)} - q'^{(1)}q'^{(2)}. \tag{11.16}$$

The basic problem of interest is as follows. If we are given the unprimed symbols, can we calculate the primed ones, or if we are given the primed symbols can we calculate the unprimed ones? In general, we cannot do this uniquely and we will need additional information in order to definitely deduce one set of probabilities from the other. We will, however, present some possible solutions to this problem, and then discuss some conditions that will lead to these solutions.

11.6.1 Method 1

Given $q^{(j)}$, $j = 1, 2$, we sum to get $q^{(\tau)}$, then calculate $p^{(\tau)}$, and then let

$$p'^{(j)} = p^{(\tau)\left(\frac{q^{(j)}}{q^{(\tau)}}\right)}, \tag{11.17}$$

which will clearly satisfy (11.15). It is not too difficult to remember this formula. We have to factor the quantity $p^{(\tau)}$ into two factors, and it is natural to take the factors as $p^{(\tau)}$ to some exponent where the exponents sum up to one. Two numbers that obviously sum to 1 are $q^{(j)}/q^{(\tau)}$, $j = 1, 2$.

For the other direction, given $q'^{(j)}$, $j = 1, 2$, we calculate $p'^{(j)}$, multiply to get $p^{(\tau)}$, then calculate $q^{(\tau)}$, and finally take

$$q^{(j)} = \frac{\log p'^{(j)}}{\log p^{(\tau)}} q^{(\tau)}. \tag{11.18}$$

For this formula we have to split up $q^{(\tau)}$ as the sum of two terms. We obtain the terms by multiplying by weights that add up to 1. Since we have a multiplicative relationship, we take logs to accomplish this.

A straightforward calculation verifies that this approach is consistent. That is, if we are given $q^{(j)}$, $j = 1, 2$, and use (11.17) to calculate $q'^{(j)}$, $j = 1, 2$, and then apply (11.18) to these numbers, we will end up with the same values of $q^{(j)}$ as we started with.

To see that (11.14) is satisfied we use the inequality

$$(1 - x)^{\alpha} \leq (1 - \alpha x), \quad \text{for } 0 \leq x \leq 1, 0 \leq \alpha \leq 1,$$

which can be demonstrated by looking at the the Taylor polynomial of degree 2 for the left hand side. From (11.17), taking $\alpha = q^{(j)}/q^{(\tau)}$,

$$p'^{(j)} = (1 - q^{\tau})^{\alpha} \leq 1 - \alpha q^{(\tau)} = 1 - q^{(j)},$$

and (11.14) follows.

The entire discussion above can be generalized to the case of m independent parts, where the failure of any one part will cause the machine to fail. Use notation as above, except that j now takes values $1, 2, \ldots, m$. Everything is the same with the obvious changes to handle m probabilities instead of two. For example, (11.15) will now read

$$\sum_{j=1}^{m} q^{(j)} = 1 - \prod_{j=1}^{m} p'^{(j)}, \tag{11.19}$$

and the solutions given by (11.17) and (11.18) will satisfy this relationship.

11.6.2 Method 2

We present another method to deduce the unprimed probabilities from the primed, which in certain cases may be more realistic than formula (11.18). However, for $m > 2$ it is not easy to invert, in order to deduce unprimed from primed. To motivate this method, we look again at the case of the joint-life statuses of Chapter 10. In fact we can directly visualize the machine point of view in this situation. The 'machine' consists of the joint-life status (xy) with the two distinct components (x) and (y). For our particular problem, suppose we are given $q'^{(1)}$ and $q'^{(2)}$, and want to compute the probability that the failure of part 1 caused the machine to fail.

Reasoning as we did in deriving (10.19), we approximate this by computing the probability that part 1 failed during the year and that at the middle of the year part 2 was still working. Making an assumption of a *uniform distribution of failures for each part*, over each year, which is analogous to our UDD assumption, we have, as in (10.19),

$$q^{(1)} = q'^{(1)} - \frac{1}{2}q'^{(1)}q'^{(2)}, \qquad q^{(2)} = q'^{(2)} - \frac{1}{2}q'^{(1)}q'^{(2)}, \tag{11.20}$$

which clearly satisfy (11.14) and (11.16).

To calculate the inverse formula, let $\Delta = q^{(1)} - q^{(2)}$. From (11.20)

$$q^{(1)} = q'^{(1)} - \frac{1}{2}\left(q'^{(1)}\right)^2 + \frac{1}{2}q'^{(1)}\Delta$$

We have a quadratic equation in $q'^{(1)}$, which can be solved by the usual formula to yield

$$q'^{(1)} = \frac{(2+\Delta) - \sqrt{(2+\Delta)^2 - 8q^{(1)}}}{2}, \tag{11.21}$$

and directly from (11.20)

$$q'^{(2)} = q'^{(1)} - \Delta.$$

Example 11.2 Given $q^{(1)} = 0.27, \quad q^{(2)} = 0.17$, find $q'^{(1)}$ and $q'^{(2)}$ by Method 2.

Solution.

$$q'^{(1)} = \frac{2.1 - \sqrt{(2.1)^2 - 2.16}}{2} = 0.3, \quad q'^{(2)} = 0.3 - 0.1 = 0.2.$$

As a check we can take these answers and apply (11.20) to get back to where we started.

$$q^{((1))} = 0.3 - (1/2).06 = 0.27, \quad q^{((2))} = 0.17.$$

Example 11.3 Given $q'^{(1)} = 0.200, q'^{(2)} = 0.488$, estimate $q^{(1)}$ and $q^{(2)}$ by both methods.

Solution. By Method 1: Since $p'^{(1)} = 0.8$ and $p'^{(2)} = 0.512$, we have $q^{(\tau)} = 0.5904$ and, from (11.18),

$$q^{(1)} = \frac{1}{4} \times 0.5904 = 0.1476, \quad q^{(2)} = \frac{3}{4} \times 0.5904 = 0.4428.$$

Solution by Method 2: From (11.20),

$$q^{(1)} = 0.2000 - 0.0488 = 0.1512, \qquad q^{(2)} = 0.4880 - 0.0488 = 0.4392.$$

As with Method 1, Method 2 can also be extended to the case of m components to deduce the unprimed from the primed, as we show later in Section 11.7.2. For example, with $m = 3$ we get

$$q^{(1)} = q'^{(1)} - \frac{1}{2}\left[q'^{(1)}q'^{(2)} + q'^{(1)}q'^{(3)}\right] + \frac{1}{3}q'^{(1)}q'^{(2)}q'^{(3)}. \tag{11.22}$$

Note that the coefficients of 1/2 and 1/3 are what is needed in order to satisfy (11.19).

With $m > 2$ there is no convenient direct method for the inverse process, and it would have to be done by numerical approximation.

11.7 Associated single-decrement tables

11.7.1 The main methods

We now return to the original setting dealing with a group of lives, to which we will apply our 'machine' model. For a definite example start with a multiple-decrement table with two decrements; cause 1 is death, and cause 2 is disability. Suppose we wish to use this table to construct a regular single-life table relating only to death, as introduced in Chapter 3. Would we be justified in simply taking $q_x^{(1)}$ as the mortality rate for age (x), in the single-life table? The answer is a definite 'no'. We stress again that $q_x^{(1)}$ is not the probability that (x) will die within the year, but rather the probability that (x) will die within the year *before* becoming disabled. The actual value of q_x should be larger than $q_x^{(1)}$. The situation is the same as the machine model in the previous section. A person could die during the year, after having already left the group by reason of disability. In other words, the mortality rate for age (x) in the single decrement table is analogous to the primed rate in the machine model, and will be denoted by $q_x'^{(1)}$.

We may also want to compute $q_x'^{(2)}$, the probability that (x) will become disabled during the year, assuming that no other causes of failure are operating. This admittedly requires a stretch of the imagination, as the possibility of death would always appear to be present. We have to imagine, however, that we are computing what these probabilities would be if we could somehow eliminate the possibility of death. The best approach is simply to view things in terms of the machine model. That is, we have to think of a person as having a 'disability component' that exists independently and could in fact continue to operate even after death.

For each cause j, the collection of values $\{q_x'^{(j)}\}$ for various values of x is known as the *associated single-decrement table for cause j*. As we have stressed above, these rates give probabilities of failure for the particular cause j, assuming that *no other* causes of decrement are operating.

The problem often arises of going from one set of rates to another. We might, for example, construct the multiple-decrement table in the first place by using our knowledge of the associated single-decrement tables. Alternatively, we might have first constructed a multiple table by actually observing the effect of the various causes acting together and wish to use that to construct the associated single-decrement tables. In order to do so we can directly apply our machine model and use the two given methods, provided we make the key assumption of that model, namely that the various causes are acting *independently*. One can certainly argue in many applications that this does not hold, but the standard actuarial model incorporates this independence assumption. The more general theory is discussed in Chapter 17.

Consider a numerical example. Suppose that in our death–disability table

$$\ell_x^{(\tau)} = 1000, \qquad d_x^{(1)} = 300, \qquad d_x^{(2)} = 100.$$

We will compute the associated single-decrement rates using method 1. From formula (11.17), we have

$$q_x'^{(1)} = 1 - 0.6^{3/4} = 0.318,$$
$$q_x'^{(2)} = 1 - 0.6^{1/4} = 0.120.$$

Suppose we are given the above single-decrement rates and want to compute the multiple-decrement table. If we use the Method 1 formula (11.18) we necessarily will get back to unprimed rates of 0.300 and 0.100 that we started with, as the reader should verify. If we use the Method 2 formula (11.20) instead, we will get

$$q_x^{(1)} = 0.318(1 - 0.06) = 0.299,$$
$$q_x^{(2)} = 0.120(1 - 0.159) = 0.101,$$

verifying as we saw in Example 11.3 that the two methods are different, but will generally give answers that are close.

11.7.2 Forces of decrement in the associated single-decrement tables

Refer back to the statement made in Section 11.4 that the multiple-decrement model is often constructed from the forces of decrement. A question that naturally arises is, how are these forces of decrement obtained? The answer is that they can be taken as the forces of decrement in the associated single-decrement tables, namely

$$\mu_x'^{(j)}(t) = -\frac{\frac{d}{dt}\,{}_tp_x'^{(j)}}{{}_tp_x'^{(j)}}.$$

Note that $\mu_x'^{(j)}(t)$ is defined differently than the quantity $\mu_x^{(j)}(t)$ as given in (11.9) but there is reason to believe that they might be the same. We observed above that we expect that $q_x'^{(j)} > q_x^{(j)}$ since the person may fail within the year from cause *j* after first failing from another cause, but this argument does not hold when we are speaking of *instantaneous* rates of failure rather than failure over a period of positive length. This by itself is not sufficient for equality – see Exercise 11.15 – and we need the independence hypothesis. In Section 17.4.5, we give a formal proof and provide an explanation of the fact that by virtue of independence,

$$\mu_x'^{(j)}(t) = \mu_x^{(j)}(t). \tag{11.23}$$

We can now summarize the main conclusions which follow from our independence hypothesis.

- Given the forces of decrement in the associated single-decrement tables, we can uniquely determine the multiple-decrement model, using (11.23) together with the procedure of Section 11.4.

- Given only the rates of decrement in the associated single-decrement tables, we cannot uniquely determine the multiple-decrement model, and must choose from different methods, the most notable being the two described in Section 11.7.1.

11.7.3 Conditions justifying the two methods

In this section we give conditions to justify the Method 1 and Method 2 introduced above, and also some additional procedures that apply in special cases. We start with the easier case.

Method 2 This will hold provided we have a *uniform distribution of decrements in each of the associated single-decrement tables*. That is, for an integer x, $0 < t < 1$, and $j = 1, 2, \ldots, m$,

$$ {}_t q_x^{\prime(j)} = t q_x^{\prime(j)}. \tag{11.24}$$

To show this, we first note that as in (8.16)

$$ {}_t p_x^{\prime(j)} = e^{-\int_0^t \mu_x^{\prime(j)}(r)dr}. \tag{11.25}$$

and then, invoking (11.23) and applying (8.20)

$$ {}_t p_x^{\prime(j)} \mu_x^{(j)}(t) = {}_t\, p_x^{\prime(j)} \mu_x^{\prime(j)}(t) = q_x^{\prime(j)}, \tag{11.26}$$

From (11.13)

$$ {}_s q_x^{(1)} = \int_0^1 {}_t p_x^{\prime(1)} (1 - {}_t\, q_x^{\prime(2)}) \mu_x^{(1)}(t) \mathrm{d}t. $$

Now substitute from (11.24) and (11.26) and integrate to obtain the Method 2 formula. The same procedure works for the general case of m decrements. We just need to include additional factors of $(1 - tq_x^{\prime(j)})$ in the integrand.

Method 1 We need a preliminary definition. We say that the *Uniform Ratio Hypothesis* (UR) holds at an integer (x) if there are constants K_j for $j = 1, 2, \ldots m$ such that for $0 < t < 1$,

$$ \frac{\mu_x^{(j)}(t)}{\mu_x^{(\tau)}(t)} = K_j. $$

As an example in the two decrement cases, let $\mu_x^{(1)}(t) = 2\mu_x^{(2)}(t)$ for $0 < t < 1$. Then UR holds at x with $K_1 = 2/3, K_2 = 1/3$.

Suppose now that UR holds at x. From (11.13) with $s = 1$,

$$ q_x^{(j)} = K_j \int_0^1 {}_t p_x^{(\tau)} \mu_x^{(\tau)} \mathrm{d}t = K_j q_x^{(\tau)}. \tag{11.27}$$

Finally, (11.23), and (11.27) show that

$$p_x'^{(j)} = e^{-\int_0^1 \mu_x'^{(j)}(r)\mathrm{d}r} = e^{-\int_0^1 \mu_x^{(j)}(r)\mathrm{d}r} = e^{-K_j\int_0^1 \mu_x^{(\tau)}(r)\mathrm{d}r} = p_x^{(\tau)K_j} = p_x^{(\tau)q_x^{(j)}/q_x^{(\tau)}},$$

which is the Method 1 formula for going from $q_x^{(j)}$ to $q_x'^{(j)}$.

There are certain assumptions that will imply UR. One obvious condition is that the forces of failure for each decrement are constant over the year from age x to age $x+1$. That is, for $0 < t < 1$ and $j = 1, 2, \ldots, m$, $\mu_x{}^{(j)}(t)$ is independent of t.

UR also holds at all ages under the assumption of a uniform distribution of deaths in the *multiple-decrement table.* That is, for an integer (x), $0 < t < 1$ and $j = 1, 2, \ldots, m$,

$$_tq_x^{(j)} = tq_x^{(j)}. \tag{11.28}$$

If (11.28) holds, we differentiate (11.13) with respect to s to conclude that for $0 < s < 1$,

$$q_x^{(j)} = {}_sp_x^{(\tau)}\mu_x^{(j)}(s).$$

Adding this equality for all j

$$q_x^{(\tau)} = {}_sp_x^{(\tau)}\mu_x^{(\tau)}(s).$$

Dividing the first equation above by the second shows that UR holds at x.

It should be noted that condition (11.28) is a somewhat unnatural assumption which differs from (11.24), and is therefore inconsistent with the usual assumptions of UDD for each single decrement. To see this, suppose that we assume (11.28) in a situation with two decrements where $q_x'^{(j)} = 0.1$, for $j = 1, 2$. Then Method 1 applies, so from (11.18), $q_x^{(j)} = 0.095$. Our assumption then gives $_{1/2}q_x^{(j)} = 0.0475$. Now we can also apply Method 1 to rates over half-year periods. From (11.17) we have $_{1/2}q_x'^{(j)} = 1 - 0.95^{1/2} = 0.0487$. However, the assumption of UDD for each decrement would give $_{1/2}q_x'^{(j)} = 0.05$.

Special Cases There are decrements that by their nature will not occur uniformly over the year, but rather at definite discrete points. A prime example is withdrawal from an insurance policy. Nobody is likely to withdraw in a middle of a premium paying period so that these can be assumed to occur only on premium due dates. If we assume that all decrements either follow such a discrete pattern or are uniformly distributed over each year, there is a fairly simple procedure to move from the primed to unprimed rates. This is illustrated by the following two examples.

Example 11.4 In a double-decrement model with decrements of (d) for death and (w) for withdrawal, we are given that deaths are uniformly distributed over each year of age in the single-decrement table, while one-third of the withdrawals in any year take place in the middle of the year and two-thirds occur at the end of the year. Given that $q_x'^{(1)} = 0.20, q_x'^{(2)} = 0.36$ calculate $q_x^{(1)}$ and $q_x^{(2)}$.

Solution. To calculate $q_x^{(2)}$ we add the probability of withdrawal before death in the middle of the year to the probability of withdrawal before death at the end of year. The probability of a withdrawal in the middle of the year is (1/3)(0.36) = 0.12. In order that this occurs before death, we require that death has not yet occurred by the middle of the year, which by the uniform distribution assumption has a probability of $1 - (1/2)(0.20) = 0.9$. Arguing similarly for the end of the year we have that

$$q^{(2)} = 0.12(0.9) + 0.24(0.8) = 0.300.$$

Since $q_x^{(\tau)} = 1 - (0.8)(0.64) = 0.488$, we know that $q_x^{(1)} = 0.188$.

As a check we can calculate the latter figure directly. The probability of death before withdrawal in the first half of the year is just the probability of death in the first half of the year, since there are no withdrawals during that time. This is equal to 0.10. The probability of death before withdrawal in the second half of the year is the probability of death in the second half, multiplied by the probability that withdrawal did *not* take place in the middle of the year. This equals $(0.1)(1 - 0.12) = 0.088$. The sum of the two probabilities is indeed equal to 0.188.

Note that in both calculations above, we used the independence assumption in order to multiply relevant probabilities.

Example 11.5 In a multiple-decrement model with three decrements, failures from causes 1 and 2 both occur uniformly over each year in the single-decrement table. For cause 3, 60% of the failures in any year occur 1/4 of the way through the year and the other 40% occur 3/4 of the way through the year. Find formulas that give the unprimed rates in terms of the primed.

Solution. To simply the notation, fix any age x and let a', b', c' denote $q_x'^{(j)}$ for $j = 1, 2, 3$, respectively, and let a, b, c denote $q_x^{(j)}$ for $i = 1, 2, 3$, respectively. Proceed as in Example 11.4 only now we need that survival from both decrements 1 and 2 occurred at the two points in question. This gives

$$\begin{aligned} c &= 0.6c'\left(1 - \tfrac{1}{4}a'\right)\left(1 - \tfrac{1}{4}b'\right) + 0.4c'\left(1 - \tfrac{3}{4}a'\right)\left(1 - \tfrac{3}{4}b'\right) \\ &= c' - \frac{9}{20}a'c' - \frac{9}{20}b'c' + \frac{21}{80}a'b'c'. \end{aligned}$$

Formula (11.30) says that

$$a + b + c = 1 - (1 - a')(1 - b')(1 - c').$$

Therefore we know by symmetry (since there is nothing to distinguish decrements 1 and 2) that

$$\begin{aligned} a &= a' - \frac{1}{2}a'b' - \frac{11}{20}a'c' + \frac{59}{160}a'b'c', \\ b &= b' - \frac{1}{2}b'a' - \frac{11}{20}b'c' + \frac{59}{160}a'b'c'. \end{aligned}$$

There is a much longer procedure for the first two decrements,which might be done as a check. To illustrate with decrement 1, we use the formula

$$a = \int_0^1 a'(1 - tb')_t p_x'^{(3)} \mathrm{d}t,$$

which follows from (11.16) after substituting from (11.26). Calculation of this will necessitate integrating over several intervals, since the formula for ${}_tp_x'^{(3)}$changes we get

$$a = a' \left[\int_0^{1/4} (1 - tb')\mathrm{d}t + \int_{1/4}^{3/4} (1 - tb')(1 - 0.6c')\mathrm{d}t + \int_{3/4}^{1} (1 - tb')(1 - c')\mathrm{d}t \right],$$

We leave it to the reader to verify that this gives the same answer as above.

11.7.4 Other approaches

We will describe two additional methods, which were popular in pre-computer days for calculating primed quantities from the unprimed. Neither of them can be easily inverted to give the unprimed in terms of the primed. They do satisfy (11.14) but their main drawback is that they do *not* satisfy the consistency relation (11.15) which must hold if we assume that the causes operate independently. On the other hand, they give approximate answers quickly and can give additional insights into the difference between the primed and unprimed rates.

Looking at the particular data given above in Section 11.7.1, we see that in order to compute $q_x'^{(1)}$ we want to estimate how many of those 100 people who became disabled during the year will die during the year after becoming disabled. Suppose we postulate that, on average, people leave from disability in the middle of the year. Since the chance of leaving by death during the year is 0.30, there should be approximately a chance of 0.15 of dying in the second half of the year. In other words, approximately 15 of the 100 disabilities would die during the year after becoming disabled and we could estimate $q'^{(1)} = 0.315$, not too far from the result obtained above by method 1.

Our final method gives a more plausible argument for decrements other than death, for as we already noted, it is difficult to imagine someone becoming disabled after they have died. The argument is as follows. The probability of death during the year should be obtained by taking the number of people who died during the year and dividing by the number of people who were in the observation period for that year. In the above example we observed 300 deaths. However, we did not really have a full 1000 people under observation for the entire year. Some of the group dropped out due to disability, and were no longer observable as potential deaths. Consider extreme cases. If all the people becoming disabled did so at the beginning of the year we would have only 900 people left. If all of them dropped out at the end of the year, then we would indeed have the full 1000 under observation. We can conclude that on average we would have the equivalent of 950 people to observe and we can estimate $q_x'^{(1)} = 300/950 = 0.316$. Similarly, we could estimate $q_x'^{(2)} = 100/850 = 0.118$. Again, we obtain answers close to those of method 1.

Both of these alternate methods can be applied to the general case of m decrements. To state the formulas it is convenient to define, for each cause j,

$$q_x^{(-j)} = \sum_{i \neq j} q_x^{(i)},$$

the probability that (x) will leave the group within the next year from some cause other than j. Our first argument above then gives the approximation

$$q_x'^{(j)} \doteq q_x^{(j)} \left(1 + \frac{1}{2} q_x^{(-j)}\right), \tag{11.29}$$

while the second argument leads to the approximation

$$q_x'^{(j)} \doteq \frac{q_x^{(j)}}{1 - \frac{1}{2} q_x^{(-j)}}. \tag{11.30}$$

These general formulas confirm that the two latter methods give different approximations, but the answers are usually close, since for a small value of a, $1 + a$ is close to $(1 - a)^{-1}$.

The reader may find it instructive to take the case of two decrements and substitute algebraically into the right hand side of (11.16) from both (11.29) and (11.30). In both cases one obtains a result that is somewhat less than the left hand side of (11.16), verifying the inconsistency. However, we see that the difference is likely to be small. It is a sum of terms each of which is close to the product of three or more $q^{(j)}$s.

Notes and references

Promislow (1991) discusses select multiple-decrement tables.

Exercises

Type A exercises

11.1 You are given the following portion of a double-decrement table. (Blanks indicate data that you must calculate from the given figures if you need them.)

x	$\ell_x^{(\tau)}$	$d_x^{(1)}$	$d_x^{(2)}$
50	–	100	300
51	700	50	–
52	470	40	–
53	320		

Find the probability that (50) will fail first from cause 2 between the ages of 51 and 53.

11.2 You are given the following portion of a double-decrement table. (Blanks are as in Exercise 11.1.)

x	$\ell_x^{(\tau)}$	$d_x^{(1)}$	$d_x^{(2)}$
50	1200	100	300
51	–	200	–
52	300		

(a) A 2-year insurance contract on (50) provides for benefits paid at the end of the year of failure if this occurs within 2 years. The benefit payable is 1 unit if failure is from cause 1, or 2 units if it is from cause 2. If the interest rate is a constant 50% per year, find the present value of the benefits.

(b) Find the associated single-decrement rates, $q_{51}^{\prime(1)}$ and $q_{51}^{\prime(2)}$, using the Method 1 approximation.

11.3 In a table with three decrements you are given that, for all $x \in [0, 100)$,

$$\ell_x^{(1)} = 10(100 - x), \qquad \ell_x^{(2)} = 20(100 - x), \qquad \ell_x^{(3)} = 30(100 - x).$$

Find the probability that (50) will fail first from cause 1 between the ages of 60 and 70.

11.4 Suppose that in a double-decrement model, you are given the associated single-decrement rates

$$q_x^{\prime(1)} = 0.3, \qquad q_x^{\prime(2)} = 0.51.$$

Compute $q_x^{(1)}$ and $q_x^{(2)}$ by Method 1 and Method 2.

11.5 A disability insurance policy provides for payments at the moment of disability should this occur within 10 years. The amount of the benefit at time t is $e^{0.10t}$. The policy is purchased by level annual premiums payable continuously for 10 years until either death or disability occurs. Nothing is paid on this policy if the insured dies before becoming disabled. If the force of disability is a constant 0.03, the force of mortality is a constant 0.06, and the force of interest is a constant 0.05, find the annual rate of premium payment.

11.6 Given $q_x^{(1)} = 0.05, q_x^{(2)} = 0.08, q_x^{(3)} = 0.10$, use Method 1 to find $q_x^{\prime(j)}$, $j = 1, 2, 3$.

11.7 Given $q_x^{\prime(1)} = 0.1, q_x^{\prime(2)} = 0.2, q_x^{\prime(3)} = 0.25$, use both Method 1 and Method 2 to find $q_x^{(j)}$, $j = 1, 2, 3$.

Type B exercises

11.8 Suppose that in a double-decrement model, $q_x^{\prime(1)} = q_x^{\prime(2)}$. Show that Method 1 and Method 2 lead to the same answer for the unprimed rates.

11.9 In a double-decrement table, with decrements d for death and w withdrawal, we have that $q_x^{(d)}$ is a constant 0.10 for all x and $q_x^{(w)}$ is a constant 0.02 for all x. The rate of interest is a constant 0.06. An insurance policy pays 1 at the end of the year of death. Nothing is paid to a withdrawing policyholder. Find the present value of the benefits on this policy.

11.10 In a multiple-decrement model with two causes of decrement, you are given associated single-decrement rates of $q_x'^{(1)} = 0.20, q_x'^{(2)} = 0.36$. Calculate the double-decrement rates $q_x^{(1)}$ and $q_x^{(2)}$ in each of the following three cases: (a) by Method 1; (b) by Method 2; (c) assuming that failures from cause 1 are uniformly distributed over the year, but that the failures from cause 2 all take place at a point three quarters of the way through the year.

11.11 In the case of two decrements, compute the difference between the right hand side and the left hand side of (11.16) in terms of $q_x^{(1)}$ and $q_x^{(2)}$, under each of (11.29) and (11.30).

11.12 Is the following statement true or false?

$$ {}_np_x^{(j)} + {}_nq_x^{(j)} = 1. $$

If false, give a correct version.

11.13 In a model with two decrements, you are given that, for all x, $\mu_x^{(1)}(t) = 0.02$ for all t and $\mu_x^{(2)}(t) = (10 - t)^{-1}$ for $0 \leq t < 10$. Find ${}_4q_x^{(2)}$.

11.14 In a model with two decrements you are given that $q_x^{(1)} = 0.1274$, $q_x^{(2)} = 0.1674$ and that there is a uniform distribution of decrements in each year for both of the single-decrement tables. Find $q_x'^{(1)}$ and $q_x'^{(2)}$.

11.15 Consider a double-decrement model, where we relax the hypothesis that the causes operate independently. Cause 1 is death, and failure from cause 2 occurs at time t if death occurs at time $t - 1$. Show that 11.23 does not hold.

11.16 In a multiple-decrement model with 4 decrements, $q_x'^{(i)} = i/10, i = 1, 2, 3, 4$. Each decrement is uniformly distributed over each year in the single-decrement table. Calculate $q_x^{(j)}$, $j = 1, 2, 3, 4$.

11.17 For an insurance contract with quarterly premiums, withdrawals will occur at times 1/4, 1/2, 3/4 and 1. Studies show that at a particular age x, one-third of the withdrawals occur at time 1/4, one-third occur at time 1, one sixth occur at time 1/2 and one sixth occur at time 3/4. Assume that deaths are distributed uniformly over the year. If $q_x'^{(d)} = 0.12$ and $q_x'^{(w)} = 0.20$, find $q_x^{(d)}$ and $q_x^{(w)}$. (Here, d denotes death and w denotes withdrawal.)

11.18 A multiple-decrement model has 3 decrements. For decrement 1, 60 % of the failures occur at time 1/3, and 40 % occur at time 2/3. For decrement 2, failures occur either at times 1/4 or 3/4 in equal numbers. For decrement 3, failures are uniformly distributed over each year. Find formulas giving the unprimed rates in terms of the primed rates.

12

Expenses and profits

12.1 Introduction

Provision for expenses and profits are two important features that we have ignored in our models up to now. Premiums paid on insurance and annuity contracts must not only provide benefits, but must also contain an extra amount to cover the expenses of operating the business. In addition the premiums must be sufficient to generate profits as a return to the investors who provide the capital necessary to start an insurance operation.

We will begin with a discussion of expenses, and the first step is to distinguish between three basic categories of regular periodic expenses. These can depend on the premiums, on the amount of the benefits, or be constant per policy. The major expense of the first type is commissions paid to the agents selling the policies. It is traditional for their compensation to take the form of certain percentages of the premiums. There are various expenses that will depend on the amount of the benefits. For example, larger policies will require additional efforts and expenses in the selection procedure to verify that the individual is a sound risk. Finally, there are expenses, such as setting up records for a new policy, that are largely independent of the size of premiums or benefits, and are fixed for each policy.

A peculiar feature of expenses for life insurance is that a large portion of these are incurred in the first year of the policy. Commissions paid on the initial premium are traditionally much higher than those paid in subsequent years (known as renewal premiums). The expenses of selection and of the setting up of policy records also occur at the beginning. One consequence of this is that the amounts to be paid upon withdrawal become an important consideration. The potential loss on withdrawal must be taken into account, since if a policyholder withdraws early, there may be no chance for the high initial expenses to be recovered through the premiums.

The expenses mentioned above will be paid at fixed times. But there are also extra expenses incurred when a loss occurs and a claim is reported to the insurer. The insurer must investigate to ensure that the claim is legitimate and then arrange to disburse the benefits. Some of these

Fundamentals of Actuarial Mathematics, Third Edition. S. David Promislow.

Companion website: http://www.wiley.com/go/actuarial

expenses would depend on the policy amount, as more care would be taken with a large policy. Others would be fixed. These are often referred to as *settlement* expenses.

For a typical life insurance or annuity policy, the mathematical treatment of expenses is quite straightforward. Suppose the insurer determines that an expense e_k will be incurred at each time $k = 0, 1, \ldots$. Then the insurer is simply providing an additional life annuity with benefit vector $\mathbf{e} = (e_0, e_1, \ldots)$. Similarly the expenses of claim settlement would be handled by an increase to the death benefit. Of course the policyholder does not receive these amounts. They are paid to those providing the goods or services involved in the expense, but they still must be provided for from the premiums.

Premiums which make provisions for expenses as well as benefits will be referred to as *expense-augmented premiums*. Some authors use the term *gross* premiums. We will however reserve this term for the premiums actually charged, which as pointed out in Section 4.6.1, can take into account factors other than the benefits and expenses. such as profits, which we discuss later on in this chapter.

Example 12.1 A whole-life policy issued at age x, with level annual premiums paid for life, provides for a death benefit paid at the moment of death. The expenses are as follows: in the first year, 70% of the initial premium, 1% of the face amount, and 30 per policy; in years 2 to 10, 10% of each premium, 0.5% of the face amount, and 10 per policy; after 10 years, 5% of each premium, 0.2% of the face amount and 5 per policy; the settlement expense is 100 per policy plus 0.5% of the face amount. Assume that the expenses in any year are paid at the beginning of the year.

Find a formula to compute the annual expense-augmented premium for a policy with a constant death benefit of 200 000.

Solution. Let G be the expense-augmented premium. We simply proceed as we did in Chapter 5, equating the present value of premiums with the present value of the death benefits plus the present value of the additional expense benefits. Of course the latter depend partly on G, which just means we have to solve a simple equation at the end.

It is best to consider two expense vectors. These are

$$\mathbf{e} = (2030, 1010_9, 405_\infty),$$

which covers the expenses depending on face amount as well as the fixed expenses, and

$$\mathbf{r} = (0.07,\ 0.1_9,\ 0.05_\infty),$$

which gives the proportion of the premium for premium-based expenses. The settlement charges mean that the amount paid on death is effectively 201 100. We then have the equation

$$G\ddot{a}_x = 201\,100\bar{A}_x + \ddot{a}_x(\mathbf{e}) + G\ddot{a}_x(\mathbf{r}),$$

from which we get

$$G = \frac{201\,100\bar{A}_x + \ddot{a}_x(\mathbf{e})}{\ddot{a}_x(0.3,\ 0.9_9,\ 0.95_\infty)}. \tag{12.1}$$

Note that the vector in the denominator is just $(1_\infty - \mathbf{r})$.

A more realistic model would also involve withdrawal. We would start with a double-decrement table with decrements d for death and w for withdrawal. We would also need a cash value vector **cv**, where cv_k would equal the cash value given to a policyholders who withdraws at time $k+1$. The subscript is chosen to correspond to the death benefit vector. (Note that we can safely assume that for a policy with premiums paid annually, withdrawal will only occur at an integer time, when a new premium is due.) We need as well a vector $\mathbf{e}^w$(standing for the *expenses of withdrawal*) where e_k^w would be the expenses incurred for withdrawal at time k, analogous to the settlement expenses added to the death benefit. Formula (12.1) would be modified to

$$G = \frac{201\ 100\bar{A}_x^{(d)} + A_x^{(w)}(\mathbf{cv} + \mathbf{e}^w) + \ddot{a}_x^{(\tau)}(\mathbf{e})}{\ddot{a}_x^{(\tau)}(\mathbf{1}_\infty - \mathbf{r})}.$$

12.2 Effect on reserves

Suppose that we compute reserves by adding expenses to the benefits and using expense-augmented premiums in place of net. The resulting quantities are known as *expense-augmented reserves*, while the reserves calculated in Chapter 6 are known as *net premium reserves*. Is it natural to ask as to how these two quantities compare. In the usual level premium policy, the expenses constitute a decreasing sequence of benefit payments that are paid for by a level addition to premiums. The extra expense charge included in the premium is less than needed to cover the expenses in early years, and more in later years. This means that typically the expense-augmented reserves will be less than the net premium reserves. To put it another way, the high initial expenses are a receivable to the insurer that will be collected from future premiums, and this causes a reduction in liabilities. Insurance regulators have usually taken the viewpoint that one should not count on these receivables since the policyholder might withdraw and never pay them. It is traditional that reserves be calculated without taking into account expenses, and using net in place of expense-augmented premiums. So even if realistic mortality and interest assumptions were used in calculating reserves, ignoring expenses causes higher reserves, and will cause losses to be shown in the early years of a policy.

Consider a simple example. Suppose that for a certain policy issued to (x) with benefits paid at the end of the year of death, $b_0 = 1000, q_x = 0.2,\ i = 0.10$, the net premium payable at time 0 is 250, the expense-augmented premium is 300, and the total expenses payable at time 0 are 100. The expense-augmented premium reserve at t time 1 would be

$$(300 - 100)\frac{1.10}{0.8} - 1000\frac{0.2}{0.8} = 25.$$

while the net premium reserve at time 1 is

$$250\frac{1.10}{0.8} - 1000\frac{0.2}{0.8} = 93.75.$$

Even if the mortality and interest matched exactly those of the reserve assumptions, the use of net premium reserves will result in a loss of 68.75 per policy in the first year. This arises from the fact that the provision for expenses is made by an extra charge of only 50 to the level premium, but this does not cover the actual expense of 100. (Of course, as we showed in Chapter 6, there will be corresponding gains in future years, when the actual expenses are less than the 50 allowed for in the premium.)

To alleviate this distortion of the profitability in the first year, many regulatory bodies permit reserves to be calculated by methods that are generally known in North America as *modified reserve systems*. (Similar procedures are popular in some European countries and are known as *Zilmerized reserves.*) Recall the term *valuation premium* to refer to premiums used in calculating reserves, which we introduced in Chapter 6. We stress again that these are not necessarily (and in fact almost never) the same as the insurer actually charges. The modified systems still call for expenses to be ignored, but they allow for lower initial valuation premiums, followed by higher valuation premiums for the later durations, calculated so that valuation premiums remain actuarially equivalent to benefits. This effectively recognizes that the insurer has a smaller premium in the first few years to provide the benefits since a large amount of these premiums must go to pay the high initial expenses.

There are many such systems. The most basic is the *full preliminary term* method. Suppose we have a life insurance policy with level premiums payable for n years, and benefits payable at the end of the year of death. The initial valuation premium is taken as $b_0 v(1) q_x$, the exact amount needed to provide the benefits for the first year, followed by a level valuation premium for all subsequent years. This will automatically make ${}_1V = 0$. This level valuation premium for years after the first will be

$$\frac{A_{[x]+1}(\mathbf{b} \circ 1)}{\ddot{a}_{[x]+1}(1_{n-1})},$$

which will provide for the benefits after the first year. This effectively means that the entire first-year valuation premium, less the amount needed for the benefits according to the reserve basis, is available to use for expenses.

There are several variations to full preliminary term. In some cases there is a recognition that for policies with high initial premiums, such as endowment insurance, or premiums payable over a limited period, this provides more expense allowance than is needed in the first year, and the first-year valuation premium is taken to be higher than $b_0 v q_x$. In other cases there may be more than one step. There will be a low premium used in the first year, a somewhat higher one for some duration, for example the next 10 years, and a still higher one after that. We will not discuss the details of the various methods here. Some of these are dealt with in the exercises.

12.3 Realistic reserve and balance calculations

As we have indicated, insurers are required to calculate reserves in accordance with the particular regulations that apply in their jurisdiction. These methods are intended mainly to ensure that policyholder will receive their promised benefits. These are often referred to as *statutory reserves*. There are however many reasons for the insurer, to calculate reserves and balances on a more realistic basis which use expense-augmented or gross premiums and expenses, rather than net premiums with expenses ignored, and which takes into account withdrawals. An acronym that is sometimes used here is that of a 'GAAP' reserve, the acronym standing for *generally accepted accounting principles*. Such a reserve would then be calculated as the present value of all future benefits, including death benefits, expenses, surrender benefits, less the present value of future premiums. It is useful to replace our basic recursion formula (6.8) with the following more general version. For simplicity we assume aggregate mortality, and benefits payable at the end of the year of death. Consider a policy with annual premiums of π_k, death benefits of b_k, cash values of cv_k, an expense of r_k per unit of premium paid at time k, fixed expenses at time k of e_k, claim settlement expenses of e_k^d in the case of death between time k and time $k+1$, and surrender expenses of e_k^w for surrender at time $k+1$. Our basic recursion (6.8), when modified for expenses and withdrawal, becomes

$$ {}_{k+1}V = [{}_kV + \pi_k(1-r_k) - e_k](1+i_k) - q_{x+k}^{(d)}\left(b_k + e_k^d - {}_{k+1}V\right) - q_{x+k}^{(w)}\left(cv_k + e_k^w - {}_{k+1}V\right), \tag{12.2} $$

and solving for ${}_{k+1}V$, we obtain the Fackler type equation

$$ {}_{k+1}V = \frac{[{}_kV + \pi_k(1-r_k) - e_k](1+i_k) - q_{x+k}^{(d)}\left(b_k + e_k^d\right) - q_{x+k}^{(w)}\left(cv_k + e_k^w\right)}{1 - q_{x+k}^{(d)} - q_{x+k}^{(w)}}. \tag{12.3} $$

We want to discuss three main quantities which arise from the above recursion. They differ in the choice of premiums and initial values.

Expense-augmented reserves. These reserves, as defined above, can be calculated from (12.3) with π_k equal to the expense-augmented premium.

Gross-premium reserves. These are simply the present value of future benefits and expenses less the present value of future *gross premiums*. They can be calculated from (12.3) with π_k equal to the gross premium.

Asset Shares. These are realistic balance calculations. They give the accumulated value of the premiums less the accumulated value of the benefits and expenses, They can be calculated from (12.3) using gross premiums, but now with an initial value of 0 at time 0. (Different notation is often used here to distinguish them from reserves. We will write AS_k in place of ${}_kV$.) The name comes from the fact that they represent that share of the insurer's assets which are attributable to the particular contract.

We will use the following simplified example to illustrate the difference between these quantities.

Example 12.2 A two-year policy on (x) provides for death benefits of 1000 paid at the end of the year of death. Premiums of 160 are paid for 2 years. There are expenses, incurred at the beginning of each year, of 20% of the premium plus 18 in year 1, and 5 in year 2. You are given that $i = 100\%$, $q_x = 0.2$, $q_{x+1} = 0.3$. There are no withdrawals. Find the expense-augmented reserves, gross-premium reserves, and asset shares.

Solution. Let G denote the expense-augmented premium. The present value of benefits and expenses is

$$ 1000[(0.5)(0.2) + (0.25)(0.8)(0.3)] + 18 + 0.2G + 5(0.5)(0.8)] = 180 + 0.2G. $$

Then $G[1 + (.5)(.8)] = 180 + 0.2G$ showing that $G = 150$.

For the expense-augmented reserves,

$$
\begin{aligned}
{}_0V &= 0, \\
{}_1V &= [0.8(150) - 18) \times 2 - 200]/0.8 = 5, \\
{}_2V &= 0.
\end{aligned}
$$

For gross-premiums reserves,

$$
\begin{aligned}
{}_0V &= 180 + 0.2(160) - 160(1.4) = -12, \\
{}_1V &= [(-12 + 0.8(160) - 18) \times 2 - 200]/0.8 = -5, \\
{}_2V &= 0
\end{aligned}
$$

For the asset shares

$$
\begin{aligned}
AS_0 &= 0, \\
AS_1 &= [(128 - 18) \times 2 - 200]/0.8 = 25, \\
{}_2AS_2 &= [25 + 160 - 5] \times 2 - 300]/0.7 = 85.71.
\end{aligned}
$$

In the following section we will revisit this example, and interpret the figures in terms of profit calculations.

Remark For simplicity, some examples involving expenses will ignore withdrawals. This does necessarily mean that withdrawals are not allowed as we stated in the previous example. The implicit assumption is that at each duration, the cash value together with any expense of withdrawal is precisely equal to the reserve on the policy. In such a case the withdrawal rates and cash values have no effect on premiums or reserves, and need not be taken into account.

12.4 Profit measurement

12.4.1 Advanced gain and loss analysis

We want to continue the profit analysis we began in Chapter 6, but now incorporating expenses and withdrawals, and also by considering a more extensive context.

We start with a very general setting. We are given a particular insurance or annuity policy, with a description of the benefits and cash values, and a set of reserves. The latter could be net premium reserves, expense-augmented reserves, gross-premium reserves, or even something else. In addition we are given a basis for all relevant factors, known as the *profit test basis*. These are interest rates, rates of mortality and withdrawal, expenses and the premiums which will be charged. We want to determine the profit we will have in a specified period, if our actual experience is according to the given profit test basis and we use the specified reserves to measure profit. There are different contexts possible. The profit test basis could be a hypothetical test basis, to see what profits will emerge under our assumptions, and the major goal is to see if the proposed premiums will achieve a desired level of such profits. Alternatively this calculation could be carried out after the fact with the profit basis reflecting the actual observed experience, and the result will be the actual profit that is achieved.

In any event, for the period running from time k to $k+1$ we have a profit of Pr_{k+1} consisting of the right hand side of (12.2) less the reserve ${}_{k+1}V$. That is

$$\begin{aligned}\mathrm{Pr}_{k+1} &= ({}_kV + \pi_k(1-r_k) - e_k)(1+i_k) - q^{(d)}_{x+k}\left(b_k + e^d_k - {}_{k+1}V\right) \\ &\quad - q^{(w)}_{x+k}\left(cv_k + e^w_k - {}_{k+1}V\right) - {}_{k+1}V, \qquad (12.4a)\\ &= ({}_kV + \pi_k(1-r_k) - e_k)(1+i_k) - q^{(d)}_{x+k}\left(b_k + e^d_k\right) \\ &\quad - q^{(w)}_{x+k}\left(cv_k + e^w_k\right) - p^{(\tau)}_{x+k}\ {}_{k+1}V. \qquad (12.4b)\end{aligned}$$

The two versions represent two possible points of view, leading to the same result, as we already indicated in Chapter 6. In (12.4a) the insurer sets up the reserve for everyone and then needs to pay only the amount of the benefits above the reserve in the case of death or surrender, while in (12.4b), the insurer pays the benefits in full but then needs to set up a reserve only for the survivors. Version (12.4b) is normally the most efficient for calculation. Yet another version is to write

$$\begin{aligned}\mathrm{Pr}_{k+1} &= (\pi_k(1-r_k) - e_k)(1+i_k) - q^{(d)}_{x+k}\left(b_k + e^d_k\right) \\ &\quad - q^{(w)}_{x+k}\left(cv_k + e^w_k\right) - \Delta_k(V), \qquad (12.4c)\end{aligned}$$

where

$$\Delta_k(V) = p^{(\tau)}_{x+k}\ {}_{k+1}V - {}_kV(1+i_k) \qquad (12.5)$$

represents the increase in reserves over the year. This version simply says that profit is what is left over after we accumulate the net amounts collected, and provide for benefit payments, expenses and the increase in reserves.

Example 12.3 Refer back to Example 12.2 and calculate Pr_k when the the underlying reserves are: (a) gross-premium reserves; (b) expense-augmented reserves. In each case, the profit test basis is the same as the reserve basis.

Solution. (a) It is clear that $\mathrm{Pr}_1 = \mathrm{Pr}_2 = 0$. So what happened to the profits? The answer is that with gross-premium reserves, and observed experience the same as the reserve basis, they all occur at time 0. The reserve of -12 at this time, means that the insurer could take 12 out of the company surplus for each such policy and still have enough left to provide for the reserves. This shows that in addition to the definition of Pr_k given above we must add the statement that

$$\mathrm{Pr}_0 = -{}_0V.$$

(b) Using (12.4(b)),

$$\begin{aligned}\mathrm{Pr}_1 &= (128 - 18)2 - 0.2(1000) - 0.8(5) = 16.\\ \mathrm{Pr}_2 &= (5 + 160 - 5)2 - 0.3(1000) = 20.\end{aligned}$$

Note the difference in the incidence of profit for the different types of reserves. With gross-premium reserves, all the profit is recognized at the beginning. With the expense-augmented reserves, the profit emerges gradually as the extra amounts above the expense-augmented premiums come in. In this case there is an extra 10 received in the first premium, but this is reduced to 8 after the percentage of premium expense. This accumulates in 1 year to 16. Similarly the extra 10 in the second premium accumulates to 20 at the end of the second year. It will instructive for the reader to verify that in the long run, that the reserve basis will not affect the actual profits received.

This example also indicates that we can view asset shares as the sum of the expense-augmented reserves plus an accumulation of the profits at interest and survivorship. For example,

$$\mathrm{AS}_2 = 0 + 8\left(\frac{4}{0.56}\right) + 10\left(\frac{2}{0.7}\right) = 85.71.$$

as calculated above.

12.4.2 Gains by source

We now elaborate on the decomposition of gains by source, as initiated in Section 6.4.1. We will in fact consider a somewhat more general situation. Suppose we have two different bases under consideration, one in which quantities are starred and the other unstarred. In the most common type of application the unstarred symbols refer to the expected basis, and the starred will refer to what actually happened, but mathematically these could be any two bases. We then have two profit calculations, Pr_k^* and Pr_k and we are interested in the difference $\mathrm{Pr}_k^* - \mathrm{Pr}_k$ which in the common situation will be the excess of actual profit over the expected profit. In the case where the reserve basis coincides exactly with the unstarred basis, we have that $\mathrm{Pr}_k = 0$ for all k and the difference will just be Pr_k^*. This was precisely the point of view we took in Chapter 6. The goal is to decompose this difference into the various gains by source. As well as the mortality and interest gains of the Chapter 6 model, we now have additional sources.

One is the gain from withdrawal, which is calculated analogously to the mortality gain as

$$\left(q_{x+k}^{(w)} - q_{x+k}^{*(w)}\right)(cv_k - {}_{k+1}V).$$

In addition we have gains from the various types of expenses. These present a complication, since they are invariably tied up with other factors. Consider for example the case of death settlement expenses. As a particular example, suppose we have $q_{x+k}^{(d)} = 0.07$, $q_{x+k}^{*(d)} = 0.05$, $e_k^d = 200$ and $e_k^{*d} = 100$. So our mortality experience is more favourable than expected and the costs of settling each claim are less than expected. We gain in both ways and in fact the total gain per policy will be $0.07(200) - 0.05(100) = 9$. We would like to divide this gain up into that attributable to mortality, and that attributable to lower settlement expenses. A somewhat surprising fact is that there seems to be no reasonable way to do this uniquely. To illustrate, consider the general situation, which applies as well to withdrawal expenses. (For simplicity in notation we will omit superscripts and subscripts.)

We have a total gain of $(qe - q^*e^*)$. Now we can write this in two distinct ways, namely

$$(q - q^*)e + (e - e^*)q^*, \text{or } (q - q^*)e^* + (e - e^*)q.$$

In either case it would be reasonable to take the first term to be the mortality gain and the second term to be the expense gain. However, there is nothing to choose between them, and they give different answers. One way to decide is to determine an order in which we will calculate the gains. If we decide to look at mortality first, it would be natural to ignore the expense differential at this point and take $(q - q^*)e$ as the mortality gain, as in the first expression. Then necessarily we would take $(e - e^*)q^*$ as the expense gain. If we decide to look at expense first, we would decompose according to the second expression. In short, for the factor considered second, the gain will be expressed with the other factor starred. So in our numerical example above, we could divide the total gain of 9 into 4 for mortality and 5 for expenses, or 2 for mortality and 7 for expenses, deciding on which decomposition we used.

Another way of viewing this is to note that we can also write the total gain in a symmetric fashion as

$$qe - q^*e^* = (q - q^*)e + q(e - e^*) - (q - q^*)(e - e^*).$$

The first term on the right is strictly mortality gain, the second term is strictly expense gain and the third term, which is a combined effect, can be allocated to one or the other.

This seems rather arbitrary and artificial but it is the best we can do if we insist on a decomposition.

Indeed we have already encountered this problem before, namely in the example involving premium difference gain in Chapter 6, a gain which is tied in with interest. Since we had previously talked about interest gain, it seemed natural to do that first, but we could have instead taken the premium difference gain as $(\pi^* - \pi)(1 + i)$, leaving the total interest gain as $({}_tV + \pi^*)(i^* - i)$.

The same problem comes with periodic expense gains which also involve interest. If we for example, focus on interest first, the total gain of $(e_k(1 + i_k) - e_k^*(1 + i^*k)$ will be decomposed into a part due to interest of $e_k(i_k - i_k^*)$ and a part due to expenses of $(e_k - e_k^*)(1 + i^*)$. (Note that earning more interest than expected will result in the first item being negative. The reader should ensure that they can explain why this is so.)

What would happen if there were a source of gain connected with *three* different factors? We'll leave it to the reader to verify that there would now be six possible ways of dividing up the gain. Theoretically this could happen. For example, a percentage of premium expense would be affected by differences in both premiums and interest. In the common situation we have described, this will not occur since the premiums payable on both the actual profit basis and the expected profit basis will be the same, namely the gross premium, and the premium difference gain will be zero.

Example 12.4 You are given the following information about a policy issued at age 50. The gross premium payable at age 65 is 2000, and a death benefit of 100 000 is payable at age 66 for death occurring between age 65 and 56. A cash value of 20 000 is paid at age 66 for policyholders surrendering at that time. Moreover ${}_{15}V = 20\ 300$, and ${}_{16}V = 22\ 500$.

The insurer's best estimate for mortality, withdrawal and interest is given by

$$q_{65}^{(d)} = 0.005, \qquad q_{65}^{(w)} = 0.008, \quad i = 0.04.$$

Anticipated expenses at the beginning of the year are 5% of the gross premium, plus 100. In addition, the expected settlement expenses are 200 for a death claim and 100 for withdrawal.

Now here is the actual experience. Out of 1000 policies at the beginning of the 16th policy year, there were 4 deaths and 10 withdrawals. The actual interest rate earned was 0.048. The actual expenses incurred were 230 per policy at the beginning of the year, including both percentage of premium and fixed expenses, 150 to settle each death claim, and 120 for each withdrawal.

For the year running from time 15 to time 16, calculate the actual profit less the expected profit, and classify this difference by source. For the latter, assume that all expense gains are calculated *after* the gains from the other sources.

Solution. Taking the actual results for the starred quantities, and using (12.3)

$$\begin{aligned}
\mathrm{Pr}^*_{16} &= (20\,300 + 2000 - 230)(1.048) - .004(10\,0150) - 0.010(20\,120) - 0.986(22\,500) \\
&= 342.56 \\
\mathrm{Pr}_{16} &= (20\,300 + 2000(1 - 0.05) - 200)(1.04) = 0.005(10\,0200) - 0.008(20\,100) \\
&\quad -0.987(22\,500) \\
&= 114.7
\end{aligned}$$

$$\text{Actual profit} - \text{Expected profit} = 227.86.$$

This can be decomposed as follows:

Gain from mortality (including the portion due to settlement expenses) $= (0.005 - 0.004)(10\,0200 - 22\,500) = 77.70$.

Gain from withdrawal (including the portion due to withdrawal expenses) $= (0.008 - 0.010)(20\,100 - 22\,500) = 4.80$.

Gain from interest $= (20\,300 + 2000 - 200)(0.048 - 0.040) = 176.80$

Gain from percentage of premium and periodic expenses $= (200 - 230)(1.048) = -\,31.44$

Death settlement expense gain $= 0.004(200 - 150) = 0.20$

Withdrawal expense gain $= 0.010(100 - 120) = -0.020$.

The expense gains above, coming after the other calculations, all use the actual experience for the other factors, as we noted in our remarks about order above. The reader should find it instructive to redo the calculations, now assuming that the expense gains are calculated first.

Note that since the insurer benefits from a withdrawal, paying out less than the reserve, the higher than expected number of withdrawals caused a positive gain.

12.4.3 Profit testing

We now return to the case where we have a proposed profit test basis, and proposed gross premiums. We want to compute the profits over several periods, and see how we can use the resulting figures.

Example 12.5 A five-year, 100 000 endowment insurance policy on (60) has benefits payable at the end of the year of death and carries annual premiums of 21 500, payable for

5 years. Initial expenses are 60% of the premium plus 200, and renewal expenses are 5% of the premium plus 100. Death settlement expenses for the first 4 years are 0.2% of the the face amount plus 50. The expense of paying any cash value, or the final payment at time 5 is 50. The profit test basis interest rate is 5%. The profit test basis mortality and withdrawal rates, reserves, and cash values are shown in the following table. Find Pr_k for $k = 0, 1, \ldots, 5$.

k	$q^{(d)}_{60+k-1}$	$q^{(w)}_{60+k-1}$	${}_kV$	cv_{k-1}	Pr_k
0			0	0	0
1	0.009	0.080	18 500	0	−8935.75
2	0.010	0.050	37 500	17 500	3636.25
3	0.012	0.030	57 500	42 500	3151.75
4	0.014	0.020	78 500	70 000	3080.75
5	–	–	100 000	0	3716.25

Solution. The values of of Pr_k are computed directly from (12.3) and are shown in the final column. As an example, here is the computation of Pr_2.

$$\begin{aligned}\mathrm{Pr}_2 &= (18\,500 + 21\,500 - 0.05(21\,500) - 100)(1.05) - 0.01(100\,250) - 0.05(17\,550)\\ &\quad - 0.94(37\,500)\\ &= 3636.25\end{aligned}$$

The vector $(\mathrm{Pr}_0, \mathrm{Pr}_1, \ldots)$ is sometimes called the *profit vector*. It is not however the vector we want in order to accomplish our goal of analyzing profits. Recall that Pr_k is the profit in the year running from time $k-1$ to time k for a policy in force at time $k-1$. The figure we want is the expected profit for that year. Accordingly we make the following definition.

Definition 12.1 The *profit signature* for a policy issued at age x is the vector Π, where $\Pi_0 = \mathrm{Pr}_0$, and for $k > 0$

$$\Pi_k = {}_{k-1}p_x^{(\tau)}\mathrm{Pr}_k.$$

In other words we multiply the yearly profits by the probability that the policy holder will survive to the beginning of the particular year.

Example 12.6 Find the profit signature for Example 12.5.

Solution. We first calculate

$$p_x^{(\tau)} = 0.911, \quad {}_2p_x^{(\tau)} = 0.85634, \quad {}_3p_x^{(\tau)} = 0.82037, \quad {}_4p_x^{(\tau)} = 0.79248.$$

A straightforward calculation then yields

$$\Pi = (0, -8935.75,\ 3312.62,\ 2698.97,\ 2527.37,\ 2945.06).$$

Our problem of analyzing profits is now reduced to the familiar problem, dealt with from the beginning of this book, of analyzing a sequence of cash flows.

For example we can summarize this cash flow sequence with a single number, by computing the present value of the profit signature according to some discount function. This gives the net present value of profits, often abbreviated as NPV. In practice the discounting is at a constant rate of interest which is called the *risk discount* rate. We can interpret this rate as one which is the desired return by an investor putting up the capital to fund an insurance enterprise. It will therefore reflect the risk involved in such an investment and will accordingly be normally higher than say the profit test rate. Note that we do not discount by survivorship in calculating the NPV. This has already been considered when calculating the profit signature.

Recall now that we have calculated this NPV according to a single policy, with a specified amount of benefits and premiums. To interpret it properly and obtain measures which can be used for comparative purposes, we must relate this to some measure of volume. A natural candidate for this is the present value of premiums.

Definition 12.2 We define the *profit margin for the policy* as the quantity

$$\frac{NPV}{\ddot{a}_x^{(\tau)}(\pi)}$$

where π is the premium vector for the policy. The discounting in the denominator is done with both interest, at the same risk discount rate used in the numerator, and with survivorship as given by the multiple-decrement table.

Example 12.7 If the risk discount rate is 0.07, find the profit margin for the policy of Example 12.5.

Solution. The denominator is just 21 500 $\ddot{a}_{60}(1_5)$ at 7% interest = 83 282.65.

The numerator is

$$(1.07^{-1})(-8935.75) + (1.07)^{-2}(3312.62) + (1.07)^{-3}(2698.97) + (1.07)^{-4}(2527.36) + (1.07)^{-5}2984.09 = 773.27$$

The profit margin is 773.27/ 83292.65 = 0.0093.

This example shows then that for the policy in question, for every dollar in premium received in sales, 0.93 cents would represent profit. This can be used for various decision problems. If it deemed lower than the company would like, then some changes must be made in the policy structure, usually by raising the premium charged, or in some case by lowering the surrender benefits if allowed by applicable legislation. The profit margin can also be compared with other policies to see which types of benefit and premium structures are most profitable.

A natural question now arises. How is the profit margin affected by the choice of reserves. Our calculations in Section 6.5 suggest that under certain conditions the reserve basis should not affect the profit margin, and this indeed is true when the discount rates used in computing the NPV are the same as those of the profit test basis.

To see this, look at formulas (12.3) and (12.5). Suppose that the single reserve figure ${}_kV$ is increased by an amount E. Then $\Delta_{k-1}V$ is increased by $p_{x+k}E$ so that

$$\mathrm{Pr}_k \text{ is decreased by } p_{x+k}E.$$

Similarly, $\Delta_k(V)$ is decreased by $E(1 + i_k)$ so that

$$\mathrm{Pr}_{k+1} \text{ is increased by } (1 + i_k)E$$

resulting in an decrease in Π_k of ${}_kp_xE$ and and increase in Π_{k+1} of ${}_kp_x(1 + i_k)E$, which will cancel out when we discount, leaving the NPV unchanged.

This is not true however in the more typical case where the risk discounting is at higher rates. Suppose that $E > 0$. The calculation above shows that discounting the increase in Π_{k+1} with a rate higher than i_k for the year running from time k to $k + 1$ will decrease the NPV. This is easily explained. The higher reserve means that more money must be set aside at time k, reducing profits at that time. This money is earning interest and will add to the profits the following year. However under the present scenario, it is not earning sufficient interest to compensate. The conclusion is that in the usual case where risk discount rates are higher than the profit test rates, increases in reserves cause reduced profitability, and decreases in reserves cause increased profitability.

Notes and references

See Bowers *et al.* (1997, Chapter 16) for a detailed discussion of various modified reserve systems.

Net premium reserves are sometimes referred to as *benefit* reserves in the literature since, they involve only benefits and not expenses. The difference between the expense-augmented reserve and the net premium reserve is sometimes referred to as the *expense reserve*. It represents the present value of the future premiums available for expenses, less the present value of the future expenses.

Some authors treat certain expenses incurred right at or just before a policy begins differently than above. They would appear as a negative contribution to Pr_0. In our treatment they are accumulated to the end of the year, and appear as a negative contribution to Pr_1, in the same way as the initial premium received at time 0, makes a positive contribution to Pr_1.

Exercises

Type A exercises

12.1 An insurance policy issued at age x provides for a payment of 100 000 at the moment of death provided this occurs within 20 years. Level annual premiums of G are payable for 10 years. The expenses are as follows: in the first year, 60% of the initial premium, 1% of the face amount and 20 per policy; after the first year, 5% of the each premium, 0.25% of the face amount and 10 per policy; the death benefit settlement expense is

50 per policy plus 0.5% of the face amount. Assume that the expenses for any year are paid at the beginning of the year. Find a formula for G.

12.2 An insurance policy issued at age x has death benefit vector $\mathbf{b}$ paid at the end of the year of death, and level annual premiums payable for n years. Assume a constant interest rate. Given $A_x(\mathbf{b}) = 330$, $A_{x+1}(\mathbf{b} \circ 1) = 280$, $\ddot{a}_x(1_n) = 11$, $\ddot{a}_{x+1}(1_{n-1}) = 8$ and $\ddot{a}_{x+k}(1_{n-k}) = 6$, find the excess of the reserve at time k calculated with net premiums over the reserve at time k calculated by the full preliminary term method.

12.3 A certain policy on (x) has gross-premium reserve of 1000 at time 10. The gross premium payable at time 10 is 100. The death benefit payable at the end of year of death for death between time 10 and time 11 is 1500. The cash value payable at time 11 is 900. Expenses for the year running from time 10 to time 11 are 20, paid at the beginning of the year. The interest rate $i = 10\%$, $q^{(d)}_{x+10} = 0.08$ and $q^{(w)}_{x+10} = 0.05$. Find the gross premium reserve at time 11.

12.4 A policy on (x) provides death benefits of 1000 paid at the end of the year of death. There is no cash value for withdrawal in the first year. For withdrawal in the second year, an amount of 30 is paid at the end of the year. You are given that

$$q_x^{(d)} = 0.1, \qquad q_{x+1}^{(d)} = 0.2, \qquad q_x^{(w)} = 0.05, \qquad q_{x+1}^{(w)} = 0.06, \qquad i = 0.07.$$

Expenses are 120 in the first year and 40 thereafter, paid at the beginning of the year. The level expense-augmented premium payable annually is 250 and the corresponding gross premium is 300. The gross-premium reserve at time 0 is -130. Find (i) the expense-augmented reserve, (ii) the gross-premium reserve, (iii) the asset share, all at the end of 2 years.

Type B exercises

12.5 A modified reserve system has a first-year premium of $b_0 v q_x$ a second-year premium of $b_1 v q_{x+1}$, followed by a level renewal premium of β. Assume constant interest and benefits at the end of the year of death. Find an expression for β, when premiums are payable for n years.

12.6 Consider a policy issued at age x with benefits payable at the end of the year of death. A modified reserve system, used in Canada for tax purposes (sometimes known as the $1\frac{1}{2}$ preliminary term), takes an initial premium to $b_0 v q_x$, a premium at time 1 of β, and a level premium of γ payable after time 1, such that β is the average of $b_1 v q_{x+1}$ and γ. Assume constant interest and benefits paid at the end of the year of death. If premiums are payable for life, show that

$$\gamma = \frac{A_{x+1}(\mathbf{b} \circ 1) + v p_{x+1} A_{x+2}(\mathbf{b} \circ 2)}{\ddot{a}_{x+1} + v p_{x+1} \ddot{a}_{x+2}}.$$

12.7 Suppose that in Example 12.3 the insurer sells 100 of the given policies. Show the the total profits accumulated with interest to time 2 are the same, regardless of whether gross premium or expense-augmented reserves are used.

12.8 Redo Example 12.4 with the expense gains calculated *before* those from mortality, withdrawal and interest.

12.9 Refer to Example 12.5

(a) Suppose that ${}_1V$ is changed to 13 000 while all other data remains the same. Calculate the new profit signature and profit margin.

(b) Redo part (a) with the risk discount rate equal to 5%.

(c) Find the profit margin, for the original problem with a risk discount rate equal to 5%.

12.10 Refer to Example 12.5. Suppose we decide to increase the premium to achieve a profit margin of 3%. What should the new premium be?

*13

Specialized topics

For the concluding chapter in Part I of this book, we cover a few specialized topics in insurance and annuities that are of current importance. We will not go into any of these topics in depth, but rather provide an introduction to some of the major features. The material here is not needed for any other chapters of the book.

13.1 Universal life

13.1.1 Description of the contract

Universal life is a type of contract that began in the 1970s, and now accounts for a substantial portion of life insurance sales. Prior to that time, the mainstay of life insurance was for the most part whole life or endowment contracts. The origin of the change could well have been the advice given by some financial advisors that people should buy term insurance instead of those products, for a much smaller premium, and then invest the difference elsewhere. Now our analysis in Section 6.4.2 shows that this is exactly what happens within the policy itself when one purchases whole life or endowment insurance. The excess premiums, not needed to provide insurance in a particular year, are invested in the savings portion. Of course, the purchaser of endowment or whole life insurance has less flexibility than the term buyer, both with regards to the relative amounts going into the insurance and savings portion, and to the types of investments and rates of return. Universal life plans were developed to allow this flexibility within a single contract. The main features involve variable premium payments, and the ability for the policyholder to participate in higher yielding investment opportunities.

In the usual type of universal life plan, premiums are not fixed in advanced and may be varied at the option of the purchaser. Each individual has in effect a separate fund that changes over time in the way we describe it by the recursion formula (6.8). Premiums paid

Fundamentals of Actuarial Mathematics, Third Edition. S. David Promislow.

Companion website: http://www.wiley.com/go/actuarial

are deposited, the fund is credited with interest or investment earnings, while expenses, and the cost of insurance, based on the net amount at risk are deducted. The policyholder need only pay sufficient premiums to ensure that the fund value will cover the cost of insurance.

In this type of plan where there is a more direct relationship between policyholders and their individual accounts, there is a more frequent demand for the type of plan that we discussed in Section 6.7. It is usual to allow two options. The purchaser can fix the amount paid at death, or they can fix the net amount at risk, so at death, the amount in their account (representing the reserve) would be returned in addition to the stipulated face amount, giving an increasing death benefit. In the usual terminology the fixed death benefit is often referred to as a Type A policy, whereas the death benefit plus account value is referred to as type B.

Some plans provide the flexibility to alter the face amount as well as the premium, with the stipulation that a request to increase the coverage would normally necessitate the type of information as required by new policyholders as to health and other conditions affecting the risk. There do arise cases however when such changes are mandated. In the United States this occurs due to regulations stipulating that in order to receive the favourable tax treatment accorded to life insurance, the death benefit must exceed the fund value by a certain minimum percentage. If the fund value gets high enough, the insurer will increase the death benefit in order to maintain this so-called *corridor requirement*.

Policyholders can lapse the policy at any time and receive their fund value less a so-called *surrender charge* that is intended mainly to account for the initial expenses, as we outlined in Section 6.6. The surrender charge will decrease with time and normally will become zero after a certain number of years, often 20 or so.

As well as the flexibility in premium payments a major attraction of universal life for the purchaser is the opportunity to earn a higher yield. The credited interest rate is usually not fixed in advance as it essentially is in traditional plans, but it is allowed to vary. There are different ways of accomplishing this. In some cases the funds of the policyholder are actually invested in certain assets, so that the account is very much like a mutual fund investment together with guaranteed benefits payable on death. Usually there is a choice of various types of assets or funds to invest in, so the policyholder can control the amount of risk they are willing to take in order to obtain higher returns. This type of contract is often referred to as *variable universal life*. In other cases the account of the policyholder need not be invested in any particular assets but interest is credited according to earnings on some reference portfolio. Such contracts are known as *equity indexed insurance*. There are normally some qualifications to the crediting of interest. There will be a stipulated minimum rate of interest that is credited regardless of what the reference portfolio does. It is also common to have limits on the other end. There may be a 'participation rate'. So for example if the reference portfolio earned 9% interest over a period and the participation rate was 80%, the the policyholder's account would be credited with 7.2%. In addition there is often a stated maximum that will be credited regardless of the actual earnings.

Universal life contracts often carry other guarantees that mitigate again unfavourable investment experience. These are referred to as *secondary guarantees*. A popular option of this type, is that as long as the insured keeps up a certain stipulated minimum rate of premium payment, the death benefit is guaranteed even though lower than expected rates of return bring the person's account below the amount needed to meet the costs of insurance. A similar provision is commonly used with endowment type contracts that have a maturity date, in which case a minimum guaranteed amount paid at maturity provided that the insured keeps up a minimum schedule of premium payments.

13.1.2 Calculating account values

For each policyholder, the amount of money in their account is calculated periodically, often monthly. For the most part this is handled easily by the basic recursion formulas that we have seen in previous chapters. There are however some aspects of terminology and special features that will be discussed. Given the account value AV_k at time k, the account value AV_{k+1} is calculated using the following familiar quantities. We assume aggregate mortality and payment of benefits at the end of the period of death.

(a) π_k, the premium paid at time k.

(b) r_k, the premium expense rate. This is a percentage, often around 5%, taken from each premium by the insurer to handle expenses. Only the remaining amount (sometimes referred to as the *allocated premium*) is credited to the account. In the so-called *unit linked policies* popular in the United Kingdom, it is common to incorporate this expense by a device termed a *bid–offer* spread. In these cases the policyholder is deemed to have their own separate fund, distinct from the general assets of the company, and the premium is used to buy a number of units of the fund. These are bought at a certain *offer price*, and sold back at a lower *bid price*. So for example if the offer price was 100 per unit and the bid price was 95 per unit, and a premium of 1000 was paid, the purchaser would own 10 units of the fund, which would have a value of 950. So the allocated premium would be effectively 950.

(c) e_k, an additional flat amount that is charged each period for expenses.

(d) c_k, the cost of insurance rate. This is just a mortality rate which will apply to the time period running from time k to time $k+1$. For a yearly calculation with an issue age of x, we would have $c_k = q_{x+k}$, for an appropriate life table. For a monthly calculation, assuming UDD, we would have $c_k = (1/12)q_{x+m}$, where $12m \le k < 12(m+1)$. It is often expressed as a rate of so much per 1000. So for example a cost of insurance rate of 15 per 1000 would mean that $c_k = 0.015$.

(e) i_k, the credited interest rate. This may be specified directly or tied to the performance of some other investments.

(f) b_k, the death benefit paid at time $k+1$ for death between time k and $k+1$.

Then the recursion is just

$$\mathrm{AV}_{k+1} = (\mathrm{AV}_k + \pi_k(1 - r_k) - e_k)(1 + i_k) - c_k\eta_k. \tag{13.1}$$

For type A policies we have $\eta_k = b_k - \mathrm{AV}_{k+1}$, if this is positive, and we solve the equation for AV_{k+1}. For type B policies where the death benefit is paid in addition to the face amount, $\eta_k = b_k$. Another possible type of arrangement is where the fund itself is returned at death, subject to a guaranteed minimum amount b_k. (As mentioned above, this may be ruled out in certain jurisdictions by taxation requirements.) In such case if $\mathrm{AV}_{k+1} \ge b_k$, we would have $\eta_k = 0$, while if $\mathrm{AV}_{k+1} < b_k$, we would have $\eta_k = b_k - \mathrm{AV}_{k+1}$.

It must be kept in mind that the account on a universal life policy should be viewed as belonging strictly to the particular policyholder, so while the account value is similar in nature to a retrospective reserve, it is not quite the same. For example, withdrawal rates and surrender values are not taken into account in this calculation as was done in some Chapter 12 formulas. To do so would mean that policyholder accounts would be augmented with positive gains from surrender, but these belong strictly to the insurer (as of course do surrender losses, should they arise).

Example 13.1 A universal life policy has a stated death benefit of 100 000, monthly premiums of 5000, monthly expenses of 100, and a further monthly expense charge of 4% of each premium. The annual credited interest rate is 8%, and the monthly cost of insurance rate in the 5th year of the policy is 30 per thousand. If the account value at the end of the 52nd month is 50 000, find the account value at the end of the 53rd month, assuming that the policy is (a) type B and (b) type A.

Solution. We first must calculate the applicable monthly interest rate which is $(1.08)^{1/12} - 1 = .00643$. Then

(a) $\mathrm{AV}_{53} = (50\,000 + 5000(0.96) - 100)(1.00643) - 0.03(100\,000) = 52\,051.72.$

(b) The 100 000 above is replaced with $100\,000 - \mathrm{AV}_{53}$, and solving the equation we just obtain the Type B amount divided by $(1 - 0.03)$.

$$\mathrm{AV}_{53} = 52\,051.72/(1 - 0.03) = 53\,661.57.$$

Another calculation that will normally be done each period is to compute the *cost of insurance* (as opposed to the cost of insurance *rate*). This is sometimes abbreviated as COI_k, for the period running from time k to $k + 1$. (Some authors prefer a different indexing method and would refer to this as COI_{k+1}. We prefer to have the index correspond to that of the cost of insurance rate c_k.) This is normally determined as at the *beginning* of each period, so if there are not sufficient funds to pay it, the policy would lapse in the absence of any guarantees to the contrary. In effect it can be viewed as the net single premium paid each period for the death benefit coverage. That is

$$\mathrm{COI}_k = (1 + i_k)^{-1} c_k \eta_k.$$

So in the above example we would have

$\mathrm{COI}_{52} = (1.00643)^{-1}(0.03)100\,000 = 2980.83$, for the type B policy and
$\mathrm{COI}_{52} = (1.00643)^{-1}(0.03)(100\,000 - 53\,661.57) = 1381.27$, for the type A policy.

There are some variations which can arise. In some circumstances the insurer may use a discount rate j_k in calculating the cost of insurance which is different from the credited interest rate i_k. In this case formula (13.1) would be modified to

$$\mathrm{AV}_{k+1} = [\mathrm{AV}_k + \pi_k(1 - r_k) - e_k - (1 + j_k)^{-1} c_k \eta_k](1 + i_k) \tag{13.2}$$

which shows directly the insurance costs being deducted at the start of the period. Clearly when $j_k = i_k$, the interest factors cancel and we just get back the same formula as before.

Notice however that in general we still have the same formula as before if we adjust the c_k. We change this to a new cost of insurance rate given by

$$c'_k = c_k \frac{1+i_k}{1+j_k}.$$

The following illustrates.

Example 13.2 Redo Example 13.1 and calculate the cost of insurance, under the assumption of a 3% discount rate for the cost of insurance.

Solution. We can do all calculations by changing c_k from 0.03 to $c'_k = 0.03(1.08)^{1/12}/(1.03)^{1/12} = 0.03012$. In the type B policy, this will change the value of AV_{53} to 52 039.72 and COI_{52} to $(1.00643)^{-1}(0.03012)100\,000 = 2992.76$. In the type A policy, this will change the value of AV_{53} to $52\,039/(1-0.03012) = 53655.83$ and COI_{52} to $(1.00643)^{-1}(0.00301)(100\,000 - 53\,655.83) = 1386.96$.

Another modification may be necessary because of the corridor requirement. In a close situation, we must test whether an account value is sufficiently low to meet this restriction. Let cor_k denote the minimum allowable ratio of b_k/AV_{k+1}. If $\mathrm{AV}_{k+1}\mathrm{cor_k} > b_k$, we must redo the calculations. The death benefit will now be $\mathrm{cor}_k F_{k+1}$ and $\eta_k = (\mathrm{cor}_k - 1)\mathrm{AV}_{k+1}$.

Example 13.3 For the policies of Example 13.1, find AV_{53} and COI_{52} assuming that $\mathrm{cor}_{53} = 1.9$.

Solution. For the type B policy the corridor requirement is clearly satisfied and the answer will remain as calculated above.

For the type A policy, we check the original answer to see that 53 661.57(1.9) = 101 957 which is above 100 000, so we must recalculate:

$$\mathrm{AV}_{53} = (50\,000 + 5000(0.96) - 100)(1.00643) - 0.03(0.9\mathrm{AV}_{53}),$$

and solving we get $\mathrm{AV}_{53} = 53\,604.40$. The death benefit will increase to 101 848.36

$$\mathrm{COI}_{52} = 1.00643^{-1}0.03(0.9 \times 53\,604.62) = 1438.07.$$

13.2 Variable annuities

A similar attempt to provide additional flexibility has become popular for certain deferred annuity contracts, which have come to be known as *variable annuities*. They correspond to the contract described in Example 5.8 to the extent that there is no survivorship accumulation prior to the annuity payments. In fact they operate in the same basic manner as we have described above for Universal Life, except that no deductions are made for the cost of insurance. In fact some purchasers use these as a vehicle to accumulate money in a high yielding investment account without any intention of converting the funds to an annuity. The annuity aspect enters into the contract however, since it common to include provisions providing for the conversion of the funds into annuities at guaranteed rates of interest and mortality.

As in the insurance case, an individual's account may be invested in particular assets or it may be equity-indexed, with the credited interest tied to some reference portfolio.

Similarly to Universal Life, their are several types of possible guarantees that have been designed to reduce the losses under bad investment experience. A common provision is a *guaranteed minimum death benefit,* whereby the account holder is promised a certain minimum return upon death, regardless of the account value. The death benefit amount could be the amount originally invested, or it could be that amount together with a certain fixed accumulation rate. A similar type of provision is the *guaranteed minimum accumulation benefit* whereby the end of a specified period, the account holder is promised a minimum account value, regardless of the investment experience. Still another option is the *guaranteed minimum withdrawal benefit.* Under this option the account holder is allowed to withdraw a minimum amount (or a minimum percentage) from their account, each year, until the original invested value is returned. Evaluation of these benefits requires a knowledge of option pricing. An introduction to this topic is found in Chapter 20.

13.3 Pension plans

Pension plans are set up by companies to provide retirement income to a group of employees. This is a vast subject and we confine ourself here to providing a survey of principal features and definitions. Pension plans can be classified into two main categories, *defined benefit* plans, abbreviated as DB, and *defined contribution* plans, abbreviated as DC, as introduced in Section 4.6.2. We now provide more detail.

13.3.1 DB plans

The usual type of DB plan provides that an employee will receive a life annuity, beginning at a specified normal retirement age, with a periodic payment K. Rules for calculating K are specified at the outset. It will normally depend on the employee's salary and years of service. The type of annuity is often a whole life annuity, although there may be options to elect a guaranteed period, or a joint-life annuity with another individual, such a spouse. The income of course will be adjusted according to the nature of the annuity selected.

A typical formula is that K will be equal to r times the average of the employee's last h years of salary times the number of years of service, for some specified r and h. As an example, suppose that the normal retirement age is 65, $r = 0.02$, $h = 3$ and that K is an annual payment. Consider an employee who is hired at age 30, retires at age 65, and whose last 3 years of salary were 89 000, 93 000, 100 000. The employee would then receive an annual pension of $0.02 \times 35 \times 94\,000 = 65\,800$.

A measure which is used to compare various plans is known as the *replacement ratio* and it is simply the ratio of the pension to the final year's salary. In the above example the replacement ratio would be 65.8%.

The quantity h will often be between 3 and 5 years. A possible variation involves plans, known as *career average earnings plan* where h is not fixed, but is equal to the complete number of years of service for each employee.

To evaluate benefits on such a plan the actuary will use an investment discount function v and a multiple-decrement table which will typically show decrements of disability, withdrawal from service, death and retirement. An additional necessary tool is a so-called *salary scale* S_x defined as follows. For some minimal age x_0, we have $S_{x_0} = 1$ and then for $y > x_0$.

$$S_y = \text{ the expected ratio of salary at age } y \text{ to salary at age } x_0.$$

The scale is then used to estimate future salaries. For example, if an employee is earning an annual salary of J at age x, then an estimate of their annual salary as age $x + h$ will be JS_{x+h}/S_x.

Pension plans must specify what benefits if any will be paid to those who leave the employee group for reasons other than retirement at the normal retirement age. Such reasons include disability, death, withdrawal from company employment and retirement at an age other than the normal one. We focus now on early retirement. Plans may specify that a person may receive pension benefits if they retire early, with some minimum criteria specified, which can depend on age and duration of employment. Early retirement means of course that the pension is paid for a longer period, and also that contributions into the plan will not be made for the remaining time to the normal date, so typically the pension income will be appropriately reduced. Rather than calculate the applicable amount of reduction in each case, it is usual to work out approximate figures, known as *actuarial reduction factors* and these are specified as part of the provisions of the plan.

In the following simplified example, provision is made for early retirement up to 2 years before the normal date, but not for other forms of decrement.

Example 13.4 Consider a DB plan which provides a yearly pension of 2% of the final 3-year average salary, times the number of years of service, for retirement at ages 63–65, but with a reduction of 5% per year should retirement occur before 65. An employee now 45 was hired at age 35. His/her present salary is 70 000. The salary scale is given by $S_x = 1.03^{x-30}$, $x \geq 30$. (This simply means that salaries are expected to increase by 3% per year.) An employee may retire at any age from 63 to 65, with a reduction in pension income of 5% per year. Find a formula for the actuarial present value of the pension benefits.

Solution. For a person retiring at age 63, the estimated 3-year final average salary is given by

$$70\,000[1.03^{15} + 1.03^{16} + 1.03^{17}]/3 = 112\,362$$

so that the estimated annual pension income, accounting for the actuarial reduction factor of 10%, is given by $0.90 \times 0.02 \times 28 \times 112\,262 = 56\,630$.

For person retiring at age 64 a similar calculation, now using a 5% reduction results in annual pension income of 63 769, and for a person retiring at age 65, the figure is 71 523.

The actuarial present value is then given by the following (the subscript r denoting the retirement decrement).

$$56\,630\,{}_{18}p_{45}^{(\tau)} q_{63}^{(r)}\, \ddot{a}_{63} + 63\,769\,{}_{19}p_{45}^{(\tau)} q_{64}^{(r)}\, \ddot{a}_{64} + 71\,523\,{}_{20}p_{45}^{(\tau)} q_{65}^{(r)}\, \ddot{a}_{65}.$$

Our solution here incorporates several simplifying assumptions. We have assumed that the pension would be paid annually, instead of the more usual monthly arrangement. We have also assumed that all employees are hired and retired on their birthdays, and that each salary increase occurs on birthdays. It is not too difficult to incorporate more realistic assumptions, using the fractional duration techniques in Chapter 7. One common provision is to assume hiring, retirements, and salary increases at the middle of each year, that is at ages $x + 1/2$, where x is an integer.

Funds to provide the income could be provided solely by the employer, or more commonly shared between the employees and the employer on some specified basis. A usual provision is that the employees will contribute each period a amount of c times their salary for that period. A typical figure might be $c = 0.05$. The employer will then contribute additional amounts which are estimated to be sufficient to provide the promised benefits. There are various methods for doing so. The subject of funding DB pensions is complex, and will not be discussed further.

13.3.2 DC plans

At one time the DB arrangement was the most common one. However periods of low interest rate earnings and improving mortality meant that many employers needed to put in more money then originally estimated in order to ensure the promised level of benefits. In some cases this became prohibitive, resulting in firms switching to the DC mode. In such a pension, the employees contribute a certain percentage of their salary, the employer adds an additional percentage, and the funds are invested and accumulated as an individual account for each participant until retirement. At the time of retirement the total accumulated contributions made on the employee's behalf are used to purchase an annuity at the then prevailing interest and mortality rates. As with the DB plan there is often a targeted goal, but this is not guaranteed. If the investment experience is unfavourable, or mortality has improved so that the cost of life annuities go up, the pension income may be short of the projected amount. Of course, things can go the other way in a DC plan. Very favourable investment returns can result in higher pensions than expected.

Funding arrangements are usually much easier for the DC plan than in the DB case. A common practice is to deposit into the account a certain fraction c of a employee's salary each year, which could be shared in some way between the employee and employer. One does not have to worry as much about benefits for withdrawal or early retirement. Withdrawing employees can be given the amount in their account, either in cash or as a deferred annuity. For those retiring early the amount can be used to buy an annuity starting on the normal retirement date.

In the following examples, we continue to make the simplifying assumptions noted above,

Example 13.5 For an employee hired at age 35 at a salary of 50 000, it is estimated that an amount of 500 000 is needed at at the normal retirement age of 65 to buy an appropriate pension. For exit from the plan before age 65 the accumulated amount of contributions with interest is returned to the employee. Find a formula to calculate the contribution rate c, assumed to be made at the beginning of each year.

Solution. The idea is similar to Example 5.8. We solve the following equation to determine c.

$$500\,000 = 50\,000\ c\ \mathrm{Val}_{30}(\mathbf{g}; v) \tag{13.3}$$

where

$$g_k = S_{35+k}/S_{35}, \quad k = 0, 1, \ldots, 34. \tag{13.4}$$

There are many variations in practice. Many plans have a so-called *vesting requirement*, which means that an employee does not receive back the employer's contributions unless they remain in the plan for a minimum period.

Example 13.6 Suppose that in the situation described above, the contributions are split, with the employee paying 40% and the employer the remaining 60%. For participants exiting the plan after 5 years, the total of all contributions with interest is returned at the end of the year of leaving. For those exiting in the first 5 years only the employee's own contributions with interest are returned. Find an equation to determine the contribution rate c.

Solution. We use a variation of formula (5.4).

$$c = \frac{500\,000 v(35)_{35}p_{30}^{(\tau)}}{\ddot{a}_{35}^{(\tau)}(\mathbf{g}) - \mathbf{A}_{35}^{(\tau)}(\mathbf{j})}.$$

where $\mathbf{g}$ is as in (13.4) and

$$j_k = \begin{cases} 0.4 \sum_{i=0}^{k-1} v(k,i) S_{35+i}/S_{35} & \text{if } 0 \le k < 5, \\ \sum_{i=0}^{k-1} v(k,i) S_{35+i}/S_{35} & \text{if } 5 \le k < 25. \end{cases}$$

Exercises

13.1 A universal life policy provides a death benefit at the end of the period of death equal to the account value, but subject to a minimum of 30 000. Monthly premiums are 2000. There is a monthly expense charge of 50 and a further charge of 3% of each premium. The annual credited interest rate is 6%. The account value at the end of the 30th month is 27 318. The monthly cost of insurance during the third year of the policy is 15 per 1000. Find AV_{31} and COI_{30}.

13.2 A universal life policy provides a death benefit at the end of the period of death of 100 000. The account value at time 20 is 2500. For the time period from time 20 to time 21, the credited interest rate is 0.006, the expense charge is 5% of any premium paid plus 100, and the cost of insurance rate is 30 per thousand.

(a) Suppose that at time 20, the policyholder makes the minimum premium payment of 500, which according to the guarantee in the contract ensures that the cost of insurance will be paid regardless of the amount in the account. Find the account value at time 21.

(b) If the contract does not carry the secondary guarantee of part (a), what is the minimum premium that the policyholder will have to pay at time 20 to ensure that the cost of insurance can be paid.

13.3 At a certain time k, two universal life policies have exactly the same account value, death benefit, premiums, expenses and cost of insurance rates. In both cases the cost of insurance discount rate is the same as the credited interest rate. One policy is type A and the other is type B. Show that the difference between the type A and type B account values at the end of the year is the same as the difference between the type B and type A COI for that period, accumulated with interest to the end of the year.

13.4 A universal life policy provides a death benefit equal to the account value plus 50 000. Monthly premiums are 3000. There is a monthly expense charge of 100 and a further charge of 4% of each premium. The annual credited interest rate is 8%. The account value at the end of the 73rd month is 49 200. The monthly cost of insurance for the following month is 20 per 1000 and is computed at an annual interest rate of 4%. There is a corridor requirement which specifies that the death benefit must be at least twice the account value. Find AV_{74} and COI_{73}.

13.5 Redo Example 13.4, only assume salary increases of 4% per year, and an actuarial reduction factor of 3% per year.

13.6 Redo Example 13.6, only assuming that an employee who dies during the first 5 years, receives at the end of the year of death, both their own and the employer's contribution accumulated with interest.

13.7 An employee starts employment in a firm at age 35 and is offered a choice of either a DC or a DB plan. In the DC plan, contributions of 15% of salary are accumulated with interest at 4% until age 65 and then used to buy a life annuity. Under the DB plan, the employee is given an annual amount, beginning at age 65 of 1.8% of their 3 years average salary times the number of years of service. If the cost of a 1-unit life annuity at age 65 is 10, and salaries increase by 4% per year, find the ratio of the annual income under the DC plan to that of the DB plan, for an employee who remains in the plan until age 65.

13.8 A DB plan provides for an annual pension of a certain percentage of the final 5 years average salary times the number of years of service upon retirement. Retirement is based on the *80 factor* which means that an employee can retire at any time that the sum of age plus years of service is greater than or equal to 80. An employee age 55 who began work at age 25, is comparing their income for retirement now, to that which they would receive if they stay for another year. They can expect a 5% increase in salary for the following year, which is the same that they received in each of the last 4 years. Find the ratio of the pension income for retirement in 1 year to that for retirement now.

Part II

THE STOCHASTIC LIFE CONTINGENCIES MODEL

14

Survival distributions and failure times

14.1 Introduction to survival distributions

Our goal in this part of the book is to introduce a stochastic model for mortality to replace the deterministic model used in Part I. This will not only provide us with a more realistic description of human mortality, but it will also have more general applications.

The basic information a prospective issuer of an insurance or annuity contract wants to know is how long the life in question will live. The insurer obviously cannot hope to answer this question exactly, since the actual future lifetime lived is random. Some people age 50, for example, will live another 40 years or more, while others will die very soon. In the deterministic model, we circumvented this issue by assuming that while we could not identify how long a particular individual would live, we could identify how many individuals of a given age would live to some other age. Clearly, however, the number of such individuals is also random. In the stochastic model we will face this randomness directly.

This and subsequent chapters will require a more advanced knowledge of probability than we have assumed so far. We follow the notation and terminology of Appendix A. For the present chapter, see in particular Sections A.4–A.8 and note that P will denote probability.

We do not need to confine ourselves to looking at the time of death of an individual. Suppose we are interested in some event that will occur once and only once at some random future time. We will refer to this event as 'failure'. The random variable T, the time of occurrence of such an event, is known as a *failure time*, and its distribution is often referred to as a *survival distribution*. At any time before 'failure' we will say we are in a state of 'survival'.

Our motivating example is the case where the event in question is the death of (x). In this case we will denote T by $T(x)$.

Fundamentals of Actuarial Mathematics, Third Edition. S. David Promislow.

Companion website: http://www.wiley.com/go/actuarial

There are, however, many other examples of interest. Suppose a manufacturer sells a product with a guarantee that it will be replaced it if fails before a certain time. In order to assess the cost of the guarantee the manufacturer wants to know the distribution of the product's failure times. We see that a guarantee can be viewed as type of insurance policy.

We have already encountered more general failure times in Chapter 10. Joint-life insurance and annuities were seen to be the same type of contracts as those for single lives except that the failure time was defined to be the time of first death. Similarly, for last-survivor annuities or second-death insurances, the failure time was the time of second death. Multiple-decrement theory provided many more examples of failure times.

14.2 The discrete case

Consider the case where failure can occur only at integer times $1, 2, 3, \dots$, so T is a discrete random variable with positive integers as values. Refer in particular to Section A.4. In place of the cumulative distribution function F, it is often more convenient to use the *survival function* s defined by

$$s(k) = 1 - F(k) = P(T > k).$$

This gives the probability that failure has not yet occurred by time k, or in other words that we are still in a state of survival at time k. If f is the probability function of T, it is clear from (A.5) that

$$f(k) = s(k-1) - s(k), \qquad s(k) = \sum_{i=k+1}^{\infty} f(i) = 1 - \sum_{i=1}^{k} f(i). \tag{14.1}$$

There is another important method of describing the distribution of T.

Definition 14.1 The hazard function of T at time k, denoted by $\lambda(k)$ is the condition probability of failure at time k given survival up to time $k - 1$. That is,

$$\lambda(k) = \frac{f(k)}{s(k-1)}.$$

This of course is defined only for those integers k such that $s(k-1) > 0$. Once $s(k-1)$ equals 0, there is no possibility of survival up to that point. (The hazard function is also known as the hazard *rate* or *intensity* function or *failure* function.)

Readers should make sure that they understand the difference between $f(k)$ and $\lambda(k)$. Both quantities give the probability of failure at time k, but from different perspectives. If, at time 0, we are asked to assess the likelihood that failure will occur at time k, the answer is just $f(k)$. As time goes on, our assessment of this likelihood must change. For example, if failure takes place before time k, then we know that the probability of failure at time k is zero. Suppose that at time $k - 1$, failure has not yet occurred, and we ask ourselves the same question. The answer now is $\lambda(k)$. The difference between the two functions is further clarified by writing the definition of λ in the form

$$f(k) = s(k-1)\lambda(k), \tag{14.2}$$

expressing the fact that we view failure at time k as arising from two events. First, there must be survival up to time $k-1$, and then, given this survival, failure must occur at time k.

In deriving an appropriate survival distribution, it is often easier to model the hazard rate rather than s or f. Given λ, we can then easily determine the other functions as follows. Equating the two different expressions for $f(k)$ given in (14.1) and (14.2),

$$s(k) = s(k-1)[1-\lambda(k)]. \tag{14.3}$$

Beginning with $s(0) = P(T > 0) = 1$, we know that $s(1) = 1 - \lambda(1), s(2) = s(1)[1-\lambda(2)] = [1-\lambda(1)][1-\lambda(2)]$, and proceeding inductively,

$$s(k) = [1-\lambda(1)][1-\lambda(2)]\cdots[1-\lambda(k)]. \tag{14.4}$$

Example 14.1 A bag contains 3 green balls and 1 red ball. A ball is drawn at random. If green, it is replaced and the draw is repeated. Failure occurs when a red ball is drawn. If this is on the kth draw we say that failure occurs at time k. Find $s(k), f(k)$ and $\lambda(k)$.

Solution. From the conditions of the problem we can immediately deduce that

$$\lambda(k) = \frac{1}{4}, \quad k = 1, 2, \ldots.$$

From (14.4),

$$s(k) = \left(\frac{3}{4}\right)^k, \quad k = 1, 2, \ldots.$$

and from (14.2),

$$f(k) = \frac{3^{k-1}}{4^k}, \quad k = 1, 2, \ldots.$$

We have the well-known geometric distribution. See Section A.11.3. As a check we can also compute $f(k)$ from the first formula in (14.1), which of course yields the same answer.

14.3 The continuous case

In most applications, the time of failure is not restricted to the integers but can be arbitrary. We model this by assuming that T is a continuous random variable with values in $[0, \infty)$.

14.3.1 The basic functions

We define the *survival function* s as in the discrete case,

$$s(t) = 1 - F(t) = P(T > t),$$

the probability of survival up to time t. It is related to the probability density function (p.d.f) by

$$f(t) = -s'(t), \qquad s(t) = \int_t^\infty f(r)\mathrm{d}r = 1 - \int_0^t f(r)\mathrm{d}r. \tag{14.5}$$

Definition 14.2 The hazard function of a continuous failure time T at time t, denoted by $\mu(t)$, is the continuous density function for failure at time t given survival up to that point. It is given by

$$\mu(t) = \frac{f(t)}{s(t)}$$

for all t such that $s(t) > 0$.

For small Δt, $\mu(t)\Delta t$ approximates the probability that T takes a value in the interval $[t, t + \Delta t]$ given survival up time t. Analogously to (14.2), we have the expression

$$f(t) = s(t)\mu(t). \tag{14.6}$$

To determine the other quantities from μ we note from (14.5) and (14.6) that

$$\mu(t) = \frac{-s'(t)}{s(t)} = -\frac{\mathrm{d}}{\mathrm{d}t}[\log s(t)].$$

Following the proof of Proposition 8.1, we deduce that

$$s(t) = \mathrm{e}^{-\int_0^t \mu(r)\mathrm{d}r}. \qquad (14.7)‡$$

Formula (14.7) seems quite different from its discrete counterpart (14.4), but if you look at it in the right way, it really is a natural continuous version. Recall that $\mathrm{e}^{-\sum a_i} = \prod \mathrm{e}^{-a_i}$. So if we think of an integral as a type of generalized sum, then e to the integral is a type of generalized product of terms of the form $\mathrm{e}^{-\mu}$, which for 'small' values of μ are close to $1 - \mu$.

14.3.2 Properties of μ

We know that the density function f must be nonnegative and satisfy $\int_0^\infty f(t)\mathrm{d}t = 1$, and that the distribution function F must be nondecreasing and satisfy $F(0) = 0, \lim_{t\to\infty} F(t) = 1$. The latter implies that the survival function s is nonincreasing and satisfies $s(0) = 1, \lim_{t\to\infty} s(t) = 0$.

What are the corresponding properties for the hazard function? The hazard function μ must be nonnegative on its domain and in addition satisfy

$$\int_0^\infty \mu(t)\mathrm{d}t = \infty.$$

To see this, suppose that the above integral had a finite value a. From (14.7) we would deduce that $\lim_{t\to\infty} s(t) = \mathrm{e}^{-a}$ which is not equal to 0, contradicting the fact that failure must occur at some time. In other words, if the integral is finite, the hazard function is not large enough to guarantee failure and there would be some chance of surviving forever. Of course, we may want to model situations where there is a chance that failure will never occur, and in this case we would want the above integral to be finite.

14.3.3 Modes

The general shape of the density function of T can often be inferred by looking at μ. In particular, we may be interested in *modes*, which are points where the density function assumes a local maximum. We assume that f is differentiable, and note from (14.5) and (14.6) that

$$f' = (s\mu)' = s'\mu + s\mu' = -f\mu + s\mu' = -s\mu\mu + s\mu' = s(\mu' - \mu^2). \tag{14.8}$$

We see then that f is increasing or decreasing according as μ' is greater than or less than μ^2. Points at which modes can occur are restricted to or those values of t for which $\mu'(t) = \mu(t)^2$, or possibly to endpoints of the domain. We will look at some particular examples later.

14.4 Examples

In this section we survey some familiar distributions that can be used to model failure times in certain situations.

The family of exponential distributions (see Section A.11.6) can be easily described as those with a *constant* hazard function. If $\mu(t) = \mu$ for all t, then from (14.6) and (14.7), $s(t) = \mathrm{e}^{-\mu t}$. Such a distribution is suitable for modelling situations where the chance of failure in the next instant remains constant regardless of the time. This is shown by the constancy of the hazard function as well as the property given in A.57. It is not suitable for modelling human mortality, where the aging process means that we would expect the hazard function to increase with time. See Section 8.10.1 where we discussed the same point in the deterministic model.

Another overly simplistic approach is to take a continuous uniform distribution (see Section A.11.4). For most applications, this is not realistic, but it has been used as a rough approximation to model human mortality. This is simply the stochastic version of Demoivre's law, introduced previously in Chapter 8. Note that it does give an increasing hazard function since

$$\mu(t) = \frac{f(t)}{s(t)} = \frac{1}{N-t}, \quad 0 \le t < N.$$

The first serious attempt to capture some of the relevant features for mortality was the *Gompertz* distribution given by

$$\mu(t) = Bc^t, \quad \text{for some parameters } B > 0 \text{ and } c > 1.$$

This was introduced by Benjamin Gompertz in 1825. His idea was that the hazard function for human mortality should increase with age at a rate proportional to itself, that is, at an exponential rate. It has been used extensively for modelling mortality in both humans and other species. For human mortality, a good fit can usually be found over the middle span of ages by choosing the parameters appropriately.

One feature of this distribution is that $\mu'(t) - \mu(t)^2 = \mu(t)[\log c - \mu(t)]$ is positive, zero, or negative, precisely when $\mu(t)$ is less than, equal to, or greater than $\log c$. Then, (14.8) shows that if $\mu(0) \geq \log c$, the density function is decreasing, while if $\mu(0) < \log c$, there is a unique mode at the point t where $\mu(t) = \log c$. The density function increases up to this point and then decreases. This does indeed seem to fit the pattern of human mortality over the middle range of ages. Starting about age 30, the probability of dying at age x will increase with x up to a certain point. There is clearly more chance of dying at age 70 than at age 30. However, at sufficiently high ages the chance of dying starts to decrease with age. For example, there is a very small chance of dying at age 110, for the simple reason that very few people will live that long in the first place. It may be enlightening to look again at (14.6), which expresses f as the product of two functions, one of which is decreasing, and one increasing.

The Gompertz distribution does not fit well with observed mortality at ages below 30 or above 70 or so, for the reasons explained at the beginning of Section 8.10.1. In 1860, a modification was advocated by Makeham. He suggested that a constant be added to the hazard function to cover causes of death that were age-independent, such as accidents. The Makeham distribution, often referred to as the Gompertz–Makeham distribution, is then given by $\mu(t) = A + Bc^t$, for some $A > 0$. The extra parameter allows for more flexibility in shape (see Exercise 14.12) but still does not capture observed mortality behaviour at very young or very old ages.

14.5 Shifted distributions

Suppose we are given a failure time T. (For convenience we will deal with the continuous case, but the conclusions for the discrete case are similar.) As remarked above, having reached a point u at which failure has not yet occurred, we must alter our assessment of the likelihood of failure at different times. To formalize this, we define a new random variable.

The random variable $T \circ u$ is equal to the time until failure occurs, as measured from time u, given survival up to time u. Therefore, it takes a value of $T - u$ when T takes a value greater than u, with probabilities conditioned on the event that T is greater than u. The survival function, density function, and hazard rate function, respectively, for $T \circ u$ are expressed in terms of the corresponding functions for T as follows. (Since we are dealing with more than one distribution here, we will insert appropriate subscripts on s,f and μ.)

$$s_{T \circ u}(t) = \frac{s_T(u+t)}{s_T(u)}. \tag{14.9}$$

$$f_{T \circ u}(t) = -\frac{\mathrm{d}}{\mathrm{d}t} s_{T \circ u}(t) = \frac{f_T(u+t)}{s_T(u)}. \tag{14.10}$$

$$\mu_{T \circ u}(t) = \frac{f_{T \circ u}(t)}{s_{T \circ u}(t)} = \mu_T(u+t). \tag{14.11}$$

Formula (14.10) provides a nice way to visualize this concept. To get the graph of the density function of the shifted random variable, we take the graph of the original density function, ignore everything to the left of u, and scale all values upward to get the total area equal to 1.

For some distributions, the shifted random variables are of the same family as the original. This makes these distributions convenient for modelling purposes.

Example 14.2 Describe the random variables $T \circ u$ when (a) T is an exponential random variable with constant hazard μ, (b) T is uniform on the interval $[0, N]$, (c) T is Gompertz–Makeham with parameters A, B and c.

Solution. In all cases it is convenient to use the hazard function and (14.11).

(a) $\mu_{T \circ u}(t) = \mu(u + t) = \mu$, showing that $T \circ u$ has the same distribution as T. This indicates the so-called *memory-less* property of exponential random variables. The time until the event in question occurs will not be affected by how long you have already waited.

(b) $\mu_{T \circ u}(t) = \mu(u + t) = 1/(N - u - t)$, showing that $T \circ u$ is uniform on the interval $[0, N - u]$.

(c) $\mu_{T \circ u}(t) = A + Bc^u c^t$, showing that $T \circ u$ is also Gompertz–Makeham with the B parameter changed to Bc^u.

14.6 The standard approximation

Associated with every continuous failure time T is a discrete failure time $\tilde{T}$. Failure according to $\tilde{T}$ will occur at the integer time k, if failure under T has occurred in the time interval $(k - 1, k]$. To be precise,

$$\tilde{T} = [T] + 1, \quad \text{where } [\cdot] \text{ denotes the greatest integer function.}$$

Let f and s denote the density function and survival function, respectively, of T. Let $\tilde{f}$ and $\tilde{s}$ denote the corresponding functions for $\tilde{T}$. Then, for all positive integers k,

$$\tilde{f}(k) = \int_0^1 f(k - 1 + t)\mathrm{d}t = s(k - 1) - s(k), \tag{14.12}$$

$$\tilde{s}(k) = s(k). \tag{14.13}$$

Suppose we want to deduce information about T by observing when failure occurs. It could be difficult or impossible to observe continuously and we may be only able to view the situation at certain discrete times, say 1, 2, 3, etc. If we observe at time k, and see that failure has occurred, we know only that failure occurred between time $k - 1$ and k. In other words, we are observing values of $\tilde{T}$. We would like therefore to infer the distribution of T from that of $\tilde{T}$. Clearly, we cannot do this exactly and must make some type of approximation. A simple method, consistent with (14.12), is just to assume that for all nonnegative integers k, and any t in the interval (0,1),

$$f(k+t) = \tilde{f}(k+1). \tag{14.14}$$

We will refer to this as *the standard approximation.* In fact, we have already used this approximation in our deterministic model, in the form of the UDD assumption for the random variable $T(x)$, as indicated by the equivalent formulation in terms of survival functions, namely,

$$s(k+t) = (1-t)s(k) + ts(k+1), \tag{14.15}$$

for an integer k and $0 < t < 1$. To see this equivalence, note that, given (14.15), we obtain (14.14) by differentiating with respect to t. Conversely, given (14.14), we have

$$\begin{aligned} s(k+t) &= s(k) - \int_0^t f(k+r)\mathrm{d}r \\ &= s(k) - \int_0^t \tilde{f}(k+1)\mathrm{d}r = s(k) - t\tilde{f}(k+1) \\ &= s(k) - t[s(k) - s(k+1)]. \end{aligned}$$

For the purpose of calculating moments, it is useful to introduce the random variable

$$R = \tilde{T} - T,$$

the duration from the time of failure until the end of the year of failure. Given a value r in the interval (0,1), consider the probability that $R \leq r$ given that $\tilde{T} = k$. This is the probability that T takes a value between $k - r$ and k, given that $\tilde{T} = k$. Using the standard approximation, this probability is given by

$$\frac{1}{\tilde{f}(k)} \int_{k-r}^{k} f(t)\mathrm{d}t = \frac{1}{\tilde{f}(k)} \int_{k-r}^{k} \tilde{f}(k)\mathrm{d}t = r.$$

This shows that under R is independent of $\tilde{T}$ and has a uniform distribution on the interval (0,1). Using standard results about the uniform distribution, we can calculate that under the standard approximation

$$E(T) = E(\tilde{T}) - E(R) = E(\tilde{T}) - \frac{1}{2}, \tag{14.16}$$

$$\mathrm{Var}(T) = \mathrm{Var}(\tilde{T}) + \mathrm{Var}(R) = \mathrm{Var}(\tilde{T}) + \frac{1}{12}, \tag{14.17}$$

where E denotes expectation and Var denotes variance.

14.7 The stochastic life table

The goal in this section is to establish a link between the stochastic model and the deterministic model for mortality. In this and subsequent chapters we will use this link to show that the questions that we answered in the deterministic model can be answered in the stochastic model in the *very same* way. In addition, we will be able to answer more questions.

The tool for doing so will be to return to the fundamental concept of a life table but look at it in a different way. As we did before, we start out with an arbitrary number ℓ_0 of newly born lives. We now let $\mathscr{L}_x$ be the number of these still alive at age x, and $\mathscr{D}_x$ be the number of these who will die between the age of x and $x+1$. These quantities look like the ℓ_x and d_x of Chapter 3, but we now recognize that these are not numbers but *random variables*. We obtain the life table by letting

$$\ell_x = E(\mathscr{L}_x), \qquad d_x = E(\mathscr{D}_x).$$

Since it is clear that $\mathscr{L}_{x+1} = \mathscr{L}_x - \mathscr{D}_x$, we have the same formula (3.1) as we had before. In other words, in our stochastic model we still can introduce a life table, but we now view the numbers as expected values rather than as an exact account of the numbers who will live or die.

Consider now the random variable $T(x)$, the time to the death of (x). We will write $s_x(t), f_x(t), \mu_x(t)$ for the survival function, density function, and hazard function respectively of $T(x)$.

For age 0, there is somewhat a different standard notation, which can cause some confusion if care is not taken. The random variable $T(0)$ is traditionally denoted by X, and the variable denoted by x rather than t, since $T(0)$ is really the *age* at death. It is common to suppress the 0 and just write $s(x)$, $f(x)$ and $\mu(x)$ for $s_0(x)$, $f_0(x)$ and $\mu_0(x)$, respectively.

A key observation is that

$$\ell_x = \ell_0 s(x). \tag{14.18}$$

The argument is the same as showing that if you flip N coins, each with a probability p of coming up heads, you can expect to get Np heads. In this case, if you take ℓ_0 people, each with a probability $s(x)$ of surviving to age (x), then you can expect to get $\ell_0 s(x)$ such survivors.

We are now ready for the main purpose of this section, which is to establish a correspondence between the survival, hazard, and density functions of $T(x)$ and the life table functions for the stochastic life table. At first, we assume aggregate mortality, as in Chapter 3, where we used a single-life table for the future mortality of all ages x, by just starting at age x. This corresponds stochastically to the statement

$$T(x) \text{ has the distribution of } T(0) \circ x,$$

which simply says that the future lifetime of (x) is the future lifetime after time x of a person who was age 0, given that they have lived to age x. There are therefore no effects of selection.

Suppose also that we calculate the quantities ${}_tp_x, {}_tq_x, \mu(x)$ starting from the life table, exactly as we did in the deterministic model. We then have from (14.10) and (14.23) that

$$s_x(t) = \frac{s(x+t)}{s(x)} = \frac{\ell_{x+t}}{\ell_x} = {}_tp_x. \tag{14.19}$$

This is certainly reasonable. Although looked at from different points of view, both of these are probabilities that a person age x will survive t years. It follows immediately that

$$F_x(t) = 1 - {}_tp_x = {}_tq_x, \tag{14.20}$$

and that the hazard rate function of $T(x)$ is given by

$$\mu_x(t) = -\frac{\mathrm{d}}{\mathrm{d}t} \log {}_tp_x = \mu(x+t), \tag{14.21}$$

where the last quantity is the force of mortality of Chapter 8. (Note that in the case of hazard rates, the notation in the deterministic model was chosen from the outset to correspond to that in the stochastic setting.) Finally,

$$f_x(t) = {}_tp_x\mu_x(t), \tag{14.22}$$

the familiar expression that we used for insurance premiums in Chapter 8.

If we assume the strictly select mortality of Chapter 9, then we cannot assume that $T(x)$ is distributed as $X \circ x$. The above correspondences would be modified as follows:

$$s_x(t) = {}_tp_{[x]}$$

and

$$f_x(t) = {}_tp_{[x]}\mu_x(t).$$

It is also instructive to compute the distribution of multiple-life failure times in terms of the quantities introduced in the discrete model. For example, consider $T(xy)$, the time of failure of the joint-life status. For this random variable, the survival function is just ${}_tp_{xy}$, the hazard function is $\mu_{xy}(t)$, and the density function is ${}_tp_{xy}\mu_{xy}(t)$.

14.8 Life expectancy in the stochastic model

Let $\tilde{T}(x)$ be the discrete random variable associated with $T(x)$ as defined in Section 14.7. It will have survival, probability and hazard functions given by

$$\tilde{s}_x(k) = {}_kp_x, \tag{14.23}$$

$$\tilde{f}_x(k) = {}_{k-1}p_x - {}_kp_x = {}_{k-1}p_x\, q_{x+k-1}, \tag{14.24}$$

$$\lambda_x(k) = q_{x+k-1}. \tag{14.25}$$

Traditional actuarial terminology defines a discrete random variable $K(x)$ known as the *curtate future lifetime*, which measures the whole number of future years to be lived by (x). It is clear that

$$K(x) = \tilde{T}(x) - 1.$$

From (14.24), (3.8), (A.12), and the fact that $s_x(0) = 1$, we can write

$$e_x = \sum_{k=1}^{\omega-x-1} s_x(k) = E[\tilde{T}(x)] - s_x(0) = E[K(x)].$$

Therefore life expectancy is indeed an expectation in the sense of probability theory. The formal definition of *complete life expectancy* in the stochastic model is

$$\mathring{e}_x = E[T(x)].$$

From (A.15) we see that this agrees with our Chapter 8 definition. Moreover, using the standard approximation and (14.16), we verify the fact that

$$E[T(x)] = e_x + \frac{1}{2}.$$

We can already see one advantage of the stochastic model. As well as computing the expected future lifetime, we can compute the variance of future lifetime.

Example 14.3 Given $\ell_{90} = 100$, $\ell_{91} = 90$, $\ell_{92} = 70$, $\ell_{93} = 40$, $\ell_{94} = 10$, $\ell_{95} = 0$, compute the expectation and variance of $T(90)$, assuming UDD.

Solution. Using either (A.7) or (A.16) we find that $E[\tilde{T}(90)] = 3.1$, $E[\tilde{T}(90)]^2 = 10.9$, so that $\text{Var}[\tilde{T}(90)] = 1.29$. Then, from (14.16) and (14.22),

$$\mathring{e}_{90} = 2.6, \qquad \text{Var}[T(90)] = 1.373.$$

14.9 Stochastic interest rates

Stochastic interest is a large topic which we will only allude to briefly. As well as the life table, the other main ingredient of the actuaries tool-kit, the discount function, should be treated stochastically. For the most part, the method of doing so in the pricing and valuation of insurance and annuity contracts has been by a simulation technique. Various scenarios of possible future interest rates are selected, together with some weighting as to how likelihood their occurrence is. Premiums or reserves are computed using each scenario, exactly as we have described, and the totality of results are analyzed. In this way, one can arrive at estimates of the expected value of the quantities, or in some cases, particularly for reserve purposes, an idea of the worst case scenarios.

A more sophisticated approach is to model $v(s, t)$ as a random variables rather than definite numbers, as we mentioned in Chapter 2. Some aspects of this idea are dealt with in Chapter 20. Numerous stochastic interest rate models have been proposed. The most popular approach is to express the force of interest $\delta(t)$ as a stochastic process rather than a function of t.

Notes and references

Standard actuarial terminology, which is based on deterministic thinking, refers to Gompertz's *law*, rather than to a Gompertz distribution. This is stated as $\mu(x) = Bc^x$. In our framework this means that $T(0)$ has a Gompertz distribution and $T(x) = T(0) \circ x$, so that $T(x)$ has a Gompertz distribution for all x. Similar remarks apply to the Makeham modification.

It was common practice at one point to use life tables that satisfied Gompertz's or Makeham's law. A major motivation was to simplify joint-life calculations, by reducing joint-life statuses to a single-life or a joint-life status with equal ages. See Exercise 14.10. There is little need for this with modern computing methods.

Additional material on Gompertz, Makeham and more general mortality distributions can be found in Brillinger (1961), Carrière (1994a) and Tennenbein and Vanderhoof (1980).

Exercises

Type A exercises

14.1 Redo Example 14.1, but now assume that each time a green ball is drawn, it is replaced with a red ball. In this case what is the most likely time of failure?

14.2 Redo Example 14.1, but now assume that each time a green ball is drawn, a new red ball is added.

14.3 A failure time T has the hazard function $\mu(t) = (1+t)^{-2}$. What is the probability that failure never occurs?

14.4 Suppose that $q_{97} = 0.5$, $q_{98} = 0.6$, $q_{99} = 0.8$, $q_{100} = 1$, and that UDD holds. Find (a) $E[T(97)]$ and (b) $\text{Var}[T(97)]$.

14.5 If Demoivre's law holds with $\omega = 100$, find the variance of $T(60)$.

Type B exercises

14.6 Consider two copying machines. Machine 1 will make 20 000 copies per month. When new, it will last T months, where T is a random variable with hazard function

$$\mu_T(t) = \frac{1}{40-t}, \quad 0 \le t < 40.$$

Machine 2 will make 15 000 copies per month. When new, it will last for S months, where S is a random variable with hazard function

$$\mu_S(t) = \frac{2}{96-t}, \quad 0 \le t < 96.$$

A prospective purchaser is trying to decide between buying a new machine 1, or a 2-year-old machine 2. Which of these two choices will produce, on average, the largest total number of copies during its lifetime?

14.7 Suppose that T has an exponential distribution with constant hazard 0.2.

(a) What is the probability that $\tilde{T} = 3$?

(b) What is the error made if we approximate the probability that $2.5 \leq T \leq 2.75$ by using the standard approximation?

14.8 Consider the distribution with density function $f(t) = \beta^2 t e^{-\beta t}$, $t \geq 0$. (This is a gamma distribution with first parameter 2.)

(a) Show that $s(t) = (1 + \beta t)e^{-\beta t}$.

(b) Find the hazard function $\mu(t)$.

(c) Describe precisely the shifted distribution $T \circ u$. (*Hint*: It is a mixture of two well-known distributions.)

14.9 For two independent lives (x) and (y), (x) is subject to a force of a mortality of $\mu_x(t) = 2/(10 - t)$ while (y) is subject to the survival function $s_y(t) = 1 - (t/10)$, in both cases for $0 \leq t < 10$. Find the p.d.f. for the following random variables.

(a) $T(xy)$, the time of failure of the first death;

(b) $T(\overline{xy})$, the time of failure of the second death.

14.10 (The law of uniform seniority)

(a) Suppose that Gompertz's law holds (see Section 14.9). Show that there is function g, defined on the nonnegative reals, such that

$${}_tp_{x:x+n} = {}_tp_{x+g(n)}, \qquad \text{for all } t > 0.$$

(b) Suppose that Makeham's law holds. Show that there is a function h, defined on the nonnegative reals, such that

$${}_tp_{x:x+n} = {}_tp_{x+h(n):x+h(n)}, \qquad \text{for all } t > 0.$$

14.11 Show that the sample table of Section 3.7 satisfies Gompertz's law (that is up to age 119 where a modification was made to get a finite value of ω).

*14.12 Suppose that T has a Gompertz–Makeham distribution with $A > 0$. Show that f_T is either (i) decreasing, (ii) increasing then decreasing or (iii) bimodal with two local maximums. Identify the conditions on the parameters that will result in each case.

14.13 For each of the following two families of distributions, decide whether or not the shifted distributions are from the same family, with changed parameters.

(a) $\mu(t) = \alpha(t + \theta)^{-1}$ for positive α and θ (Pareto);

(b) $\mu(t) = kt^r$ for positive k and r (Weibull).

15

The stochastic approach to insurance and annuities

15.1 Introduction

In this chapter we deal with a perfectly general failure time T, and develop the stochastic approach for calculating premiums and reserves for insurance and annuity contracts based on T. These will all be calculated as expected values of appropriate random variables. In the particular case where $T = T(x)$, we will show, using the correspondence established in Chapter 15, that these agree with the results that we obtained in the deterministic model. The advantage of the stochastic approach is that we can augment these expected values with other quantities, such as variances.

Throughout the chapter, f, s and μ will denote the density, survival and hazard functions respectively of T. We let $\tilde{T}$ be the associated discrete failure time, $\lfloor T \rfloor + 1$. We will let $\tilde{f}$ and λ denote the probability function and hazard function respectively of $\tilde{T}$. We denote the survival function of $\tilde{T}$ with the same symbol s. This will not cause confusion since the value at any integer is the same in both cases.

In some cases the values of T are bounded (e.g., the values of $T(x)$ are bounded by $\omega - x$), but this is not necessary. The upper limit on integrals or sums will be written as ∞ to cover all possibilities.

The approach differs somewhat from the deterministic case and it is important to understand the distinction. We suppose, as we did before, that we have a fixed deterministic investment discount function v. We will consider contracts that pay certain benefits provided that failure has not occurred, and/or benefits at the time of failure. We will want to determine present values of these benefits, and these will all be computed with respect to the discount function v. In this model, we do not have the interest and survivorship discount function that we had before. However, the present values that we obtain will not be definite numbers but rather *random variables*, since they will depend on the unknown value that T assumes.

Fundamentals of Actuarial Mathematics, Third Edition. S. David Promislow.

Companion website: http://www.wiley.com/go/actuarial

Premiums and reserves will be calculated as *expectations* of these random variables, and it is in calculating these expectations that the probabilities of life and death are taken into account. These expectations are commonly known as *actuarial present values*, abbreviated as APV.

Throughout this chapter and the next, we adopt the following notation, which will ensure that our theory is applicable both to the net premium situation, as discussed in Chapters 4–6 and to the gross premium situation as discussed in Chapter 12. The premium symbol π_k will denote the *total inflow* received at time k, which will be the net premium in a model which ignores expenses, or in the expense situation will be the gross- or expense-augmented premium less the expenses paid at time k. Similarly the benefit b_k paid for failure at time k will include any costs involved with making the benefit payment. However to keep things simple we confine our attention to a single decrement case where we have only one cause of failure. In particular then, in the context of life insurance we are ignoring withdrawals, implicitly making the assumption discussed in the second remark after Example 12.2.

15.2 The stochastic approach to insurance benefits

We deal here with contracts that provide benefits upon failure, and consider both the discrete and continuous cases.

15.2.1 The discrete case

We consider a contract with failure benefit vector **b** paying b_{k-1} at time k for failure between time $k-1$ and time k. Let Z denote the present value of the benefits with respect to v. Then Z is a function of the random variable $\tilde{T}$, since when $\tilde{T} = k$, the value of Z is $b_{k-1}v(k)$. We can therefore write

$$Z = b_{\tilde{T}-1}v(\tilde{T}).$$

Letting $A_{\tilde{T}}(\mathbf{b};v)$ denote $E(Z)$, we have

$$A_{\tilde{T}}(\mathbf{b};v) = \sum_{k=1}^{\infty} b_{k-1}v(k)\tilde{f}(k) = \sum_{k=0}^{\infty} b_k v(k+1)\tilde{f}(k+1). \qquad (15.1)‡$$

As before, we will often suppress the v and write this as $A_{\tilde{T}}(\mathbf{b})$. Note that when $T = T(x)$, formula (15.29) shows that we obtain $A_x(\mathbf{b})$ as calculated in (5.1).

For many applications we want more information about Z than just its expected value; at the very least, we would want to calculate Var(Z). This is easily done, since $Z^2 = b^2_{\tilde{T}-1}v(\tilde{T})^2$. In other words, to calculate the second moment, we simply square the benefits, and square the discount function. This leads to

$$\mathrm{Var}(Z) = A_{\tilde{T}}(\mathbf{b}^2;v^2) - A_{\tilde{T}}(\mathbf{b};v)^2. \qquad (15.2)‡$$

15.2.2 The continuous case

Consider a contract that pays $b(t)$ at the moment of failure, should failure occur at time t. We will let $\overline{Z}$ denote the present value of the benefits and denote the expected value of $\overline{Z}$ by $\overline{A}_T(b; v)$, often shortened to $\overline{A}_T(b)$. When $T = t$, $b(t)$ is paid at time t, so

$$\overline{Z} = b(T)v(T)$$

and

$$\overline{A}_T(b; v) = \int_0^{\infty} b(t)v(t)f(t)\mathrm{d}t. \qquad (15.3)‡$$

When $T = T(x)$, we see from (15.27) that this is the same quantity as we obtained in (8.18).

The variance is computed as we did in the discrete case:

$$\mathrm{Var}(\overline{Z}) = \overline{A}_T(b^2; v^2) - \overline{A}_T(b; v)^2. \qquad (15.4)‡$$

Example 15.1 Suppose that $\mu(t)$ is a constant 0.04 and the force of interest is a constant 0.07. An insurance contract pays a benefit of $\mathrm{e}^{0.03t}$ at time t, should failure occur at that time. Find the expectation and variance of $\overline{Z}$, the present value of the benefits.

Solution. This is an exponential failure time, so $f(t) = 0.04\mathrm{e}^{-0.04t}$. We have $b(t) = \mathrm{e}^{0.03t}$ and $v(t) = \mathrm{e}^{-0.07t}$. Substituting into (15.3),

$$E(Z) = \int_0^{\infty} 0.04\mathrm{e}^{0.03t}\mathrm{e}^{-0.07t}\mathrm{e}^{-0.04t}\mathrm{d}t = \frac{0.04}{0.04 + 0.07 - 0.03} = \frac{1}{2}.$$

Now $b(t)^2 = \mathrm{e}^{0.06t}$ and $v(t)^2 = \mathrm{e}^{-0.14t}$, so calculating as above,

$$E(Z^2) = \frac{0.04}{0.04 + 0.14 - 0.06} = \frac{1}{3}, \qquad \mathrm{Var}(Z) = \frac{1}{3} - \frac{1}{4} = \frac{1}{12}.$$

15.2.3 Approximation

Suppose now that we know only the distribution of $\tilde{T}$, and wish to approximate $\overline{A}_T(b)$. If interest and benefits are constant over each year, we can use the standard approximation to obtain the same i/δ adjustment as we obtained before for $T(x)$. We will give an alternate stochastic derivation of this result. We first assume constant interest, so that $v(t) = v^t$ for some constant v. Let b_k denote the constant value of $b(t)$ over the time interval k to $k + 1$, and let $\mathbf{b} = (b_0, b_1, \ldots)$. Recall the random variable $R = \tilde{T} - T$ introduced in Section 15.7. Then

$$E(\overline{Z}) = E[b(T)v^T] = E[b(T)v^{\tilde{T}-R}] = E[b(T)v^{\tilde{T}}]E[v^{-R}],$$

where we invoke independence for the last equality. When $\tilde{T} = j, b(T) = b_{j-1}$, so that $E[b(T)v^{\tilde{T}}]$ is just $A_{\tilde{T}}(\mathbf{b})$. Moreover, $v^{-R} = (1+i)^R$, and since R is uniform on [0,1].

$$E[v^{-R}] = \int_0^1 (1+i)^u \mathrm{d}u = \frac{i}{\delta},$$

leading to $\bar{A}_T(b) = (i/\delta)A_{\tilde{T}}(\mathbf{b})$.

The general formula, where interest can vary from year to year is then given by

$$\bar{A}_T(b) = A_{\tilde{T}}\left(\frac{\mathbf{i}}{\delta} * \mathbf{b}\right). \tag{15.5}$$

where $\mathbf{i}/\delta$ is now a vector.

15.2.4 Endowment insurances

In the deterministic model, endowment insurances were viewed as a kind of hybrid, consisting of both insurance and annuity components. In our stochastic model, we can view these strictly as insurances. Define the failure time

$$T = \min\{T(x), n\},$$

so that failure occurs at the death of (x), or at time n if earlier. Insurances based on T are precisely n-year endowment insurances. When benefits are paid at the moment of death, T is an example of a random variable that has both a continuous part and a discrete part (see Section A.5). It is continuous over the interval $[0, n)$, on which it has a density function f_x satisfying

$$P(a \leq T \leq b) = \int_a^b f_x(t)\mathrm{d}t, \qquad \text{provided } b < n.$$

This density will just be the restriction of the density function for $T(x)$ to the interval $[0, n)$, and it integrates to $F_x(n)$ over this interval. The remaining probability is all concentrated on the point n, since T will take this value whenever $T(x)$ takes a value of n or greater, an event with probability $s_x(n)$. Expectations of a general function g of T must be calculated as a sum of two terms,

$$E[g(T)] = \int_0^n g(t)f_x(t)\mathrm{d}t + g(n)s_x(n).$$

In particular, consider an n-year endowment insurance with death benefit function b defined on $[0, n]$. That is, $b(t)$ is paid at the moment of death if this occurs before n and $b(n)$ is paid at time n if the insured is then alive (so that failure occurs at time n). Then

$$\bar{A}_T(b) = E[v(T)b(T)] = \int_0^n b(t)v(t)f_x(t)\mathrm{d}t + b(n)v(n)s_x(n),$$

which is easily seen to agree with the present value obtained in the deterministic model.

From (15.4) we calculate the variance of the benefits as

$$\int_0^n [b(t)v(t)]^2 f_x(t)\mathrm{d}t + [b(n)v(n)]^2 s_x(n) - \left[\bar{A}_T(b)\right]^2 .$$

Some care is needed when applying the approximation (15.5). This is only valid for the continuous part of the random variable T. In the present example of the endowment insurance, Equation (15.5) would be modified to

$$\bar{A}_T(b) = A_T\left(\frac{\mathbf{i}}{\boldsymbol{\delta}} * \mathbf{b}\right) + b(n)v(n)s(n).$$

The last term does not get multiplied by the interest adjustment, since in both the end-of-the-year-of-death and moment-of-death cases the final amount of $b(n)$ is paid to survivors at exact time n.

Example 15.2 A two-year-endowment insurance on (x) has benefits of 100 if death occurs in the first year or 80 if death occurs in the second year, and a pure endowment of 80 if the insured is alive at time two. You are given that $q_x = 0.2$, $q_{x+1} = 0.3$, and the interest rate is a constant 100%. Find the expectation and variance of the benefits when (a) benefits are payable at the end of the year of death; (b) benefits are payable at the moment of death.

Solution. (a) Let Z denote the present value of the benefits in the end of the year of death case. As we noted in Example 6.2, we do not even need the value of q_{x+1}. In fact we could just notice that Z takes the value 50 with probability 0.2 and 20 with probability 0.8. It therefore has a mean of 26 and a variance of $30^2(0.2)(0.8) = 144$.

(b) Let $\overline{Z}$ denote the present value of the benefits in the moment of death case. Now we do need the value of q_{x+1} since it makes a difference if the person dies in the second year, in which case 80 is paid at the moment of death, or if they survive, in which case 80 is paid at the end of the year. Begin again with the calculation of $E(Z)$ and $E(Z^2)$ but now split into the insurance and pure endowment amounts.

$$E(\overline{Z}) = 14.8 + 80(1/4)(0.56) = 14.8 + 11.2.$$

$$E(\overline{Z}^2) = 596 + 6400(1/16)(0.56) = 596 + 224.$$

Now for $i = 1$, $\delta = \log(2)$ so

$$E(\overline{Z}) = \left(\frac{1}{\log 2}\right) 14.8 + 11.2 = 32.55$$

Squaring v means that δ is doubled and the i changes to

$$(1+i)^2 - 1 = 2i + i^2.$$

which equals 3 when the original value is 1. So, for the second moment calculation we take $i = 3, \delta = \log(4)$, to get

$$E(\overline{Z}^2) = \left(\frac{3}{\log 4}\right) 596 + 224 = 1513.77$$

$$\mathrm{Var}(\overline{Z}) = 1513.77 - 32.55^2 = 454.27$$

The variance is much larger in (b), due to the variation in the present value of the benefits, depending on when the person dies during the year, which is significant in view of the high interest rate.

15.3 The stochastic approach to annuity benefits

15.3.1 Discrete annuities

Consider a contract with annuity benefit vector $\mathbf{c}$ that pays c_k at time k provided that failure has not yet occurred. Let Y denote the present value of the benefits with respect to the investment discount function v, and let $\ddot{a}_{\tilde{T}}(\mathbf{c}; v)$ (usually shortened to $\ddot{a}_{\tilde{T}}(\mathbf{c})$) denote $E(Y)$. (The subscript $\tilde{T}$ reflects the fact that in view of the yearly payments, this quantity will depend only on the value of $\tilde{T}$, rather than the exact value of T).

There are two methods for computing expectations and variances of annuities in the stochastic setting. This first has the advantage of simplifying variance calculations. Recall that in the deterministic model we often viewed insurances as annuities. It is now helpful to view annuities as insurances in a certain sense. (Readers who attempted Exercise 5.22 will have encountered this idea previously.) To be precise, we want to express Y as a function of $\tilde{T}$. Let g be the function defined on the positive integers by

$$g(k) = c_0 + c_1 v(1) + c_2 v(2) + \cdots + c_{k-1} v(k-1).$$

We can also write $g(k) = \ddot{a}({}_k\mathbf{c}; v)$ using the notation introduced in Section 2.10.1.

Suppose that $\tilde{T}$ takes the value k. Then payments will have been made at all integer times from 0 to $k-1$, so Y will take the value $g(k)$. We can therefore write

$$Y = g(\tilde{T}),$$

and it follows that

$$\ddot{a}_{\tilde{T}}(\mathbf{c}) = \sum_{k=1}^{\infty} g(k)\tilde{f}(k) \tag{15.6‡}$$

and

$$\mathrm{Var}(Y) = \sum_{k=1}^{\infty} g(k)^2 \tilde{f}(k) - \ddot{a}_{\tilde{T}}(\mathbf{c})^2. \tag{15.7‡}$$

Expression (14.6) is known as the *aggregate payment formula*, since we consider the function $g(k)$ which is the aggregate present value received for failure at time k.

Example 15.3 For a 4-year life annuity on (60). The sequence of annuity benefits is 1, 2, 3, 4 beginning at age 60. You are given $q_{60} = 0.1,\ q_{61} = 0.2,\ q_{62} = 0.3$ and $i = 100\%$. Find $E(\mathrm{Y})$ and Var(Y).

Solution. Calculate recursively $g(1) = 1, g(2) = 1 + 2 \times \frac{1}{2} = 2, g(3) = 2 + 3 \times \frac{1}{4} = 2.75$, $g(k) = 2.75 + 4 \times \frac{1}{8} = 3.25$ for all $k > 3$.

The distribution of $\tilde{T}$ is given by $\tilde{f}_{60}(1) = 0.1,\ \tilde{f}_{60}(2) = 0.180,\ \tilde{f}_{60}(3) = 0.216,\ s_{60}(3) = 0.504$. We can now calculate

$$E(Y) = 0.100 + 2(0.180) + 2.75(0.216) + 3.25(0.504) = 2.692,$$
$$\mathrm{Var}(Y) = 0.100 + 2^2 \times 0.180 + (2.75)^2 \times 0.216 + (3.25)^2 \times 0.504 - 2.692^2 = 0.530.$$

Remark Note carefully that when the final payment of an annuity is at time k_0, then Y takes the value of $g(k_0 + 1)$ with probability $s(k_0)$ since $g(k) = g(k_0 + 1)$ for all $k > k_0$.

Remark There is an important point here which is often overlooked. Equation (15.6) applied to life annuities says that we can view the present value of a set of cash flows with respect to the interest and survivorship function as the expected value of the present value of payments received with respect to an interest only function. Some people have mistakenly assumed that similarly, the accumulated value of a life annuity is the same as the expected value of the payments received, accumulated with interest. This is not true and in fact the equality does not hold for values at any time other than 0. We have

$$Val_k(\mathbf{c}, y_x) = y_x(k)^{-1}\ddot{a}_x(\mathbf{c}),$$

while the expected value at time k with respect to v, of the payments received from $\mathbf{c}$ is $v(k)^{-1}\ddot{a}_x(\mathbf{c})$, which will be less than the first quantity for all $k > 0$.

The second method to calculate annuity present values is more closely related to what we did in the deterministic model. We define random variables

$$I_k = \begin{cases} 1, & \text{if } \tilde{T} > k, \\ 0, & \text{if } \tilde{T} \le k. \end{cases}$$

The contract pays a present value of $c_k v(k)$ for each integer k such that $\tilde{T} > k$, or equivalently, such that $I_k = 1$. We can therefore write

$$Y = \sum_{k=0}^{\infty} c_k v(k) I_k. \tag{15.8}$$

This is analogous to viewing the annuity as consisting of several pure endowments. Since $E(I_k) = s(k)$, taking expectations, gives

$$\ddot{a}_T(\mathbf{c}) = \sum_{k=0}^{\infty} c_k v(k) s(k). \tag{15.9‡}$$

Expression (15.9) is known as the *current payment formula*, since it looks at each payment received, discounts it with interest, and multiplies it by the probability of receiving it. In the case where $T = T(x)$, this is precisely the formula obtained in (4.1).

The equality of the expressions in (14.6) and (15.9) can be seen from (A.16). (Condition (A.14) will almost always hold. It is automatic when benefits are positive since g will be increasing, or when there are finitely many benefit payments, even if negative, since g will be bounded.)

Note that the I_k are not independent. To calculate variances from the current payment approach we use the following facts.

$$\mathrm{Var}(I_k) = s(k)(1 - s(k)). \tag{15.10}$$

For $j < k$, we have $I_j I_k = I_k$, so that

$$\mathrm{Cov}(I_j, I_k) = E(I_j I_k) - E(I_j)E(I_k) = s(k)(1 - s(j)). \tag{15.11}$$

To simplify notation let

$$r_k = c_k v(k) s(k), \qquad u_k = c_k v(k)(1 - s(k)).$$

Then from (A.24), (15.10) and (15.11),

$$\mathrm{Var}(Y) = \sum_k r_k u_k + 2 \sum_{j<k} r_k u_j. \tag{15.12}$$

This will normally be a much more involved calculation than that given by the aggregate payment approach.

15.3.2 Continuous annuities

Consider a contract that makes payments continuously at the annual rate of $c(t)$ at time t, provided T has not yet occurred. We will denote the actuarial present value by $\overline{a}_T(c; v)$ (often shortened to $\overline{a}_T(c)$).

Define the function $\overline{g}$ by

$$\overline{g}(t) = \int_0^t c(r) v(r) \mathrm{d}r.$$

If T fails at time s, the present value of the benefits received will be precisely $\overline{g}(s)$, so that

$$\overline{Y} = \overline{g}(T).$$

It follows that

$$\bar{a}_T(\mathbf{c}) = E(\overline{Y}) = \int_0^\infty \bar{g}(t)f(t)\mathrm{d}t \tag{15.13‡}$$

and

$$\mathrm{Var}(\overline{Y}) = \int_0^\infty \bar{g}(t)^2 f(t)\mathrm{d}t - [\bar{a}_T(c)]^2. \tag{15.14‡}$$

For the alternate current payment formula we use (A.15). From the fundamental theorem of calculus, $g'(t) = c(t)v(t)$ and, moreover, $g(0) = 0$. If (A.14) holds (which as in the discrete case will almost always be the case) we can write the current payment formula

$$\bar{a}_T(c) = \int_0^\infty c(t)v(t)s(t)\mathrm{d}t, \tag{15.15‡}$$

which agrees with (8.9) when $T = T(x)$. There is no convenient variance formula analogous to (15.12) for continuous annuities, since we do have the representation as a sum of indicator random variables that was present in the discrete case.

Example 15.4 An annuity provides continuous benefits at the rate of 1 per period, provided failure has not occurred. The failure time T has an exponential distribution with constant hazard μ. The force of interest is a constant δ. If $\overline{Y}$ is the present value of the benefits, find $\mathrm{Var}(\overline{Y})$.

First solution. Since

$$\bar{g}(t) = \int_0^t \mathrm{e}^{-\delta r}\mathrm{d}r = \frac{1 - \mathrm{e}^{-\delta t}}{\delta},$$

We have

$$E(\overline{Y}^2) = \int_0^\infty \left[\frac{1 - \mathrm{e}^{-\delta r}}{\delta}\right]^2 \mu \mathrm{e}^{-\mu r}\mathrm{d}r = \frac{\mu}{\delta^2}\int_0^\infty [1 - 2\mathrm{e}^{-\delta r} + \mathrm{e}^{-2\delta r}]\mathrm{e}^{-\mu r}\mathrm{d}r$$
$$= \frac{\mu}{\delta^2}\left[\frac{1}{\mu} - \frac{2}{\mu + \delta} + \frac{1}{\mu + 2\delta}\right] = \frac{2}{(\mu + \delta)(\mu + 2\delta)}.$$

From (15.13), or simply referring back to Section 8.10, we see that $E(\overline{Y}) = 1/(\mu + \delta)$ so that

$$\mathrm{Var}(\overline{Y}) = \frac{2}{(\mu + \delta)(\mu + 2\delta)} - \frac{1}{(\mu + \delta)^2} = \frac{\mu}{(\mu + \delta)^2(\mu + 2\delta)}.$$

Note that, as a check, for $\delta = 0$, this reduces to μ^{-2}, the variance of the given exponential distribution (see Section A.11.6) which must be true since $\overline{Y} = T$ at zero interest.

A simplification. We can simplify and also obtain a more general formula by the following trick:

$$\overline{Y}^2 = \frac{1 - 2e^{-\delta T} + e^{-2\delta T}}{\delta^2} = \frac{2}{\delta}\left[\left(\frac{1 - e^{-\delta T}}{\delta}\right) - \left(\frac{1 - e^{-2\delta T}}{2\delta}\right)\right].$$

Upon taking expectations and recalling that $E[e^{-\delta T}] = \mu/(\mu + \delta)$, we get the same answer as above.

*15.4 Deferred contracts

Consider any contract based on the failure time T, in which no payments are to be made in the first m years regardless of whether failure occurs or not. A typical example is a deferred annuity, but this can be a perfectly general contract, continuous or discrete, including death benefits, annuity benefits or both. Let Y be the present value of the benefits, and let Y' be the value of the benefits at the starting time m. As mentioned previously, there is no real need for any special mathematical treatment of such, as we simply take zero entries or zero values in the benefit vector or function. It is sometimes convenient, however, to express $\mathrm{Var}(Y)$ in terms of $\mathrm{Var}(Y')$. The purpose of this section is to derive such a formula.

The idea is to notice that Y takes the value 0 with probability $1 - s(m)$ and $v(m)Y'$ with probability $s(m)$. It follows from basic probability theory that

$$E(Y) = s(m)E(v(m)Y') = v(m)s(m)E(Y'),$$

a formula that we are familiar with from the deterministic model.

It similarly follows that

$$E(Y^2) = v(m)^2 s(m)E(Y'^2) = v(m)^2 s(m)[\mathrm{Var}(Y') + E(Y')^2].$$

Substituting for $E(Y)$ in terms of $E(Y')$ in the expression $\mathrm{Var}(Y) = E(Y^2) - E(Y)^2$ gives

$$\mathrm{Var}(Y) = v(m)^2 s(m)\mathrm{Var}(Y') + v(m)^2 s(m)(1 - s(m))E(Y')^2. \tag{15.16}$$

This gives a familiar decomposition of the variance of Y. The first term measures the uncertainty arising from actual benefits once they start at time m, and the second term measures the uncertainty arising from the fact that failure may or may not occur prior to time m.

15.5 The stochastic approach to reserves

We will illustrate the stochastic approach to reserves with the discrete model. The results are easily adapted to the continuous case. We have vectors $\mathbf{b}$ and π, where b_k is paid at time $k + 1$ if failure occurs between time k and $k + 1$, and π_k is paid at time k if failure has not yet occurred. We are adopting the point of view (used before in Chapter 6) of treating annuity benefits as negative premiums, so that π_k represents a net inflow at time k, consisting of the premium received, less expenses paid at time k if these are being considered, and also less any annuity payments made at time k. We will therefore not need to refer to a separate annuity benefit vector.

Fix an integer duration r and suppose that failure has not yet occurred by that time.

Definition 15.1 The *prospective loss at time r,* denoted by ${}_rL$, is the value at time r of the future net cash flows to be *paid out.*

These net cash flows are the future benefits to be paid minus the future net inflows to be received. The definition here is different than that used in the deterministic model. The benefits are the actual benefits paid on the contract, and not the individual share of the total death benefits. As we stressed in Section 14.1, the discounting is with regard to the investment discount function only. The value will depend of course on when failure occurs, so ${}_rL$ will be a random variable. It is a function of the shifted random variable $\tilde{T} \circ r$, since we have assumed survival up to time r.

We next determine this function. Suppose that failure occurs at time $r + t$ where $k \leq t < k + 1$, for some integer k. Then $\tilde{T} \circ r$ will assume a value of $k + 1$. The contract will pay out a failure benefit of b_k at time $k + 1$. Offsetting this, the net inflows would have been collected from time r to time $r + k$, so that in this case the value of ${}_rL$ will be

$$b_{r+k}v(r, r + k + 1) - [\pi_r + \pi_{r+1}v(r, r + 1) + \cdots + \pi_{r+k}v(r, r + k)].$$

In terms of random variables, we can write

$${}_rL = b_{r+\tilde{T}\circ r-1}v(r, r + \tilde{T} \circ r) - [\pi_r + \pi_{r+1}v(r, r + 1) + \cdots + \pi_{r+\tilde{T}\circ r-1}v(r, r + \tilde{T} \circ r - 1)].$$

Remark We have used the standard terminology here, but the reader should be aware that the word 'loss' in the above definition has a completely different connotation than that discussed in Section 6.4.1. It is not a definite number, measuring the loss brought about by deviation from the expected pattern of mortality or interest, but rather it is a random variable, giving the difference between what is collected and what is paid out under the contract at the different random times of failure. It should also be noted that ${}_rL$ is of the few random variables we encounter that can take negative values, which occur when there is a 'gain'.

Definition 15.2 In this stochastic model, we define the *reserve at time r,* denoted as before by ${}_rV$, simply as

$${}_rV = E({}_rL).$$

So ${}_rV$ is the expected value of the prospective loss, which is the expected present value of the future benefits less the expected present value of the future inflows. It is clear that in the case of $T = T(x)$ we obtain the same value as in the deterministic model.

Example 15.5 Let $T = T(60)$. We are given $q_{60} = 0.1,\ q_{61} = 0.2,\ q_{62} = 0.3,\ i = 100\%$. Suppose we have a 3-year term insurance policy on (60) with $b_0 = 200,\ b_1 = 200,\ b_2 = 100,\ \pi_0 = 20,\ \pi_1 = 20,\ \pi_2 = 10$.

(a) Find the distribution of ${}_0L$.

(b) Find the distribution of ${}_1L$.

(c) From your answer to (b), calculate ${}_1V$.

Solution.

(a) The distribution of ${}_0L$ is as follows:

k	Value of ${}_0L$ when $\tilde{T} = k$	$P(\tilde{T} = k)$
1	$200 \times \frac{1}{2} - 20 = 80$	0.100
2	$200 \times \frac{1}{4} - \left(20 + 20 \times \frac{1}{2}\right) = 20$	0.180
3	$100 \times \frac{1}{8} - \left(20 + 20 \times \frac{1}{2} + 10 \times \frac{1}{4}\right) = -20$	0.216
≥ 4	$0 - \left(20 + 20 \times \frac{1}{2} + 10 \times \frac{1}{4}\right) = -32.5$	0.504

(b)

k	Value of ${}_1L$ when $\tilde{T} \circ 1 = k$	$P(\tilde{T} \circ 1 = k)$
1	$200 \times \frac{1}{2} - 20 = 80$	0.20
2	$100 \times \frac{1}{4} - \left(20 + 10 \times \frac{1}{2}\right) = 0$	0.24
≥ 3	$0 - \left(20 + 10 \times \frac{1}{2}\right) = -25$	0.56

(c) ${}_1V = 80 \times 0.20 + (-25 \times 0.56) = 2$. The reader should verify this agrees with the answer that we would get from the methods of Chapter 6.

15.6 The stochastic approach to premiums

For this section only, it is convenient to adopt a different convention. Assume now that expenses and annuity payments are included as benefits, so that the inflows π_k are just the premiums collected.

15.6.1 The equivalence principle

The random variable ${}_0L$ is usually denoted just as L. It is the present value of all future benefits less the present value of all future premiums. Recall that in the deterministic model we calculated net premiums by setting the present value of benefits equal to the present value of premiums. The stochastic version of this concept is to set premiums so that

$$E(L) = 0.$$

This is known as the *equivalence principle*.

15.6.2 Percentile premiums

The stochastic viewpoint allows us to incorporate other features of the random variable L rather than just its expectation, when computing premiums. For example, it would be highly desirable for an insurer to avoid a positive loss. Of course, it is impossible to ensure that this will always hold. If the insured dies shortly after purchasing a life insurance policy, then L will almost certainly be positive. Suppose, however, we set a threshold α, and then set premiums as small as we can, so that the probability will be at least α that the loss will be nonpositive. Typically α will be a number close to 1 such as 0.95 or 0.99. These are sometimes referred to as *percentile premiums*.

This sounds reasonable, but there are major problems to such an approach. In the first place, the method does not take into account the important *right tail* of the distribution, consisting of those values of L which are greater than the specified percentile. If there are very large claims present, the method may not provide sufficient premiums to cover the risk. This is evidenced by the fact that the method can produce premiums that are actually less than the equivalence principle premium. For an extreme example of this, suppose we have an n year term policy and the probability of (x) dying within n years is less than $1 - \alpha$. We could achieve the stated goal by charging premiums of 0, which is absurd.

On the other hand, for high values of α the method can produce premiums that seem inordinately high when compared with the equivalence principle premium. See Example 15.7 below.

Still another instance of the pathological behaviour produced by percentile premiums is illustrated by the following example.

Example 15.6 On a single premium contract, the present value of the benefits takes the value 0 with probability 0.94, 1 with probability 0.01 and 3 with probability 0.05.

(a) Find the percentile premium for $\alpha = 0.95$.

(b) Suppose the insurer plans to sell 2 contracts. Find the smallest possible premium for each, such that there is a probability of at least 0.95 that the total premiums will cover the total benefits on both contracts.

Solution.

(a) This is obviously 1, which gives exactly a 0.95 probability of covering the benefits.

(b) Using independence, direct calculation shows that the probability of the total present value of benefits being less than or equal to z is 0.9025 for $z = 2$ or 0.9965 for $z = 3$. The smallest total premium needed is then 3 resulting in a charge of 1.5 per policy.

The conclusion of this example is exactly contrary to what one would expect, and indicates a major flaw in the percentile premium approach. As the number of policies increases, the risk should decrease due to diversification. Unfavourable occurrences on one contract can be offset by favourable ones on another. The premium per policy should therefore decrease rather than increase.

Despite the drawback inherent in percentile premiums, there may be times when one want to compute these, possibly for comparison with premiums produced by other methods. Here is a procedure for the case where benefits are paid at the end of the year of failure. We want to find the smallest possible level premium π payable for h years, such that the probability of a nonpositive L is greater than or equal to some given number α. We consider only the case where L decreases as the value of T increases. This is fairly typical. As the time of failure increases, the discount factor reduces the present value of the benefit paid. In addition, the later the occurrence of failure, the more premiums will be collected. Both of these factors tend to decrease L. (Of course, the required condition may not hold in the case where benefits are increasing rapidly.) The procedure is as follows:

1. Let k_0 be the largest positive integer k such that $s(k) \geq (\alpha)$.
2. Solve for π so that L is 0 when $\tilde{T} = k_0 + 1$. That is,

$$b_{k_0} v(k_0 + 1) = \pi \ddot{a}(1_r; v),$$

where r is the minimum of $k_0 + 1$ and h. Since L is decreasing with k, $L \leq 0$ implies that $\tilde{T} \geq k_0$. Therefore,

$$\alpha \leq s(k_0) \leq P(L \leq 0).$$

Moreover, if we take any smaller premium π', then for $\tilde{T} = k_0 + 1$, L will be positive, so that

$$P(L \leq 0) \leq s_{k_0+1} < \alpha.$$

by our choice of k_0.

Example 15.7 Refer back to Example 15.5. Find the level percentile premium π payable for 3 years if $\alpha = 0.8$.

Solution. We take $k_0 = 1$ since $P(\tilde{T} > 1) = 0.90 > 0.8$, while $P(\tilde{T} > 2) = 0.72 < 0.8$. When $\tilde{T} = 2, L = 200 \times \frac{1}{4} - \pi(1 + \frac{1}{2})$, so that $\pi = 33.33$. This is much higher than the equivalence principle level premium of 13.31.

15.6.3 Aggregate premiums

An analysis of the examples above indicate that the percentile premium method can give unreasonable results with highly asymmetric distributions, where there is a large probability of a value in one of the tails. This approach however is a reasonable one to use in order to compute the total premiums for a large group of policies, where the distribution of total benefits is likely to be more heavily concentrated about the mean. Suppose the insurer issues several contracts of the same type and want to be fairly certain, that in the aggregate, the total premiums collected will cover the total benefits. There are two major questions that could be considered. First, the premiums may be fixed and we want to determine the number of contracts to sell to achieve the desired confidence. Second, the number of contracts may be fixed and the problem is to determine the premium to charge. In general, exact calculations are prohibitive, but we can get approximate answers by assuming that the totality of all losses has a normal distribution (see Section A.11.5).

It is convenient to introduce the following quantity, which we will discuss in some generality.

Definition 15.3 For a random variable X with a positive mean, the *coefficient of variation* of X is the quantity

$$CV(X) = \frac{\sqrt{\mathrm{Var}(X)}}{E(X)}.$$

It is clear from (A.8) and (A.10) that for any constant c,

$$CV(cX) = CV(X),$$

showing that, unlike variance or standard deviation, this is a measure of variation that is independent of the particular units. So for example, if we measure loss in US dollars and then change the units to pounds sterling, the variance will be quite different, but the coefficient of variation will remain the same.

There are some other facts about this quantity that we want to derive. Suppose $S = X_1 + X_2 + \cdots + X_N$ where these are independent random variables each distributed as X. Then

$$\mathrm{Var}(S) = N\ \mathrm{Var}(X), \qquad E(X) = NE(X),$$

showing that

$$CV(S) = \frac{CV(X)}{\sqrt{N}}, \tag{15.17}$$

illustrating the expected diversification effect that occurs when we take an independent sum of random variables. We expect to reduce variation, as some high values are offset by low values.

Next, suppose that X is a normal random variable with positive mean, and we want the probability that X takes a value greater than or equal to 0. This is clearly independent of any units, so we should be able to express it in terms of the coefficient of variation. Indeed, suppose X has mean μ and standard deviation σ, so that $X = \mu + \sigma Z$ where Z is the standard normal. Then, $P(X \geq 0) = P(Z \geq -\mu/\sigma)$ which by symmetry is equal to $P(Z \leq \mu/\sigma)$. We can write

$$P(X \geq 0) = \Phi[CV(X)^{-1}], \tag{15.18}$$

where Φ is the cumulative distribution function of the standard normal.

Consider now the first question given above. Suppose we have a contract with a prospective loss at time 0 of L, where $E(L) < 0$ and we want to determine the smallest value of N, so that if N independent contracts are sold, the probability is at least α that premiums will cover benefits.

Let S denotes the aggregate loss on all contracts and apply our results above to $-S$, the aggregate gain. We want a probability of α that $-S$ will be positive. Assume that $-S$ is normal. Invoke (15.18) with $X = -S$, and take Φ^{-1} of each side, resulting in

$$\Phi^{-1}(\alpha) = CV(-S)^{-1}.$$

Then apply (15.17) with $X = -L$ to get $\Phi^{-1}(\alpha) = \sqrt{N}/CV(-L)$, so that

$$\sqrt{N} = \Phi^{-1}(\alpha)CV(-L). \tag{15.19‡}$$

We see then that the required number of contracts increases as the confidence level goes up, and also as the uncertainty in L goes up, which is just what we expect to happen.

For the second question above where N is fixed, consider a single premium contract with present value of benefits $\overline{Z}$. Let the required premium be of the form $(1 + \theta)E(\overline{Z})$. The quantity θ is known as the *relative security loading* – the words *risk loading* and *contingency loading* are also used. (So if $\theta = 0.20$ for example, it means that the premium is calculated by adding 20% to the equivalence principle premium.) Now in this case

$$-L = (1 + \theta)E(\overline{Z}) - \overline{Z},$$

so that $\text{Var}(-L) = \text{Var}(\overline{Z})$, and $E(-L) = \theta E(\overline{Z})$. Therefore

$$CV(-L) = \frac{CV(\overline{Z})}{\theta}. \tag{15.20}$$

Substituting from (15.19) we can write

$$\theta = \frac{\Phi^{-1}(\alpha)CV(\overline{Z})}{\sqrt{N}}. \tag{15.21‡}$$

This is certainly reasonable as it shows that the loading increases with the tolerance level, and the uncertainty in the benefits, but decreases as more contracts are sold.

Example 15.8 Suppose the force of interest is a constant 0.06 and that $\mu(t) = 0.04$ for all t. An insurer sells 100 contracts, each of which pays 1000 upon failure. Assuming a normal approximation, how much should be charged as a single premium on each contract, so that there is a 95% chance that premiums will be sufficient to cover the aggregate claims for this group of 100 contracts.

Solution. If $\overline{Z}$ = the benefit for a contract paying one unit at failure. Then

$$E(\overline{Z}) = \frac{\mu}{\mu + \delta} = 0.4, \qquad E(\overline{Z}^2) = \frac{\mu}{\mu + 2\delta} = 0.25.$$

The formula for the second moment reflects the fact that squaring the discount function is equivalent to doubling the force of interest. It follows that $\text{Var}(\overline{Z}) = 0.09$ and $CV(\overline{Z}) = 0.75$. From (15.21), with $\alpha = 0.95$, $\theta = 1.645(0.75)/10 = 0.123375$. For each 100-unit contract the equivalence principle premium is 400 and so the premium charged should be $400(1.123375) = 449.35$.

Example 15.9 Suppose the contract in Example 15.8 is sold with premiums paid continuously for life at a level rate, where the security loading $\theta = 0.10$. Assuming a normal approximation, how many contracts must be sold, so there is a 95% chance that total premiums will cover total benefits?

Solution. Since our formulas are independent of units we can ignore the 1000 and consider a 1-unit contract. The main problem is to determine $CV(-L)$. This is not always so easy when we no longer have a single premium, but it can be done for level benefits and premiums. This is carried out in Chapter 16, and we refer the reader to formulas (16.5) and (16.6) with $t = 0$. The equivalence principle premium is 0.04, so the premium $\pi = 0.044$ and $\pi/\delta = 11/15$. From formula (16.6) $\sqrt{\text{Var}(L)} = (26/15))0.3 = 0.52$. From formula (16.5) $E(L) = (26/15)(0.04) - 11/15 = -0.04$. So

$$CV(-L) = \frac{0.52}{0.04} = 13.$$

From (15.19) $\sqrt{N} = 1.645 \times 13$, so $N = 457.32$. This means that 458 contracts must be sold.

15.6.4 General premium principles

There are many other possibilities for computing premiums to allow for the random nature of the loss. These are known in general as *premium principles*. Formally, a premium principle is a function H, which assigns to each random variable X representing a loss, a number $H(X)$, which is the premium to be charged for accepting the risk X. This definition is applicable to contracts purchased by a single premiums, rather than periodic premiums.

The equivalence principle premium is given by $H(X) = E(X)$. The *standard deviation* principle is given by $H(X) = E(X) + \beta\sqrt{\text{Var}(X)}$, for some positive β. Dividing this by $E(X)$, we can see from (15.21) that the standard deviation principle can be interpreted (assuming a normal approximation) as setting premiums so that there is a certain probability, which will depend on N, that premiums will cover losses. Another example is the *variance* principle given by $E(X) + \beta\mathrm{Var}(X)$, for some positive β. (Exercise 15.25 points out a flaw in this principle.) There is extensive treatment in the actuarial literature comparing the properties of these and other premium principles. We will not elaborate further here, but more comments can be found in Chapter 22.

15.7 The variance of $_rL$

We have seen that in our stochastic model, the reserve represents the expected value of the amount that we need in order to meet future obligations. It is useful to have more information about the distribution of this random variable. In this section we present a formula for calculating the variance of $_rL$ under the setup of Section 14.2. Of course, if we have computed the exact distribution of $_rL$, as we did in Example 15.5 we could calculate the variance directly. The present method will allow us to calculate the variance without knowing the exact distribution, as long as we know all the reserves after time r. Moreover, it gives us a decomposition of this variance into that attributable to each of the future years. In the actuarial literature, this result is known as *Hattendorf's theorem.*

We start by taking $r = 0$, so we compute the variance of $L = {}_0L$. To simplify the task we look at the net cash outflow year by year. Let C_k denote the value at time k of the net amount paid out in the year from k to $k+1$. That is, C_k is equal to the value at time k of the failure benefit paid out at the end of the year, less the inflow at time k. So C_k is a random variable taking three possible values depending on whether failure has occurred before time k, during the year $(k, k+1)$ or at time $k+1$ or later. If failure occurs before time k, nothing is paid and nothing is received. If failure occurs during the year $(k, k+1)$, then b_k is paid at time $k+1$ and π_k is received at time k. If failure occurs after time $k+1$, nothing is paid out and a premium of π_k is received at time k. The following table summarizes the possible values of C_k and the respective probabilities:

Time of failure	Value of C_k	Probability
Before time k	0	$1 - s(k)$
Between time k and k+1	$v(k, k+1)b_k - \pi_k$	$\tilde{f}(k+1) = s(k)\lambda(k+1)$
At time $k+1$ or later	$-\pi_k$	$s(k+1) = s(k)(1 - \lambda(k+1))$

The random variable L is related to the values of C_k by

$$L = \sum_{k=0}^{\infty} v(k)C_k. \tag{15.22}$$

A fundamental property is that

$$C_jC_k = -\pi_jC_k, \quad j < k. \tag{15.23}$$

To verify (15.23), observe that it is obvious if $C_k = 0$, since then both sides are 0. If C_k is not equal to 0, it means that failure did not occur before time k and therefore must have occurred at time $j + 1$ or later. We see from the table above that C_j is equal to $-\pi_j$.

We will first derive a variance formula for the case of a contract with zero reserves. This means that the premium paid at any time k is just enough to cover the death benefit paid at the end of that year, so that

$$\pi_k = v(k, k+1)b_k\lambda(k+1), \quad \text{for all } k. \tag{15.24}$$

The significance of this case is that, as we show below, the random variables C_h for different values of h are uncorrelated (although they are not independent). Moreover, we have a fairly easy expression for the variance of each C_h, so that (15.22) gives us a formula for Var(L).

Substituting from (15.24), we see that the second and third values of C_k as listed above are respectively $v(k, k+1)b_k[1 - \lambda(k+1)]$ and $-v(k, k+1)b_k\lambda(k+1)$. It follows immediately that under assumption (15.24),

$$E[C_k] = 0, \quad \text{for all } k, \tag{15.25}$$

so that from (15.23), if $j < k$,

$$\mathrm{Cov}(C_j, C_k) = -\pi_j E(C_k) = 0. \tag{15.26}$$

Moreover, after some algebraic manipulation,

$$\mathrm{Var}[C_k] = E\left[C_k^2\right] = v(k, k+1)^2 b_k^2 s(k)\lambda(k+1)(1 - \lambda(k+1)). \tag{15.27}$$

Invoking the fact the variance of a sum of uncorrelated random variables is the sum of the variances, and noticing that $s(k)(1 - \lambda(k+1)) = s(k+1)$, we deduce from (15.22) that

$$\mathrm{Var}(L) = \sum_{k=0}^{\infty} v(k+1)^2 b_k^2 s(k+1)\lambda(k+1).$$

We next consider the case of a general duration r, still keeping the assumption of zero reserves. In this case, ${}_rL$ will just be ${}_0L$ for the policy on $(x + r)$ with the benefits and premiums starting at age r, and with probabilities conditioned on the survival of (x) to age $x + r$. The s and λ in the formula above will be those associated with the distribution of $\tilde{T} \circ r$. The result, using (15.10) and the discrete counterpart of (15.8), is

$$\mathrm{Var}({}_rL) = \frac{1}{s(r)} \sum_{k=0}^{\infty} v(r, r+k+1)^2 b_{r+k}^2 s(r+k+1)\lambda(r+k+1).$$

Finally, we consider the general case where we remove the restriction given by (15.24). We split the policy into two parts, the risk portion and the savings portion, in exactly the same way as we did in Section 6.4.2. This leads to a corresponding split of the prospective loss into that portion of this loss attributable to the risk portion and that attributable to the savings portion. But the prospective loss for the savings portion must be a constant, since it is completely independent of the mortality experience. (In fact this constant is just the present value, with respect to interest, of the pure endowment paid out at the end, less the savings portions of the premiums. Since the fund is reduced to zero, this present value must be what you started with, which is ${}_rV$.) The means that the variance of the prospective loss is the same as the variance of the prospective loss on the risk portion. Therefore, the only change from the previous formula is to replace the original death benefits by the benefits attributable to the risk portion, namely the net amount at risk. Our final formula is

$$\mathrm{Var}({}_rL) = \frac{1}{s(r)} \sum_{k=0}^{\infty} v(r, r+k+1)^2 \eta_{r+k}^2 s(r+k+1)\lambda(r+k+1). \tag{15.28‡}$$

This leads to a backwards recursion formula. Splitting off the first term,

$$\mathrm{Var}({}_rL) = v(r, r+1)^2 \frac{s(r+1)}{s(r)} [\eta_r^2 \lambda(r+1) + \mathrm{Var}({}_{r+1}L)]. \tag{15.29}$$

This is useful if T is bounded, as is true with $T(x)$. We can start the recursion with a value of 0 for the variance of the loss at the final duration.

Example 15.10 Refer back to Example 15.5. Use (14.23) to find the variance of ${}_1L$.

Solution. We will first calculate the product of the three mortality factors in (14.23). In the case where $T = T(x)$, this product for index k is ${}_{k+1}p_{x+r}\, q_{x+r+k}$. For $k = 0$ we get 0.8×0.2, and for $k = 1$ we get $0.8 \times 0.7 \times 0.3$. We also calculate that ${}_2V = 5$, so that $\eta_1 = 195$. Then

$$\mathrm{Var}({}_1L) = \frac{1}{4} \times 195^2 \times 0.16 + \frac{1}{16} \times 100^2 \times 0.168 = 1626.$$

We can verify this answer directly from the solution to Example 15.5, where we have the complete distribution of ${}_1L$.

15.8 Standard notation and terminology

The only major item we have not yet discussed is the use of an upper left subscript 2 to denote a quantity that is computed at a force of interest that is double that of the standard one or, equivalently, at a discount function that is the square of the standard one. This was developed for variance formulas. Its use is, however, restricted to cases where the failure benefits are either 0 or 1. In that case, squaring the benefits leaves them unchanged, and the second moments are calculated by just squaring the discount function. For example, the variance for the benefits paid on a 1-unit, end-of-the-year-of-death, n-year term insurance would be written as

$${}^2A^{1}_{x:\overline{n}|} - \left(A^{1}_{x:\overline{n}|}\right)^2.$$

Notes and references

To be precise, one should specify what convergence means in Equation (15.8) in the case that the sum is infinite. We did not worry about that here, since the important result for our purposes is (15.9) which, as we noted, is valid in almost all cases.

Young (2004) discusses a variety of premium principles and their properties.

Exercises

Type A exercises

15.1 A failure time T is uniformly distributed on the interval [0,10]. An insurance contract pays $e^{0.06t}$ at the moment of failure if this occurs at time t. The force of interest is a constant 0.04. Find the expectation and variance of the present value of the benefits.

15.2 A 3-year life annuity on (x) and (y) provides for annuity benefits of 1 at time 0, 2 at time 1 and 4 at time 2, provided that both individuals are alive. You are given that $p_x = 0.8,\ {}_2p_x = 0.6,\ p_y = 0.7,\ {}_2p_y = 0.5$ and the interest rate is a constant 100%.

(a) Find the expectation and variance of the present value of the benefits.

(b) A single premium for the contract is found by adding a 20% relative risk loading to the present value. What is the probability that the premium will be sufficient to provide for the benefit payments on a single contract?

15.3 A life annuity contract provides for continuous payments at the annual rate of $e^{-0.05t}$ at time t. The force of mortality is a constant 0.1, and the force of interest is a constant 0.05. If $\overline{Y}$ is the present value of the benefits, find $E(\overline{Y})$ and $\mathrm{Var}(\overline{Y})$.

15.4 A 10-year term insurance policy provides for a payment of $10 - t$ at the moment of the first death of (x) and (y), should this occur at time t. The mortality of (x) and (y) is as given in Exercise 14.9. Assuming an interest rate of 0, find the expectation and variance of the present value of the benefits.

15.5 A failure time T is uniformly distributed on the interval [0,10]. There is a constant interest rate of zero. An annuity provides for continuous payments at the rate of t at time t, provided that failure has not yet occurred. Find the expectation and variance of the present value of the benefits.

15.6 A 4-year pure endowment contract on (60) provides for 1000 paid at age 64 if (60) is then alive. Nothing is paid if death occurs before age 64. This is purchased by four level annual premiums of 100. You are given that $q_{60} = 0.1, q_{61} = 0.2, q_{62} = 0.25, q_{63} = 0.30$, and the interest rate i is a constant 25%. Write down the exact distribution of ${}_1L$. Use this to find ${}_1V$.

15.7 A certain product has a failure time of T. The approximating discrete random variable $\tilde{T}$ has a constant hazard function $\lambda(k) = 0.3$. (In other words, there is a 30% chance that the product will fail each year, given that it was still working at the beginning of the year.) An insurer agrees to pay 100 at the end of the year of failure should the product fail within 2 years. In return it collects a premium of 40 now and a second premium of 20 at time 1, if failure did not occur in the first year. The interest rate is 25%.

(a) Find $E(L)$, $\text{Var}(L)$, $\text{Var}({}_1L)$.

(b) Find the probability that, for a given contract, the premiums collected will be sufficient to pay the benefits.

15.8 A 3-year endowment insurance on (70) provides death benefits of 100 payable at the end of the year of death, if this occurs within 3 years, plus a pure endowment of 100 at age 63 if the insured is then alive. This is purchased by two premiums of 30 payable at age 60 and 61. You are given that $q_{60} = 0.2, q_{61} = 0.4$, and the interest rate is a constant 25%. Give the complete distribution of the random variable L and use this to find ${}_0V$.

15.9 An insurer charges a single premium of $E(X) + 0.1\sqrt{\text{Var}(X)}$ for a contract with present value of benefits equal to X. Use a normal approximation.

(a) How many contracts must be sold so that the probability is 95% that premiums cover benefits?

(b) What is the probability that premiums cover benefits if 225 contracts are sold?

15.10 A failure time has a hazard rate of

$$\mu(t) = \frac{2}{20 - t}.$$

A contract provides for a benefit of 1 at time 10 provided that failure occurs before time 10 (so, for example, if failure occurs at time 6, the benefit is not paid until 4 years later). You are given that $v(10) = 0.6$. Find the expectation and variance of the present value of the benefits.

15.11 Consider a 3-year life annuity on (60). The sequence of payments (beginning at time 0) is 4, 2, 1. You are given that $q_{60} = 0.2,\ q_{61} = 0.4,\ q_{62} = 0.5$. The interest rate is a constant 25%. The insurer sells 100 such contracts, charging $(1 + r)$ times the equivalence principle single premium. What should r be so that there is at least a 95% chance that premiums will cover the benefits? (Use a normal approximation.)

15.12 A 2-year term insurance contract sold to two lives, (70) and (71), provides for benefits payable at the end of the year of the first death. The benefit paid is 40 in the first year and 50 in the second year. You are given that $q_{70} = 0.4, q_{71} = 0.5, q_{72} = 0.6$, and the interest rate is 100%. Let $\overline{Z}$ denotes the present value of the benefits.

(a) Find $E(\overline{Z})$ and $\text{Var}(\overline{Z})$.

(b) Suppose the contract is to be paid for by level annual premiums of P payable for 2 years. What should P be if the equivalence principle is used, and what is the resulting probability that $L \leq 0$?

(c) What is the smallest amount that P could be if we want a probability of at least 0.25 that the loss L will be less than or equal to 0?

15.13 A 3-year endowment insurance on (70) provides death benefits of 400 payable at the end of the year of death, if this occurs within 3 years, plus a pure endowment of 200 at age 73, if the insured is then alive. This is purchased by three level premiums of 40. You are given that $q_{70} = 0.2, q_{71} = 0.4, q_{72} = 0.5$, and the interest rate is a constant 100%.

(a) Give the complete distributions for the random variable L, and the random variable ${}_1L$. Use the latter to find the reserve at time 1.

(b) What should the level premium be if the insurer wants the smallest premium such that the probability of a positive loss is less than 25%?

15.14 Use Equation (15.12) to calculate the variance in Example 15.3.

15.15 You are given that $q_x = 0.1, \quad q_{x+1} = 0.2, \quad q_y = 0.2, \quad q_{y+1} = 0.3$, and the rate of interest is a constant 25%. Find the variance of the present value of the benefits in each of the following contracts

(a) A payment of 100 is made at the end of the year of the second death, provided this occurs within 2 years.

(b) Three annual payments of 100 are made, the first at time 0, provided that at least one of (x) or (y) are alive.

15.16 A failure time T has a uniform distribution on [0,30]. The force of interest is a constant 5%. A single premium insurance contract pays 1 unit at the moment of failure. It is desired to have at least a 95% probability that premiums will cover benefits on a large group of contracts. Answer the following assuming a normal approximation.

(a) Suppose the security loading is 20%. How many contracts should be sold?

(b) It is estimated that 100 contracts will be sold. What should the security loading be?

15.17 Redo Example 15.2(b) only now assuming that the interest rate is 25% in the first year and 50% in the second year.

Type B exercises

15.18 A failure time has a constant hazard rate of μ and the force of interest is a constant δ. A contract pays $e^{\gamma t}$ at the moment of failure for failure at time t, where $\gamma < \mu + \delta$. Find the expectation and variance of the present value of the benefits as a function of μ, δ and γ.

15.19 Suppose that the force of mortality μ is a constant 0.04 and the force of interest is a constant 0.06.

(a) An insurer sells to one individual a single-premium, whole-life policy, with 100 payable at the moment of death, and also sells, to an independent life, a life annuity with benefits payable continuously at the rate of c per year. If W denotes the present value of the benefits on the two contracts combined, find Var(W) as a function of c.

(b) Suppose now that the two contracts in (a) are sold to the same individual. What is the variance of W in this case? Is it less than, greater than or equal to the variance in (a)? Explain why. For what values of c, if any, will the variance be equal to 0? Explain.

15.20 A failure time T has a p.d.f. of $f(t) = (12 - t)/72$ for $0 \leq t < 12$. An annuity contract provides for continuous payments at the annual rate of 1, which stop at failure, or at time 6 if earlier. The interest rate is 0. Let $\overline{Y}$ denote the present value of the benefits.

(a) Express $\overline{Y}$ as a function of T.

(b) Calculate $E(\overline{Y})$ and $\mathrm{Var}(\overline{Y})$.

(c) The single premium charged for this contract is the expected value plus 20%. What is the probability that the premium will cover the benefits?

15.21 For a certain failure time T, an insurance contract pays benefits at the end of the year of failure. You are given that $s_T(9) = 0.8,\ s_T(10) = 0.7\ E({}_{10}L) = 70, \mathrm{Var}({}_{10}L) = 1000$. The amount payable at time 10 for failure between time 9 and time 10 is 100. The interest rate is a constant 20 %. Find $\mathrm{Var}({}_9L)$.

15.22 A discrete failure time T has a constant hazard rate of $\lambda(k) = 1/2$. There is a constant interest rate of 100%. An insurance contract pays a benefit of 1 unit at the end of the year of failure. Level annual equivalence principle premiums are payable prior to failure. Using formula (15.28), show that for any nonnegative integer r, $\mathrm{Var}({}_rL) = 1/15$.

15.23 Consider two independent lives $(x), (y)$, where $T(x)$ is uniform on [0,1] and $T(y)$ is uniform on [0,2].

(a) Calculate the p.d.f. of the last survivor failure time $T(\overline{xy})$.

(b) An annuity provides continuous payments for as long as either (x) or (y) is alive. The rate of payment at time t is $3t^2$. The rate of interest is zero. If $\overline{Y}$ denotes the present value of the benefits for this annuity, find $E(\overline{Y})$ and $\mathrm{Var}(\overline{Y})$.

15.24 An endowment insurance on (x) provides for 1000 at the end of the year of death, provided this occurs within 4 years, and 1000 at time 4 if (x) is still alive. Net level annual premiums are payable for 4 years. You are given that $q_x = 0.05, q_{x+1} = 0.08, q_{x+2} = 0.10$. Interest rates are 5% for the first 2 years and 6% thereafter. Find $\mathrm{Var}({}_1L)$ in two ways: first, by finding the exact distribution of ${}_1L$; and second, by finding ${}_2V$ and ${}_3V$ and then using Equation (15.28).

15.25 A failure time takes the values 1 or 2 each with probability 1/2. The interest rate is 0. Insurance contract A pays 5 at failure. Insurance contract B pays 1 if failure occurs at time 1, and 5 if failure occurs at time 2. Show that using the variance premium principle with $\beta = 1$, the premium for contract B is higher than that for contract A.

Spreadsheet exercise

15.26 Modify the spreadsheet of Chapter 6 to calculate $\mathrm{Var}({}_rL)$.

16

Simplifications under level benefit contracts

16.1 Introduction

The calculation of variances and other distributional features simplifies considerably when we have level benefits and constant interest. In fact, we can write down formulas for exact distributions of the major random variables of interest. Throughout this chapter, we consider the following setup. We have a general failure time T. We will consider insurances paying a level amount upon failure, either at the end of the year of failure or at the moment of such, and we will consider annuities paid prior to the failure of T with either a level payment or continuous payments at a level rate. In addition, we assume a constant force of interest δ.

By taking T to be $T(x)$, this will apply to level benefit, whole-life insurances and to level benefit whole-life annuities. By taking $T = \min\{T(x), n\}$, this will apply to level benefit n-year endowment policies and to level benefit n-year temporary life annuities. Our assumption does *not* apply to term insurance, even when there is a level benefit during the term, since the benefit drops to zero after the expiration of the contract. However, in Section 16.5 we do illustrate that the calculation of exact distributions is possible for term or deferred insurances with a level death benefit paid over the benefit period.

16.2 Variance calculations in the continuous case

It is convenient to begin with a continuous failure time T.

Fundamentals of Actuarial Mathematics, Third Edition. S. David Promislow.

Companion website: http://www.wiley.com/go/actuarial

16.2.1 Insurances

Consider an insurance policy paying 1 at the moment of failure. The discount function is given by $v(t) = v^t = e^{-\delta t}$. Let $\bar{Z}$ be the present value of the benefits. In this case there is little simplification, and we know from the previous chapter that

$$\bar{A}_T = E(\bar{Z}) = E(v^T), \qquad \mathrm{Var}(\bar{Z}) = E(v^{2T}) - (\bar{A}_T)^2. \tag{16.1}$$

16.2.2 Annuities

Consider an annuity with continuous payments at the rate of 1 per year, made prior to the occurrence of failure. If $\bar{Y}$ is the present value of the benefits, then

$$\bar{Y} = \bar{a}(1_T) = \frac{1 - v^T}{\delta} = \frac{1 - \bar{Z}}{\delta}, \tag{16.2‡}$$

$$\bar{a}_T = E(\bar{Y}) = \frac{1 - E(\bar{Z})}{\delta}, \qquad \mathrm{Var}(\bar{Y}) = \frac{\mathrm{Var}(\bar{Z})}{\delta^2}. \tag{16.3‡}$$

For the case where $T = \min\{T(x), n\}$, we have already seen the first part of (16.3). This was the continuous 'endowment identity' given at the end of Section 8.8.

16.2.3 Prospective losses

Consider a contract that pays 1 unit at the moment of failure and has continuous level premiums at an annual rate of π payable prior to failure. (As a practical matter, this means that we are dealing with a net premium model that ignores expenses, which are unlikely to be level.) The prospective loss at time t is

$${}_tL = \bar{Z} \circ t - \pi(\bar{Y} \circ t), \tag{16.4}$$

where $\bar{Z} \circ t = v^{T \circ t}$ and $\bar{Y} \circ t = (1 - \bar{Z} \circ t)/\delta$. This leads to

$${}_tL = \left(1 + \frac{\pi}{\delta}\right) v^{T \circ t} - \frac{\pi}{\delta}. \tag{16.5}$$

This greatly simplifies the calculation of the variance of the prospective loss. Compare the following with (14.23).

$$\mathrm{Var}({}_tL) = \left(1 + \frac{\pi}{\delta}\right)^2 \mathrm{Var}(v^{T \circ t}). \tag{16.6‡}$$

16.2.4 Using equivalence principle premiums

Suppose that π is an equivalence principle premium. Then $\pi E(\bar{Y}) = E(\bar{Z})$, so that from (16.3),

$$\bar{\pi} = \frac{1}{\bar{a}_T} - \delta. \tag{16.7}$$

Taking $t = 0$ in (16.6),

$$\mathrm{Var}(L) = \frac{\mathrm{Var}(v^T)}{(\delta \bar{a}_T)^2} = \frac{\mathrm{Var}(\bar{Z})}{(1 - \bar{A}_T)^2}, \tag{16.8}$$

showing that

$$\mathrm{Var}(\bar{Z}) < \mathrm{Var}\,(L).$$

This reflects the fact that there is more risk involved in selling an insurance contract where premiums are payable over the entire life of the contract as opposed to the single-premium case. For failure occurring early, the insurer not only loses interest but will have collected relatively small amounts in premiums.

For a final formula, express $\bar{Z} \circ t$ in terms of $\bar{Y} \circ t$ in (16.4). Then

$${}_tL = (1 - (\pi + \delta)(\bar{Y} \circ t))$$

If π is an equivalence principle premium, we can substitute from (16.7) and take expectations to give a simple formula for the reserve at time t.

$${}_t\bar{V} = 1 - \frac{\bar{a}_{T \circ t}}{a_T}. \tag{16.9}$$

(For $T(x)$, we obtained the discrete version of this in (6.18).)

16.3 Variance calculations in the discrete case

We now consider the case where failure benefits are paid at the end of the year of failure, and annuity benefits and premiums are payable yearly. All of the formulas in Section 16.2 have discrete counterparts, which for the most part are obtained by replacing T by $\tilde{T}$, $\bar{A}$ by A, a by $\ddot{a}$ and δ by d. We will leave the formal derivations to the reader, but will list the formulas with the corresponding equation numbers as in Section 16.2, only with a prime to denote the discrete case.

If Z denotes the present value of the benefits for an insurance paying 1 unit at the end of the year of failure,

$$A_{\tilde{T}} = E(Z) = E(v^{\tilde{T}}), \qquad \mathrm{Var}(Z) = E(v^{2\tilde{T}}) - (A_{\tilde{T}})^2. \tag{16.1$'$}$$

If Y denotes the present value of an annuity paying 1 unit yearly provided that failure has not occurred,

$$Y = \ddot{a}(1_T) = \frac{1 - v^{\tilde{T}}}{d} = \frac{1 - Z}{d}, \tag{16.2$'$)‡ $$

$$E(Y) = \frac{1 - E(Z)}{d}, \qquad \mathrm{Var}(Y) = \frac{\mathrm{Var}(Z)}{d^2}. \tag{16.3$'$)‡ $$

Now consider a contract which pays 1 unit at the end of the year of failure and has level annual premiums of π payable prior to failure. Then for any positive integer k,

$$ {}_kL = \left(1 + \frac{\pi}{d}\right) v^{\tilde{T} \circ k} - \frac{\pi}{d}, \qquad (16.5') $$

$$ \mathrm{Var}({}_kL) = \left(1 + \frac{\pi}{d}\right)^2 \mathrm{Var}(v^{\tilde{T} \circ k}). \qquad (16.6')‡ $$

Suppose that π is an equivalence principle premium. Then

$$ \pi = \frac{1}{\ddot{a}_{\tilde{T}}} - d, \qquad (16.7') $$

$$ \mathrm{Var}(L) = \frac{\mathrm{Var}(v^{\tilde{T}})}{(d\ddot{a}_{\tilde{T}})^2} = \frac{\mathrm{Var}(v^{\tilde{T}})}{(1 - A_{\tilde{T}})^2}, \qquad (16.8') $$

$$ {}_kV = 1 - \frac{\ddot{a}_{\tilde{T} \circ k}}{\ddot{a}_{\tilde{T}}}. \qquad (16.9') $$

One could possibly consider expenses when using the above simplified prospective loss formulas if the difference is only in the first year, as the following example shows.

Example 16.1 An insurance policy provides 1000 at the end of the year of death with level premiums payable for life. There are expenses in the first year of 60% of the premium plus 50, and in the subsequent years of 15% of the premium plus 20. In addition, there is a death benefit settlement expense of 50. You are given that $E(Z) = 0.4$ and $\mathrm{Var}(Z) = 0.10$ where Z is the present value of 1 unit of death benefit. The rate of discount is a constant 0.06. Find Var (L) where L is calculated using expense-augmented premiums, and including all expenses.

Solution. We first compute the expense-augmented premium G as

$$ G\ddot{a}_x = 0.45G + 0.15G\ddot{a}_x + 30 + 20\ddot{a}_x + 1050A_x $$

so that

$$ G = \frac{1050A_x + 30 + 20\ddot{a}_x}{.85\ddot{a}_x - 0.45}. $$

We know that $\ddot{a}_x = (1 - A_x)/d = (1 - 0.4)/6 = 10$, and substituting in the above, $G = 80.75$. This means that the total inflow after the first year is $0.85(80.75) - 20 = 48.64$.

Now consider a policy where the total inflow was 48.64 in every year. The value of L from that contract would differ from that on the one in question only by a constant amount, namely the extra amount in the first year due to higher expenses. Therefore, the variance of L would be the same, and we can use formula (16.6′), with $\pi = 48.64$, and the 1 replaced by the death benefit of 1050. From (16.6′),

$$ \mathrm{Var}(L) = \left(1050 + \frac{48.64}{0.06}\right)^2 0.10 = 346\,208. $$

16.4 Exact distributions

In this section, we calculate the exact distributions for $\bar{Z}$, $\bar{Y}$ and L. In each case, we will derive distribution functions. These will be given for values between the greatest lower bound and least upper bound of the values. (We know that F takes the value 0 for arguments less than the greatest lower bound , and 1 for arguments greater than the least upper bound.) It is convenient here to introduce some new notation. For any random variable X, let

$$\hat{F}_X(x) = P(X < x), \quad \hat{s}_X(x) = P(X \geq x).$$

Of course, when X is continuous, $\hat{F} = F$ and $\hat{s} = s$.

The distribution functions we want are all easily expressed in terms of the distribution of T. Let N denote the least upper bound of the values of T. In the case that $N = \infty$ (as for example when T is exponential), the term v^N in the formulas below will be equal to 0.

16.4.1 The distribution of $\bar{Z}$

The distribution of $\bar{Z}$ is given by

$$F_{\bar{Z}}(z) = P(e^{-\delta T} \leq z) = P\left(T \geq -\frac{\log z}{\delta}\right) = \hat{s}_T\left(-\frac{\log z}{\delta}\right), \quad v^N < z < 1. \tag{16.10}$$

16.4.2 The distribution of $\bar{Y}$

First note that arguing as in (16.10) gives

$$\hat{F}_Z(z) = s_T\left(\frac{-\log(z)}{\delta}\right).$$

Using (16.2)

$$F_{\bar{Y}}(y) = P(\bar{Z} \geq 1 - \delta y) = 1 - \hat{F}_{\bar{Z}}(1 - \delta y).$$

Substituting from above,

$$F_{\bar{Y}}(y) = 1 - s_T\left(-\frac{\log(1-\delta y)}{\delta}\right) = F_T\left(-\frac{\log(1-\delta y)}{\delta}\right), \qquad 0 \leq y < \frac{1-v^N}{\delta}. \tag{16.11}$$

16.4.3 The distribution of L

The minimum value of L will be

$$\frac{-\pi}{\delta} + \left(1 + \frac{\pi}{\delta}\right) v^N,$$

occurring for failure at time N, and its maximum value will be 1, occurring for failure at time 0. Using (16.5),

$$\begin{aligned} F_L(u) &= P\left[\left(1+\frac{\pi}{\delta}\right)\mathrm{e}^{-\delta T}\right] \le u + \frac{\pi}{\delta} \\ &= P\left[\mathrm{e}^{-\delta T} \le \left(\frac{\delta u + \pi}{\delta + \pi}\right)\right] \\ &= \hat{s}_T\left[-\frac{1}{\delta}\log\left(\frac{\delta u + \pi}{\delta + \pi}\right)\right], \qquad \frac{-\pi}{\delta} + \left(1+\frac{\pi}{\delta}\right)v^N \le u \le 1. \end{aligned} \tag{16.12}$$

For the more general case, $F_{{}_tL}(u)$ is given by the same formula, but with $\hat{s}_{T\circ t}$ replacing $\hat{s}_T$ and v^{N-t} replacing v^N.

16.4.4 The case where T is exponentially distributed

In the particular case where T is exponential with constant hazard function μ, we know that $\hat{s}_T(z) = s_T(z) = \mathrm{e}^{-\mu z}$ and $N = \infty$. The above formulas simplify to

$$F_{\bar{Z}}(z) = z^{\mu/\delta}, \qquad 0 < z < 1, \tag{16.13}$$

$$F_{\bar{Y}}(y) = 1 - (1-\delta y)^{\mu/\delta}, \qquad 0 \le y < \frac{1}{\delta}, \tag{16.14}$$

$$F_L(u) = \left(\frac{\delta u + \pi}{\delta + \pi}\right)^{\mu/\delta}, \qquad -\frac{\pi}{\delta} \le u \le 1. \tag{16.15}$$

It is of interest to observe that if $\mu = \delta$ in the exponential case, the exponent in the formulas above equals 1 so that $\bar{Z}$, $\bar{Y}$ and L are all *uniform* random variables.

Example 16.2 A company decides to add 20% to its equivalence principle premiums as a protection against unfavourable experience. In each of the following cases, find the probability that premiums will cover claims. Suppose that T is exponential with $\mu = 0.04$ and that $\delta = 0.06$.

(a) A single-premium annuity providing continuous payments at the annual rate of 1 prior to failure.

(b) A contract paying 1 unit at failure, with level premiums payable continuously prior to failure.

Solution. (a) The equivalence principle premium is $1/(\mu + \delta) = 10$. The actual premium charged will be 12. From (16.14),

$$P(Y \le 12) = 1 - 0.28^{2/3} = 0.57.$$

(b) The equivalence principle annual premium rate is just $\mu = 0.04$, so the actual premium rate charged is 0.048. The probability that premiums cover claims is just

$$P(L \leq 0) = \left(\frac{0.048}{0.108}\right)^{2/3} = 0.58.$$

16.5 Some non-level benefit examples

It is also possible to obtain exact distributions in some simple cases involving non-level benefits such as term or deferred insurance.

16.5.1 Term insurance

Consider a contract that pays 1 at failure, provided failure occurs within n years. In order to handle failure times that are not continuous, we adopt the convention that a benefit is paid for failure at exact time n. We will compute the distribution of $\bar{Z}$, the present value of the benefits. The minimum *positive* value of $\bar{Z}$ is $e^{-\delta n}$ occurring for death at time n. Since nothing is paid for death strictly after time n, which occurs with probability $s_T(n)$, we can easily obtain the term distribution from the whole life distribution. Pick up the probability mass to the *left* of $z = e^{-\delta n}$ and set it down as a point mass at the point 0, as illustrated in Figure 16.1. From this, utilizing formula (16.10), we can read off this distribution of $\bar{Z}$.

$$F_{\bar{Z}}(z) = \begin{cases} s_T(n), & 0 \leq z < e^{-\delta n}, \\ \hat{s}_T\left(\dfrac{-\log z}{\delta}\right), & e^{-\delta n} \leq z \leq 1. \end{cases} \tag{16.16}$$

Example 16.3 Redo part (b) of Example 16.2, but now assuming n-year term insurance for $n = 15$ and 10. Level premiums are payable continuously for n years.

Solution. The equivalence principle premium rate is still $\mu = 0.04$ so the actual premium rate charged is still 0.048 and $\pi/\delta = 0.8$. For the contract in Example 16.2, the value of L when $T = n$ is $1.8e^{-0.06n} - 0.8$. For $n = 15$, this is negative. This means that L becomes negative at some point prior to expiration of the term contract, so the probability that premiums cover claims is 0.58, exactly the same as it was in constant benefit case. When $n = 10$, the value of L for the contract in Example 16.1 is positive. Therefore, in order that L in this example be negative, it is necessary that $T \geq 10$, so that no benefits are paid. The probability of this is $e^{-0.6} = 0.67$.

16.5.2 Deferred insurance

A similar example is provided by *deferred insurance*. Consider a contract that pays 1 at failure, provided failure occurs after time n. We now adopt the convention that nothing is paid for failure at exact time n. In view of this convention, there need not be a maximum value of $\bar{Z}$ but $e^{-\delta n}$ is certainly an upper bound, since any benefits will be paid at a time later than n.

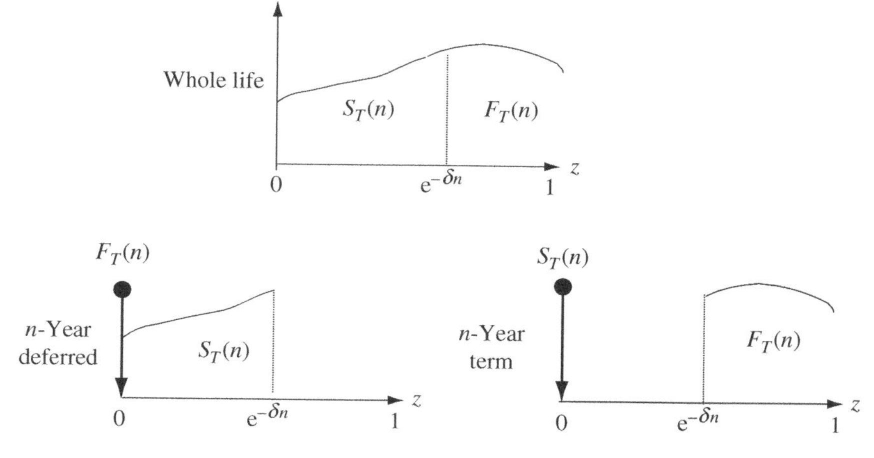

Figure 16.1 Graph of $f_{\bar{Z}}(z)$ for various types of insurance

Refer again to Figure 16.1. Since nothing is paid for death at or before time n, which occurs with probability $F_T(n)$, the probability mass to the *right* of $e^{-\delta n}$ is picked up and set down as a point mass at 0.

Utilizing (16.10), we can write

$$F_{\bar{Z}}(z) = \begin{cases} F_T(n) & z = 0, \\ F_T(n) + \hat{s}_T\left(\dfrac{-\log z}{\delta}\right) & 0 < z < e^{-\delta n}, \\ 1 & e^{-\delta n} \leq z \leq 1. \end{cases}$$

16.5.3 An annual premium policy

We next investigate a more complicated case where we compute the exact distribution of L in an annual premium policy. Consider a failure time T that is unbounded (i.e. $N = \infty$). We consider insurances which have premiums payable continuously at a level annual rate π for the duration of the contract. We will compare the distribution of L for a contract that pays 1 unit on failure or at time n if earlier, and a contract that pays 1 unit at failure provided this occurs within n years. These then correspond respectively to endowment and term insurance. To simplify the notation, let

$$b_0 = -\frac{\pi}{\delta} + \left(1 + \frac{\pi}{\delta}\right) v^n, \qquad c_0 = -\frac{\pi}{\delta} + \frac{\pi}{\delta} v^n.$$

In both cases, the value of L is given by (16.5) (with $t = 0$), provided that T takes a value less than n, which corresponds to L taking a value greater than b_0. So for any interval (a, b) with $a > b_0$, we have that $P(a < L \leq b)$ is the same in both cases and this value can be calculated directly from (16.12).

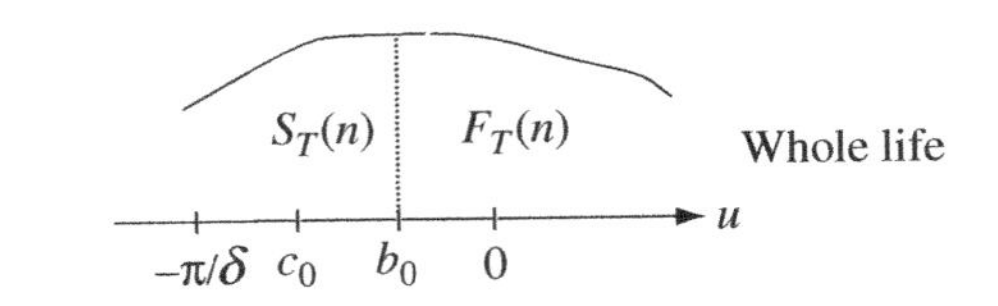

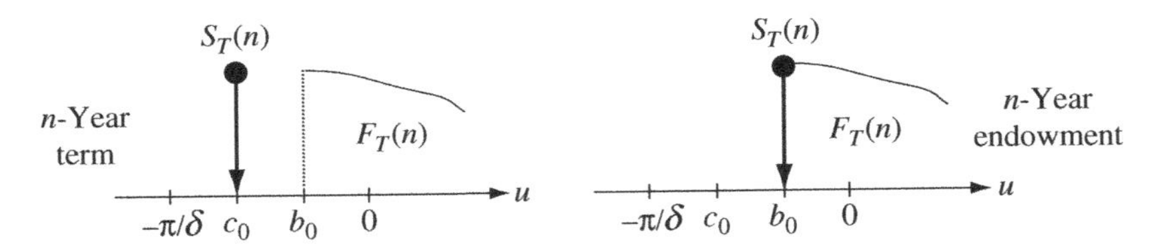

Figure 16.2 Graph of $f_L(u)$ for various types of insurance

Consider the remaining probability of $s_T(n)$. For the endowment contract, this will all be concentrated at the single point b_0. For the term contract, it will all be concentrated at the single point c_0, which lies in the interval $(-\pi/\delta, b_0)$. As n increases to ∞ both b_0 and c_0 approach $-\pi/\delta$ and the distribution approaches the whole-life case as given in (16.12).

The various density functions are compared in Figure 16.2. The probability mass to left of the point b_0 on the whole-life graph is picked up and set down as a point mass at b_0 for the endowment policy, or at c_0 for the term policy.

Exercises

Type A exercises

16.1 For a certain failure time T, an insurance contract pays 1 at the moment of failure, and has level premiums payable continuously prior to failure at the annual rate of 0.06. You are given that

$$\text{Var}(\bar{Z}) = 0.064, \qquad \text{Var}(\bar{Y}) = 10,$$

where $\bar{Z}$ is the present value of the failure benefits, and $\bar{Y}$ is the present value of an annuity contract with continuous payments at the annual rate of 1, payable prior to failure. Find Var(L).

16.2 For a certain failure time T, an insurance contract pays 1 at the moment of failure. Level equivalence principle premiums are payable prior to failure. If $E(\bar{Z}) = 0.6$ and $\text{Var}(L) = 2$, find $\text{Var}(\bar{Z})$.

16.3 A whole-life insurance contract provides for 1 unit payable at the moment of death. Level premiums are payable continuously for life at the annual rate of 0.08. The force of mortality is a constant 0.05, and the force of interest is a constant 0.10. How many contracts must be sold in order that there is at least a 95% chance that the total premiums on all these contracts will cover the total benefits? Use a normal approximation.

16.4 An insurer sells 100 whole-life insurance policies each providing for 1 unit payable at the moment of death. Level premiums are payable continuously for life. The force of interest is a constant 0.06, and the force of mortality is a constant 0.04. What should the rate of premium payment be on each policy, in order that there is a 95% chance that total premiums on all 100 policies will cover the total benefits on all policies? Use a normal approximation.

Type B exercises

16.5 (See Exercise 13.8.) A continuous failure time has a density function, $f_T(t) = \beta^2 t e^{-\beta t}$ (a gamma distribution with first parameter 2). The force of interest is a constant δ. Find expressions in terms of β and δ for: (a) $\bar{A}_T$; (b) $\bar{a}_T$; (c) the net annual rate of premium payment when premiums are payable continuously prior to failure, for an insurance paying 1 at failure and (d) the reserve at time k for the contract in (c).

16.6 An insurance contract, based on the failure time T, pays 1 unit at the moment of failure provided this occurs within 5 years. Nothing is paid for failure after that time. The force of interest is a constant 0.1. If T has the hazard function

$$\mu_T(t) = \frac{2}{10 - t},$$

find the probability that the present value of the benefits is strictly positive, but less than or equal to $e^{-0.3}$.

16.7 Suppose T is uniform on [0, 10], and $\delta = 0.05$. A contract pays 1 unit at failure. Level premiums are payable continuously for 10 years. The premiums charged are the net premiums plus 20%.

(a) Find the probability that $L \leq 0$.

(b) Suppose now that, instead of 1.2 times equivalence principle premiums, the company charges premiums at the rate of 0.05 per year. What is the probability that $L \leq 0$?

16.8 A failure time T is uniformly distributed on the interval [0, 20]. The force of interest is a constant 0.05.

(a) A deferred insurance contract provides for a payment of 1 at the moment of failure provided that this occurs after time 5. If $\bar{Z}$ is the present value of the benefits, find the 80th percentile of $\bar{Z}$. That is, find the point z such that $P(\bar{Z} \leq z) = 0.80$.

(b) Find the 80th percentile of $\bar{Z}$ when the contract is a term insurance policy that pays 1 unit if failure occurs in the first 5 years.

16.9 Consider a deferred term insurance contract. It provides for 1 unit payable at the moment of failure provided that this occurs between N and $2N$ years from now. Nothing is paid if failure occurs in the first N years or after $2N$ years. You are given that the force of failure μ and the force of interest δ are constants such that $\mu = \delta/2$. Moreover, you are given that $e^{-N\delta} = 0.36$. If $\bar{Z}$ is the present value of the benefits, calculate $F_{\bar{Z}}(z)$ for all nonnegative values of z.

16.10 An insurance contract provides for $e^{\delta t}$ payable at failure, should this occur at time t. Net level premiums are payable continuously prior to failure. The force of interest is a constant δ. Show that

$$L = \frac{v^T - \bar{A}_T}{1 - \bar{A}_T}.$$

16.11 An insurance policy provides a benefit of 1 at the moment of failure provided this occurs after 10 years. The hazard rate of the failure time is a constant 0.10 and the force of interest is a constant 0.05. If $\bar{Z}$ is the present value of the benefits, find the probability that: (a) $0 < \bar{Z} \leq 0.5$, (b) $0 \leq \bar{Z} \leq 0.5$ and (c) $0 \leq \bar{Z} \leq 0.7$.

16.12 An insurance policy provides a benefit of 1 at the moment of failure plus a pure endowment of 1 at time 10. This is purchased by level premiums payable continuously for 10 years. The premium is the net premium plus 10%. There is a constant force or mortality of 0.05 and a constant force of interest of 0.05. Find (a) $P(-1/3 < L \leq 1/2)$ and (b) $P(-1/4 \leq L \leq 1/2)$.

16.13 Let T be any failure time, and assume constant interest.

(a) Show that

$$E[\bar{a}(1_T)]^2 = \frac{2}{\delta}(\bar{a}_T - {}^2\bar{a}_T),$$

where the superscript 2 on the right-hand side indicates calculation at a force of interest equal to 2δ or equivalently with v replaced by v^2.

(b) Is the following formula true? If not, give a correct version.

$$E[\ddot{a}(1_{\tilde{T}})]^2 = \frac{2}{d}(\ddot{a}_{\tilde{T}} - {}^2\ddot{a}_{\tilde{T}}).$$

16.14 Modify the formula in Section 16.5.2 in the case that there is a benefit paid for failure at exact time n.

17

The minimum failure time

17.1 Introduction

Suppose that $T_1, T_2, \ldots, T_m$ are failure times defined on the same sample space. In this chapter, we investigate the random variable

$$T = \min\{T_1, T_2, \ldots, T_m\}.$$

In other words, T is the time of the first failure to occur among the m different failure times that are possible. In particular, this will recast the material of Chapters 10 and 11 into a stochastic framework, and show that both joint-life theory and multiple-decrement theory are special cases of this general problem. In addition, it will provide more rigorous arguments for some of the results of those chapters that were obtained in an intuitive fashion. Finally it will deal with the important cases where the failure times need not be independent.

In the joint-life case, where we have a group of m lives numbered $1, 2, \ldots, m$, we can take T_i to be the future lifetime of the ith life, so that T is the failure time of the joint m-life status. In the multiple-decrement context, we can take T_i to be the time of failure from cause i in the *associated single-decrement* setting, so it is the failure time of the ith cause, assuming no other causes of failure are operating. Then, the random variable T is the time of failure in the multiple-decrement model. (In the machine analogy of Section 11.6, T_i would be the failure time of the ith part.)

17.2 Joint distributions

We wish to expand somewhat on our brief description for the case $m = 2$ given in Appendix A. Suppose that each T_i is continuous. There are various ways of describing the joint distribution.

Fundamentals of Actuarial Mathematics, Third Edition. S. David Promislow.

Companion website: http://www.wiley.com/go/actuarial

We can do so by the joint density function $f_{T_1T_2,\ldots,T_m}(t_1, t_2, \ldots, t_m)$, or alternatively by the joint distribution function

$$F_{T_1,T_2,\ldots,T_m}(t_1, t_2, \ldots, t_m) = P[T_1 \le t_1, T_2 \le t_2, \ldots, T_m \le t_m],$$

or by the joint survival function

$$s_{T_1,T_2,\ldots,T_m}(t_1, t_2, \ldots, t_m) = P[T_1 > t_1, T_2 > t_2, \ldots, T_m > t_m].$$

To simplify the notation, we will often omit the subscripts and just write f, F or s when no confusion arises.

The reader is cautioned that $s(t_1, t_2, \ldots, t_m) \neq 1 - F(t_1, t_2, \ldots, t_m)$ when $m > 1$.

As in the one-dimensional case, we integrate to obtain the distribution or survival functions from the density function, and differentiate to go in the other direction. We must, however, use multiple integrals and partial derivatives. For example, with $m = 3$,

$$F(t_1, t_2, t_3) = \int_0^{t_1} \int_0^{t_2} \int_0^{t_3} f(u, v, w) dw\, dv\, du,$$

and similarly

$$s(t_1, t_2, t_3) = \int_{t_1}^{\infty} \int_{t_2}^{\infty} \int_{t_3}^{\infty} f(u, v, w) dw\, dv\, du.$$

Now differentiate the latter expression with respect to t_1. The fundamental theorem of calculus tells us to replace the variable u in the integrand with t_1 and affix a minus sign since t_1 is a lower limit. The integrand consists of the second two integrals. The result is

$$\frac{\partial}{\partial t_1} s(t_1, t_2, t_3) = -\int_{t_2}^{\infty} \int_{t_3}^{\infty} f(t_1, v, w) dw\, dv \tag{17.1}$$

After two more iterations of the procedure, we have

$$f(t_1, t_2, t_3) = (-1)^3 \frac{\partial}{\partial t_1} \frac{\partial}{\partial t_2} \frac{\partial}{\partial t_3} s(t_1, t_2, t_3).$$

Similarly, we can derive

$$f(t_1, t_2, t_3) = \frac{\partial}{\partial t_1} \frac{\partial}{\partial t_2} \frac{\partial}{\partial t_3} F(t_1, t_2, t_3).$$

Analogous expressions hold for general m, which appears as the exponent of -1 in the first formula.

Note that the individual distributions are easily obtained from the the joint distribution or survival functions by

$$s_{T_i}(t) = s_{T_1,T_2,\ldots,T_m}(0, 0, \ldots, t, \ldots, 0),$$
$$F_{T_i}(t) = F_{T_1,T_2,\ldots,T_m}(\infty, \infty, \ldots, t, \ldots, \infty),$$

where the t is in the ith position, and ∞ indicates that you take limits.

17.3 The distribution of T

17.3.1 The general case

It is a simple matter to deduce the distribution of T from the joint distribution. Clearly, the minimum will take a value greater than t if and only if each T_i takes a value greater than t. Therefore,

$$s_T(t) = s(t, t, \dots, t). \tag{17.2}$$

(The reader should note that $F_T(t) \neq F(t, t, \dots, t)$.)

17.3.2 The independent case

Let the density function, survival function, and hazard rate function of T_i be denoted respectively by f_i, s_i, μ_i. Things become much easier to deal with when the random variables T_i are independent. We can then readily write down the relevant functions for T in terms of the corresponding functions for T_i. For example, from (17.2) we obtain

$$s_T(t) = s_1(t)s_2(t) \cdots s_m(t). \tag{17.3}$$

By taking logs and differentiating, we can find a similar relationship involving hazard functions.

$$\mu_T(t) = \sum_{i=i}^{m} \mu_i(t). \tag{17.4}$$

We have already encountered a particular case of (17.4) in (10.12).

Example 17.1 Suppose that T_1 and T_2 are independent and both have the hazard function

$$\mu(t) = \frac{2}{1-t}, \quad 0 \le t < 1.$$

Find the probability that the minimum value of these two random variables will be less than or equal to 1/2.

Solution. We note that $s_T(t) = (1-t)^4$, for $0 \le t < 1$, either by first noting that, for $i = 1, 2, s_i(t) = (1-t)^2$ or by noting that $\mu_T(t) = 4/(1-t)$. The desired probability is $F_T(1/2) = 1 - s_T(1/2) = 15/16$.

17.4 The joint distribution of (T, J)

17.4.1 The distribution function for (T, J)

Suppose that there is zero probability of the simultaneous occurrence of two or more failure times $T_1, T_2, \dots, T_m$. We can then define the random variable J as the index of the random variable giving the minimum. For an example with $m = 3$, suppose $T_1 = 7$, $T_2 = 5$, and $T_3 = 10$. Then T would take the value 5, and J would take the value 2, since the minimum occurs for T_2.

In many applications, we are interested in the joint distribution of T and J. This can be described in various ways. One method is by the joint distribution function,

$$F_{T,J}(t,j) = P(T \leq t \text{ and } J = j).$$

This joint distribution has the somewhat unusual feature that the random variable T is normally continuous, while J is discrete. Therefore, unlike the usual notation for distribution functions, the second variable in $F_{T,J}(t,j)$ is not cumulative, but refers to one specific index.

We now consider the problem of deducing the joint distribution (T, J) from the joint distribution of $T_1, T_2, \dots, T_m$.

If we are given the joint density function, then $F_{T,J}(t, j)$ is calculated by integrating over a suitable region – see (A.17). Take $m = 2$. Then $F_{T,J}(t, 1)$ is the probability that T_1 takes a value less than or equal to t, and that T_2 takes any value that is greater than that taken by T_1. This is given by the double integral

$$F_{T,J}(t, 1) = \int_0^t \int_u^\infty f_{T_1,T_2}(u, v)\mathrm{d}v\, \mathrm{d}u, \tag{17.5}$$

and similarly

$$F_{T,J}(t, 2) = \int_0^t \int_v^\infty f_{T_1,T_2}(u, v)\mathrm{d}u\, \mathrm{d}v. \tag{17.6}$$

Example 17.2 The joint distribution of T_1 and T_2 is given by

$$f_{T_1,T_2}(s, t) = \begin{cases} 6(s - t)^2, & 0 \leq s \leq 1, 0 \leq t \leq 1, \\ 0, & \text{elsewhere.} \end{cases}$$

Find $F_{T,J}(t,j)$ for $j = 1, 2$.

Solution.

$$F_{T,J}(t, 1) = \int_0^t \int_u^1 6(u - v)^2 \mathrm{d}v\, \mathrm{d}u = 2\int_0^t (1 - u)^3 \mathrm{d}u = \frac{1 - (1 - t)^4}{2}, \quad \text{for } 0 \leq t \leq 1.$$

By symmetry, we must have

$$F_{T,J}(t, 2) = \frac{1 - (1 - t)^4}{2}, \quad 0 \leq t \leq 1.$$

Since T_i is bounded above by 1 for $i = 1, 2$, we necessarily have

$$F_{T,J}(t, i) = F_{T,J}(1, i) = 1/2, \quad i = 1, 2, \quad t > 1.$$

The sum of $F_{T,J}(t, 1)$ and $F_{T,J}(t, 2)$ must of course equal $F_T(t) = 1 - s_T(t)$. It will be instructive for the reader to verify this by drawing a picture in the plane, showing that the union of the regions of integration in (17.5) and (17.6), and the region corresponding to $s_T(t)$, is the entire positive quadrant.

In the general case, we will need m-dimensional integrals to compute $F(t,j)$ from the density function. However, if we already have the joint survival function, the computation can be simplified. To illustrate, consider the case with $m = 3$. Then, reasoning as above,

$$F_{T,J}(t, 1) = \int_0^t \int_u^\infty \int_u^\infty f(u, v, w) \mathrm{d}w \, \mathrm{d}v \, \mathrm{d}u.$$

From (17.1) the inner two integrals can be written compactly as a partial derivative. We have

$$F_{T,J}(t, 1) = \int_0^t \sigma(u) \mathrm{d}u,$$

where

$$\sigma(u) = -\frac{\partial}{\partial t_1} s(t_1, t_2, t_3), \quad \text{evaluated at } t_1 = t_2 = t_3 = u.$$

In the general case,

$$F_{T,J}(t,j) = \int_0^t \sigma_j(u) \mathrm{d}u, \tag{17.7}$$

where

$$\sigma_j(u) = -\frac{\partial}{\partial t_j} s(t_1, t_2, \ldots t_m), \quad \text{evaluated at } t_1 = t_2 = \cdots = t_m = u.$$

Example 17.3 Suppose that $m = 2$ and the joint survivor function is given by

$$s_{T_1,T_2}(u, v) = \frac{1}{2}[(1 - u)^4 + (1 - v)^4 - (u - v)^4], \quad 0 \le u \le 1, 0 \le v \le 1.$$

Find $F_{T,J}(t, 1)$.

Solution. We could take two derivatives to calculate $f(u, v) = 6(u - v)^2$, and apply (17.5). (This is in fact the same distribution as in Example 16.2.) Note, however, that (17.5) would just 'undo' the calculation of the second derivative by integrating. For this form of the distribution, it is easier to apply (17.7). On the given region

$$-\frac{\partial}{\partial u} s(u, v) = 2[(1 - u)^3 + (u - v)^3], \qquad \sigma_1(u) = 2(1 - u)^3,$$

so

$$F_{T,J}(t,1) = \int_0^t 2(1-u)^3 \mathrm{d}u = \frac{1-(1-t)^4}{2}, \quad 0 \le t \le 1,$$

verifying the previous example.

17.4.2 Density and survival functions for (T, J)

We can also define the distribution of (T, J) by the joint density function $f_{T,J}(t,j)$. This is the function satisfying

$$F_{T,J}(t,j) = \int_0^t f_{T,J}(s,j)\mathrm{d}s, \qquad f_{T,J}(t,j) = \frac{\mathrm{d}}{\mathrm{d}t} F_{T,J}(t,j). \tag{17.8}$$

The function f is interpreted in the normal way. Namely, for 'small' $\Delta t, f_{T,J}(t,j)\Delta t$ is approximately the probability that the first failure will be from cause j and that it will take place in the time interval from t to $t + \Delta t$. The precise statement, following from the first expression in (17.8), is that the probability that the first failure will be from cause j and will take place between time a and time b is given by

$$F_{T,J}(b,j) - F_{T,J}(a,j) = \int_a^b f_{T,J}(t,j)\mathrm{d}t.$$

We define the joint survival function for (T, J) by thinking of survival as we did at the end of Section 11.2.2. Let

$$s_{T,J}(t,j) = P(T > t \text{ and } J = j) = \int_t^\infty f_{T,J}(s,j)\, \mathrm{d}s.$$

This is the probability that failure will occur after time t due to cause j.

Figure 17.1 shows a typical graph of $f(t,j)$ for $m = 2$. Note that the mass is concentrated on parallel sheets. If we cut the jth sheet by the plane $T = t$, the area of the left portion will be $F(t,j)$ and the area of the right portion will be $s(t,j)$.

All functions pertaining to the random variable T alone are obtained by summing over all j, as we noted above for F in the case $m = 2$. That is,

$$F_T(t) = \sum_{j=1}^m F_{T,J}(t,j), \tag{17.9}$$

$$f_T(t) = \sum_{j=1}^m f_{T,J}(t,j), \tag{17.10}$$

$$s_T(t) = \sum_{j=1}^m s_{T,J}(t,j). \tag{17.11}$$

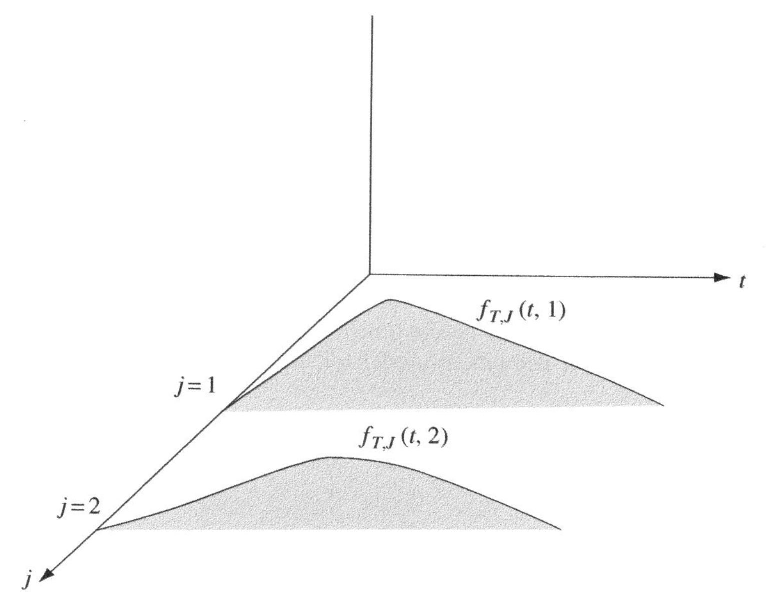

Figure 17.1 The graph of $f_{T,J}(t,j)$

17.4.3 The distribution of J

To obtain the distribution of J from the joint distribution, we compute the other marginal, which can be expressed in various ways. If $f_J(j)$ denotes the probability that $J = j$, then

$$f_J(j) = \int_0^\infty f_{T,J}(t,j)\mathrm{d}t = s_{T,J}(0,j) = \lim_{t\to\infty} F_{T,J}(t,j) \tag{17.12}$$

In Figure 17.1, $f_J(j)$ is the area of the jth sheet. Note that $F_{T,J}(t,j) + s_{T,J}(t,j)$ is not equal to 1, but rather to $f_J(j)$.

Example 17.4 Take $m = 2$. If T_1 is uniform on [0,1], T_2 is uniform on [0,2], and T_1 and T_2 are independent, find the distribution of J.

Solution. The joint density function takes a constant value of 1/2 on the rectangle $0 \le s \le 1, 0 \le t \le 2$, and is 0 elsewhere. The maximum value of T is the minimum of the respective maximums of the T_i, which in this case is 1. Therefore,

$$f_J(1) = F_{T,J}(1,1) = \int_0^1 \int_u^2 (1/2)\mathrm{d}v\,\mathrm{d}u = \frac{3}{4},$$

$$f_J(2) = F_{T,J}(1,2) = \int_0^1 \int_v^1 (1/2)\mathrm{d}u\,\mathrm{d}v = \frac{1}{4}.$$

17.4.4 Hazard functions for (T, J)

Definition 17.1 The hazard function for (T, J) is given by

$$\mu_{T,J}(t,j) = \frac{f_{T,J}(t,j)}{s_T(t)}.$$

This is a conditional density. For small Δt, $\mu_{T,J}(t,j)\Delta t$ is approximately the probability that failure will occur first from cause j in the time interval from t to $t + \Delta t$, given that failure *from any cause* has not yet taken place before time t.

In the Chapter 11 multiple-decrement model for a life age x, $\mu_{T,J}(t,j)$ corresponds to $\mu_x^{(j)}(t)$.

Given the hazard rates, we can obtain the joint distribution of (T, J) by the same method as employed in Chapter 11. From (17.10) and the definition of $\mu(t,j)$,

$$\mu_T(t) = \sum_{j=1}^{m} \mu_{T,J}(t,j), \tag{17.13}$$

and then, from (15.7),

$$s_T(t) = \mathrm{e}^{-\int_0^t \mu_T(r)\mathrm{d}r}.$$

We know that $f_{T,J}(s,j) = s_T(s)\mu_{T,J}(s,j)$ and, from the first expression in (17.5),

$$F_{T,J}(t,j) = \int_0^t s_T(s)\mu_{T,J}(s,j)\mathrm{d}s. \tag{17.14}$$

The last formula is easily explained intuitively. For the event in question to occur, there must be some point s, before t, for which failure from any cause has not yet occurred, and then failure will occur from the jth cause at time s Although (17.14) has this intuitive appeal, it is not necessarily useful for computing $F_{T,J}(t,j)$ as we may not know the joint hazard rates until we have already computed $F_{T,J}(t,j)$ and $f_{T,J}(t,j)$. It is, however, an important formula in the independent case to which we now turn.

17.4.5 The independent case

As in Section 17.2, we can simplify calculations when the T_i are independent, and deduce the distribution of (T, J) directly from the individual distributions of each T_i. Since $s_j'(t) = -s_j(t)\mu_j(t)$, (17.3) shows that

$$\sigma_j(t) = \prod_{i \neq j} s_i(t)s_j'(t) = s_T(t)\mu_j(t)$$

and then from (17.7),

$$F_{T,J}(t,j) = \int_0^t s_T(u)\mu_j(u)\mathrm{d}u, \quad \text{for independent } T_i. \tag{17.15}$$

Note, as a comparison to (17.14.), that (17.15) can be used directly to compute $F_{T,J}(t,j)$ in the independent case, when we know μ_i from the individual distributions. Moreover, differentiating and dividing by $s_T(t)$ verifies that in the independent case

$$\mu_{T,J}(t,j) = \mu_j(t) \tag{17.16}$$

for $j = 1, 2, \ldots m$ and all t for which $s_T(t) \geq 0$, This provides the promised proof for the result stated in in formula (11.23) .

The result is easily explained intuitively. Looking at the machine model of Section 11.6, for example, both quantities in (17.16) give a conditional density for failure of part j at time t. In the case of $\mu_{T,J}(t,j)$, the condition is that all parts have survived up to time t and in general this may give information regarding the failure time of part j. Suppose, for example, that the parts are connected so that part 2 cannot fail until part 1 does, and then it fails five seconds later. (The same idea in the multiple decrement model provided the simple counter-example of Exercise 11.15.) In the independent case, however, we obtain exactly the same information as if we told only that part j has survived up to time t, which is precisely the condition applicable to $\mu_j(t)$.

We have already encountered special cases of (17.15). One example is (10.28). The hazard rate $\mu_x(t)$ corresponds to $\mu_1(t)$ which equals $\mu_{T,J}(t, 1)$ since we postulated independence. Another example is (11.13).

Example 17.5 Suppose that the T_i are independent and exponential with constant hazard μ_i. (a) Find $F_{T,J}(t,j)$. (b) Find $f_J(j)$. (c) Show that T and J are independent.

Solution.

(a) Let $\mu = \mu_1 + \mu_2 + \cdots + \mu_m$. Formula (17.4) shows that T is exponential with constant hazard μ, and, from (17.15),

$$F_{T,J}(t,j) = \int_0^t \mathrm{e}^{-s\mu}\mu_j \mathrm{d}s = \frac{\mu_j}{\mu}(1 - \mathrm{e}^{-\mu t}).$$

(b)

$$f_J(j) = \lim_{t\to\infty} F_{T,J}(t,j) = \frac{\mu_j}{\mu}.$$

(c) We just note that

$$F_{T,J}(t,j) = F_T(t)f_J(j).$$

The solution to (b) gives an important result that has many applications. It says that for independent exponential failure times, the probabilities of first failure are proportional to the hazard rates. This makes sense, since the higher the hazard rate, the lower the mean, which is the reciprocal of the hazard rate, and therefore more likelihood of occurring first.

Example 17.6 Take $m = 2$. Suppose T_1 and T_2 are independent, T_1 is uniform on $[0, a]$, and T_2 is uniform on $[0, b]$, where $0 < a \leq b$. Find $F_{T,J}(t, 1)$ and $F_{T,J}(t, 2)$.

Solution. Since $s_T(t)\mu_1(t) = s_2(t)s_1(t)\mu_1(t) = s_2(t)f_1(t)$, we can write

$$F_{T,J}(t, 1) = \int_0^t \left(1 - \frac{s}{b}\right)\frac{1}{a}ds = \frac{t}{a} - \frac{t^2}{2ab} = F_1(t) - \frac{1}{2}F_1(t)F_2(t), \quad 0 < t < a.$$

Similarly,

$$F_{T,J}(t, 2) = \frac{t}{b} - \frac{t^2}{2ab} = F_2(t) - \frac{1}{2}F_1(t)F_2(t), \quad 0 < t < a.$$

Since failure must take place before time a, for any $t \geq a$,

$$F_{T,J}(t, 1) = F_{T,J}(a, 1) = 1 - \frac{a}{2b}, \qquad F_{T,J}(t, 2) = F_{T,J}(a, 2) = \frac{a}{2b}$$

As a check, note that $F_{T,J}(t, 1) + F_{T,J}(t, 2) = F_1(t) + F_2(t) - F_1(t)F_2(t)$, which must be true, since for T to take a value less than t means that at least one of T_1 and T_2 takes a value less than t. Also note that the answer here could have been written down immediately, since the distributions satisfy the stochastic version of the condition given in (11.24), and therefore we can apply Method 2 of Chapter 11.

17.4.6 Nonidentifiability

We motivate the idea of this section by an example. First note that the joint distributions given in Examples 17.1 and 17.2 are easily seen to be different. One way is to note that T_1 and T_2 are not independent in the latter.

Example 17.7 Calculate $F_{T,J}(t, j)$ for the distribution given in Example 17.1

Solution. We could do this by (17.15), but it is easier to note that, by symmetry, $F_{T,J}(t, 1) = F_{T,J}(t, 2)$ and the two must sum to $F_T(t) = 1 - s_T(t)$. We can conclude directly from Example 17.1 that

$$F_{T,j}(t, 1) = F_{T,J}(t, 2) = \frac{1 - (1 - t)^4}{2}, \quad 0 \leq t \leq 1,$$

and necessarily

$$F_{T,J}(t, 1) = F_{T,J}(t, 2) = \frac{1}{2}, \quad 1 \leq t.$$

Compare the above result with Example 17.2. The somewhat surprising conclusion is that two completely different distributions for (T_1, T_2) have led to exactly the same distribution for (T, J). This is known as the *nonidentifiability problem* and it has statistical implications. Suppose we want to make inferences about the joint distribution of the T_i by observing failure times. In many cases, all we can possibly observe is the joint distribution (T, J). An example is when the random variables represent the time of death from various causes. Once death occurs, we know the time and the cause, but no further observation of the subject is possible. Our example above shows that it is impossible to uniquely determine the joint distribution of the random variables that give rise to a given (T, J). We need additional information in order to obtain a unique solution. One instance when this occurs is in the independent case.

Theorem 17.1 *Given any joint distribution for (T, J) there is a unique joint distribution of $(T_1, T_2, \ldots, T_m)$ such that the (T_i) are mutually independent and induce the given distribution of (T, J).*

Proof. Uniqueness follows immediately, since, given independence, we know the joint distribution if we know the distribution of each T_i, and (17.16) implies that each T_i is necessarily a random variable with hazard function $\mu_{T,J}(t, i)$, which is determined uniquely from (T, J).

For the existence, given any joint distribution function F for (T, J), we let T_i be a random variable with hazard function $\mu_{T,J}(t, i)$. This collection of independent T_i in turn generates a joint distribution function $\hat{F}_{T,J}$. From (17.15),

$$\hat{F}_{T,J}(t,j) = \int_0^t s_T(s)\mu_{T,J}(s,j)\mathrm{d}s,$$

which equals $F_{T,J}(t,j)$ as shown by (17.14). □

To illustrate the use of this theorem, suppose you are told that

$$F_{T,J}(t, i) = \frac{1-(1-t)^4}{2}, \quad 0 \le t \le 1, \quad F_{T,J}(t, 1) = 1, \quad t > 1$$

for $i = 1, 2$ and asked to identify the joint distribution of (T_1, T_2). You cannot do this without further information, for it could be either the joint distribution of Example 16.1 or that of Example 16.2, or indeed several other possibilities. However, if you are given the additional information that T_1 and T_2 are independent, then you know that it must be the distribution of Example 16.1.

17.4.7 Conditions for the independence of T and J

Another question of interest is to determine when T and J are independent. In Example 17.5, we saw that this occurred with constant hazard functions. We present here a more general criterion. Define the ratios

$$K(t,j) = \frac{\mu(t,j)}{\mu_T(t)}$$

for all j, and all t, such that $s_T(t) > 0$.

Theorem 17.2 *$K(t,j) = P(J = j|T = t)$. Therefore, T and J are independent if and only if $K(t,j)$ is independent of t.*

Proof. We have

$$f_{T,J}(t,j) = s_T(t)\mu(t,j) = K(t,j)s_T(t)\mu_T(t) = K(t,j)f_T(t).$$

So

$$P(J=j|T=t) = \frac{f_{T,J}(t,j)}{f_T(t)} = K(t,j).$$

□

The condition of this theorem is sometimes expressed by saying that the hazards for the individual causes are fixed proportions of the total hazard.

17.5 Other problems

There are several other questions regarding the joint distribution of $(T_1, T_2, \dots, T_m)$ that can be answered by similar techniques to those in Section 17.3. That is, we find a certain probability by integrating failure times over a suitable region of m-dimensional space. As an example, we illustrate the method for a problem analogous to Example 10.7 of Section 10.9. Take $m = 2$, and consider the probability that both causes of failure will occur within a specified duration of each other. That is, for some fixed n, we want the probability that $(|T_1 - T_2| \leq n)$. It will normally be easier to compute this as

$$1 - P(|T_1 - T_2| > n) = 1 - [P(T_2 > T_1 + n) + P(T_1 > T_1 + n)].$$

Each term is found by integrating the joint density function over a suitable region in the plane. For example,

$$P(T_2 > T_1 + n) = \int_0^\infty \int_{u+n}^\infty f_{T_1,T_2}(u,v)\mathrm{d}v\mathrm{d}u.$$

Example 17.8 Find $P\left(|T_1 - T_2| \leq n\right)$, when T_1, T_2 are independent, and both are exponential with hazard functions μ_1 and μ_2, respectively.

Solution. The integral above reduces to

$$\frac{\mu_1}{\mu_1+\mu_2}\mathrm{e}^{-n\mu_2},$$

so the final answer is

$$1 - \frac{\mu_1}{\mu_1+\mu_2}\mathrm{e}^{-n\mu_2} - \frac{\mu_2}{\mu_1+\mu_2}\mathrm{e}^{-n\mu_1}.$$

17.6 The common shock model

In many applications, we have a group of objects whose future lifetimes are generally independent, except that they are all subject to a common hazard, which will result in the failure of all, should it occur. In the case of human lives, this could be a natural disaster such as a hurricane. In the case of machine parts, it could be something like an electrical problem that affects all components at once. The presence of the common shock introduces dependence into what would otherwise be independent future lifetimes.

To model the general situation, we have $m+1$ independent, continuous random variables, $(T_1^*, T_2^*, \ldots, T_m^*, Z)$, and for each i we let

$$T_i = \min(T_i^*, Z).$$

The interpretation is that T_i^* is the time until failure of the ith object for reasons other than the common shock, and Z is the time until the common shock occurs. It follows then that T_i will be simply the time until failure of the ith object, since such failure will occur at either time T_i^* or time Z, whichever is earlier.

In the remainder of this section, we will confine ourselves to the case where $m = 2$. Quantities referring to T_i^* will have a superscript *.

We are interested in questions about the joint distribution (T_1, T_2), which involves dependent random variables. However, in many cases, we can answer these questions by considering the independent collection (T_1^*, T_2^*, Z). We will illustrate with several examples. A key fact to note is that

$$T = \min(T_1, T_2) = \min(T_1^*, T_2^*, Z),$$

since both give the time of first failure.

Example 17.9 Find a formula for the probability that both objects will survive to time t.

Solution. This is just $s_1^*(t)s_2^*(t)s_Z(t)$.

Example 17.10 What is the probability that failure will occur as a result of the common shock?

Solution. This is just $P(J = 3)$ in the joint distribution of (T, J), where J takes the values 1, 2, 3 and T is the minimum of T_1^*, T_2^* and Z.

Example 17.11 What is the probability that the second failure will occur before time t?

Solution. We divide this up into two mutually exclusive cases. It will always occur if $Z \leq t$. If $Z > t$, we need both T_1^* and T_2^* less than or equal to t. The probability is

$$F_Z(t) + s_Z(t)F_1^*(t)F_2^*(t) = 1 - s_Z(t)[s_1^*(t) + s_2^*(t) - s_1^*(t)s_2^*(t)].$$

Other problems are not so straightforward and require special attention. The joint distribution of (T_1, T_2) is quite different from the typical two-dimensional continuous distribution. It still is continuous, but it has a mass of positive probability all concentrated on a single line, namely the diagonal, since the occurrence of the common shock will cause failure from both causes, leading to a failure point of the form (t, t). In determining the probability that (T_1, T_2) lies in some region A, we will in general have to break A up into three pieces, the part that is above the diagonal, the part that is below the diagonal, and the part that is on the diagonal.

We adopt the convention that the value of T_1 is on the horizontal axis. For the part of the plane above the diagonal, $\{(u, v) : u < v\}$, we use the joint density function

$$f_1^*(u)f_2(v),$$

since the only way T_1 can take a value $u < v$ is if T_1^* took the value u. In other words, failure from cause 1 at time u did not occur from the common shock, since if it did, then failure from cause 2 would also have occurred at time u and could not have occurred at the later date v.

Similarly, for the part of the positive quadrant below the diagonal, $\{(u, v) : v < u\}$, we use the joint density function

$$f_1(u)f_2^*(v).$$

Since $T_i = \min(T_i^*, Z)$, the densities $f_i, i = 1, 2$, are easily calculated as

$$f_i(t) = -(s_i^* s_z)'(t) = f_i^*(t)s_z(t) + s_i^*(t)f_z(t). \tag{17.17}$$

Failure on the diagonal arises if and only if the occurrence of the common shock occurs before the other two causes. We use the one-dimensional density function,

$$f_{T,J}(t, 3) = s_1^*(t)s_2^*(t)f_Z(t),$$

and project the diagonal onto the line. That is, to find the probability that failure took place at a point (t, t), where $a \leq t \leq b$, we integrate this density from a to b.

Example 17.12 Suppose T_1^*, T_2^*, and Z are exponential with hazard functions μ_1, μ_2, and ρ, respectively. Consider the event that T_1 and T_2 are both less than or equal to n. This can be subdivided into three cases according as (a) $T_1 < T_2$, (b) $T_2 > T_1$, (c) $T_1 = T_2$. Find the probability of each case.

Solution.

(a) Note first that, from (17.17), we can calculate

$$f_2(t) = (\mu_2 + \rho)e^{-(\mu_2+\rho)t},$$

so that, using the above-diagonal joint density, the required probability is

$$\mu_1(\mu_2 + \rho) \int_0^n \int_u^n e^{-\mu_1 u} e^{-(\mu_2+\rho)v} dv du,$$

which equals

$$-\mathrm{e}^{-n(\mu_2+\rho)} + \frac{\mu_1}{\mu_1+\mu_2+\rho} + \frac{\mu_2+\rho}{\mu_1+\mu_2+\rho}\mathrm{e}^{-n(\mu_1+\mu_2+\rho)}.$$

(b) Similarly, the required probability in this case is

$$-\mathrm{e}^{-n(\mu_1+\rho)} + \frac{\mu_2}{\mu_1+\mu_2+\rho} + \frac{\mu_1+\rho}{\mu_1+\mu_2+\rho}\mathrm{e}^{-n(\mu_1+\mu_2+\rho)}.$$

(c) The required probability is

$$\int_0^n \rho \mathrm{e}^{-\rho t}\mathrm{e}^{-\mu_1 t}\mathrm{e}^{-\mu_2 t}\mathrm{d}t = \frac{\rho(1-\mathrm{e}^{-n(\mu_1+\mu_2+\rho)})}{\mu_1+\mu_2+\rho}.$$

The sum of these three cases is

$$1-\mathrm{e}^{-n(\mu_1+\rho)} - \mathrm{e}^{-n(\mu_2+\rho)} + \mathrm{e}^{-n(\mu_1+\mu_2+\rho)},$$

as we can verify from the general formula given in Example 16.12.

17.7 Copulas

This section, like the previous one, is concerned with situations where there is a lack of independence. We present a general method that is often used to deal with this. Attention is confined to the case $m = 2$.

A joint distribution (T_1, T_2) can be thought of as having two ingredients. One is the distributions of the two-component random variables, and the other is the way in which these are linked together. The latter can be described by a device known as a *copula*, which can then be applied to an arbitrary pair of individual distributions. The copula provides a means of dealing with these two ingredients separately. To elaborate, we start with the observation that whenever T_1 and T_2 are independent, we immediately recover the joint distribution from the individual distributions by the rule

$$F_{T_1,T_2}(t_1, t_2) = F_{T_1}(t_1)F_{T_2}(t_2). \tag{17.18}$$

We can then ask whether we can replace the multiplication on the right-hand-side of (17.18) by other transformations, and still obtain a joint distribution – that is, if I denotes the unit interval [0,1], whether we can find a function C from $I \times I$ to itself, so that we obtain a legitimate joint distribution by the rule

$$F_{T_1,T_2}(t_1, t_2) = C\big(F_{T_1}(t_1), F_{T_2}(t_2)\big). \tag{17.19}$$

We need some restrictions on the function C. Take any point s in I. If $T_1 > s$, so that $F_{T_1}(s) = 0$, then, for all t, $F_{T_1,T_2}(s,t) = 0$. The same holds for T_2, leading to the condition that for all u, v in I,

$$C(0, v) = C(u, 0) = 0. \tag{17.20}$$

If $T_1 \leq s$, so that $F_{T_1}(s) = 1$, then for all t, $F_{T_1,T_2}(s,t) = F_{T_2}(t)$, leading to the condition that for all u, v in I,

$$C(1, v) = v, \qquad C(u, 1) = u. \tag{17.21}$$

Another requirement stems from the fact that probabilities cannot be negative. For any sub-rectangle $R \subseteq I \times I$, the probability that (T_1, T_2) lies in R is just the sum of the values of F_{T_1,T_2} on the northeast and southwest corners, minus the sum of the values on the other two corners. Since this is nonnegative, it follows that

$$C(u_2, v_2) + C(u_1, v_1) - C(u_1, v_2) - C(u_2, v_1) \geq 0. \tag{17.22}$$

whenever $u_1 \leq u_2$ and $v_1 \leq v_2$.

These are the only conditions we need and we can now state the formal definition.

Definition 17.2 A *copula* is a function C from $I \times I$ to I satisfying (17.20)–(17.22).

It can be shown that if C is a copula, then (17.19) gives a valid joint probability distribution for any T_1 and T_2. Conversely (and harder to show), for any joint distribution (T_1, T_2) there is a copula C such that F_{T_1,T_2} is given by (17.19).

If T_1 and T_2 are both uniform distributions on I, then $F_{T_i}(u) = u$ for $i = 1, 2$, and all $u \in I$, from which it follows that

$$F_{T_1,T_2}(u, v) = C(u, v).$$

This shows that as an alternate definition, we can simply define a copula as a distribution function of a joint distribution involving two random variables that are uniform on I.

The following are three simple examples of copulas:

1. $C(u, v) = uv$. This is just the copula for an independent distribution, as mentioned;
2. $C(u, v) = \min(u, v)$;
3. $C(u, v) = \max(u + v - 1, 0)$.

Copulas 2 and 3 are extreme in the sense that for any copula C and for all $u, v, \in I$,

$$\max(u + v - 1, 0) \leq C(u, v) \leq \min(u, v).$$

They are also extreme in the following sense. Consider all possible joint distributions for a given T_1 and T_2. In many cases, one is interested in the sum $T_1 + T_2$. For example, an insurer sells two insurance contracts and T_i denotes the claim on the ith policy, or a person buys two stocks and T_i is the value of the ith stock at some future date. We may want to compare all possible joint distributions as to their degree of risk. We will not go into the details of comparing joint distributions as to risk here. (In Section 22.4, we do introduce this idea for single variable distributions). However, it seems clear that risker possibilities arise when when large values of one random variable tend to go with large values of the other so there is a tendency for either both values to be large or both to be small. The less riskier possibilities arise when large values of one tend to go with small values of another, so there is a possibility for bad results in one case to be balanced by good results in the other. It can be shown, that under some natural risk comparing criteria, copula 2 will give the most risky joint distribution and copula 3 the least risky joint distribution. This point is illustrated further in Exercise 17.14.

We deal only briefly with problems of choosing a copula to model a given situation. In many cases, the modeler likes to choose a copula from a parametric family, and select the parameter to suit certain conditions. A popular choice for this is Frank's family of copulas given by

$$C_\Theta(u, v) = \frac{1}{\Theta} \log\left(1 + \frac{(\mathrm{e}^{\Theta u} - 1)(\mathrm{e}^{\Theta v} - 1)}{\mathrm{e}^{\Theta} - 1}\right),$$

where Θ can be any nonzero real number. It can be shown, using L'Hôpital's rule, that for all $u, v, \in I$,

$$\lim_{\Theta \to 0} C_\Theta(u, v) = uv,$$

so that the smaller the parameter is in absolute value, the greater the extent of independence between the two random variables, with full independence occurring for $\Theta = 0$.

An interesting feature which some copulas, but not all, have is that

$$C(u, v) = u + v - 1 + C(1 - u, 1 - v). \tag{17.23}$$

It is straightforward to verify this property for copulas 1–3 above. It is true, but harder to verify, that this holds for Frank's family. The significance of (17.23) is that

$$s_{T_1,T_2}(t_1, t_2) = 1 - F_{T_1}(t_1) - F_{T_2}(t_2) + F_{T_1,T_2}(t_1, t_2) = C\big(s_{T_1}(t_1), s_{T_2}(t_2)\big).$$

In other words, the same transformation rule can be applied to either distribution or survival functions.

Example 17.13 Suppose that Demoivre's law holds with $\omega = 100$. Consider two lives (60) and (70). Find the probability that both will be alive at the end of 10 years, assuming each of the three basic copulas given above, for the joint distribution of $T(60)$ and $T(70)$.

Solution. The individual survival probabilities are 3/4 and 2/3, so the survival of the joint-life status is in the respective cases:

1. 1/2 as we known already from Chapter 10;

2. $\min\{3/4, 2/3\} = 2/3$. This copula applied to a joint-life status just means that the younger life will die at exactly the same time as the older;

3. $(3/4 + 2/3 - 1) = 5/12$.

For a particular application of copulas, refer back to the nonidentifiability problem of Section 17.4.6. Instead of assuming independence, we might postulate a certain copula C and then ask for a joint distribution with the chosen copula that gives rise to the given distribution (T, J). In many cases this will be unique.

Notes and references

Nelsen (1999) is a good general source for additional material on copulas, including full derivations of the unproved results that we have given. Frees and Valdez (1998) discuss various actuarial applications of copulas. Carriérre (1994b) discusses the application of copulas to the nonidentifiability problem in multiple-decrement theory. For methods of comparing two-variable distributions for risk, see Shaked and Shantikumar (2007), Chapter 6.

Exercises

Type A exercises

17.1 A joint distribution is given by

$$f_{T_1,T_2}(s,t) = \begin{cases} 3s, & 0 \le s \le t \le 1, \\ 3t, & 0 \le t \le s \le 1, \\ 0, & \text{elsewhere.} \end{cases}$$

Find $F_{T,J}(t, 1)$ and $F_{T,J}(t, 2)$.

17.2 A joint survival function is given by

$$s(u, v) = \begin{cases} 1 - (3/2)(u^2 + v^2) + (1/2)u^3 + (3/2)uv^2, & 0 \le v \le u \le 1, \\ 1 - (3/2)(u^2 + v^2) + (1/2)v^3 + (3/2)vu^2, & 0 \le u \le v \le 1. \end{cases}$$

Find $F_{T,J}(t, 1)$.

17.3 Suppose that T_1 and T_2 are independent with p.d.f.'s

$$f_1(r) = 2e^{-2r}, \qquad f_2(r) = 3e^{-3r}.$$

(a) Find $F_{T,J}(t, 1)$ and $F_{T,J}(t, 2)$.

(b) Find the distribution of J.

17.4 Suppose that T_1 and T_2 are independent. T_1 has an exponential distribution with constant hazard rate μ. T_2 is uniform on $[0, a]$. Find (a) $F_{T,J}(t, 2)$, (b) $P(J = 2)$.

17.5 The failure times T_1 and T_2 are independent, and have respective hazard functions

$$\mu_1(t) = \frac{3}{2-t}, 0 \le t < 2, \qquad \mu_2(t) = \log(2).$$

Find the probability that the minimum value of these two random variables will be less than or equal to 1.

17.6 A machine is subject to two independent causes of failure. The time of the first cause of failure is uniformly distributed on the interval [0, 4]. The time of the second cause of failure is uniformly distributed on the interval [0, 5]. Find the probability that: (a) the machine will fail from cause 1 before time 3; (b) the machine will eventually fail from cause 2.

17.7 Two failure times T_1 and T_2 have a joint distribution given by the joint density function

$$f_{T_1,T_2}(u,v) = \begin{cases} 4(v-u)^2, & 0 \le u \le v \le 1, \\ 4(u-v), & 0 \le v < u \le 1, \\ 0, & \text{elsewhere.} \end{cases}$$

Find: (a) $F_{T,J}(t,1)$ and $F_{T,J}(t,2)$; (b) the distribution of J; (c) $\mu(t,1)$ and $\mu(t,2)$.

17.8 Two failure times T_1 and T_2 have a joint distribution given by the joint density function

$$f_{T_1,T_2}(u,v) = \begin{cases} (8/3)v^2, & 0 \le u \le v \le 1, \\ (8/3)uv, & 0 \le v < u \le 1, \\ 0, & \text{elsewhere.} \end{cases}$$

(a) Find $F_{T,J}(t,1)$ and $F - T, J(t,2)$.

(b) Find the distribution of J.

(c) Suppose that $\hat{T}_1$ and $\hat{T}_2$ are two independent variables whose joint distribution leads to the same distribution of (T, J) as you found in part (a). What are the hazard functions of $\hat{T}_1$ and $\hat{T}_2$?

Type B exercises

17.9 For the joint distribution given in Example 17.2, find $\mu(t,j)$ and $\mu_j(t)$ for $j = 1, 2$. Show that these are not the same.

17.10 For the joint distribution given in Exercise 17.2, find $\mu(t,j)$ and $\mu_j(t)$ for $j = 1, 2$. Show that these are not the same.

17.11 Consider independent failure times T_1 and T_2 where the hazard rate of T_1 is $\alpha/(1-t), 0 \le t < 1$, and the hazard rate of T_2 is $\beta/(1-t), 0 \le t < 1$. Answer the following in terms of α and β.

(a) Find $F_{T,J}(t,1)$.

(b) Find $P(J = 2)$.

(c) An insurance contract provides for a benefit at the moment of the first failure, provided that this is due to 'cause 1' (i.e. provided that $T_1 < T_2$). The amount of the benefit for failure at time t is $(1 - t)e^{0.1t}$. The force of interest δ is a constant 0.10. Find the expected present value of the benefits.

17.12 Two lives age (x) and (y) are subject to the common shock model. $T^*(x)$ has a constant hazard rate of 0.06, $T^*(y)$ has a constant hazard rate of 0.04, and Z has a constant hazard rate of 0.02. The force of interest is a constant 0.05.

(a) Calculate $\bar{A}_x(1_{10})$ and write it as a sum of three terms, namely, the expected present value of benefits when: (i) (x) dies at a time strictly before the death of (y); (ii) (x) dies at time strictly after the death of (y); (iii) (x) dies as result of the common shock.

(b) Calculate $\bar{A}_{\overline{xy}}$.

17.13 You are given a common shock model (T_1^*, T_2^*, Z) where T_1^* has the survival function $s(t) = 1 - 0.1t$, T_2^* has a constant hazard function of 0.04, and Z has a constant hazard function of 0.02. Find the probabilities that (a) $T_1 < T_2$, (b) $T_2 < T_1$, (c) $T_1 = T_2$.

17.14 Suppose that T_1 and T_2 each take the values 0 with probability 1/2, and 1 with probability 1/2. Calculate the probability function for the joint distribution of (T_1, T_2) under each of the following copulas: (a) $C(u, v) = \min(u, v)$, (b) $C(u, v) = \max(u + v - 1, 0)$, (c) $C(u, v) = uv$, (d) Frank's copula with values of $\theta = 0.01, 50, -50$. What happens when θ approaches ∞ or $-\infty$?

17.15 A joint life insurance on (x) and (y) has death benefits which are constant over each year. Assume that either copulas 2 or 3 of Section 17.7 apply to the joint distribution of $T(x)$ and $T(y)$. Show that unlike the independent case, the term R in equation (10.15) is equal to 0.

PART III

ADVANCED STOCHASTIC MODELS

18

An introduction to stochastic processes

18.1 Introduction

The purpose of this chapter is to provide background for the remaining chapters in the book. We make much use of the concept of conditioning, so the reader may wish to review Sections A.2 and A.8 of Appendix A.

A stochastic process is the tool used to model a quantity varying randomly in time. The following are the essential ingredients. We have an index set T, which gives the points of time that we are interested in. Normally, T will either be the nonnegative integers, $0, 1, 2, \ldots$ (discrete time) or the whole nonnegative line $[0, \infty)$ (continuous time). In both cases, we will sometimes have a maximum time horizon that we are interested in, in which case, ∞ will be replaced by a finite N). We also need a sample space with a probability measure P, and for each t in T, a random variable X_t defined on this space. The random variable X_t gives the value at time t, of the random quantity that we are trying to model. A *stochastic process* can then be defined formally as a collection of random variables X_t defined for each t in a set T.

We will illustrate briefly by considering the price of a certain stock. Anyone who has been involved with the stock market can attest that this is indeed a quantity that varies in time, and is subject to all kinds of random influences. Let time be discrete and refer to days. Suppose the stock is selling for 100 per share now, and we know that each day it will either increase by 20% or decrease by 20%. Therefore, X_0 takes a value of 100 with probability 1, while X_1 will take a value of either 80 or 120, X_2 will take one of three possible values, 144, 96, or 64, and so on. The details are shown in Figure 18.1. Of course, to complete the model, we have to specify probabilities. We will not do this quite yet, as we want to first discuss the complications that arise.

The first principle to observe is that we are not just interested in the distribution of each X_t but also in all the possible joint distributions. To illustrate, suppose we want to know the

---

Fundamentals of Actuarial Mathematics, Third Edition. S. David Promislow.

Companion website: http://www.wiley.com/go/actuarial

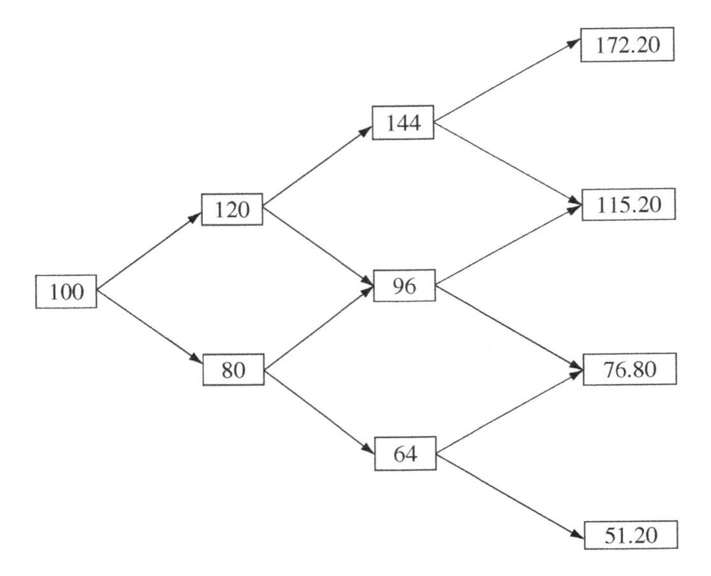

Figure 18.1 Evolution of stock price

probability that the stock will be priced at 115.20 at time 3. From Figure 18.1, we can identify three mutually exclusive 'path segments' leading from 100 at time 0 to 115.20 at time 3. The desired probability can then be written as

$$P(X_0 = 100, X_1 = 120, X_2 = 144, X_3 = 115.20) + P(X_0 = 100, X_1 = 120, X_2 = 96, X_3 = 115.20) + P(X_0 = 100, X_1 = 80, X_2 = 96, X_3 = 115.20).$$

Many of the questions we ask of stochastic processes are, like the above, concerned with realizations. A *realization* (sometimes called *sample path* or *scenario*) of a stochastic process is a function defined on the index set T, sending t to x_t, which represents a possible outcome of the process, namely that in which the random variable X_t takes the value x_t for all t in T. For example, in our example above, the realization $x_t = 100 \times 1.2^t$ represents the outcome whereby the stock continually moves upward.

It is often useful to view a stochastic process as a model for assigning probabilities to realizations. A technical difficulty arises, however. Normally T is infinite, and X_t takes at least two values, except possibly for X_0. This means there are *uncountably* many realizations, and so each single realization will normally have probability 0. We must therefore deal with infinite sets of realizations, as these can have positive probability. (This is exactly analogous to the situation with any continuous random variable X where we know that the probability that X takes a particular value of x is always 0, but we are interested in the probability that X takes values in some infinite set.)

Note that in our stock example, an event such as $(X_0 = 100, X_1 = 120, X_2 = 144, X_3 = 115.20)$, which we referred to above, is not by itself a realization (unless our index set T were just $\{0, 1, 2, 3\}$). Rather, it is an infinite set of realizations, consisting of all those with the given value of x_t for $t = 0, 1, 2, 3$, and including all possible values for $t > 3$.

The reader should realize at this point that if the index set T is in fact finite, then a stochastic process is formally nothing more than a multi-dimensional *joint distribution*.

18.2 Markov chains

18.2.1 Definitions

Throughout this section, we will assume that our index set T is the nonnegative integers and that each X_t is discrete.

To answer relevant questions about a stochastic process, we want to be able to compute all finite joint distributions. Normally, however, we do not do so directly but deduce these joint distributions through some other information that we can determine about our model. Suppose we are at time k, and we want to predict what will happen to our quantity at time $k+1$. This can be quite complicated as it could depend on the entire past history of what happened up to time k. In many applications, this is simplified, since the only relevant part of this past history for this prediction is the actual value at time k, and we get no further information from looking at the values before that time. The following formulates this precisely.

Definition 18.1 A discrete-time stochastic process with discrete random variables is called a *Markov chain*, if given any finite sequence $x_0, x_1, x_2, \ldots, x_{k+1}$, where x_i is a possible value of X_i,

$$P(X_{k+1} = x_{k+1}|X_k = x_k, X_{k-1} = x_{k-1}, \ldots, X_0 = x_0) = P(X_{k+1} = x_{k+1}|X_k = x_k). \quad (18.1)$$

To illustrate, consider the stock example given above. Suppose that we decide that each day the stock will move up with probability of $2/3$ and move down with probability $1/3$. This is clearly a Markov chain since both the left hand side and right hand side of (18.1) are $2/3$ if $x_{k+1} = 1.2x_k$, $1/3$ if $x_{k+1} = 0.8x_k$, and 0 in all other cases. Suppose, however, that we decide that these probabilities will hold only when the stock has made two different movements on the previous two days. On the other hand, we decide that if the stock moves up two days in a row, it signifies a trend, and the probability of an upward move on the next day changes to $3/4$, while if the the stock moves down two days in a row, it signifies a pessimistic attitude, and the probability of an upward move on the next day is only $3/5$. This would no longer be a Markov chain, since the probability of X_{k+1} is clearly influenced by the values of X_{k-2} and X_{k-1} as well as that of X_k. For example, taking $k = 3$,

$$P[X_4 = 92.16|X_3 = 115.20, X_2 = 96, X_1 = 80] = \frac{1}{4},$$

but

$$P[X_4 = 92.16|X_3 = 115.20, X_2 = 96, X_1 = 120] = \frac{1}{3}.$$

In a Markov chain, these would both have to equal the same number, namely $P[X_4 = 92.16|X_3 = 115.20]$.

The important feature of the Markov property is that it allows us to compute all relevant probabilities once we know the probabilities of each 'branch' of the tree-like structure, as we have drawn in Figure 18.1. To elaborate, we introduce some fundamental notation.

For any integers $k \leq n$, let

$$p_{xy}(k, n) = P(X_n = y | X_k = x). \tag{18.2}$$

The probability of any branch is of the form $p_{xy}(k, k+1)$ and we can multiply the probabilities of each branch to get the probability of any path. Finally, we can compute the general probability $p_{xy}(k, k+n)$ by adding the probabilities on all paths that lead from a value of x at time k to a value of y at time n.

Consider our stock example with the (2/3, 1/3) probabilities of the respective up or down move. We can now easily answer our question, and find the probability that the stock price is 115.20 at time 3, that is $p_{100,115.20}(0, 3)$ There are 3 paths leading from the starting value of 100 to the value of 115.20 at time 3. Each of them consists of two up moves and one down move, and will have probability equal to $\frac{2}{3} \times \frac{2}{3} \times \frac{1}{3} = \frac{4}{27}$, and so the total probability in question is 12/27.

In many common applications (as in this one), there is the further simplifying feature that $p_{x,y}(k, k+1)$ is independent of k, and can be denoted by just p_{xy}. In this case, we say that we have *stationary* transition probabilities. (This feature is also referred to as *time homogeneity*.) In connection with stochastic processes, the word *stationary* can be thought of as referring to a process 'without a watch'. In the present context, it means that whenever the evolving quantity takes a value of x, the probability that it takes a value of y at the next stage is always the same, regardless of the particular time.

18.2.2 Examples

We now look at some other examples of Markov chains. One of the most famous is the *random walk*. An indecisive person goes for a walk but cannot decide whether to go east or west. A coin is flipped and the person goes 1 unit east if a head comes up or 1 unit west if a tail comes up. After each move, the coin is flipped again and the procedure repeated. Suppose the probability of a head coming up is p. Letting X_k refer to his position east of the starting point at time k, we have the process shown in Figure 18.2.

What is the probability of being 1 unit to the west of our starting point at time 3, that is, at position -1? This is similar to the stock question asked above. We have three possible path segments leading to -1 at time 3, each with probability $p(1-p)^2$, so the answer is $3p(1-p)^2$.

This same model applies to many situations. Imagine a gambler repeatedly playing a game with an even money payoff, such as betting on black at roulette. In any single play, this person either wins 1 unit with probability p or loses 1 unit with probability $1 - p$. If we let X_k denote the total winnings after k plays (a negative amount signifying a loss), then we have exactly the random walk process as described above.

Many of the processes that we will deal with can be viewed in this gambling context. It will be convenient to make a slight adjustment. Instead of keeping track of the amount won or lost, we will keep track of the total fortune of the gambler starting from an initial fortune of u. This is just a matter of adding u to each entry. For example, if the gambler starts play with a fortune of 10 units, then the diagram of the process would be as shown in Figure 18.3.

Consider a situation where the wager at each stage is more complicated than a simple even money bet. Suppose that the return to the gambler on this bet is some discrete random variable G. (In the case above, G simply took the value 1 with probability p and -1 with probability $1 - p$.) The gambler starts with a fortune of u, and we want to consider the stochastic process

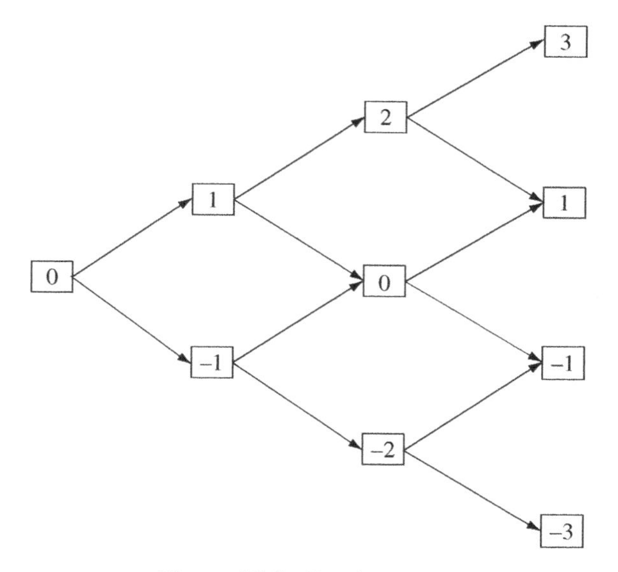

Figure 18.2 Random walk

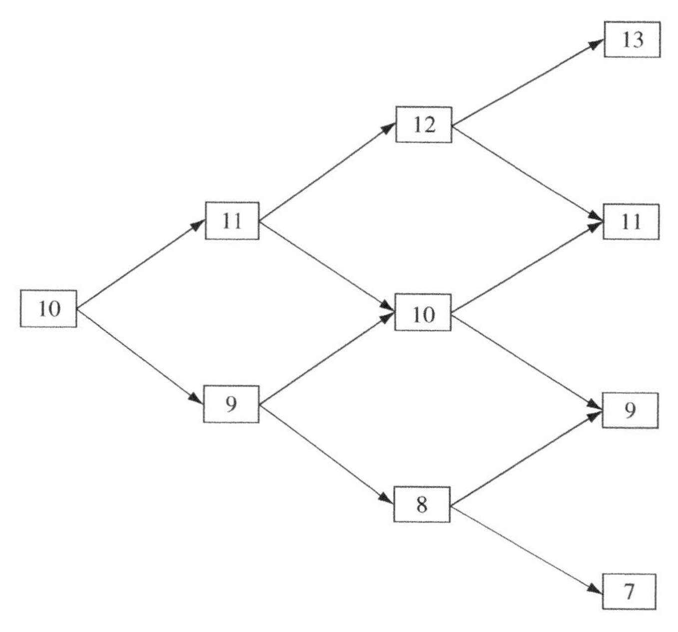

Figure 18.3 Gambler's fortune

U_n which equals the fortune of the gambler at time n. Letting G_k be the return at time k, we can write

$$U_n = u + G_1 + G_2 + \cdots + G_n, \tag{18.3}$$

where the G_is are independent and each distributed as G. This is clearly a Markov chain with transition probabilities given by

$$p_{x,y}(k, k+1) = P(G = y - x),$$

and since the right-hand side is independent of k, we see also that this process has stationary transition probabilities.

Note that it would still be a Markov chain, but not necessarily a stationary one, if the gambler were to change the bet at various times, as long as this was done independently of the past history of winnings. An example of this would be for the gambler to decide that after five turns at the roulette table, regardless of what happens, he will switch to blackjack. In this case, the distributions of the various G_n are not all the same. If, however, the gambler decided that he would play roulette until he had five consecutive losses and then switch to blackjack, the resulting process would *not* have the Markov property.

We return to the process given by (18.3) in Chapter 23 where it will play an important role in modelling the surplus of an insurer.

18.3 Martingales

In this section, we briefly introduce another important class of stochastic processes, which we will make much use of later.

Definition 18.2 A discrete-time stochastic process is a *martingale* if

$$E(X_{k+1}|X_k = x_k, X_{k-1} = x_{k-1}, \ldots, X_0 = x_0) = x_k,$$

for all k and values $x_0, x_1, \ldots, x_k$. In other words, at any time, the expected value of our quantity at the next time period is exactly what it is now.

Note that in the case of a Markov process, the requirement above simplifies to

$$E(X_{k+1}|X_k = x_k) = x_k.$$

Is our stock price process with the $2/3$ and $1/3$ probabilities a martingale? The answer is no, since

$$E(X_{k+1}|X_k = x) = \frac{2}{3} \times 1.2x + \frac{1}{3} \times 0.8x = \frac{3.2x}{3} > x.$$

A process like this, for which the expectation of the future random variable is always greater than or equal to the present one, is known as a *submartingale*. It would only be a martingale if the $2/3$ were replaced by $1/2$. Indeed, we would not expect a stochastic process for stock prices to be a martingale, since it would mean that there was no tendency for the price to increase. However, the reason for purchasing stock in the first place, given the inherent risk, is the expectation of capital gains as the stock increases in value.

What about the process given by (18.3)? We have

$$E(U_{k+1}|U_k = x) = E(U_k + G_{k+1}|U_k = x) = x + E(G).$$

This process will be a martingale if and only if $E(G) = 0$, that is, if and only if the bet made by the gambler is a fair bet. The martingale concept was in fact introduced originally as the model of a fair game. In the usual casino games, $E(G) < 0$, and we would not have a martingale. Such a process, where the expected value of the future random variable is always less than or equal to the present one, is known as a *supermartingale*. A martingale is therefore both a supermartingale and a submartingale. (To remember the terminology, note that the modifier applies to the *current* value. So a *sub*martingale means that at any time, the current value of the process is *under* the expectation of the future value.)

18.4 Finite-state Markov chains

Sometimes, it is convenient to relabel with integers the possible values that can be taken by the random variables in a Markov chain. We refer to these integers as the *states* of the system and say that the system is in state i at time k if X_k takes the value i. In fact we often just take N, the number of the state, as the underlying random variable in place. of X, so that our fundamental probabilities $p_{ij}(k, n)$ will be the probability of being in state j at time n given that the process was in state i at time k. In this section we consider Markov chains with a finite number of states and develop matrix methods for investigating their properties. In this chapter we confine our attention to the stationary case. Nonstationary chains are needed in Chapter 19 and will be introduced at the beginning of that chapter.

Starred sections contain somewhat more advanced material and these can be omitted at first reading (although we do refer back to one of the results in one of the Chapter 23 examples).

18.4.1 The transition matrix

Suppose we have a stationary Markov chain with N states. We will number them $\{0, \ldots , N - 1\}$. (Some authors use the numbers 1 to N instead). By the stationary condition, the quantity $p_{ij}(k, k + 1)$ the probability of moving from state i to state j in one step, is, independent of k, and we can write it as just p_{ij}. The matrix $\mathbf{P}$, with entries of p_{ij} in the ith row and j column is known as the *transition matrix* of the chain.

Note that the only condition required for an $N \times N$ matrix to be the transition matrix of some Markov chain is that all entries are nonnegative, and each row sums to 1.

Remark The reader should keep in mind that we have started the indexing at 0, so the top row (or left column) of the matrix will be considered as row (or column) number 0, *not* number 1.

Example 18.1 A box contains two balls, either of which can be red or yellow. At each stage, a ball is chosen at random and replaced with a ball of the opposite colour. Let X_n be the number of red balls in the box at time n. Find the transition matrix.

Solution. Here we have a three-state Markov chain. We will number the states by the number of red balls in the box. If there are 0 or 2 red balls, we are sure to move to state 1, while if there is 1 red ball, we move to either state 0 or 2, with equal probability. Therefore the transition matrix is

$$\mathbf{P} = \begin{pmatrix} 0 & 1 & 0 \\ 1/2 & 0 & 1/2 \\ 0 & 1 & 0 \end{pmatrix}.$$

18.4.2 Multi-period transitions

We now wish to calculate the general probabilities $p_{ij}(k, k+m)$ which by stationarity will be equal to $p_{ij}(0, m)$. We start by calculating this for $m = 2$. Suppose we move from state i to state k at time 1 and then from state k to state j time 2. We know from our previous discussion on Markov chains that the probability of this two-step move occurring is just $p_{ik}p_{kj}$. Summing this over all states k, we get the probability of moving from i to j after two periods as

$$\sum_{k=1}^{N} p_{ik}p_{kj}, \tag{18.4}$$

which, by the ordinary rules of matrix multiplication, is just $\mathbf{P}^2(i,j)$ the (i,j)th entry of $\mathbf{P} \times \mathbf{P} = \mathbf{P}^2$. The same argument can be repeated to show the important fact that in a stationary chain,

$$p_{ij}(0, m) \text{ is equal to the entry in row } i \text{ and column } j \text{ of } \mathbf{P}^m \tag{18.5}$$

(Note that $\mathbf{P}^0$ is just the identity matrix, usually denoted by I, with entries of 1 on the main diagonal, and entries of 0 elsewhere.)

18.4.3 Distributions

Given a finite-state Markov chain, what is the distribution of X_n? This is a basic question that we ask about any stochastic process. The distribution can be given as an N-dimensional vector π_n, whose ith entry, denoted by $\pi_n(i)$, equals $P(X_n = i)$. This will depend of course on π_0, the vector giving the initial distribution at time 0. (In certain applications, we might know the value of X_0, in which case, π_0 will simply be a vector with a single entry of 1, and other entries equal to 0.) Starting with $n = 1$, we calculate

$$P(X_1 = j) = \sum_i \pi_0(i)p_{ij},$$

which is just the jth entry of the vector obtained by multiplying the vector π_0 (viewed as a $1 \times N$ matrix) on the right by $\mathbf{P}$. The same argument holds for any n, and we can conclude

$$\pi_n = \pi_0\mathbf{P}^n. \tag{18.6}$$

Example 18.2 In Example 18.1, suppose we start with a uniform distribution, that is, there is a $1/3$ chance that X_0 takes each of the values $0, 1, 2$. What is the probability that $X_{101} = 1$?

Solution. Computing large powers of matrices is often done by calculating eigenvalues, but we can avoid that here. After a couple of multiplications, we see that $\mathbf{P}^3 = \mathbf{P}$. Therefore $\mathbf{P}^5 = \mathbf{P}^3\mathbf{P}^2 = \mathbf{P}\mathbf{P}^2 = \mathbf{P}$, and similarly, *any odd power* of $\mathbf{P}$ equals $\mathbf{P}$. From (18.6),

$$\pi_{101} = (1/3, 1/3, 1/3)\begin{pmatrix} 0 & 1 & 0 \\ 1/2 & 0 & 1/2 \\ 0 & 1 & 0 \end{pmatrix} = (1/6, 2/3, 1/6).$$

The probability that X_{101} equals 1 is $2/3$.

Remark Some writers define the transition matrix $\mathbf{P}$ by taking the entry in row i and column j as the probability of moving from state j to state i. In this case, columns, rather than rows, add to 1. The vectors π_n are written as column vectors and in (18.6), we multiply π_0 on the *left* to get π_n.

*18.4.4 Limiting distributions

The distributions π_n will normally continue to change with time, but in many applications, we would like to show that there is some limiting distribution π which is independent of the initial state at time 0. That is, for any initial state j, and all states i, the resulting probabilities $\pi_n(i)$ will converge to $\pi(i)$ as n approaches ∞. This would enable us to predict with reasonable accuracy the probabilities of being in various states, provided the process has been continuing for a sufficiently long time. We consider here the problem of finding this limiting distribution, *provided it exists*. The last provision is necessary, as indicated by the following examples where a limiting distribution does not exist.

Consider the chain of Example 18.1. If the process is in state 0 or 2, it will move to state 1 in one transition. If it is in state 1, it will move to state 0 or 2 in one transition. So if we start, say in state 0, the process is sure to be in state 0 or 2 at even times, and in state 1 at odd times. Clearly, no limiting distribution can exist. This is an example of what is known as a *periodic* chain of period 2. In general, this means we can divide the set S of all states into two disjoint subsets, S_1 and S_2, such that in any one transition, all states in S_1 move to S_2 and all states in S_2 move to S_1. More generally, we could have chains of period d where we can find d pairwise disjoint subsets, such that from each subset, we cycle through the other sets in a fixed order, and return to the original set after d transitions.

Another example where a limiting distribution will not exist is the chain with transition matrix

$$\mathbf{P} = \begin{pmatrix} 1/2 & 1/2 & 0 \\ 1/2 & 1/2 & 0 \\ 0 & 0 & 1 \end{pmatrix}.$$

In this case, if we start in state 2 at time 0, then we remain in state 2 forever. If we start in state 0 or 1, we will reach a limiting distribution, which is to be in state 0 or 1 with equal probabilities.(In fact, this limiting distribution is achieved exactly at time 1). The problem here is that the limiting distributions varys according to the initial state. This is an example of what is known as a *reducible chain*. This means we can divide the set S of all states into two nonempty disjoint subsets S_1 and S_2 such that any state in S_1 transfers to another state in S_1

after one transition and any state in S_2 transfers to another state in S_2 after one transition. We can really consider such a Markov chain as two separate chains, one comprising the states in S_1 and the other comprising the states in S_2.

A simple condition that ensures the existence of a limiting distribution is that there is a positive integer n such that all entries of $\mathbf{P}^n$ are positive. That is, given any states i, j (not necessarily distinct), there is some chance of getting from i to j in n transitions. A proof of this result is beyond the scope of this book. It is, however, not difficult to see that this condition rules out both periodic and reducible chains. See Exercise 19.12.

If we know that a limiting distribution π exists, then there is a straightforward procedure for finding it from the matrix $\mathbf{P}$. We know that $\pi = \lim_{n\to\infty} \pi_n$ (in the sense of converging at each i as described above). By continuity considerations, we must have $\pi\mathbf{P} = \lim_{n\to\infty} \pi_n\mathbf{P}$. From (18.6), we know that $\pi_n\mathbf{P} = \pi_0\mathbf{P}^n\mathbf{P} = \pi_0\mathbf{P}^{n+1} = \pi_{n+1}$. But clearly π_{n+1} also converges to π. This establishes that

$$\pi = \pi\mathbf{P}. \tag{18.7}$$

Note that a solution π of Equation (18.7) need not be a limiting distribution, since none may exist. It is true that once we have a distribution π satisfying this equation, we will remain with that distribution forever, but it is possible that from some initial states we will never converge to π. Example 18.1 is an illustration. We have exactly one solution to (18.7), namely $\pi = (1/4, 1/2, 1/4)$, but as we have shown, we will not approach this for all initial distributions. (It is of interest to note that this is the distribution we would get if we started the process by choosing the colour of each ball randomly.)

*18.4.5 Recurrent and transient states

The states of a stationary Markov chain can be divided into two classes. Given any state j, let f_j be the probability that, starting in state j, the process will return to that state.

Definition 18.3 State j is said to be *transient* if $f_j < 1$. In this case, there is some chance that the process will never return.

State j is said to be *recurrent* if $f_j = 1$. In this case, the process is sure to return.

Example 18.3 Classify the states of the Markov chain with transition matrix

$$\mathbf{P} = \begin{pmatrix} 3/4 & 1/4 & 0 \\ 1/2 & 1/2 & 0 \\ 1/4 & 0 & 3/4 \end{pmatrix}.$$

Solution. State 2 is transient, since starting at state 2, there is a positive chance of moving to state 0. From there, only states 0 and 1 can be reached, and there is no return to state 2.

We will show that states 0 and 1 are recurrent. Suppose we are in state 0. We first observe that it is certain that we will eventually get to state 1. For any positive integer n, the probability of staying in state 0 forever is certainly less than the probability of staying in state 0 for the next n transitions, which is $(3/4)^n$. As this quantity approaches 0, we see that the probability of staying in state 0 forever is 0, and so with probability 1 we will eventually move to state 1. Similarly, if we are in state 1, we are certain to move to state 0. Therefore, if we are in state 0, we are certain to move to state 1 and then back to state 0 Arguing in the same way for state 1, we see that both states are recurrent.

This is a fairly simple case, and in general it might not be so easy to classify the states from the matrix. We can, however, adapt the argument in Example 18.3 to prove a general result, which can be of great help in the classification.

Definition 18.4 We say a state j is *reachable* from a state i if there is a positive probability of eventually moving from state i to j. In terms of the matrix $\mathbf{P}$, this condition can be stated as $P^n(i,j) > 0$ for some n.

Theorem 18.1 *If i is recurrent and j is reachable from i, then j is recurrent.*

Proof. Starting in state j, we are certain to eventually reach state i, for if not, then there would be a positive probability of going from i to j and never returning to i, contradicting the fact that i is recurrent. Now, starting in state i, let α be the probability that we will return to state i without ever hitting state j. Since j is reachable from i, we must have that $\alpha < 1$. The probability that starting in state i we will make n return visits to i without ever hitting state j is α^n, which approaches 0 as n goes to ∞. Therefore, the probability is 0 that, starting in i, we never reach state j, which means that we are in fact *certain* to eventually reach state j. To conclude, starting in state j, we are certain to reach state i and certain to come back to state j from state i, showing that state j is recurrent. □

Starting in a recurrent state, we must always remain in recurrent states. What happens if we start in a transient state?

Theorem 18.2 *In a finite-state stationary Markov chain, there is at least one recurrent state. Moreover, starting from any transient state, we must eventually reach a recurrent state.*

Proof. Given a transient state j, consider all realizations of the process for which there is at least one occurrence of j. The probability that there will be exactly one such occurrence, given that there is at least one, is $1 - f_j$, the probability of never returning to j. The probability that there will be exactly two such occurrences, given that there is at least one, is just $f_j(1 - f_j)$. The process must return once, and then never return again. Continuing, we see that the number of occurrences of j, less 1, given that there is at least 1, has a geometric distribution.(See Section A.11.3) Now a geometric distribution is a proper frequency distribution with no probability of assuming the value ∞. Our conclusion is that the probability of infinitely many occurrences of any transient state j is 0. Consider the event that the chain never visits a recurrent state. If this occurs, since there are only finitely many transient states, one of them must appear infinitely often, but as we have seen the probability of this is 0. This means that with certainty, we must reach a recurrent state. □

We can apply the last two theorems to Example 18.3. Once we see that state 3 is transient, we know by Theorem 18.2 that either state 1 or state 2 is recurrent, and then by Theorem 18.1 (or even by symmetry) they both must be recurrent.

Note that Theorem 18.2 need not hold for an infinite-state chain. For a trivial example, take the random walk where the probability of moving to the right is 1. All states are transient.

Some transient states are 'less transitive' than others. That is, the expected time spent in that state is longer. Indeed, for many applications we may be interested in this expected time spent in a certain transitive state j, that is, the expected number of n for which $X_n = j$. (There is no point in asking this question for a recurrent state, since, by definition, the process is in the state for infinitely many values of n.) Of course, this expectation may depend on the starting state i. Once again this question is of interest only for a transitive starting state i, since for a recurrent starting state, we know by Theorem 18.1 that the answer is 0. We calculate this expectation by the familiar trick of looking at *indicator* random variables (see the end of Section A.5). Fix such a starting state i and consider the random variable I_n that takes the value 1 if $X_n = j$, or 0 if $X_n \neq j$. Then the expected number of visits to state j is just $E[\sum_{n=0}^{\infty} I_n] = \sum_{n=0}^{\infty} E(I_n)$, since A.22 extends to infinite sums for nonnegative random variables. Since $E(I_n) = p_{ij}(0, n)$, we have

$$\text{Expected visits to state } j \text{ starting from state } i = \sum_{n=0}^{\infty} p_{ij}(0, n). \tag{18.8}$$

The right hand side of (18.8) may seem quite formidable to calculate. It is possible to use eigenvalues of the transition matrix $\mathbf{P}$ to shorten this, (see Section 19.3.4) but there is a much simpler method, based on an idea that is useful in many contexts. Recall the formula for an infinite geometric progression. For a number x, of absolute value less than 1,

$$(1 - x)^{-1} = 1 + x + x^2 + \cdots .$$

Similarly, it can be shown that for a matrix $\mathbf{Q}$ with sufficiently small entries, the matrix $\mathbf{I} - \mathbf{Q}$ is invertible, and

$$(\mathbf{I} - \mathbf{Q})^{-1} = \mathbf{I} + \mathbf{Q} + \mathbf{Q}^2 + \cdots .$$

Suppose that the transient states in our matrix are numbered $0, \dots . m - 1$ and take $\mathbf{Q}$ to be the $m \times m$ submatrix consisting of the first m rows and first m columns. Using Theorem 18.1, we can show that for i, j between 0 and $m - 1$, the (i, j)th entry of $\mathbf{Q}^n$ is equal to $p_{ij}(0, n)$ with the final result that for any two transient states i, j not necessarily distinct,

$$\text{Expected visits to state } j \text{ starting from state } i = \text{the } (i, j)\text{th entry of } (\mathbf{I} - \mathbf{Q})^{-1} \tag{18.9}$$

Example 18.4 (Random walk with absorbing barriers) Consider a random walk on four consecutive points $1, 2, 3, 4$ on a line. Starting at either 2 or 3, the process moves right with probability $2/3$ or left with probability $1/3$. Whenever the process gets to either 1 or 4, it remains there forever. Classify the four states as to recurrent or transitive, and for each pair (i, j) of transitive states find the expected number of times the process will be in j starting from state i.

Solution. Points 2 and 3 are clearly transient. Points 1 and 4 are clearly recurrent, in fact a particular type of recurrent state, known as an *absorbing* state, which means that once there, you never leave. Numbering point 2 as state 0 and point 3 as state 1, for the transient state submatrix $\mathbf{Q}$, we have

$$\mathbf{Q} = \begin{pmatrix} 0 & 2/3 \\ 1/3 & 0 \end{pmatrix}, \quad \mathbf{I} - \mathbf{Q} = \begin{pmatrix} 1 & -2/3 \\ -1/3 & 1 \end{pmatrix}, \quad (\mathbf{I} - \mathbf{Q})^{-1} = \begin{pmatrix} 9/7 & 6/7 \\ 3/7 & 9/7 \end{pmatrix}.$$

The final matrix gives the expected number of visits. For example, starting at point 2, the expected number of times the process visits point 3 is $6/7$.

18.5 Introduction to continuous time processes

We now turn our attention to processes where the index set T takes all values in the interval $[0, \infty)$. We will also allow continuous rather than discrete random variables.

One goal is to define continuous-time analogue for the process given in the discrete case by (18.3). In Chapter 23, we apply this to modelling the surplus of an insurer. Another goal is to introduce *Brownian motion* which can be viewed as a type of continuous-time random walk, which we apply in Chapter 20.

Notation For continuous-time processes, we will write the time variable in brackets rather than as a subscript. That is, we write $X(t)$ in place of X_t.

We write $X \sim Y$ where X and Y are any two random variables to mean that they have the same distribution.

We begin with a few general definitions that capture some of the features we introduced in the discrete time setting. Given a stochastic process $X(t)$ and two times $s < t$, the random variable $X(t) - X(s)$ is called an *increment* of the process, since it gives the increase in the value over the period running from time s to time t.

Definition 18.5 We say that the process has *independent increments* if the increments over disjoint time intervals are independent.

This constitutes a strong version of the Markov property. Given times $s < t$, we can write $X(t) = X(s) + X(t) - X(s)$. This shows that the value of $X(t)$ can certainly depend on the value of $X(s)$, but we get no additional information from looking at times before s, since both $X(t) - X(s)$ and $X(s)$ are independent of what happened in the interval $[0, s)$.

Definition 18.6 We say that the process has *stationary increments* if the distribution of any increment depends only on the length of the time interval and not the particular starting point. That is, given any $s, t, h > 0$, we require that

$$X(s + h) - X(s) \sim X(t + h) - X(t).$$

This constitutes a strong version of the assumption of stationary transition probabilities that we made for discrete Markov chains.

18.6 Poisson processes

Suppose we have a particular 'event' that we are interested in, occurring repeatedly and randomly in time. This could be an insurance claim or the arrival of a person at a queue, or a number of other possibilities. A *counting process* is a stochastic process $N(t)$ that counts the number of such 'events' that have occurred up to time t. Formally, it is just any continuous-time process that takes nonnegative integer values, and such that all realizations are increasing. A particular realization can then always be drawn as an increasing step function. We will always assume that in any counting process, $N(0) = 0$. In other words, the counting starts at time 0 before any events have occurred. A major application in this text will be to insurance claims, which is discussed in Chapter 23.

We will confine attention to the particular case of Poisson processes defined as follows.

Definition 18.7 A counting process $N(t)$ is called a *Poisson process* with rate λ if it has stationary and independent increments and if, for all $h > 0$,

$$N(h) \sim \text{Poisson}(\lambda h).$$

It follows from the stationary increment assumption that, given any $t > 0$, the random variable $N(t + h) - N(t) \sim \text{Poisson}(\lambda h)$. In other words, a Poisson process is simply a counting process with independent increments, such that the number of occurrences in any time interval is a Poisson distribution, with parameter proportional to the length of the interval.

When should we choose a Poisson process to model a counting situation? There is another characterization of Poisson processes that gives some insight into answering this question. Suppose we assume that the increments are indeed stationary and independent. The alternate formulation says essentially that we will get a Poisson process if it is 'highly unlikely' to have more than one event occurring in a 'sufficiently small' time interval. Therefore, if you feel that this is the case for the particular event you are trying to model, you can be justified in choosing the Poisson process. To derive this characterization, we must first give precise meaning to the phrases 'highly unlikely' and 'sufficiently small'. This is done conveniently through the 'little o' notation, which we will now review.

We say that a function f defined on an interval $[0, b]$ is $o(h)$ if $\lim_{h\to 0} f(h)/h = 0$. This means that f is getting small rapidly as h gets small, more rapidly than h itself. For example, $f(h) = h^2$ is $o(h)$, while $f(h) = \sqrt{h}$ is not. We often write the symbol $o(h)$ for such a function. For example, we would write

$$e^{\beta h} = 1 + \beta h + o(h),$$

as can be seen from the Taylor series expansion. It is clear that, given two functions f and g that are both $o(h)$, their sum $f + g$ is $o(h)$, and for any constant c the function cf is $o(h)$ as well. We can now state the desired characterization.

Theorem 18.3 *A counting process that has stationary and independent increments is a Poisson process with rate λ if and only if the following hold:*

(i) $P(N(h) = 1) = \lambda h + o(h)$,

(ii) $P(N(h) \geq 2) = o(h)$.

Partial proof. One direction is clear. For a Poisson process with rate λ, we have $P[N(h) = 1] = \lambda h e^{-\lambda h} = \lambda h[1 - \lambda h + o(h)] = \lambda h + o(h)$. Similarly, we have $P[N(h) = 0] = e^{-\lambda h} = 1 - \lambda h + o(h)$. Therefore, $P[N(h) \geq 2] = 1 - P[N(h) = 0] - P[N(h) = 1] = 1 - (1 - \lambda h + o(h)) - (\lambda h + o(h)) = o(h)$.

The converse is the more difficult part. We assume conditions (i) and (ii) and must show that $N(h)$ is Poisson. There are various methods of doing this and we will not elaborate further. A nice proof can be found in Ross (2010, Theorem 5.1) where this is done by calculating the moment generating function of $N(h)$.

18.6.1 Waiting times

Given any counting process, $N(t)$, there is an associated *waiting-time* process, $W_n, n = 1, 2, \ldots$, where W_n is the time between the $(n-1)$th event and the nth event. We take the 0th event as occurring at time 0. So, for example, if the first event occurs at time 1, the second at time 1.7, the third at time 2.3, we would have $W_1 = 1$, $W_2 = 0.7$, and $W_3 = 0.6$. For some problems, it is more convenient to deal with W_n rather than $N(t)$. Note that the waiting-time process is a discrete-time process (although the index does not refer to times exactly) with continuous random variables, while the counting process is a continuous-time process with discrete random variables.

A natural question is to investigate the waiting-time process for a Poisson process, and this is easily answered. Let λ be the rate of the process. Suppose that $W_1 = w_1, W_2 = w_2, \ldots, W_{n-1} = w_{n-1}$. What is the distribution of W_n? If $s = w_1 + w_2 + w_{n-1}$, the $(n-1)$th occurrence was at time s, and in order for W_n to be greater than or equal to w, we require that there be no occurrences in the interval $(s, s+w]$, which, by the stationary increment assumption, has the same probability as no occurrences in the interval $(0, w]$. This probability is just $e^{-\lambda w}$, which is easily recognized as the survival function of an exponential distribution. We conclude that the W_n are independent and each is distributed as $\text{Exp}(\lambda)$.

Another random variable that is of interest in many applications is T_n, the time of the nth occurrence (sometimes called the nth *arrival time*). Clearly, $T_n = W_1 + W_2 + \cdots + W_n$, and it is easy to see using moment generating functions that T_n is distributed as $\text{Gamma}(n, \lambda)$.

18.6.2 Nonhomogeneous Poisson processes

In many counting processes, the stationarity assumption is not realistic, as we can expect the rate of occurrence to vary with time. To model this, we use a type of process that is similar to a Poisson process, except the rate λ is no longer a constant but rather a function of t.

Definition 18.8 A counting process is called a *nonhomogeneous Poisson process* with intensity function $\lambda(t)$ if it has independent increments and, for all $t > 0$,

(i) $P(N(t+h) - N(t) = 1) = \lambda(t)h + o(h)$,

(ii) $P(N(t+h) - N(t) \geq 2) = o(h)$.

It can then be shown for $s < t$, the increment $N(t) - N(s) \sim \text{Poisson}(\phi(s, t))$, where

$$\phi(s, t) = \int_s^t \lambda(r)\mathrm{d}r.$$

We omit the proof.

Note that when $\lambda(t)$ is a constant λ, then $\phi(s, t)$ is just equal to $(t-s)\lambda$, and we have exactly the same conclusion that we had before in the case of a regular Poisson process.

18.7 Brownian motion

18.7.1 The main definition

We now look at another continuous-time stochastic process which has several applications. This is intended as an introduction and we will not supply all proofs.

A stochastic process $X(t), 0 \leq t < \infty$, is called a *Brownian motion* process with variance parameter σ^2, for some $\sigma > 0$, if it satisfies the following four conditions:

(i) $X(0) = 0$;

(ii) the process has independent and stationary increments;

(iii) for each $t > 0$, $X(t)$ is normally distributed with mean 0 and variance $\sigma^2 t$;

(iv) the realizations, $t \to x_t$ are continuous functions of t.

Note that on a formal basis the definition is similar to that of a Poisson process, although the nature of the processes are quite different. Both have independent and stationary increments. In the Poisson case, the increments have a Poisson distribution with expectation proportional to the length of the interval, and in the Brownian motion case, they have a normal distribution with variance proportional to the length of the interval. In the Poisson case, the realizations are step functions which involve sudden jumps, in contrast to the continuous realizations of Brownian motions.

A *standard* Brownian motion is one in which $\sigma^2 = 1$. We will denote this by $B(t)$. For the Brownian motion $X(t)$ with variance parameter σ^2 it is clear that

$$X(t) = \sigma B(t).$$

18.7.2 Connection with random walks

A Brownian motion can be viewed as a *continuous* random walk, where moves are made at each instant of time. Consider first our simple random walk of section 18.2.2. We start at 0 on the real line, and each time unit, we move one unit either to the right or left, with equal probability. Let us compute the distribution of $X(t)$, our position at time t. It is easy to see that this is related to a binomial distribution. Precisely

$$X(t) \sim 2\text{Bin}(t, 1/2) - t, \tag{18.10}$$

from which it is clear that (see (A.41))

$$E(X(t) = 2t(1/2) - t = 0, \quad \text{Var}(X(t) = 4t(1/4) = t.$$

We see that our simple random walk has the mean and variance of condition (iii) in the definition of a standard Brownian motion.

Let us now speed up our random walk by making moves more frequently. Instead of moving each time unit, let us move every $1/m$ of a time unit where m is some positive integer. Of course, we also want to change the length of each move. One may think that instead of moving 1 unit at each step, we should now move $1/m$ of a unit. However, this will not preserve the variance condition that we want. To do so, we need to make much larger moves. A key fact is that we will need to make the length of each move equal to $1\sqrt{m}$. So for example, if $m = 4$, we would make moves of length $1/2$ at times $1/4, 2/4, 3/4$, etc, while if $m = 100$, we would make move of length $1/10$ at time $1/100, 2/100$ etc. Let $X^{(m)}(t)$ denote our position at time t under this new arrangement. It is not hard to see that this will be the same as the position at time mt of the simple case with moves of 1 unit each time period, only multiplied by the length of each move, which is $1/\sqrt{m}$. From (18.10)

$$X^{(m)}(t) \sim \frac{2}{\sqrt{m}}\text{Bin}(mt, 1/2) - \frac{mt}{\sqrt{m}},$$

and we see as above that

$$E(X^{(m)}(t) = 0, \text{Var}(X^{(m)}(t) = t.$$

The standard Brownian motion $B(t)$ can be viewed as a limiting case of $X^{(m)}(t)$ as m approaches ∞.

*18.7.3 Hitting times

For many stochastic processes we are interested in the time that it first reaches a certain point. These random times are known as *hitting times*. We will deduce hitting time distributions for Brownian motion, by invoking one of its key properties, known as the *symmetry principle*. This simply says that if the process is at a point b at time s, then at any later time t the distribution must be symmetric about b. In other words, given that $B(s) = b$, then for any numbers $h < k$.

$$P(b + h \leq B(t) \leq b + k) = P(b - k \leq B(t) \leq b - h) \tag{18.11}$$

This feature is clear for the simple Random walk, where we are just as likely to move to the right as to the left, and therefore is true for Brownian motion by viewing it as a limiting case of random walks.

Let $B(t)$ be a standard Brownian motion, and for any point a, we let

$$T_a = \inf\{t : T(t) = a\}$$

To compute the distribution of T_a, we will derive a relationship between the distribution of T_a and distribution of $B(t)$. Suppose first that $a > 0$. Then by Equation (A.30)

$$\begin{aligned} P[B(t) \geq a] = {} & P[B(t) \geq a | T_a \leq t] P[T_a \leq t] \\ & + P[B(t) \geq a | (T_a > t] P[T_a > t]. \end{aligned}$$

We note first that the second term on the right equals 0. Indeed, If $T_a > t$ then the first time the process hits a is after time t, and the value at time t could not be greater than or equal to a, since if so, by continuity of the realizations, we would have needed to reach a sometime before time T.

Consider now the first term. The quantity $P(B(t) \geq a | T_a \leq t) = 1/2$, since $T_a \leq t$ means that the process hit a at some time s prior to t and by the symmetry principle it is just as likely to be below as above. Formally, we are applying (18.11) with $h = 0, k = \infty$. This enables us to state the desired relationship

$$P(T_a \leq t) = 2P(B(t) > a).$$

Since $B(t)$ is normal with mean 0 and variance t, we know that $P(B(t) > a) = 1 - \Phi(a/\sqrt{t})$ where Φ is the c.d.f of the standard normal. Moreover, for $a < 0$, the distribution of T_a is, by symmetry, the same as that T_{-a} since we start the process at 0. Our final conclusion is then

$$P(T_a \leq t) = 2[1 - \Phi(|a|/\sqrt{t})].$$

This enables us to find the distribution of another random variable of interest. Let $M(t)$ be the maximum value that B assumes in the interval [0,t]. This will be of course a nonnegative random variable, since we start at 0. For $a > 0$, we use continuity again to see that $M(t)$ will be larger than or equal to a if and only if the process reached a sometimes in the interval $[0, t]$ which means that $T_a \leq t$. So

$$P(M(t) \geq a) = 2[1 - \Phi(|a|/\sqrt{t})].$$

This is an intuitively appealing result, since it says that for any $a > 0$, the probability that a standard Brownian motion takes a value greater than a at any time before time t is exactly twice the probability that it takes a value greater than a at time t.

*18.7.4 Conditional distributions

Given two times s and t, what is the conditional distribution of B_t given that X_s took a certain value b? We consider the two cases.

If $s \leq t$, this is relatively straightforward. We can imagine the process just starting all over again at time s except with a starting value of b rather than 0. It is then easy to deduce that

$$\text{If } t \geq s,\ (B(t)|B(s) = b) \text{ is normal with mean } b \text{ and variance } (t - s). \tag{18.12}$$

What if $s > t$? Now we are asking for the distribution of a random variable conditional on future values and the procedure is more complex. Could we perhaps get the same answer except with variance $= (s - t)$? Obviously, not since this would be false for $t = 0$ where we know by definition that B_t takes the value 0 with certainty. It turns out that we have to multiply these quantities by t/s.

We will work with the density functions and let f_r stand for $f_{B(r)}$. Now given that $B(s) = b$, we will have $B(t) = a$ provided that the increment $B(s) - B(t)$ takes a value of $b - a$. By stationarity, this increment is distributed as B_{s-t}. Then

$$f_t(a|X_s = b) = \frac{f_t(a)f_{s-t}(b - a)}{f_s(b)},$$

where we invoke the hypothesis of independent increments for the numerator. Let $r = t/s$. Plugging in the various normal densities and doing some algebra, the left hand side reduces to

$$\text{K}\exp\left(\frac{-(a - rb)^2}{r(s - t)}\right)$$

for some constant K and we can then deduce that in place of (18.12) we have the following.

If $t < s$, $(B(t)|B(s) = b)$ is normal with mean rb and variance $r(s - t)$, where $r = t/s$.

18.7.5 Brownian motion with drift

A Brownian motion process is a martingale. We have not defined this precisely for a continuous time process, but the idea is similar to that of the discrete time case. If a Brownian motion process takes the value b at some point of time, then the expected value at any future point of time will be b, as the symmetry principle shows. One often wants to model situations where this is not the case and the quantity under study is growing in value. A simple model of this type is a Brownian motion with a drift coefficient μ. This is a process which still has stationary and independent increments, $X(t)$ is still normally distributed with variance $\sigma^2 t$, but now, the mean of $X(t)$ is μt rather than 0.In other words, on the average the quantity grows as a rate of μ per time period, so in any time period of length h, the quantity can be expected to increase by μh. To state this precisely, $X(t)$ is a Brownian motion with drift coefficient μ and variance parameter σ^2 if and only if

$$X(t) \sim \sigma B(t) + \mu t.$$

18.7.6 Geometric Brownian motion

We are often interested in quantities which grow at a constant *relative* rather than *absolute* rate of growth. An example is stock prices. We might, for example expect a stock to grow in value at a rate of say 5% per year. A possible model in this case is *geometric Brownian motion*. This is a positive valued process $X(t)$ such that $Y(t)$, the logarithm of $X(t)$ is a Brownian motion with drift parameter μ and variance parameter σ^2. That is,

$$X(t) = e^{Y(t)} = e^{\mu t + \sigma B(t)}.$$

Note that each $X(t)$ will have a log-normal distribution.

When $\sigma = 0$, $X(t)$ is simply the amount of a quantity growing exponentially in time at rate μ. So a geometric Brownian motion can be viewed as a process that has an underlying pattern of exponential growth, with a rate of growth that is not constant but is itself subject to random changes as given by a Brownian motion. This is a commonly used model for stock prices in modern finance theory and we will so apply it in Section 20.18.

A quantity which is useful in many applications is the conditional expectation of $X(t)$, giving that the value of $X(s) = a$ for some $s < t$. To compute this, we first write

$$X(t) = e^{Y(t)} = e^{Y(s)}e^{(Y(t)-Y(s))} = X(s)e^{(Y(t)-Y(s))}.$$

We know that $Y(t) - Y(s)$ is normal with mean $(t - s)\mu$ and variance $(t - s)\sigma^2$. So using the expression given in Equation (A.58)

$$E(X(t)|X(s) = a) = ae^{\mu(t-s)+(t-s)\sigma^2/2}.$$

Notes and references

Ross (2010) has a good basic introduction to stochastic processes, covering more material than we do here. The section on Poisson processes is based largely on Chapter 5 of that book. Hoel *et al.* (1972) give a somewhat more advanced treatment. Kemeny and Snell (1963) provide extensive coverage of finite-state Markov chains. A proof for the sufficient condition given for the existence of a limiting distribution can be found there in Theorem 4.12. Mikosch (1998) has more material on Brownian motion, with applications to some of the material we discuss in Chapter 20.

Exercises

18.1 A certain stock has a price of 100 at time 0. At any time k, let M_k denote the average of the prices at times $0, 1, \ldots, k$. The price at time $k+1$ will be either $M_k, M_k + 20$, or $M_k - 20$, each with probability 1/3. What is the distribution of the stock price at time 2? What is the expected value at time 2? Is this process a Markov process, martingale, submartingale, supermartingale?

18.2 Suppose that r black balls and r white balls are distributed equally among two urns. At each trial, a ball is chosen randomly from each urn and put in the other urn. Let X_n be the number of black balls in urn 1 at time n. Describe the transition function for this Markov chain.

18.3 The price of a certain stock is either 9, 10 or 11. If the price is 9 on any day, the next day's price will either be 9 or 10 with equal probability. If the price is 11, the next day's price will be either 11 or 10 with equal probability. If the price is 10, the next day's price will be either 9, 10, or 11 with equal probability.

(a) Model this as a Markov chain and write down the transition matrix.

(b) If the price is 9 on Monday, what is the probability that it will be 11 on Friday?

*(c) Find the proportion of the time, in the long run, that the price will be each of the three possible values.

*18.4 Consider the Markov Chain on states 1 to 5 with transition matrix

$$\begin{pmatrix} 2/3 & 0 & 0 & 1/3 & 0 \\ 0 & 1/2 & 0 & 0 & 1/2 \\ 1/4 & 0 & 3/4 & 0 & 0 \\ 1/2 & 0 & 0 & 1/2 & 0 \\ 0 & 3/4 & 0 & 0 & 1/4 \end{pmatrix}.$$

Decide whether each state is transient or recurrent.

*18.5 Consider the Markov chain on states 1 to 4 with transition matrix

$$\begin{pmatrix} 1/3 & 1/3 & 1/3 & 0 \\ 1/3 & 1/3 & 0 & 1/3 \\ 1/3 & 0 & 1/3 & 1/3 \\ 0 & 0 & 1/2 & 1/2 \end{pmatrix}.$$

Decide whether each state is transient or recurrent.

*18.6 Consider the Markov chain on states 1 to 4 with transition matrix

$$\begin{pmatrix} 0 & 0.8 & 0.2 & 0 \\ 0 & 0 & 0.3 & 0.7 \\ 0.5 & 0.5 & 0 & 0 \\ 0 & 0 & 0 & 1 \end{pmatrix}.$$

Decide whether each state is transient or recurrent.

*18.7 For a two-state Markov Chain with transition matrix

$$\begin{pmatrix} 1-p & p \\ q & 1-q \end{pmatrix},$$

find the limiting distribution in terms of p and q

*18.8 An indecisive diner enters a restaurant and is taken to a round table with five chairs. Unable to decide which chair is best, the diner switches every minute, moving clockwise with probability p and counterclockwise with probability $1-p$. We can then consider a Markov chain X_n, the number of the chair occupied at time n, in minutes. Show that the condition ensuring a limiting distribution holds in this case. What is the limiting distribution? What can you say if there are four chairs rather than five?

*18.9 A particle moves around a square with four vertices, numbered $0, 1, 2, 3$ going clockwise. It moves clockwise with probability $2/3$ and counterclockwise with probability $1/3$. Motion stops when the particle reaches vertex 3. Find the expected amount of time the vertex will spend at vertex 2, starting from each of vertices, $0, 1, 2$.

*18.10 Show that for either a periodic or reducible Markov chain, given any integer n, there exists some ordered pair (i,j) such that the entry in row i and column j of $\mathbf{P}^n = 0$.

*18.11 A golfer has a probability of $(1-0.1n)$ of making an n-foot putt, where n is less than 10. He adopts the following routine for practicing putts from three to seven feet. He begins with a three foot putt. Anytime he makes a putt, he tries one which is a foot longer and any time he misses, he goes back to a foot shorter putt. If he misses on the three-foot putt or makes the seven-foot putt, he repeats that distance.

(a) If he continues his session for a very long time, what is the most likely distance that he will finish with? Estimate the probability of finishing at that distance.

(b) Repeat part (a) only assuming now that he is practicing putts from four to eight feet.

18.12 In a Poisson process, the probability that exactly one event will occur in any given hour is $3e^{-3}$.

(a) What is the probability that exactly two events will occur in any 20-minute period?

(b) Suppose you start observing the process at some point of time. What is the probability that it will be less than 10 minutes until an event occurs?

(c) Take a unit of time to be 1 hour. Starting at time 0, find the expectation and variance of the time of the fifth occurrence.

18.13 Suppose $N(t)$ is a Poisson process such that $P[N(2) = 1] = 6e^{-6}$.

(a) Find $P[N(1) = 2]$.

(b) Identify the distributions (including parameters) of W_n, the associated waiting-time process, and T_n, the time of the nth arrival.

(c) Find $P[(N(1) = 1 \text{ and } N(3) = 3]$.

18.14 Vehicles pass a certain marker on a highway in accordance with a Poisson process at the rate of 48 per hour; 25 % of the vehicles on the road are trucks and 75 % are cars. Suppose you observe this marker at a particular point of time.

(a) What is the probability that exactly five vehicles will pass in the first 15 minutes?

(b) Given that exactly two trucks passed in the first 5 minutes, what is the probability that the fourth truck will pass in the first 10 minutes?

(c) Given that exactly two trucks passed in the first 5 minutes, what is the expected value and the variance of the number of vehicles that passed in the first 10 minutes?

18.15 The following gives information about three counting processes, $N^1(t), N^2(t), N^3(t)$. Give reasons why each of these cannot be a Poisson process.

(a) $P[N^1(2) - N^1(1) = 10] = 0.5, P[N^1(3) - N^1(2) = 10] = 0.4$

(b) $P[N^2(1) = 1] = 0.5$ $P[N^2(1) = 1 \text{ and } N^2(2) = 2] = 0.2$

(c) $P[N^3(1) = 0] = 0.5, \quad P[N^3(2) = 1] = 0.3$

18.16 Consider a Poisson process $N(t)$ with rate 2 per time period. You are given that $N(1.8) = 4$. Using this information, answer each of the following.

(a) What is the probability that $N(2.3) \leq 6$?

(b) What is the probability that the fifth occurrence will be before time 2?

(c) What is the expected time and the variance of the seventh occurrence?

18.17 At a subway station, eastbound trains and northbound trains arrive independently, both according to a Poisson process. On average, there is one eastbound train every 12 minutes and one northbound train every 8 minutes. Suppose you arrive at the subway station at a certain point of time and start observing trains.

(a) What is the probability that exactly two eastbound trains will arrive in the first 24 minutes?

(b) What is the probability that exactly two eastbound trains will arrive in the first 24 minutes and exactly three eastbound trains will arrive in the first 36 minutes?

(c) What is the expected waiting time, in minutes, until the first train (of either type) arrives?

(d) What is the probability that the first train to arrive is eastbound, and the next two are northbound?

(e) What is the probability that it will take at least 20 minutes for two northbound trains to arrive?

18.18 Events occur according to a Poisson process. Suppose that the expected number of occurrences per hour is 3, and let T_n denote time in hours of the nth occurrence. Find the expected value and variance of T_3 given that the first occurrence was at time 1 and the second occurrence was at time 2.

18.19 In a Poisson process, the probability is 0.60 that after an occurrence of the event, it will take at least 2 months until the next event. Find the probability that exactly four claims will occur within any 5-month period.

18.20 An insurer finds that out of a certain group of insured drivers, the accident rate over each 24-hour period rises from midnight to noon, and then declines until the following midnight. They decide that the number of accidents can be modelled by a nonhomogenous Poisson process where the intensity at time t is given by $[1/6 - (12 - t)^2/1152]$, where t is the number of hours since midnight.

Find: (a) the expected number of daily accidents; (b) the probability that there will be exactly one accident between 6:00 a.m. and 6:00 p.m.

18.21 Suppose that $X(t)$ is a Brownian motion with variance parameter 0.03. If $X(2) = 3$, find the probability that $X(5) > 3.5$.

*18.22 For a certain security, each 1 unit invested now will yield the random amount $e^{0.2B(t)}$ at time t where $B(t)$ is a standard Brownian notion. What is the probability that sometime within the next five time periods, an initial investment will have doubled in value.

*18.23 The price of a stock at time t is siven by

$$S(t) = 15e^{\mu t+0.3B(t)},$$

where $B(t)$ is a standard Brownian motion. The quantity μ is unknown. An investor buys the stock at time 3 and sells it for 20 a share at time 5. What is the probability that the investor lost money on this transaction?

19

Multi-state models

19.1 Introduction

Multi-state models are an attempt to look at a variety of life insurance and annuity contracts in a unified manner, by making use of Markov processes. As motivation, take an individual now age x and consider a two-state Markov chain, where the person is in state 0 (alive) or state 1 (deceased) at any time. Life insurance contracts provide benefits upon transfer from state 0 to state 1, while life annuity contracts provide benefits as long as the process remains in state 0.

More generally, consider a multiple-decrement model for (x) with m causes of failure. We can consider a chain with $m + 1$ states. State 0 means that (x) has not succumbed to any cause and is often referred to as the *active* state. State j refers to having succumbed first to cause j. The insurance benefits discussed in Chapter 11 can be viewed as payments upon transfer from state 0 to other states.

For still another example, consider a joint-life contract issued to (x) and (y). We can now take a chain with four states as illustrated in Figure 19.1. The arrows indicate that there are possible transitions from state 0 into state 1 or state 2, occasioned by the death of (y) or (x) respectively, and then further transitions from state 1 or state 2 into state 3 when the second death occurs. The dotted line, showing a transition directly from state 0 to state 3, would not be present in our original model, but would be there if we wanted to consider a common shock model as discussed in Section 17.6. A joint life insurance can be considered as two contracts, one paying benefits upon transfer from state 0 to state 1, and the other paying benefits upon transfer from state 0 to state 2. A general two-life annuity can be considered as three separate contracts, where the ith contract, $i = 0, 1, 2$, pays benefits provided the process is in state i. This can be generalized to contracts involving n lives where we will have 2^n states.

We can of course imagine more general patterns of transition. We may wish to investigate a more enriched multiple-decrement model where individuals can transfer between states several times. A disabled person might recover and re-enter the main group of lives. In our original model we ignored what happened to a life once it left the group for any cause, but we

Fundamentals of Actuarial Mathematics, Third Edition. S. David Promislow.

Companion website: http://www.wiley.com/go/actuarial

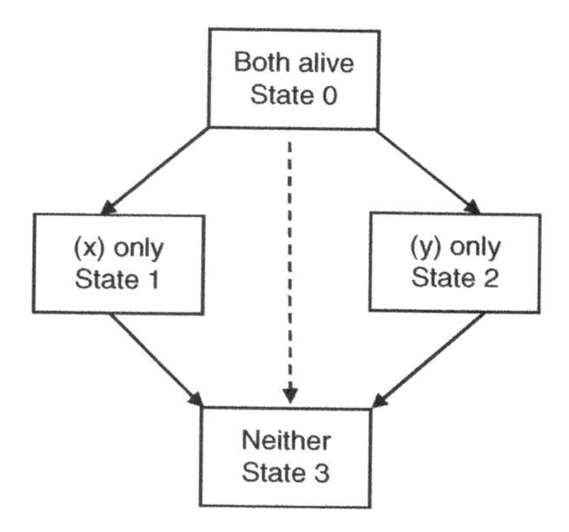

Figure 19.1 A two life multi-state model

may wish to model the fact that someone leaving for a cause other than death will subsequently die. There are many examples of insurance and annuity benefits applicable to this general case. When disability is one of the decrements, we may have a contract that pays benefits when a person becomes disabled, and then further benefits when a disabled person dies, and possibly additional payments that continue during disability. Indeed, a common provision in many life insurance policies is a *disability premium waiver* clause, which mean the person does not have to pay premiums during the time they are disabled. We view this as an annuity providing payments during a state of disability, which cease when there is a transfer back to an active state. These are only some of the possibilities, and we invite the reader to think of additional applications.

In this chapter, we will investigate the general multi-state model. We first discuss the discrete-time model, where transitions between states can occur only at integer times. Following that we deal with the more complicated continuous-time model, where we allow for transitions at arbitrary times.

19.2 The discrete-time model

19.2.1 Non-stationary Markov Chains

The transitions in multi-state models are normally age-dependent. To handle this, we need to extend the concepts and notation that we introduced in Section 18.4.1 so that they apply to the non-stationary case.

For a non-stationary Markov Chain, in place of the single transition matrix $\mathbf{P}$, we need for each nonnegative integer n, a matrix $\mathbf{P}_n$ whose ijth entry is $p_{ij}(n, n+1)$

The main quantities of interest are the probabilities $p_{ij}(k, n)$ of being in state j at time n conditional upon being in state i at time k. In deriving (18.5), we do not need the fact that the matrices are the same, and the argument given is easily adapted to show that

$$p_{ij}(k, k+n) \text{ is the } ij\text{th entry of } \quad \mathbf{P}_k\mathbf{P}_{k+1} \cdots \mathbf{P}_{k+n-1}, \tag{19.1}$$

which reduces to (18.5) in the stationary case. Equation (18.6) now takes the form

$$\pi_{k+n} = \pi_k \ \mathbf{P}_k\mathbf{P}_{k+1} \cdots \mathbf{P}_{k+n-1}.$$

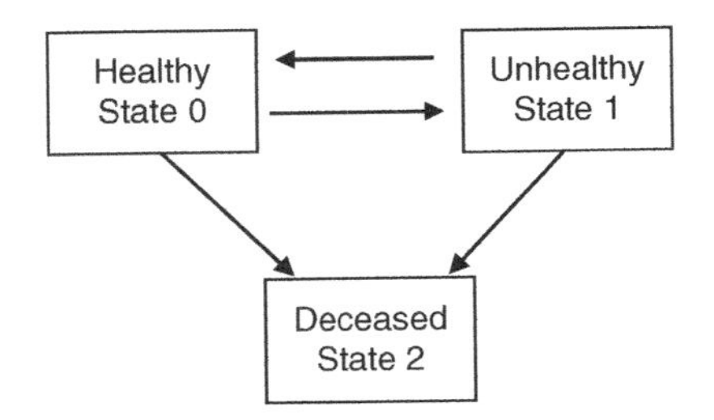

Figure 19.2 The healthy-unhealthy-deceased model

For example, in our simple life-death chain the transition matrices are given by

$$\mathbf{P}_n = \begin{pmatrix} p_{x+n} & q_{x+n} \\ 0 & 1 \end{pmatrix}.$$

In the multiple decrement example, the matrix $\mathbf{P}_n$ will be similar to that above, except the first row will be $(p_{x+n}^{(\tau)}, q_{x+n}^{(1)}, q_{x+n}^{(2)}, \ldots, q_{x+n}^{(m)})$, and each other row will have 1 on the main diagonal and 0s elsewhere.

We leave is as an exercise for the reader to write down the matrix $\mathbf{P}_n$ in the joint two life case.

We now introduce a multi-state model which is not covered by our previous cases of multiple lives or multiple-decrement theory, and to which we will refer to frequently in the rest of the chapter. It allows for a *two-way transition* between certain states, which is really where the generality of the multi-state model is the most helpful. This is the chain as depicted in Figure 19.2 where a life can be healthy or unhealthy, but we now allow for recovery from the unhealthy state.

There are various applications, The unhealthy state could refer to being disabled, or sick, or a number of other possibilities for a state which we wish to distinguish somehow from the main group of lives, but which can transfer back to the main group.

Example 19.1 Consider the model of Figure 19.2 and suppose that transition matrices for times 0, 1, 2 are given as follows;

$$\mathbf{P}_0 = \begin{pmatrix} 0.7 & 0.2 & 0.1 \\ 0.2 & 0.6 & 0.2 \\ 0 & 0 & 1 \end{pmatrix} \quad \mathbf{P}_1 = \begin{pmatrix} 0.5 & 0.3 & 0.2 \\ 0.1 & 0.6 & 0.3 \\ 0 & 0 & 1 \end{pmatrix} \quad \mathbf{P}_2 = \begin{pmatrix} 0.4 & 0.3 & 0.3 \\ 0.1 & 0.5 & 0.4 \\ 0 & 0 & 1 \end{pmatrix}.$$

What is the probability that a person who is healthy at time 0 will be unhealthy at time 2 and deceased at time 3?

Solution.

$$\mathbf{P}_0\mathbf{P}_1 = \begin{pmatrix} 0.37 & 0.33 & 0.30 \\ 0.16 & 0.42 & 0.42 \\ 0 & 0 & 1 \end{pmatrix},$$

Using formula (19.1), the required probability equals

$$p_{01}(0,2)p_{12}(2,3) = 0.33 \times 0.4 = 0.132.$$

19.2.2 Discrete-time multi-state insurances

To model the general contract of this type, we will take a Markov chain with states numbered from 0 to N. Suppose the process begins in state a at time 0. Fix any two states i and j, and consider a contract which pays a benefit of b_k at time $k+1$, provided a transfer from state i to state j occurs between time k and time $k+1$. That is, the process was in state i at time k and state j at time $k+1$.

Notation If we have several different transfer benefits we will distinguish by writing b_k as $b_k^{(ij)}$. In general, for quantities like benefits, premiums, reserves and expenses, we will let superscripts refer to states, and maintain our previous usage of subscripts to refer to time. This differs from our convention for probabilities where we use subscripts to refer to states, and write times in brackets.

A convenient method of discussing multi-state benefits in the context of the stochastic model is to make use of indicator random variables. For each nonnegative integer k, let I_k be the random variable that takes the values of 0 or 1 respectively, accordingly as a transfer from i to j has not or has occurred between time k and $k+1$. The present value of the benefits on the contract is given by

$$Z = \sum_{k=1}^{\infty} b_k v(k+1) I_k. \tag{19.2}$$

Since

$$E(I_k) = p_{ai}(0,k)p_{ij}(k,k+1), \tag{19.3}$$

the actuarial present value is

$$\sum_{k=0}^{\infty} b_k v(k+1) p_{ai}(0,k) p_{ij}(k,k+1). \tag{19.4}$$

We again have our familiar pattern with a sum of three-termed factors. In fact, we can consider each summand as a four-termed factor where the probability of receiving a payment is itself the product of two factors, the probability of transition to state i by time k and the probability of transition from state i to state j in the next period.

Example 19.2 Consider the chain given in Example 19.1. A 3-year contract written on a healthy life provides for benefits at the end of the year of death. The death benefit is 3 if the person dies while healthy and 2 if the person dies while unhealthy. If interest is a constant 5%, find the actuarial present value.

Solution. We can view this as two separate contracts, one paying 3 for transfer from state 0 to state 2, and the second paying 2 for transfer from state 1 to state 2. We use the matrices in

Example 19.1 and formula (19.1). For the first contract,

$$p_{02}(0,1) = 0.1, \quad p_{00}(0,1)p_{02}(1,2) = 0.14, \quad p_{00}(0,2)p_{02}(2,3) = 0.37 \times 0.3 = 0.111,$$

so that

$$\text{APV} = 3[(0.1)(1.05)^{-1} + 0.14(1.05)^{-2} + 0.111(1.05)^{-3}] = 0.9543.$$

For the second contract,

$$p_{01}(0,1)p_{12}(1,2) = 0.06, \quad p_{01}(0,2)p_{12}(2,3) = 0.33 \times 0.4 = 0.132$$

so that

$$\text{APV} = 2[0.06)(1.05)^{-2} + 0.132(1.05)^{-3}] = 0.3369.$$

Therefore, the total APV is 1.2912.

Example 19.3 Take the same chain as in the previous example. Consider a contract that pays 1 at the end of any year of transfer from being healthy to being unhealthy, provided this occurs within the next 3 years. Find the expected value and variance of the benefits.

Solution. We now have

$$p_{01}(0,1) = 0.2, \quad p_{00}(0,1)p_{01}(1,2) = 0.21 \quad p_{00}(0,2)p_{01}(2,3) = 0.37 \times 0.3 = 0.111,$$

so that

$$\text{APV} = 0.2(1.05)^{-1} + 0.21(1.05)^{-2} + 0.111(1.05)^{-3} = 0.4768.$$

The variance calculation can not be done conveniently by adapting a version of formula (15.4). The situation here is different from the multiple decrement model where benefits are paid upon transition to an absorbing state. In this case, there can be benefits paid at more than one time. We must instead use the approach taken in the current payment formula for annuity variances. Note however that the covariance calculations will be more complicated than that given by (15.11). Letting I_k take the value 1 if transfer from the healthy state to the unhealthy state occurs between time k and $k+1$, we have

$$E(I_0) = 0.2, \quad E(I_1) = 0.21, \quad E(I_2) = 0.111, \quad E(I_0 I_1) = E(I_1 I_2) = 0,$$

while

$$E(I_0 I_2) = p_{01}(0,1)p_{10}(1,2)p_{01}(2,3) = 0.006,$$

which is the probability of becoming unhealthy, recovering, and then becoming unhealthy again, thereby being paid at both time 1 and time 3. We can then calculate the covariances as

$$\text{Cov}(I_0 I_1) = -0.042, \quad \text{Cov}(I_0 I_2) = -0.0162 \quad \text{Cov}(I_1 I_2) = -0.02331,$$

leading to the calculation of the variance as

$$(0.2)(0.8)(1.05)^{-2} + (0.21)(0.79)(1.05)^{-4} + (0.111)(0.889)(1.05)^{-6}$$
$$-2[0.042(1.05)^{-3} + 0.0162(1.05)^{-4} + 0.02331(1.05)^{-5} = 0.2195.$$

Suppose we wished to modify the contract of the previous example so that it paid off only on the *first* occurrence of becoming unhealthy. In our simple example, this could be done readily enough by changing the probability of a benefit payment at time 3 to 0.105, subtracting 0.006, which we have calculated above as the probability of the second occurrence. To handle the problem in general, we need a more systematic approach. Consider a contract paying benefits upon the first transition from a state i_0 to a state j_0. A method that works here (as well for many other problems) is to add additional states.

The easiest way is to add a single new absorbing state z, and all transfers from state i_0 to state j_0 are redirected to z, and the contract is modified to pay upon a transfer from i_0 to z. To state this precisely, let p^* denote probabilities in the augmented chain. To simplify the notation, fix a time n and write p_{ij} and p^*_{ij} for $p_{ij}(n, n+1)$ and $p^*_{ij}(n, n+1)$ respectively. Then p^* is the same as p except that

$$p^*_{i_0j_0} = 0, \quad p^*_{i_0z} = p_{i_0j_0}, \quad p^*_{zz} = 1, \quad p^*_{zi} = 0\,, i \neq z, \quad p^*_{iz} = 0, \; i \neq z, i_0$$

There is a more involved method which has the advantage that it preserves relationships between other states that might be needed in more complex problems. This is to add $N+1$ new states $i^*, i = 0, \ldots, N$, where state i^* can be considered as a sort of 'clone' of state i. Transitions between un-starred states are as before except that transitions from i_0 to j_0 are directed to j_0^* in place of j_0. Except for that, there are no transitions between the two types of states, starred and un-starred. Transitions between starred states are as given by the corresponding transitions between their clones. In this case, $p^*_{i_0j_0} = 0$ and for all ordered pairs (i,j) not equal to (i_0,j_0),

$$p^*_{ij} = p_{ij}, \quad p^*_{i_0j_0^*} = p_{i_0j_0}, \quad p^*_{i^*j^*} = p_{ij}, \quad p^*_{i^*j} = 0, \quad p^*_{ij^*} = 0. \tag{19.5}$$

For example, if our original matrix for a two state model is given by

$$\mathbf{P} = \begin{pmatrix} a & 1-a \\ b & 1-b \end{pmatrix},$$

and we take $i_0 = 0, j_0 = 1$, the the matrix of the augmented chain, using the order $0, 1, 0^*, 1^*$ is

$$\mathbf{P}^* = \begin{pmatrix} a & 0 & 0 & 1-a \\ b & 1-b & 0 & 0 \\ 0 & 0 & a & 1-a \\ 0 & 0 & b & 1-b \end{pmatrix}.$$

In addition to this, the provisions of the insurance are modified to pay upon a transfer from state i_0 to state j_0^*. After the first occurrence of such an event, there can never be a return to state i_0, so subsequent occurrences are ruled out.

Remark Solving multi-state problems with large matrices can involve a great deal of calculation. There are a number of computing packages that can assist with this. One convenient example is the MMULT function of Excel®.

19.2.3 Multi-state annuities

We now look at annuity contracts associated with a chain. Suppose again that the individual begins in state a at time 0. For a fixed state i, we can consider a contract that pays c_k at time k provided the person is in state i at that time. Let I_k now be the indicator random variable that takes the value of 1 or 0 according as the process is or is not in state i at time k. Then the present value of the benefits is the random variable

$$Y = \sum_{k=0}^{\infty} c_k v(k) I_k,$$

from which we calculate immediately

$$\text{APV} = \sum_{k=0}^{\infty} c_k v(k) p_{ai}(0, k). \tag{19.6}$$

To derive variances, we calculate for all $m < n$

$$E(I_m I_n) = p_{ai}(0, m) p_{ii}(m, n).$$

From this, we calculate the covariances and then proceed as in Example 19.3.

Example 19.4 Consider again the chain given in Example 19.1. A contract on a healthy life provides for four yearly payments beginning at time 0. The amount of the payment is 1 if the person is healthy, or 2 if the person is unhealthy. If interest is a constant 5%, find the actuarial present value.

Solution. We view this as two separate annuities. Contract 1 is for state 1 and contract 2 is for state 2. In addition to the matrix $\mathbf{P}_0\mathbf{P}_1$ calculated in Example 19.1, we need

$$\mathbf{P}_0\mathbf{P}_1\mathbf{P}_2 = \begin{pmatrix} 0.181 & 0.276 & 0.543 \\ 0.106 & 0.258 & 0.636 \\ 0 & 0 & 1 \end{pmatrix}.$$

The APV of contract 1 is

$$1 + \frac{0.7}{1.05} + \frac{0.37}{1.05^2} + \frac{0.181}{1.05^3} = 2.1586,$$

and that of contract 2 is

$$2\left(\frac{0.2}{1.05} + \frac{0.33}{1.05^2} + \frac{0.276}{1.05^3}\right) = 1.4564,$$

giving a total APV of 3.6150.

Example 19.5 Suppose the contract in Example 19.2 is to be purchased by three level annual premiums of π, which are payable only if the person is healthy. Calculate π.

Solution. Equating values of premiums and benefits,

$$\pi(1 + (1.05)^{-1}0.7 + (1.05)^{-2}0.37) = 1.2912,$$

which gives $\pi = 0.6449$.

There are other types of annuity contracts that can be handled by the trick of adding states. Suppose for example, a contract provides periodic annuity benefits starting when the process enters state i_0 (or starting at time 0 if the process begins in state i_0), but which stop completely when the process leaves state i_0, even if there is a subsequent return. The easiest approach here is to add an absorbing state z and all transitions from state i_0 to other states are directed to state z, ensuring the process never returns to i_0. If we want to preserve information about other transitions, the method is to add a clone for each state as we did in the insurance case. Transitions from state i_0 to any state $j \neq i_0$ are redirected to the clone j^*, also ensuring that the process will never return to state i_0.

19.3 The continuous-time model

In a realistic situation, transitions between states need not occur at discrete intervals, but can happen at any point of time, and the benefits payable upon transition from one state to the another will be paid at the moment of transition. To model this situation, we need to consider *continuous-time Markov processes*. We still keep a finite number of states, but let time vary continuously. The definition parallels that which we make in Section 18.2, only we must consider all possible points of time. (We will follow the notational usage introduced in Section 18.5 of writing the time variable in brackets.)

Definition 19.1 A continuous-time Markov process is a stochastic process $X(t) : 0 \leq t < \infty$, where each $X(t)$ is discrete, with the property that given any sequence of times, $0 \leq t_0 < t_1 < \cdots < t_n < t_{n+1}$, and any sequence $(x_0, x_1, x_2 \ldots x_n, x_{n+1})$ where x_i is a value of $X(t_i)$ we have

$$P\big(X(t_{n+1})=x_{n+1}|X(t_n)=x_n, X(t_{n-1})=x_{n-1}, \ldots, X(t_0)=x_0\big)=P\big(X(t_{n+1})=x_{n+1}|X(t_n)=x_n\big)$$

As before, for times $s \leq t$ and any two states i, j we let

$$p_{ij}(s,t) = P(X(t) = j|X(s) = i),$$

which is the probability of reaching state j at time t when starting in state i at time s.

19.3.1 Forces of transition

With the continuous time framework, we are able to define analogues of the force of mortality in our original life-death model, and the forces of decrement in the multiple-decrement model. As we have seen in earlier examples, the process is often specified by giving these forces, and then the central problem is to use these to calculate the desired probabilities.

Definition 19.2 We define the *force of transition from state i to state j at time t* as follows. If $i \neq j$

$$\mu_{ij}(t) = \lim_{h \to 0^+} \frac{p_{ij}(t, t+h)}{h},$$

while

$$\mu_{ii}(t) = \lim_{h \to 0^+} \frac{p_{ii}(t, t+h)) - 1}{h}.$$

Note that μ_{ii} is always nonpostive, which may seem a bit strange at first, but it is easily explained if we consider, for example, the multiple-decrement case where state 0 refers to an active life. The negativity of μ_{00} reflects the action of the other forces of decrement, which cause transfer *out* of state 0.

An important fact is the following.

Theorem 19.1

$$\sum_{j=0}^{N} \mu_{ij}(t) = 0, \textit{ for all } i \textit{ and } t.$$

Proof. Fix any i and t. Refer to the little o notation introduced in Section 18.3. From the definitions above, for all states j and $h > 0$,

$$p_{ij}(t, t+h) = h\mu_{ij}(t) + o(h), i \neq j, \tag{19.7}$$

and

$$p_{ii}(t, t+h) = h\mu_{ii}(t) + o(h) + 1. \tag{19.8}$$

Summing over all j gives

$$1 = h \sum_{j=0}^{N} \mu_{ij}(t) = o(h) + 1.$$

Subtract 1 from each side, divide by h and take a limit as h approaches 0 to establish the conclusion. □

We now want to relate the forces of transition to the transition probabilities. This is much more complicated in our general setting than in the simple cases we looked at previously.

Theorem 19.2 (Kolmogorov forward equations) *For any states i, j and times $s < t$*

$$\frac{\partial}{\partial t} p_{ij}(s, t) = \sum_{k=0}^{N} p_{ik}(s, t)\mu_{kj}(t) \tag{19.9}$$

Proof. We start by deriving the discrete version of the above formula. Following the reasoning used in Equation (18.4), we can deduce that for states i, j times $s < t$ and $h > 0$

$$p_{ij}(s, t+h) = \sum_{k=0}^{N} p_{ik}(s, t)p_{kj}(t, t+h), \tag{19.10}$$

reflecting the fact that in order to reach state j from state i in the time interval $s, t+h$ we must first reach some state k by time t and then go from state k to state j in the next h time units. The system of equations given by (19.10) is known as the *Chapman-Kolmogorov equations*.

Substituting from (19.7) and (19.8) into the second factor of the summand,

$$p_{ij}(s, t+h) = h \sum_{k=0}^{N} p_{ik}(s, t)\mu_{kj}(t) + p_{ij}(s, t) + o(h).$$

Subtracting $p_{ij}(s, t)$ from both sides, dividing by h and taking a limit gives (19.9). □

We can give an intuitive explanation of (19.9) paralleling that given in the discrete case. View each side as a type of density function for reaching state j at time t when starting from state i at time s. The right side reflects the fact that in order to accomplish this, we must first reach some state k at time t and then at that instant of time, transfer from state k to state j (or in the case that $k = j$, not transfer back into some other state).

We view the statement of the theorem as a system of differential equations in the variable t for a fixed value of s. The word 'forward' arises since our transition is moving forward in time. There is a related system, the corresponding *backwards equations*, which will be developed in Exercise 19.6.

There will normally be many solutions to a system of differential equations of this type. There will however often be a unique solution if we specify appropriate initial conditions. In our context, these conditions are the obvious requirements that for all t,

$$p_{ij}(t, t) = \begin{cases} 0, & i \neq j \\ 1, & i = j. \end{cases}$$

We have of course encountered versions of the Kolmogorov equations before. Take $s = 0$. The simplest case with $N = 1$ gives Equation (8.15). In the multiple decrement model, the Kolmogorov equations produce Equation (11.10). These statements may not be completely obvious since the notation is somewhat different. It will be instructive for the reader to verify these claims.

There is a nice matrix formulation of (19.9). Fixing s, t we let P denote the transition matrix for the time interval (s, t). This is the matrix with an entry in the ith row and jth column of $p_{ij}(s, t)$, as we had in the discrete case. We also have a corresponding matrix M in which the entry in the ith row and jth column is $\mu_{ij}(t)$. M is often referred to as the *intensity* matrix. We let P' denote the matrix in which each entry is the partial derivative with respect to t of the corresponding entry in P. The Kolmogorov forward equations together with the initial conditions can then be expressed as

$$P' = PM, \tag{19.11}$$

and

$$P = I \quad \text{for } s = t. \tag{19.12}$$

The following is a simple but useful consequence of the forward equations. It generalizes a result from our life-death model, when we have a state, such as that of being alive, which cannot be entered from any other state. That is, for all $j \neq i$ and $s < t$, we have $p_{ji}(s, t) = 0$. We will refer to such a state as being *anti-absorbing*.

In the following theorem, we adopt a formally weaker hypothesis involving forces rather than probabilities, which by the second statement turns out to be equivalent.

Theorem 19.3 *Let i be a state such that $\mu_{ji}(t) = 0$ for all $j \neq i$, and all t. Then, for $s < t$,*

$$p_{ii}(s,t) = e^{\int_s^t \mu_{ii}(r)dr},$$

and

$$p_{ji}(s,t) = 0, \text{ for } j \neq i.$$

Proof. From Theorem 19.2 we get

$$\frac{\partial}{\partial t} p_{ji}(s,t) = p_{ji}(s,t)\mu_{ii}(t), \tag{19.13}$$

a familiar differential equation which we first encountered in (8.15) and which is easily solved to give

$$p_{ji}(s,t) = K(s)e^{\int_s^t \mu_{ii} r \, dr}$$

for some function $K(s)$ independent of t, Since $p_{ii}(s,s) = 1$, and $p_{ji}(s,s) = 0$ for $j \neq i$, we must have $K(s) = 1$ if $j = i$ or 0 if $j \neq i$ and the conclusion follows. □

Another quantity of interest is the so-called *sojourn* probability for a state. We define

$$p_{\overline{ii}}(s,t) = P[X_r = i, s \leq r \leq t | X_s = i].$$

Note that the sojourn probability will in general differ from $p_{ii}(s,t)$, as the corresponding event requires the process to remain in state i for an entire interval of time, rather than just at the endpoints. The two probabilities will however be the same for an anti-absorbing state.

Since sojourn probabilities are featured prominently in the next section, we will attempt to find a way to evaluate them. Let

$$q_i(s,t) = p_{ii}(s,t) - p_{\overline{ii}}(s,t)$$

This is the probability that starting in state i at time s, the process is back in state i at time t, having left state i sometime during the interval. Let

$$v_i(t) = \lim_{h \to 0^+} \frac{q_i(t, t+h)}{h}.$$

Now we employ once again the technique of adding states. Given a process and a state i, we add a single new state i^* which is a clone of state i. Transitions out of state i^* follow the same pattern as transitions out of state i, while transitions that originally came into state i will be diverted to the cloned state i^*. Now to say that the new process is in state i at a time s and also at a later time t means that it must have stayed there throughout as it cannot re-enter. In addition, the forces of transition out of state i remain the same in both processes. So we have that

$$p^*_{ii}(s,t) = p_{\overline{ii}}(s,t), \quad \mu^*_{ij} = \mu_{ij},\ j \neq i, i*$$

We do have to consider the quantity μ^*_{ii*}. Now

$$p^*_{ii*}(s,t) = q_i(s,t),$$

since the new process will move from state i to state i^* precisely when the original process moves from state i back to state i having left state i at some point. From Definition 19.2,

$$\mu^*_{ii*}(t) = v_i(t).$$

Applying Theorem 19.3 to state i in the new process then leads to the formula

$$p_{\overline{ii}}(s,t) = e^{\int_s^t (\mu_{ii}(r) - v_i(r))dr}.$$

This formula does not however allow one to calculate the term on the left exactly since v_i also involves sojourn probabilities. All we can say without further assumptions is that

$$p_{\overline{ii}}(s,t) \leq e^{\int_s^t \mu_{ii}(r)dr}.$$

Suppose, however, that we have a process such that, for all states i and $t, h > 0$

$$q_i(t, t+h) = o(h). \tag{19.14}$$

This assumption says that it is highly unlikely that in a small time interval there is a move out of a state and back into it. It is the same idea that underlies a Poisson process as discussed in Section 18.3, where it was highly unlikely to have two occurrences of an event in a small time interval. When this assumption holds, we have that $v_i(t) = 0$ for all t and we can now state the following theorem.

Theorem 19.4 *For a process satisfying (19.14),*

$$p_{\overline{ii}}(s,t) = e^{\int_s^t \mu_{ii}(r)dr}.$$

Remark We can view the first statement in Theorem 19.3 as the special case of Theorem 19.4, when q_i is actually equal to 0.

An alternative direct proof of Theorem 19.4 is outlined in Exercise 19.18.

To summarize our conclusions in this section we have shown that given the forces of transition, we can in theory, solve a system of differential equations to arrive at the required probabilities that we are interested in for our applications. However, in practice, the solution can be difficult or impossible to obtain exactly. In the following subsections we look at various possibilities for obtaining or approximating the probabilities from the forces.

19.3.2 Path-by-path analysis

Assume then that we are given $\mu_{ij}(t)$ for all i, j and t, and we want to determine the probability that starting in state i at any time s, we will be in state j at time t. What we can do in a relatively straightforward manner is the following. Given any path going from state i to state j, we can derive a formula for the probability that starting in state i at time s, the process will be in state j at time t, having following the prescribed path exactly. Precisely, we are given a path π which is a sequence of states, $i_0, i_1, \dots, i_M$ where $i_0 = i$ and $i_M = j$. We then want the probability that the process will make a transition from state i to i_1 followed by a transition from i_1 to i_2, followed by a transition from i_2 to i_3 and so on, continuing in such a fashion, without visiting any other states, and finally reaching state j at or before time t, and remaining in state j until time t. Let us denote such a probability by $p^{\pi}_{i,j}(s, t)$. These can be calculated by integration, using only the forces and sojourn probabilities. So assuming that (19.14) holds, we have reduced the problem to one of evaluating integrals.

To illustrate, consider the simplest possible case of a path of length 1, that is $\pi = (i, j)$. Then arguing much the same as we did in the single life and multiple decrement models we can write

$$p^{\pi}_{i,j}(s,t) = \int_s^t p_{\overline{ii}}(s,r)\mu_{i,j}(r)p_{\overline{jj}}(r,t)dr.$$

The above formula parallels formula (11.13) except that in that case all non-active states are absorbing, so one cannot get out of them, which means that the third factor in the integrand does not appear as it is equal to 1. To recap, the formula above shows that in order to go from i to j along the one-step path, three events have to occur. First, the process must remain in state i until some time r between time s and time t. This event has probability $p_{\overline{ii}}(s, r)$. It cannot first go to another state and return, since that would constitute a different path from the given one. Secondly, there must be a transition to state j at time r. We can think of the probability of this as being $\mu_{i,j}(r)dr$. Finally, the process must remain in state j from time r to time t. It cannot leave state j and return as this would similarly constitute a different path. We get the required probability by integrating the product of these probabilities over all times r between s and t.

As the path length increases the formula gets more complicated. Consider a two-step path, $\pi = (i, k, j)$. Now we need a double integral. The formula is

$$p^{\pi}_{i,j}(s,t) = \int_s^t \int_{r_1}^t p_{\overline{ii}}(s,r_1)\mu_{ik}(r_1)p_{\overline{kk}}(r_1,r_2)\mu_{kj}(r_2)p_{\overline{jj}}(r_2,t)dr_2\, dr_1.$$

This looks extremely complex, but it simply records what are now five steps to meet the required conditions. The process must remain in state i until some time r_1, then transfer to state k, then remain in state k until some time r_2, then transfer to state j, and then finally remain in state j until time t.

A similar calculation that arises for a given path Π, is to find the probability that starting in state i at time s, the process will transfer to state j before time t having followed precisely the prescribed path. We get the same multiple integral as described above, except the final sojourn probability is omitted, since the process need only enter, but not necessarily remain in the final state on the path. Of course, when the final state is absorbing, the two problems are identical.

Example 19.6 Consider a heathy-unhealthy-deceased model as introduced in Figure 19.2. The forces of transition are constant with $\mu_{01} = 2$, $\mu_{02} = 1$, $\mu_{10} = 1$, $\mu_{12} = 3$. Find the probability that a healthy life will become unhealthy, and then die without recovering before time 1.

Solution. We have $\mu_{00} = -3$, $\mu_{11} = -4$. The path in question is $\pi = (0, 1, 2)$ and the required probability is

$$p_{02}^{\pi}(0, 1) = \int_0^1 \int_{r_1}^1 e^{-3r_1} 2e^{-4(r_2 - r_1)} 3dr_2 dr_1.$$

The double integral evaluates to

$$\frac{6}{4}\left[\frac{1 - e^{-3}}{3} - e^{-4}(e - 1)\right] = 0.4279.$$

In general, a path of length n will require an n-dimensional integral with an integrand consisting of $2n + 1$ factors, so this can quickly become computationally infeasible. Aside from that there is another difficulty. Usually, what we really want is $p_{ij}(s, t)$. In some cases this might be obtained by adding up $p_{i,j}^{\pi}(s, t)$ for all possible paths π from i to j. However, even in the simplest of models, there could well be an infinite number of such paths and this procedure will not work. This will normally occur whenever there is a positive probability of two-way transitions, as in the healthy-unhealthy-deceased model where healthy people can recover. A person could go from a healthy state to the deceased state, by getting sick and dying, or getting sick, then recovering, then getting sick again and dying, or recovering twice before death, or any of a number of infinite possibilities. We need other approaches to calculating transition probabilities, which we discuss in the next two subsections.

19.3.3 Numerical approximation

There are various ways to find a numerical approximation to the solution of the Kolmogorov equations. We will describe one approach that is fairly simple to implement. It is similar to Euler's method, described in Section 8.9, for Thiele's equation. For a fixed time s, the goal is to approximate the probabilities $p_{ij}(s, t)$. Pick a small interval of time h, and change our unit of time, so that each original time unit is now $1/h$ units. (So, for example, if our original time unit was a year, and $h = 1/12$, this changes the time unit to months.). We then essentially define a discrete time multi-state model, which will in generally be nonstationary. We will define one step transition probabilities for this model by simply looking at formulas (19.7) and (19.8), which formed the basis of the Kolmogorov equations, only omitting the $o(h)$ terms in each case. Since these terms becomes very small as h does, we can hope that for small enough h, we obtain a good approximation. We also switch the origin of time so that the original time s is now time 0 and the original time $s + mh$ for an integer m is now time m. We then take our approximating discrete non-stationary Markov chain to be that in which the transition probabilities, denoted by $\tilde{p}$, are given by

$$\tilde{p}_{ij}(m, m + 1) = h\mu_{ij}(s + mh), i \neq j,$$
$$\tilde{p}_{ii}(m) = h\mu_{ii}(s + mh) + 1.$$

We then obtain transition matrices ($\tilde{\mathbf{P}}_m$) which have for an (i,j)th entry, the quantity $\tilde{p}_{ij}(m, m+1.)$

Another way of looking at this is that we simply get the matrix ($\tilde{\mathbf{P}}_m$) by taking the intensity matrix $\mathbf{M}(s+mh)$, multiplying all non-diagonal entries by h and adjusting the diagonal entries so that the rows add to 1.

When $t = s + mh$ for some integer m we can then obtain approximations by

$$p_{ij}(s,t) \doteq \text{ the } (i,j)\text{th entry of } \tilde{\mathbf{P}}_0\tilde{\mathbf{P}}_1 \cdots \tilde{\mathbf{P}}_{m-1}.$$

Example 19.7 In a chain with three states, you are given that

$$\mu_{01}(t) = \mu_{10}(t) = 0.09(1+t),\ \mu_{02}(t) = 0.18(1+t),\ \mu_{12}(t) = 0.09,\ \mu_{20}(t) = \mu_{21}(t) = 0.$$

Find an approximation to $p_{01}(1,2)$ using the method described above with $h = 1/3$.

Solution. We have

$$\mathbf{M}_{(1)} = \begin{pmatrix} 0.46 & 0.18 & 0.36 \\ 0.18 & 073 & 0.09 \\ 0 & 0 & 1 \end{pmatrix}.$$

so that

$$\tilde{\mathbf{P}}_0 = \begin{pmatrix} 0.82 & 0.06 & 0.12 \\ 0.06 & 0.91 & 0.03 \\ 0 & 0 & 1 \end{pmatrix}$$

and similarly

$$\tilde{\mathbf{P}}_1 = \begin{pmatrix} 0.79 & 0.07 & 0.14 \\ 0.07 & 0.90 & 0.03 \\ 0 & 0 & 1 \end{pmatrix}, \quad \tilde{\mathbf{P}}_2 = \begin{pmatrix} 0.76 & 0.08 & 0.16 \\ 0.08 & 0.89 & 0.03 \\ 0 & 0 & 1 \end{pmatrix}.$$

Then the approximation to $p_{01}(1,2)$ is the entry in row 0 and column 1 of $\tilde{\mathbf{P}}_0\tilde{\mathbf{P}}_1\tilde{\mathbf{P}}_2$, which is 0.15131.

19.3.4 Stationary continuous time processes

One case where we can make some progress in solving the Kolmogorov equations directly is that of a *stationary* process as defined in Chapter 18. We then have that each $\mu_{ij}(t)$ is a constant, denoted by μ_{ij}. Moreover, the process will be determined by the quantities $p_{ij}(t) = p_{ij}(0,t)$ since the stationary condition implies that $p_{ij}(s,t) = p_{ij}(t-s)$.

As an example, we can write down completely the solution for the general two-state stationary process. Suppose the intensity matrix is

$$\mathbf{M} = \begin{pmatrix} -\mu & \mu \\ \nu & -\nu \end{pmatrix}.$$

Let $\rho(t) = e^{-t(\mu+\nu)}$. Then the transition matrix $\mathbf{P}(t)$, which has the entry of $p_{ij}(t)$ in the ith row and jth column, is given by

$$\mathbf{P}(t) = (\mu+\nu)^{-1}\begin{pmatrix} \mu\rho(t)+\nu & -\mu\rho(t)+\mu \\ -\nu\rho(t)+\nu & \nu\rho(t)+\mu \end{pmatrix} \tag{19.15}$$

We can verify that $\mathbf{P}$ satisfies equations (19.11) and (19.12) by direct calculation, noting that $d\rho(t)/dt) = -(\mu+\nu)\rho(t)$, and that $\rho(0) = 1$.

For the general stationary process, solutions to $\mathbf{P}' = \mathbf{PM}$ can be generated from *eigenvectors* of $\mathbf{M}$. These are non-zero vectors $\mathbf{a} = (a_0, a_1, \ldots, a_N)$ for which

$$\mathbf{aM} = \lambda\mathbf{a}$$

for some constant λ known as an *eigenvalue* of $\mathbf{M}$. (The left hand side above is a matrix multiplication in which we view $\mathbf{a}$ as a $1 \times N+1$ matrix. According to the convention we have used for the transition probabilities, the vector appears on the left.) To obtain eigenvectors, we first find the eigenvalues as those constants for which the determinant of $(\lambda\mathbf{I} - \mathbf{M}) = 0$. This involves solving a polynomial equation in λ. We can then solve for the resulting eigenvectors. Examples will follow.

Note now that given any such eigenvector $\mathbf{a}$ and any fixed i, the matrix $\mathbf{P}$ that has all zero entries except for an ith row of $(a_0e^{\lambda}t, a_1e^{\lambda}t \ldots a_Ne^{\lambda}t)$ is a solution to $\mathbf{P}' = \mathbf{PM}$. This follows since the ijth entry of $\mathbf{PM}$ will be $\lambda a_je^{\lambda t}$ which is the derivative of the corresponding entry in $\mathbf{P}$.

Note next that given any finite number of solutions to $\mathbf{P}' = \mathbf{PM}$, the linearity implies that any linear combination of these solutions will also be a solution. We therefore seek a linear combination that will satisfy the initial conditions. This can always be done in the case that we can find N linearly independent eigenvectors. We illustrate with a simple example.

Example 19.8 For the data as given in Example 19.6, find the probabilities of the following events.

(a) A life now disabled will be active at time 0.5.

(b) A life now active will be deceased at time 1.

Solution. The intensity matrix is

$$\mathbf{M} = \begin{pmatrix} -3 & 2 & 1 \\ 1 & -4 & 3 \\ 0 & 0 & 0 \end{pmatrix}.$$

We can see immediately that 0 is an eigenvalue with an eigenvector of $(0, 0, 1)$.

We can also note that since the rows of $\mathbf{P}$ add up to 1, the rows of $\mathbf{M}$ add to 0, and the last row of $\mathbf{P}$ is $(0, 0, 1)$, it is sufficient to solve the reduced system $\mathbf{P}^{r'} = \mathbf{P}^r\mathbf{M}^r$ in which the superscript r indicates a matrix with the row N and column N removed. So we can consider eigenvectors of

$$\mathbf{M}^r = \begin{pmatrix} -3 & 2 \\ 1 & -4 \end{pmatrix}.$$

The determinant of $\lambda\mathbf{I} - \mathbf{M}^r$ is $\lambda^2 + 7\lambda + 10$ giving eigenvalues of -5 and -2. Solving, we find respective eigenvectors of $(1, -2)$ and $(1, 1)$. (Note that eigenvectors are not unique as they can be multiplied by any non-zero constant.).

To satisfy the initial conditions, we want the first row of $\mathbf{P}^r$ to be the particular linear combination of the vectors $\mathbf{a} = e^{-5t}(1, -2)$ and $\mathbf{b} = e^{-2t}(1, 1)$ which gives the vector $(1, 0)$ when $t = 0$. We solve this to get a first row vector of $(1/3)\mathbf{a} + (2/3)\mathbf{b}$. For the second row, we want a linear combination that gives the vector (0,1) when $t = 0$. We solve this to get a second row vector equal to $(-1/3)\mathbf{a} + (1/3)\mathbf{b}$. Transition probabilities are then given by

$$p_{00}(0,t) = \frac{1}{3}e^{-5t} + \frac{2}{3}e^{-2t}, \quad p_{01}(0,t) = -\frac{2}{3}e^{-5t} + \frac{2}{3}e^{-2t},$$

$$p_{10}(0,t) = -\frac{1}{3}e^{-5t} + \frac{1}{3}e^{-2t}, \quad p_{11}(0,t) = \frac{2}{3}e^{-5t} + \frac{1}{3}e^{-2t}.$$

The answers to the particular questions asked are

(a) $p_{10}(0, 0.5) = 0.0953$.

(b) $1 - p_{00}(0, 1) - p_{01}(0, 1) = 0.8218$.
(The answer to part (b) is of course larger than that of Example 19.6 since here we allow for an arbitrary number of recoveries from being sick, as well as death from a healthy state.

19.3.5 Some methods for non-stationary processes

The above procedure can also be used solve the problem of determining probabilities from the forces, when these are 'piecewise constant'. We can find transition matrices for each time interval on which the forces are constant, and then use the discrete time technique. The following simple example illustrates the procedure. It involves only two time intervals but the method can easily be extended.

Example 19.9 For a two-state model, we have forces of transition given by

$$\mu_{01}(0,t) = \begin{cases} 1, & \text{if } 0 \le t < 2, \\ 2 & \text{if } 2 \le t < 3. \end{cases}$$

$$\mu_{10}(0,t) = \begin{cases} 2, & \text{if } 0 \le t < 2, \\ 3 & \text{if } 2 \le t < 3. \end{cases}$$

Find the probability that the process will be in state 1 at time 2.5 given that it is is state 0 at time 0.

Solution.

$$p_{01}(0, 2.5) = p_{00}(0, 2)p_{01}(2, 2.5) + p_{01}(0, 2)p_{11}(2, 2.5).$$

From the matrix (19.15), we have

$$p_{00}(0,2) = \frac{2+e^{-6}}{3}, \quad p_{01}(0,2) = \frac{1-e^{-6}}{3} \quad p_{01}(2,2.5) = \frac{2-2e^{-2.5}}{5},$$
$$p_{11}(2,2.5) = \frac{2+3e^{-2.5}}{5}$$

Substituting, the required probability is 0.39446.

One method for handling a perfectly general chain is to approximate it by a piecewise continuous one, and then use the method outlined above. To do so, one can choose a small time unit, and approximate the forces of transition by replacing them with those that are constant on each time interval, possibly using the midpoint value.

19.3.6 Extension of the common shock model

As mentioned in the introduction, multiple life theory, multiple decrement theory, or the more general model discussed in Chapter 17 can all be looked at in a multi-state framework. In the standard cases where failure times are independent, this does not normally provide any advantage over the conventional treatments that we have described previously. The multi-state approach, however, can be useful in modelling certain types of dependence. As an illustration, we will revisit the common shock model of Section 17.5 and discuss some possible refinements. We confine the analysis to the case of two types of failure, the first with failure time equal to $\min(T_1^*, Z)$ and the second with failure time equal to $\min(T_2^*, Z)$. Failure for the two types is dependent due to the common shock but there will be additional sources of dependence when T_1^* and T_2^* are themselves not independent. An example of this is the 'broken-heart' syndrome mentioned in Section 10.2 where one type of failure can hasten failure of the other type.

We consider a multi-state model with four states, as given in Figure 19.1, but adapted to cover general failure times. State 0 means that neither type of failure has occurred, state 1 means that only the second cause of failure has occurred, state 2 means that only the first type of failure has occurred, and state 3 means that both types of failure have occurred. (The states 1 and 2 then are labelled by the number of the surviving type.) Let $\mu_i(t)$ denote the hazard function for T_i^*, and $\rho(t)$ denote the hazard function for Z.

Suppose that forces of failure are of the form.

$$\mu_{01}(t) = \mu_2(t), \quad \mu_{02}(t) = \mu_1(t), \quad \mu_{03}(t) = \rho(t),$$
$$\mu_{13}(t) = \mu_1(t) + \rho(t) + \epsilon_1(t), \quad \mu_{23}(t) = \mu_2(t) + \rho(t) + \epsilon_2(t)$$

for some functions $\epsilon_1(t)$ and $\epsilon_2(t)$. When these are 0, we have precisely the common shock model. In this more general setting, we use $\epsilon_i(t)$ to build in some dependence between T_1^* and T_2^* by letting the distributions change after the first failure. The 'broken-heart syndrome' would call for positive values of $\epsilon_i(t)$. On the other hand, there may be cases where we assume that the prospects for failure type i improves after the first failure, and we would reflect this with a negative $\epsilon_i(t)$.

Probabilities in this model can all be calculated by the procedure of Section 19.3.2 since there are at most three paths between any two states. Moreover, since no state can be re-entered once the process leaves it, we can use Theorem 19.4 for the sojourn probabilities.

For the following examples we assume that all forces are *constant*.

Example 19.10 Calculate the probability that at time n, the first type of failure will have occurred but not the second.

Solution. From Theorem 19.1,

$$\mu_{00} = -(\mu_1 + \mu_2 + \rho), \quad \mu_{22} = -(\mu_2 + \rho + \epsilon_2).$$

We want

$$p_{02}(0, n) = \int_0^n p_{00}(s)\mu_{02}(s)p_{22}(s, n)ds = \mu_1 \int_0^n e^{-(\mu_1+\mu_2+\rho)s}e^{-(\mu_2+\rho+\epsilon_2)(n-s)}\, ds.$$

The integral is easily evaluated to give

$$p_{02}(0, n) = \mu_1 \left[\frac{e^{-(\mu_2+\rho+\epsilon_2)n} - e^{-(\mu_1+\mu_2+\rho)n}}{\mu_1 - \epsilon_2}\right]. \tag{19.16}$$

As a check, suppose that $\epsilon_1 = \epsilon_2 = \rho = 0$, so we simply have two independent exponential failure times. The answer reduces to

$$e^{-n\mu_2}(1 - e^{-n\mu_1}),$$

which is just the probability that the first type of failure occurred before time n multiplied by the probability that the second type did not occur before time n.

Example 19.11 Find the probability that the second cause of failure occurs before time n and after the first cause of failure.

Solution. If Π is the path (0,2,3) we want,

$$p_{03}^{\Pi}(0, n) = \int_0^n p_{02}(0, t)(\mu_2 + \rho + \epsilon_2)dt,$$

(a special case of formula (19.17) in the next section.) Substituting from (19.16) and integrating, the answer is

$$\frac{\mu_1(\mu_2 + \rho + \epsilon_2)}{\mu_1 - \epsilon_2}\left[\frac{1 - e^{-n(\mu_2+\rho+\epsilon_2)}}{\mu_2 + \rho + \epsilon_2} - \frac{1 - e^{-n(\mu_1+\mu_2+\rho)}}{\mu_1 + \mu_2 + \rho}\right]$$

When $\epsilon_2 = 0$ we recover the answer of Example 17.12 (a).

19.3.7 Insurance and annuity applications in continuous time

Consider an insurance contract that provides for two fixed states j and k, payments at the moment of transfer whenever a transfer occurs from state j to state k. The amount paid for a transfer at time t will be b_t (denoted by $b_t^{(jk)}$ if we wish to distinguish between several transfer benefits). Suppose the process is in state i at time 0. Then the actuarial present value parallels the result we saw in the multiple decrement case and is given by

$$\int_0^\infty b_t v(t) p_{ij}(0,t)\mu_{jk}(t)dt.$$

If benefits are to be paid on only the first transfer from state j to state k, then this can be handled by adding new states, exactly as we did in the discrete case.

Similarly, for a contract that provides payments made continuously at the periodic rate of c_t (written as $c_t^{(j)}$ if needed to distinguish states) at time t, provided the process is in state j the present value is given

$$\int_0^\infty c_t v(t) p_{ij}(0,t)dt.$$

Some annuities may make use of the sojourn probabilities. Suppose, for example, that at time 0, the process is in state i, and payments at the periodic rate of c_t are paid as long as the process remains in state i. All payments stop upon the first exit from state i. The present value is given by

$$\int_0^\infty c_t v(t) p_{\overline{ii}}(0,t)dt.$$

It is possible to fit the continuous time insurances and annuities into a stochastic model, and calculate variances as well as expected values, as we did in the discrete case. We will not pursue this here. In general, this will require integration of functions whose values are random variables rather than definite numbers. This presents both theoretical and computational difficulties.

More general probabilities can be deduced from the insurance formulas by the standard method we described in Section 10.8.4 of taking zero interest and benefit functions that take the value 1 or 0. For example suppose the process is in state i. The probability that within n years it will at some point make a transfer from state j to state k is given by

$$\int_0^n p_{ij}(0,t)\mu_{jk}(t)dt. \tag{19.17}$$

Example 19.12 An insurance contract based on the model in Example 19.8 provides for a benefit at time t of $e^{.04t}$ provided that a person now healthy dies while unhealthy. The force of interest is a constant 5%. Find the actuarial present value.

Solution. This is given directly by

$$\int_0^\infty e^{0.04t}e^{-0.05t}p_{01}(0,t)\mu_{12}(t)dt = 3\int_0^\infty e^{-.01t}\left(\frac{2e^{-2t}-2e^{-5t}}{3}\right)dt = 0.596.$$

The following is an example which can be thought of as a generalization of contingent insurances. Suppose we have a path of states $\Pi = i_0 i_1 \ldots i_M$ as described in Section 19.3.2. At time 0, the process is in state i_0 and an insurance contract provides a payment of 1 at the moment that the process enters state i_M, having previously followed precisely the path Π. For example, consider the chain covering the two lives (x) and (y) as given in the introduction and the path $\Pi = \{0, 1, 3\}$. The contract based on this path is the second death-contingent insurance paying upon the death of (x) if this occurs after the death of (y).

Example 19.13 Suppose that the force of interest and all forces of transition are constant. Find a formula for $\bar{A}_\Pi$, the actuarial present value of the insurance based on the path Π.

Solution. To simplify notation, let $v_j = \mu_{i_j i_{j+1}}$ and $\mu_j = -\mu_{i_j i_j}$ Then, following the explanation for the calculation of probabilities in section 19.3.2, we can set up the following multiple integral in which r_j denotes the time of transfer from state i_{j-1} to state i_j.

$$\bar{A}_\Pi = \int_0^\infty \int_{r_1}^\infty \cdots \int_{r_n}^\infty v_0 v_1 \ldots v_{M-1} e^{\mu_0 r_1} e^{\mu_1 (r_2 - r_1)} \ldots e^{\mu_M (r_N - r_{M-1})} e^{-\delta r_M} dr_N dr_{M-1} \cdots dr_1.$$

The integration can be carried out in a straightforward manner to give

$$\bar{A}_\Pi = \frac{v_0 v_1 \ldots v_M}{(\mu_0 + \delta)(\mu_1 + \delta) \cdots (\mu_{M-1} + \delta)}.$$

The formula is quite easy to remember and to compute from. The numerator is the product of all the forces of transition along the path. The denominator is a product of factors, one for each state on the path, except for the final one. Each such factor is the sum of all outgoing forces from that state plus δ. We can take $\delta = 0$ to give the probability that the path Π will be followed.

Note that Formula (8.28), Example 10.5 and Formula (10.25) with substitution from these first two are special cases of the above.

19.4 Recursion and differential equations for multi-state reserves

The basic idea of calculating reserves for multi-state contracts is the same as we have seen, except that the reserve must be calculated for each state, (an idea we have already encountered in Section 10.4.2). We let ${}_kV^{(j)}$ denote the reserve at time k when in state j. Given that the process is in a certain state, the reserve at any time is as usual, the present value of future benefits less the present value of future premiums. Recursion and differential equation formulas can become more complicated.

We start first by looking at the discrete time case, where we assume transfer benefits are payable at the end of the period of transfer.

Example 19.14 Consider the healthy-unhealthy-deceased model, and suppose that we have a stationary chain with transition matrix

$$\mathbf{P} = \begin{pmatrix} 0.6 & 0.2 & 0.2 \\ 0.2 & 0.6 & 0.2 \\ 0 & 0 & 1 \end{pmatrix}.$$

An insurance contract provides for payments of 2 when a healthy life becomes disabled, 3 when a healthy life dies, and 4 when a disabled life dies. All benefits are paid at the end of the year of the change of state. This is paid for by a single premium. Assume the interest rate $= 0$. Find the reserves at time k, for $k > 0$.

Solution. This can be done by recursion, but we will first do it in a more complicated way, so that the ease of the recursion formula method can be better appreciated. We could consider this as three separate contracts, but it is just as easy to do it all at once. In view of the stationarity, the reserve at time k for $k > 0$ will be independent of the time. It will however depend on the state at time k. Let $a = {}_kV^{(0)}$ and $b = {}_kV^{(1)}$. In view of the single premium, a will just equal the present value of future benefits given the state is 0. Therefore, summing over all possible transfer times j, we have

$$a = 2\sum_{j=0}^{\infty} p_{00}(0,j)p_{01} + 3\sum_{j=0}^{\infty} p_{00}(0,j)p_{02} + 4\sum_{j=0}^{\infty} p_{01}(0,j)p_{12}. \tag{19.18}$$

Now, in this case, the matrix $\mathbf{P}$ is of a particular simple form. Namely

$$\mathbf{P} = \begin{pmatrix} \frac{c+d}{2} & \frac{c-d}{2} & 1-c \\ \frac{c-d}{2} & \frac{c+d}{2} & 1-c \\ 0 & 0 & 1 \end{pmatrix}.$$

It is not hard to see that

$$\mathbf{P}^k = \begin{pmatrix} \frac{c^k+d^k}{2} & \frac{c^k-d^k}{2} & 1-c^k \\ \frac{c^k-d^k}{2} & \frac{c^k+d^k}{2} & 1-c^k \\ 0 & 0 & 1 \end{pmatrix},$$

In the present case, we have $c = 0.8, d = 0.4$. Substituting in (19.18), we easily sum the infinite geometric progressions to get

$$a = 2\frac{1}{2}\left(\frac{1}{0.2}+\frac{1}{0.6}\right)0.2 + 2\frac{1}{2}\left(\frac{1}{0.2}+\frac{1}{0.6}\right)0.2 + 4\frac{1}{2}\left(\frac{1}{0.2}-\frac{1}{0.6}\right)0.2 = \frac{14}{3},$$

We will leave it to the reader to verify similarly that $b = 13/3$.

We now look at general recursion formulas. Suppose we have a contract providing for each i, j, benefits of $b_k^{(ij)}$ at time $k+1$ providing there is a transfer from state i to state j between time k and time $k+1$. If we want to include expenses of payment, these can be incorporated into the transfer benefits. We can assume that for each state i, we have $b_k^{(ii)} = 0$, since payments for remaining in a state will be handled by annuity type benefits. We let $\pi_k^{(i)}$denote the *net* payment collected at time k assuming the process is in state i. These will be the premium less any annuity payments paid out and less any expenses if applicable. Then if the system is in state i at time k,

$$ {}_{k+1}V^{(i)} = \left({}_kV^{(i)} + \pi_k^{(i)}\right) - \sum_{j\neq i}(p_{ij}(k,k+1)\left(b_k^{(ij)} + {}_{k+1}V^{(j)} - {}_{k+1}V^{(i)}\right) \qquad (19.19) $$

Note that this is somewhat more complicated than previous recursion formulas, where the transfer was always to the deceased state, for which no reserves are required. In this more general case, upon transfer from state i to state j, we pay out the net amount at risk, and then also must set up the reserve required for state j, necessitating the extra term of ${}_{k+1}V^{(j)}$ above.

The following rearranged version of the recursion formula is instructive as well as being well suited to computation. For each state i,

$$ \left({}_kV^{(i)} + \pi_k^{(i)}\right)(1+i_k) = \sum_j p_{ij}(k,k+1)\left(b_k^{(ij)} + {}_{k+1}V^{(j)}\right). $$

This simply says that the amount accumulated at the end of a period must provide for all the benefits paid out at that time, as well as all the reserves that are needed in the various states.

Solving this general recursion to get reserves is not as straightforward as the cases we have previously looked at where reserves were all 0 except for one state. In general, we need to solve a system of linear equations rather than a single equation.

Example 19.15 Solve the previous example by recursion.

Solution. Substitute in (19.19) to get

$$ a = a - 0.2(2+b-a) - 0.2(3-a), \quad b = b - 0.2(a-b) - 0.2(4-b), \qquad (19.20) $$

which simplifies to

$$ 0.4a - .0.2b = 1, \quad -0.2a + 0.4b = 0.8, $$

and this is solved to give

$$ a = 14/3, \quad b = 13/3. $$

Consider a variation on the above where a level premium of π is paid as long as the person is healthy. We now need a π on the right hand side of the first equation in (19.20) and solving will lead to

$$ a = \frac{14-10\pi}{3} \quad b = \frac{13-5\pi}{3}. $$

Interestingly enough, we can now calculate the equivalence principle premium. In view of the stationarity, this must be the value of π which makes $a = 0$. This will be $\pi = 7/5$ and the corresponding reserve for the unhealthy lives will be 2. The value of $7/5$ can also be verified directly by dividing the benefit present value of $14/3$ by the value of an annuity of 1 per year while the life is healthy. This is calculated as in Example 19.19 by summing the appropriate geometric progression to give $(1/2)(1/.2 + 1/.6) = 10/3$. The disabled state reserve of 2 can always be verified since $2 = 0.6(2) + 0.2(4)$, showing that 2 will provide for the same reserve for those who remain disabled and a death benefit of 4 for those who die.

In the continuous case, we have Thiele's differential equations for the multi-state reserves. Consider a contract providing benefits, including accompanying expense, of $b_t^{(ij)}$ at time t for transfer from state i to state j. Net payments are are collected continuously, where the periodic rate of payment at time t is $\pi_t^{(i)}$ when the system is in state i. We then have a system of differential equations, one for each state i, as follows.

$$\frac{d}{dt}{}_tV^{(i)} = \delta_t\, {}_tV^{(i)} + \pi_t^{(i)} - \sum_{j\neq i} \mu_{ij}(t)\left(b_t^{(ij)} + {}_tV^{(j)} - {}_tV^{(i)}\right)$$

which we can also write as

$$\frac{d}{dt}{}_tV^{(i)} = \delta_t\, {}_tV^{(i)} + \pi_t^{(i)} - \sum_{j} \mu_{ij}(t)\left(b_t^{(ij)} + {}_tV^{(j)}\right), \qquad (19.21)$$

by invoking Theorem 19.1.

19.5 Profit testing in multi-state models

We now revisit Section 12.4.3 and adapt it to the multi-state case. As we did in that section, we will include notation for all the expenses, rather than incorporating them into other symbols. Suppose we have a general insurance contract based on a discrete time multi-state model providing for benefits of $b_k^{(ij)}$ at time $k+1$ for transfer from state i to state j between time k and $k+1$ and annuity benefits of $c_k^{(i)}$ paid at time k when in state i. We have a profit test basis at time k consisting of premiums $\pi_k^{(i)}$, percentage of premium expenses of $r_k^{(i)}$, periodic expenses of $e_k^{(i)}$, all paid at time k when in state i, and in addition a transition matrix $\mathbf{P}_k$, interest rates i_k, and transfer expenses of $e_k^{(ij)}$ paid at time $k+1$ for a transfer from state i to state j between time k and time $k+1$. The profit testing procedure will conform precisely to the principles outlined in the single-life case. The profit for the period running from time k to $k+1$, for a policy in state i at time k is given by

$$\mathrm{Pr}_{k+1}^{(i)} = \left({}_kV^{(i)} + \pi_k^{(i)} - c_k^{(i)} - e_k^{(i)}\right)(1+i_k) - \sum_j (p_{ij}(k,k+1)\left[(b_k^{(ij)} + e_k^{(ij)} + {}_{k+1}V^{(j)}\right]$$

Note that this is just a straightforward generalization of the formula (12.4b) with the modification that we must provide reserves for the states transferred into, as noted in the Section 19.4.

Example 19.16 An insurance policy issued to a healthy life provides for death benefits at the end of the year of death of 100 000 if death occurs while healthy or 70 000 if death occurs while unhealthy. In addition there are yearly payments of 30 000 provided the insured is unhealthy. You are given reserves as follows, with state 0 being healthy and state 1 being unhealthy.

$$ {}_5V^{(0)} = 25\,000, \quad {}_5V^{(1)} = 75\,000, \quad {}_6V^{(0)} = 30\,000 \quad {}_6V^{(1)} = 60\,000 $$

The profit test basis has annual premium of 15,000 paid while healthy, an interest rate of 10%, periodic expenses paid each year of 200 if healthy or 400 if unhealthy, a transition matrix for the year running from time 5 to time 6 as the matrix $\mathbf{P}_0$ of Example 19.1, and expenses for paying a death claim of 1% of the face amount. Find the profit Pr_6 both when the policyholder is healthy at the end of 5 years and when the policy holder is unhealthy at the end of 5 years.

Solution.

$$ \mathrm{Pr}_6^{(0)} = (25\,000 + 15\,000 - 200)(1.1) - 0.7(30\,000) - 0.2(60\,000) - 0.1(100\,000(1.01) = 680. $$

$$ \mathrm{Pr}_6^{(1)} = (75\,000 - 30\,000 - 400)(1.1) - 0.2(30\,000) - 0.6(60\,000) - 0.2(70\,000(1.01) = -7080 $$

19.6 Semi-Markov models

In this section, we discuss briefly the problem that arises when the Markov condition does not hold. This is a frequent occurrence, since this will happen whenever the probabilities of movement from one state to another depend on the length of time elapsed since entering the current state. These are known as *semi-Markov models*. A typical example is found in the healthy-unhealthy-deceased model. The likelihood of either recovery or death for a unhealthy person will certainly depend on how long they have been in the unhealthy state, a fact which was not considered in our previous models.

We will not discuss these in detail. One method of approach is to approximate the semi-Markov model by a Markov chain, by dividing states up into several sub-states. For example, in place of a single state h for being unhealthy we could have states $h_1, h_2, \ldots, h_k$ where the only transitions with positive probability between this collection of states is from h_i to state h_{i+1}. Therefore, a higher subscript is indicative of a longer period of being unhealthy. The probabilities of transition to both recovery and death could then differ between the sub-states.

Notes and references

For a more detailed account of multi-state theory than we have presented here, see Norberg (2008).

Transition probabilities, which involve four variables, can be written in a myriad of ways, and various choices of notation are found in different works. In much of the actuarial literature, the standard form of notation is maintained. The basic symbol used is ${}_tp_x^{ij}$ to refer to the probability that a life now age x and in state i will be in state j at time t. In the notation of this chapter, we would consider the age (x) fixed and write the above probability as $p_{ij}(0, t)$, following the typical usage found in much of the probability theory literature. The actuarial notation has the advantage of following classical actuarial usage for probabilities, but seems restricted to models which refer to a body of lives, and not directly applicable to more general situations.

See Jones (1997) for an interesting application of multi-state models.

Exercises

19.1 Redo Example 19.2, but assume now that in the first year, the probability that a healthy person will become ill is 0.3 rather than 0.2, and the probability that they will remain healthy is 0.6 rather than 0.7.

19.2 Consider a Markov chain with transition matrices for the first three time periods given by

$$\mathbf{P}_0 = \begin{pmatrix} 1/3 & 1/3 & 1/3 \\ 1/3 & 1/3 & 1/3 \\ 1/3 & 1/3 & 1/3 \end{pmatrix}, \quad \mathbf{P}_1 = \begin{pmatrix} 1/2 & 1/2 & 0 \\ 0 & 1/2 & 1/2 \\ 1/2 & 1/2 & 0 \end{pmatrix}, \quad \mathbf{P}_2 = \begin{pmatrix} 1/2 & 1/4 & 1/4 \\ 1/2 & 1/4 & 1/4 \\ 1/2 & 1/4 & 1/4 \end{pmatrix}.$$

Interest rates are given by $i_0 = 0.05$, $i_1 = 0.06$, $i_2 = 0.07$. Find the APV of each of the following contracts.

(a) At time 0, the process is in state 0. A contract provides for payments at the end of the period of a transfer from state 0 to state 1 if this occurs within three periods. The payment is 100 in the first year, 200 in the second year, and 300 in the third year. Note that more than one payment can be made.

(b) At time 0 the process is in state 0. A contract provides for a payment of 1000 at time k, $k = 0, 1, 2, 3$, provided the process is in state 0 at time k.

19.3 A transition matrix for a three-state homogeneous Markov chain is given by

$$\mathbf{P} = \begin{pmatrix} 0.4 & 0.3 & 0.3 \\ 0.7 & 0.2 & 0.1 \\ 0.2 & 0.7 & 0.1 \end{pmatrix}$$

A process starts in state 0. Annuity contract 1 provides for periodic payments, provided the process is in state 1. Annuity contract 2 provides for periodic payments which starts at the end of the period in which process enters state 2, but stop completely upon exit from state 2 and do not begin again, even upon subsequent return. For each contract, find the probability that a payment is made at time 3.

19.4 A transition matrix for a four-state homogeneous Markov chain is given by

$$\mathbf{P} = \begin{pmatrix} 0.1 & 0.2 & 0.3 & 0.4 \\ 0.3 & 0.1 & 0.3 & 0.3 \\ 0.5 & 0.2 & 0.1 & 0.2 \\ 0.4 & 0.3 & 0.2 & 0.1 \end{pmatrix}.$$

At time 0, the process is in state 0. A contract provides for payments at the end of the year of the *first* transition to state 3 from any other state, provide this occurs within 4 years. The amount of the payment is 100 if the transition is from state 0, 200 if the transition is from state 1, or 300 if the transition is from state 2. Level net annual premiums are paid beginning at time 0, and continuing until the time of transition to state 3. The interest rate is a constant 5%.

(a) Find the annual premium.

(b) Find the reserve at time 2, assuming the process is at that time in (i) state 0, (ii) state 1, (iii) state 2.

19.5 Verify that the Kolmogorov forward equations with $s = 0$ give Equation (8.15) in the case of a single life, or Equation (11.10) in the multiple decrement model.

19.6 Derive the Kolmogorov backwards equations. For times $s \leq t$,

$$\frac{\partial}{\partial s} p_{i,j}(s,t) = -\sum_{k=1}^{N} \mu_{i,k}(s) p_{k,j}(s,t).$$

19.7 Redo Examples 19.8 and 19.12 now given that $\mu_{01} = 1$, $\mu_{02} = 3$, $\mu_{10} = 2$, $\mu_{12} = 3$.

19.8 For a two-state chain, derive the matrix (19.15) directly by using eigenvalues and eigenvectors.

19.9 In a two-state model, the forces of transition are given by $\mu_{01}(0,t) = 1$ for all t while

$$\mu_{10}(0,t) = \begin{cases} 1 & \text{if } 0 \leq t < 1, \\ 3 & \text{if } 1 \leq t. \end{cases}$$

Find the probability that the process will be in state 0 at time 2.4 given that it is in state 1 at time 1.4.

19.10 A stationary Markov chain has the following transition matrix.

$$\mathbf{P} = \begin{pmatrix} 0.6 & 0.1 & 0.3 \\ 0.2 & 0.5 & 0.3 \\ 0.1 & 0.3 & 0.6 \end{pmatrix}.$$

Interest is a constant 5%. Find the APV of the following contracts based on this chain, both issued when the process is in state 0.

(a) An temporary annuity provides 100 per year, beginning when the process enters state 1 and stopping when the process leaves state 1. Payments do not resume upon re-entry to state 1. The last possible payment is at the end of 4 years.

(b) A 7-year term insurance pays 100 at the end of the year of the *first* transfer from state 0 to state 2, provided that that prior to this time there was a transfer from state 1 to state 0. If a transfer from state 0 to state 2 occurs before a transfer from state 1 to state 0, then nothing is paid.

19.11 Consider the chain of Figure 19.1. Suppose that (x) and y are two independent lives, where x is subject to a constant force of mortality a and (y) is subject to a constant force of mortality b.

(a) Write down the reduced intensity matrix M^r, (as defined in Example 19.8).

(b) Write down the reduced transition matrix $P^r(t)$ directly from Chapter 10 formulas.

(c) Find the eigenvalues and eigenvectors of M^r.

(d) Verify that the rows of $P^r(t)$ are linear combinations of vectors of the form $e^{\lambda t}(a_1, a_2, a_3)$ where λ is an eigenvalue of M^r and (a_1, a_2, a_3) is the corresponding eigenvector.

19.12 Consider the chain of Figure 19.1. We do not necessarily assume independence. You are given the constant intensities $\mu_{01} = b$, $\mu_{02} = a$, $\mu_{13} = c$, $\mu_{23} = d$.

(a) Given any positive u, v, find the probability that (x) survives u years and (y) survives v years, in terms of a, b, c, d.

(b) Show that $T(x)$ and $T(y)$ are independent if and only $a = c$ and $b = d$.

19.13 Consider a multi-state model for the three lives $(x), (y), (z)$ where we have eight states described below, where the stipulated lives are those still living,

State 0: all; State 1: $(x), (y)$ only; State 2: $(x), (z)$ only; State 3: $(y), (z)$ only; State 4: (x) only; State 5: (y) only; State 6: (z) only; State 7: None.

You are given that $\mu_{01} = 0.02$, $\mu_{02} = 0.01$, $\mu_{03} = 0.04$, $\mu_{34} = 0.02$, $\mu_{35} = 0.06$, $\mu_{36} = 0.05$, $\mu_{67} = 0.10$ and the force of interest is a constant 0.05.

(a) Find the actuarial present value of an insurance policy on the three lives which pays 1 at the moment of death of (z) provided that (x) dies first and (y) dies second.

(b) Find the probability that the lives die in the order $(x), (y), (z)$.

19.14 A contract is based on a three-state chain. There are benefits at the end of the year of transfer, of 100 for transfer from state 0 to state 2, and 50 for transfer from state 1 to state 2. Annual premiums of 10 are paid while the process is in state 0. There are annuity payments of 5 made each year when the process is in state 1. You are given that $i_5 = 0.10$ and the transition matrix

$$\mathbf{P}_5 = \begin{pmatrix} 0.6 & 0.1 & 0.3 \\ 0.2 & 0.7 & 0.1 \\ 0.2 & 0 & 0.8 \end{pmatrix}.$$

If

$${}_5V^{(0)} = 50, \quad {}_5V^{(1)} = 35 \quad {}_5V^{(2)} = 70,$$

find ${}_6V^{(i)}$, for $i = 0, 1, 2$.

19.15 Redo Example 19.16 with the following changes in the given data:

(i) The transition matrix for the year running from time 5 to time 6 is now given by the matrix $\mathbf{P}_1$ of Example 19.1.

(ii) There is a payment at the end of the year of 20 000 when a healthy life becomes unhealthy, and the yearly income for unhealthy lives is reduced to 15 000.

19.16 A insurance on two independent lives (x) and (y) provides for a death benefit at the moment of the second death. The death benefit is $e^{0.05t}$ for death at time t. Premiums are payable until the second death, The premium is level while both are alive and reduces to one half of the initial amount upon the first death. The life (x) is subject to a constant force of mortality of 0.10 and (y) is subject to a constant force of mortality of 0.15. The force of interest is a constant 0.10. Refer to the two-life model as given Figure 19.1.

(a) Using the method in Chapter 10, calculate for each of the first three states the annual premium and the reserve at time t.

(b) Write down the Thiele differential equations.

(c) Verify that your answers to (a) satisfy the equations in (b).

19.17 In a certain four state chain, the forces of transition are given by

$$\mu_{01}(t) = 0.1 + 0.02t, \quad \mu_{12} = 0.2 + 0.01t, \quad \mu_{03}(t) = 0.1 + 0.03t, \quad \mu_{13} = 0.2 + 0.02t.$$

For all other i, j with $i \neq j$, $\mu_{ij}(t) = 0$. The force of interest is a constant 0.06.

The following double integral represents the net single premium for a certain insurance contract.

$$\int_0^3 \int_0^t e^{-(0.6+0.09t+0.02s)}(0.1 + 0.02s)(0.2 + 0.01t)ds\, dt.$$

Describe the benefits on the contract.

19.18 Prove Theorem 19.4 by observing first that

$$p_{\overline{ii}}(s, t + h) = p_{\overline{ii}}(s, t)\, p_{\overline{ii}}(t, t + h).$$

20

Introduction to the Mathematics of Financial Markets

20.1 Introduction

This chapter will introduce some basic concepts of modern mathematical finance. One goal is to cover the fundamentals of option pricing. This has become an important tool in actuarial mathematics, since many insurance and annuity contracts today contain the so-called 'embedded options', which we discussed in Section 13.2. For the most part, we carry this out in a discrete setting, but we do move into the continuous-time approach briefly in order to introduce the Black–Scholes–Merton formula. Another major objective in this chapter is to revisit the basic quantity of a discount function which we introduced early on. In the first part of the book, we treated this as a deterministic function, but a more realistic approach would be to consider $v(s, t)$ as a random variable, reflecting the stochastic nature of investment returns that we discussed in Chapter 14. In particular, we seek a version of the key identity, Formula (2.1) in this stochastic setting.

A prerequisite for this chapter is the starred Section 2.12. We assume familiarity with concepts discussed in that section such as as short selling, forward contracts, and arbitrage.

20.2 Modelling prices in financial markets

A *financial market* is an institution designed to facilitate the trading of financial assets, such as stocks or bonds. Certain individuals wish to buy such assets, while others wish to sell them, and the financial market provides a forum to bring together the various parties. The potential holder of an asset quotes a desired selling price, known as the *asking price*, while the potential buyer quotes a desired buying price, known as the *bidding price*. When bidding and asking prices are equal it establishes a price for the asset, and a sale can be made. Our goal in this section is to develop a stochastic model for the evolution of prices.

Fundamentals of Actuarial Mathematics, Third Edition. S. David Promislow.

Companion website: http://www.wiley.com/go/actuarial

For most of this chapter, we assume a relatively simple framework, which will allow us to present the main ideas without too many technical mathematical difficulties. We assume a discrete-time model. That is, trading of assets will take place at integer times $0, 1, 2, \ldots$. The time period can be arbitrary, and we can think of it as possibly a very short time, (say an hour or even a minute) which could then constitute an approximation to the more realistic continuous-time setting. We adopt a finite-time horizon, with time N as the last date we are interested in. Finally, we assume that the price of any asset can take on only finitely many possible values.

Suppose we have $M + 1$ assets traded in our market, numbered from 0 to M. We will let

$$S_j(n) = \text{the price of the } j\text{-th asset at time } n.$$

We consider each such price as a random variable. Therefore, our financial market is modelled by $M + 1$ discrete time stochastic processes $S_j(n)$, where $j = 0, 1, 2, \ldots M$, and $n = 0, 1, \ldots N$.

We will single out a particular asset, often referred to as a *bank account*, for asset numbered 0. To describe this, we will first need to postulate in our model a nonnegative quantity r called *the risk-free rate of interest*, which is the interest rate that we can obtain on a risk-free investment as described in Section 2.12.

We can then define this asset by

$$S_0(0) = 1, \quad S_0(n) = (1 + r)^n.$$

In other words, this is an asset which accumulates at the risk-free interest rate. It is a stochastic process in which each random variable takes a single value with probability 1. For simplicity, we are at first adopting a constant risk-free interest rate. The definition could be based on a more general discount function and we comment on this below in Section 20.13.

We will make the same idealized assumptions that we made in Section 2.12. That is, we postulate that for each asset, any real number of the units can be bought at any trading date. Through short selling if necessary, this includes negative quantities. We also assume that there are no transaction costs such as commissions.

Note that the existence of the bank account means that we are assuming that all participants in our market can freely borrow at the risk-free rate.

Another simplifying assumption made throughout is that none of our assets provide any payments at intermediate dates, such as dividends on stocks or coupons on bonds. They provide funds only upon sale or maturity.

20.3 Arbitrage

An initial observation is that in the typical financial market, the various asset prices do not move independently. If asset i moves up in price, asset j may have a tendency to move up, or possibly to move down, or be certain to move up or down. It can be quite complicated to model all dependencies, but the *no-arbitrage* principle will often enable us to reduce the possibilities. In our stochastic models, this requires a more complicated definition than the one we gave in Section 2.12.

We first define the concept of a *trading strategy.* This is roughly a description you, as an investor in the market, would give to an assistant before leaving for a holiday on a remote desert island where you cannot be reached. You would specify the number of units of each asset to be held at each time period. At each trading date after the initial portfolio is established at time 0, certain assets in the existing portfolio would be sold and others bought to achieve the stipulated amounts. These amounts could depend on the entire past and present history. The description could be enormously complicated or quite simple. For example, a trading strategy might be as follows:

Start with an initial portfolio of 1 unit each of asset 1 and asset 2. Keep this intact until the first time that the price of asset 1 is above 40 per unit and the price of asset 2 is below 30 per unit. At that time, sell all units of asset 1 and use the proceeds to buy shares of asset 2. These are then held without further trading.

Note that the amounts to be held at time n can depend on all the prices of all assets at or before time n, but not after. It would not be a feasible trading strategy to specify that a a certain asset should be sold at time 2, if the price of some other asset were below 40 at time 3.

To formalize this somewhat, we can represent the asset holdings at any time n by a vector

$$\alpha(n) = (\alpha_0(n), \alpha_1(n) \dots \alpha_M(n))$$

where $\alpha_j(n)$ is the number of units of asset j held at time n.

The entries of this vector are random, depending on the prices up to time n. So a trading strategy $\mathcal{T}$ is formally a vector of these random vectors.

$$\mathcal{T} = (\alpha(0), \alpha(1) \dots \alpha(N-1)),$$

where each $\alpha(r)$ is a function of the values of $S_j(k)$ for $j = 0, 1, \dots M,\ k = 0, 1, 2 \dots r$.

For any trading strategy and any time n, we will have a portfolio consisting of a certain number of units of each of our $M + 1$ assets. The portfolio at time n will then then have a value $V(n)$ obtained by multiplying the number of units of each asset by the price of that asset at time n, and summing. That is

$$V(n) = \sum_{j=0}^{M} \alpha_j(n) S_j(n),$$

a random variable depending on all prices as well as the trading strategy as followed up to time n. Of course $V(0)$ is a definite number as it is the cost of setting up the initial portfolio at time 0 when all prices are known.

For any trading strategy, there is a reverse strategy which involves holding at each time, the negative of the number of units held in the original strategy. In other words, one sells in place of buying and buys in place of selling. Formally, if a trading strategy is given by the the vector $\mathcal{T}$, the reverse strategy is given by $-\mathcal{T}$. If V^* denotes values for the reverse strategy it is clear that $V^*(n) = -V(n)$ for all n.

Here is another important concept.

Definition 20.1 A trading strategy is said to be *self-financing* if for any trading date after time 0 and before time N, the total price of all the assets sold on a given trading date exactly equals the total price of all the assets bought on that date, so no additional infusion or withdrawal of capital is required.

For a self-financing strategy, the value of the portfolio at any intermediate trading date is the same before and after trading.

We can now summarize the procedure we will be following in subsequent discussions. We set up an initial portfolio at time 0 for a cost of $V(0)$, dictate a self-financing trading strategy, retreat to the dessert island, where no additional outlays of cash are required and none are received. Finally, the portfolio is liquidated at time N for proceeds of $V(N)$, a random variable which depends on both the trading strategy and the evolution of prices. We let P denote the probability measure for $V(N)$.

The key definition of this section can now be given in terms of the starting value $V(0)$ and the ending value $V(N)$.

Definition 20.2 The financial market admits *arbitrage* if there exists a self-financing trading strategy such that

$$V(0) = 0, \quad V(N) \geq 0, \text{ and } \quad P[V(N) > 0] > 0.$$

A financial market which does not admit arbitrage is said to be *arbitrage-free*.

In other words, an arbitrage opportunity is one where starting with a zero investment, we cannot possibly lose by the end of the trading period, and we have at least some chance of making a gain. Note that the arbitrage opportunity does not guarantee a positive gain. One can think of it as being given a lottery ticket for free. We cannot lose anything, and there is some chance of profiting. It is important to note that cases where there is a very small probability of loss do not constitute an arbitrage under this definition. The avoidance of the loss must be absolutely certain.

Remark In the definition of arbitrage, we could replace the condition on $V(N)$ by $V(N) \leq 0$ and $P[V(N) < 0] > 0$, since, if this holds, the reverse strategy will satisfy the original condition. This looks a bit strange at first, but it simply says that if there is a strategy for which we are sure not to gain, then the reverse strategy is sure not to lose.

An important consequence of the above is the following.

Theorem 20.1 *In an arbitrage-free financial market, if there is a self-financing trading strategy for which $V(n)$ is a constant c for some n, then*

$$c = V(0)(1 + r)^n.$$

Proof. Modify the strategy by holding $-V(0)$ units of the bank account at time 0, so that the new strategy has initial value 0. If necessary, modify the strategy further to stipulate that everything should be settled at time n, and the proceeds (possibly negative) left to accumulate in the bank account at the risk-free rate until time N. The new strategy will have the constant value of $c - V(0)(1 + r)^n$ at time n, and this must be equal to 0. If not, there would be a sure chance of having either a positive or negative amount at time N, which would imply an arbitrage opportunity, by Definition 20.2 and the remark following this definition. □

This was a reasonably simple result, but there is an important message behind it. It says that in the absence of arbitrage, if we can find a self-financing trading strategy which eliminates risk at some point, then our initial investment must accumulate at the risk-free rate up to that point.

Remark It is true that any trading strategy can be converted into a self-financing one by using the bank account. An excess of the sales over purchases can be placed in the bank account, while excesses of purchases over sales can be handled by borrowing. However this gives a different strategy with a different amount held in the bank account at time N, and therefore a different value of $V(N)$. The self-financing hypothesis is therefore essential in the definition of arbitrage.

For our first example, we consider a very simple financial market. We will take $N = 1$, so a trading strategy involves simply specifying the initial portfolio. Our financial market has, in addition to S_0, a single risky asset S_1 consisting of a stock. We can assume, changing units if necessary, that the price of a unit of the stock is 1 at time 0. Suppose that the price of the stock at time 1 can only take two possible values, u or d (standing for 'up', 'down' respectively) with $d < u$, each with positive probability. We call this a *binomial* model to reflect the two possible values at time 1.

Theorem 20.2 *The above financial market is arbitrage free if and only if*

$$d < (1 + r) < u. \tag{20.1}$$

Proof. Consider any trading strategy with $V(0) = 0$. If α is the number of units of stock in the initial portfolio. we must have $-\alpha$ units of the bank account. Then we will have either $V(1) = \alpha u - \alpha(1 + r)$ or $V(1) = \alpha d - \alpha(1 + r)$. Suppose (20.1) holds. If $\alpha < 0$ then the first such value will be negative and the second will be positive, while the reverse holds if $\alpha > 0$. If $\alpha = 0$, both values are 0. An arbitrage opportunity cannot exist.

Conversely if (20.1) is not true, then at least one of two possibilities holds. Suppose $d \geq (1 + r)$. We create an arbitrage opportunity by choosing $\alpha > 0$ which makes both values of $V(1)$ nonnegative,with at least one positive. The other possibility is that $u \leq (1 + r)$, in which case we similarly create an arbitrage opportunity by taking $\alpha < 0$. □

Note that the converse statement is intuitively obvious. If the inequality does not hold, then we can create an arbitrage opportunity by either buying a stock which is sure to yield more than the risk-free return, or short selling a stock which is sure to yield less than the risk-free return.

Another pertinent fact to notice in the definition of arbitrage is that the condition does not depend on the particular values of P but only on whether such values are positive or zero. We are therefore led to make use of the following standard definition of probability theory.

Definition 20.3 Two probability measures P and Q on a sample space S are said to be *equivalent* if for all $A \subseteq S$, we have $P(A) = 0$ if and only if $Q(A) = 0$. (For readers familiar with the concept of an equivalence relation, one can readily verify that this is a legitimate such relation.)

It is clear from the definition that a financial market is arbitrage-free with respect to P if and only if it is arbitrage-free with respect to any equivalent probability measure Q.

20.4 Option contracts

Given a financial market, we can do more than just buy or sell the existing assets. We have already seen one possibility, which is to enter into forward contracts. Another possibility is *option contracts*. These are in one sense similar to forward contracts since they are both transactions which involve trading of assets at a future date for prices that are specified now. There are major differences however, for in an option, unlike the forward contract, one party is not obligated to complete the transaction, but has the option to do so, and will only exercise this option if it is advantageous. There are two basic types of option contracts, known as *calls* and *puts*. A buyer of a call option has the right to buy a specified asset at a specified future time, known as the *expiration date* or *exercise date*, for a specified price, known as the *strike price* or *exercise price*, if they should choose to do so. The call option buyer has a similar motivation to a speculator taking a long position in a forward contract. They hope for a rise in price, so that they can buy the asset at a price which is lower than prevailing at the time of purchase. If the price of the asset at the expiration date is below the strike price, the option will not be exercised. A buyer of a put option has the right to sell a specified asset on the expiration date for a specified strike price. The put option buyer has a similar motivation to the speculator taking a short position in a forward contract. They hope for a fall in price so that they can sell the asset for more than it is worth at the time of sale. In this case, if the price of the asset at the expiration date is above the strike price, the option will not be exercised. Unlike the forward contract, the call and put buyers are not on opposite sides. For each of them, there must be another party who sells or (as it is commonly said) writes the option, and agrees to complete the transaction should the option holder so elect. Now if the option is exercised, the option writer is necessarily selling or buying at an unfavourable price, and they are compensated for this by the option price which they receive from the option holder at the time the agreement is entered into. The option writers of course hope that options will not be exercised, so they profit by the full amount of the option price, and do not have to engage in an unfavourable transaction. Determining option prices is complicated, and will form much of the material of this chapter.

It should be noted that what we have described are more properly known as *European* options, which specify that the option can only be exercised on the one specific expiration date. We will assume all options we discuss are of this nature unless specified otherwise. Another type of contract, known as an *American* option, allows for the exercise of the option at any time before or on the expiration date. These are more complicated and will be dealt with briefly in Section 20.7.

Although the underlying assets for calls and put are normally taken as financial instruments like stocks, as will be the case in our treatment, the basic idea of an option arises in many diverse contexts. For example, buying insurance on an asset like a house, is essentially buying a type of put option. You are protecting yourself from a drop in value, not from market variation in this case, but rather from physical damage. Similarly, the guarantees for variables annuities (as discussed in Section 13.2) which protect your account against unfavourable investment experience constitute put options. For another example, suppose that you take out a long-term loan or mortgage, and the lender gives you the right to repay in full at any time without penalty. In effect, you have been given a call option. In this case, you are protected from a rise in the cost of repayment, which will occur if interest rates decline. (Refer to the discussion in Section 2.10.3.)

In essence, protection against declines in the value of an asset that you own, while allowing you the full benefit of increases in value, can be viewed as being given a put option. Protection against increases in the value of an asset that you may wish to acquire in the future, while allowing you the full benefit of decreases in value, can be viewed as being given a call option. Note that in contrast, forward contracts protect you from unfavourable declines or increases, but they do not allow the parties to reap the full upside benefits, since the transactions must be completed with the agreed upon prices.

20.5 Option prices in the one-period binomial model

In this section, we show how the no-arbitrage principle allows us to calculate an option price for the one-period binomial market where condition (20.1) holds. We illustrate with a particular example. Suppose that a share of the stock is selling for 108 at time 0 and at time 1 it will be either 132 or 99, each with positive probability. Assume a risk-free interest rate of 10%.

Consider a call option on the stock with an expiration date of time 1 and a strike price of 110. What should the price per unit of this option be? At first, one may think that there is no way to determine this exactly, and that it could take on many possible values. After all, the option is just another asset with its price being determined by the amounts bid and asked by the various market participants. The worth of this asset, however, is directly tied to the performance of the stock, so it should be clear that its price must be related in some way to the stock price. Such an asset is often termed a *derivative security*, since its value is derived from that of another security.

To help determine the price, we take the following point of view. Purchasers of call options are not normally interested in actually taking possession of the stock at maturity. They simply want to buy it at the strike price, and sell it immediately for the higher market price if available. If the market price is below the strike price, the option is worthless and they receive nothing. The option then is just another asset S_2 with $S_2(1) = 132 - 110 = 22$ if the stock price goes up, or $S_2(1) = 0$ if the stock price goes down. The problem is to determine $S_2(0)$.

Those well-versed in the actuarial models we discussed in earlier chapters may well think that we can determine $S_2(0)$ by simply taking a discounted expected value, as we did with several other similar sounding problems. That is, we simply take the price as $22vp$ where v is the discount factor for one period, and p is the probability that the stock goes up. We will first illustrate why one cannot solve the problem this way, and after that, we will, paradoxically, illustrate why one *can* do it this way.

The first problem is that one is not given p as part of the model. All that we postulated about the probability measure P was that both of the possible outcomes at time 1 have positive probability. Indeed, there may not be any reasonable choice for a single value of p. The many different participants in the market may well have completely different assessments of this figure. It is not unusual to find two experts commenting on a particular stock, where one claims it is the best buying opportunity to come along in the last decade, and the other predicts imminent bankruptcy of the firm.

The second problem is that one is not given v. Now the reader may take issue with this statement since we postulated a risk free rate of 10% a few paragraphs back, so it appears as if v is simply $(1.10)^{-1}$. Use of this rate would imply that the buyer is looking for an expected return of 10% on their investment. However, 10% is the return for a perfectly *risk-free investment*. Investing in a call option is far from being risk-free. If the stock price at expiry is below the strike price, the entire investment is lost. It is to be expected that a rational option purchaser will want a return in excess of 10% as compensation for taking on the risk. (Recall that we discussed the same concept when introducing the risk discount rate in Section 12.4).

We will now solve the puzzle, and show that regardless of the assessment of p or of the desired yields of different individuals, the price of this option can only be 12. The reason is that one can in fact *replicate* the option for an initial investment of 12. That is, by investing only in the bank account and the stock, one can produce an outcome at the expiration date, which matches exactly the payouts of the option. This is done by buying 2/3 of a share of stock at time 0, which will cost 72. We can put in 12 cash, and borrow the additional 60. If the stock price is 132 per share at time 1, we sell our 2/3 of a share for 88, pay off the loan balance which is now 66, leaving us with 22. If the stock price is 99, we sell our 2/3 of a share for 66, and pay off the loan, leaving us with nothing extra. We have therefore exactly replicated the option for the price of 12. It is is clear that no one would pay more than 12 to buy this option. Similarly, nobody would sell the option for a price of less than 12, since instead they could reverse the above strategy and be in the same position at time 1 as if they had written the option, but they would have have received 12 at time 0.

Here is another point of view, which ties in with our previous definition of arbitrage. If we enlarge our financial market by adding the option as another asset S_2, then we must take $S_2 = 12$ to make this enlarged market arbitrage-free. To take a definite example, suppose the option price is 13. We will construct an arbitrage opportunity. Take the trading strategy which has as initial portfolio $\alpha(0) = (-59, 2/3, -1)$. The reader can verify that $V(0) = 0$. Now $S_0(1) = -64.90$, so If $S_1(1) = 132$, then $S_2(1) = 22$ and $V(1) = 1.1$. If $S_1(1) = 99$ then $S_2(1) = 0$ and again $V(1) = 1.1$. We leave it to the reader to find an arbitrage opportunity if the option price is below 12.

Let us now go back to the proposed solution of of $22vp$ as an option price, which we criticized a few paragraphs above. If we in fact use the risk-free rate and therefore take $v = 1.10^{-1}$, we will get the correct answer by using $p = 0.6$. Is there someway we could have discovered this probability of 0.6 beforehand? The answer is yes. Let us suppose that there exists a so called *risk-neutral* individual, that is one who ignores the risk and is happy to accept an expected 10% return on any investment, regardless of the degree of safety involved. Let p be the particular probability of rise in the stock price, which would be assumed by such a risk-neutral person. In order that this person would be willing to pay 108 for a share of stock, we should have that

$$108 = 1.10^{-1}[132p + 99(1 - p)]$$

and solving we have indeed that $p = 0.6$.

We have now discovered the important principle of *risk-neutral valuation*. The assignment here of 0.6 and 0.4 to the events of the stock going up or down, respectively, is known as a *risk-neutral* probability measure. It is the probability that must be assigned by a risk-neutral individual in order to justify buying the stock at the market price. Note that we are not saying that such a person necessarily exists, and indeed have stressed that most investors would be unlikely to possess such an attitude. We are only saying that if one did exist, the price of the underlying asset would necessarily imply a unique probability assessment for that individual. The principle then says that if we use the risk-free interest rate, along with the risk-neutral probability measure, then we can indeed value options by following the usual actuarial approach of taking a discounted expected value.

Note carefully that the risk-neutral probability measure need not be the same or indeed have any particular relation to the original measure P, other than being equivalent in the sense defined above. Even if we had specified values for P, these would have had no effect on the resulting option price. The fact that one can risklessly replicate the option means that only the risk-neutral probability and the risk-free interest rate need be considered.

This is a puzzling observation at first, and for those who are still skeptical, we will look into the situation a little further. We mentioned above that the probability p was not even specified as part of the model, but let us suppose it is. In fact, suppose that instead of a stock with uncertain returns, we have two lotteries each depending on the same random draw. A ball is drawn randomly from an urn containing two white balls and one red. The payoff from lottery 1 at time 1 is 132 if a white ball is drawn, or 99 if a red is picked. The payoff from lottery 2 at time 1 is 22 if a white ball is drawn or 0 if a red is drawn. So the true underlying value of p is now indisputable as $2/3$. If the price for a lottery 1 ticket is 108, and we make the assumption that we can buy or sell any fraction of lottery 1 tickets, then the price for a lottery 2 ticket must be 12, by exactly the same argument as given above, regardless of the known value of p. What does this imply for people who participate? Buyers of a ticket in lottery 1 are in effect earning an expected return of $[(2/3)132 + (1/3)99)]/108 - 1 = 12.04\%$. There is a reasonable extra return over the risk-free rate, to compensate for the risk taken on. Buyers of a ticket in lottery 2 are in effect earning an expected return of $[(2/3)22/12] - 1 = 22.22\%$, a much higher return, which compensates for the greater risk in lottery 2 when the entire stake could be lost. Indeed for any value of p above 0.6, there will be a return above the risk-free rate in lottery 1 and an even higher return in lottery 2. It is only for the risk-neutral value of p equal to 0.6, for which the expected returns on both lotteries will coincide with the risk-free rate.

Going back to our original example with the stock, is it possible that an investor who assesses the probability of an upward movement as being less than 0.6 would still pay 108 per share, thereby earning an expected return of less than the risk-free rate? This may seem irrational, but it is no more so than the behaviour of a vast number of people who buy lottery tickets or gamble in casinos at highly unfavourable odds. (For more on this topic, see Example 22.2.)

We next derive a general formula. Suppose that (20.1) holds. As we did above, we can set up an equation to solve for p, the risk-neutral probability that the upward move will occur. Taking $S_1(0) = 1$, this is

$$1 = (1 + r)^{-1}[pu + (1 - p)d], \tag{20.2}$$

which we solve to obtain

$$p = \frac{(1 + r) - d}{u - d}, \quad 1 - p = \frac{u - (1 + r)}{u - d}. \tag{20.3}$$

Note that condition (20.1) ensures that $0 < p < 1$.

The above procedure allows us to uniquely price, not only call options, but a general derivative security in this market, which pays an amount A if an upward move occurs or B if a downward move occurs. We do this in one of two ways: first, we can find a replicating initial portfolio consisting of α units of the stock and β units of the bank account by solving the equations

$$\alpha S_1(0)u + \beta(1 + r) = A, \quad \alpha S_1(0)d + \beta(1 + r) = B. \tag{20.4}$$

Then

$$\text{Price} = \alpha S_1(0) + \beta, \tag{20.5}$$

which is the cost of establishing the replicating portfolio. Secondly, and usually easier, we can bypass finding the replicating portfolio and just take the price as the discounted expected value of the the payoff with respect to the risk-free interest rate and risk-neutral measure. That is,

$$\text{Price} = (1+r)^{-1}[pA + (1-p)B], \tag{20.6}$$

where p is as given in formula (20.3). The reader can verify that both methods lead to the same answer.

Example 20.1 For the example given at the beginning of this section, find the price of a put option with a strike price of 110.

Solution. If the upward move occurs, the holder tears up the option. If the downward move occurs, the holder buys the stock for 99, and sells it for 110. So this is a derivative security with $A = 0, B = 11$. Directly from Equation (20.6), we have that the price is $1.10^{-1}(0.4 \times 11) = 4$. Alternatively, solve (20.4) to derive the replicating portfolio given by $\alpha = -1/3$, $\beta = 40$, and use (20.5) to get the same answer. To see this directly, we replicate the option for a cost of 4 by selling $1/3$ of a share short, receiving 36, letting the total of 40 accumulate to 44 at time 1. This allows one to just cover the short position if the stock is up, or cover the short position and have 11 left over if the stock is down.

There is in this case yet another way to obtain the answer. In fact, we develop a general formula relating puts and calls.

Theorem 20.3 (Put-call Parity) *Let γ denote the cost of a call option and π denote the cost of a put option on the same stock with a current price of $S(0)$, the same strike price of K, and same expiration date N. Then*

$$S(0) + \pi - \gamma = K(1+r)^{-N}. \tag{20.7}$$

Proof. Suppose an investor at time 0 adopts the following trading strategy. Buy one unit of stock, sell one call option, buy one put option, and hold these without further trading up to the expiration date N. If $S(N) > K$, the put will expire worthless, the call will be exercised by the other party, so that the investor must give up the stock for a price of K. If $S(N) < K$, the call will expire worthless, the investor will exercise the put and sell the stock for a price of K, while if $S(N) = K$, both options are worthless and the value is just the stock price. Whatever happens, the value at time N of the portfolio will be K. Since $V(0)$ is just the left side of Equation (20.7), the formula follows from Theorem 20.1. □

The proof shows in fact that this theorem is true for a general arbitrage-free market, and does not depend on the binomial assumption. In our present example, we know $\gamma = 12$, $N = 1$, $S(0) = 108$ and $K = 110$, and we can immediately calculate that $\pi = 4$.

20.6 The multi-period binomial model

The model of the last section is clearly too simple to be representative of reality. As a further extension, we keep the binomial feature, but allow the prices to evolve over several periods.

We assume that the price of the stock evolves each period as described in the one-period model above. That is, if the value is s at the beginning of a period, the price at the end will be either su or sd where $d < u$. So for an initial price of $S(0)$ at time 0, the price at time 1 will be either $S(0)u$ or $S(0)d$, and the price at time 2 will be either $S(0)u^2, S(0)ud$, or $S(0)d^2$, etc. We can represent this by what is known as a *binomial tree*. See, for example, Figure 18.1 which is an example with $u = 1.2, d = 0.8$.

We can now consider any general contingent claim, which will be a payoff at time N which can depend on the entire history of up and down movements in the stock price. To formalize this, consider the sample space Ω consisting of all paths in the binomial tree. Each such path can be labeled by an N-termed sequence formed of the entries U and D, where U denotes an upwards branch and D a downwards branch, so there are 2^N paths altogether. We postulate that there is a probability measure P on Ω, but we need not specify anything about it except that $P(\omega) > 0$ for all $\omega \in \Omega$.

We now formally define the general type of derivative security we are interested in.

Definition 20.4 A *contingent claim* is a contract which provides a payment at time N which is dependent on the particular outcome in our underlying sample space. It is modelled by a random variable X, where for $\omega \in \Omega$, $X(\omega)$ is the payment for outcome ω.

For example, a call option with strike price K and expiration date N on a stock with current price $S(0)$ is a contingent claim given by

$$X(\omega) = [S(0)u^m d^{(N-m)} - K]_+$$

whenever ω is a sequence with m upward movements and $N - m$ downward movements. (For any real number t, the symbol t_+ denotes $\max\{t, 0\}$.)

A contingent claim can be more complicated than the options we have described up to now. Consider, for example, a *lookback* option on a stock which will return at expiry the maximum value of the stock over the period from time 0 to time N. So looking for example at Figure 18.1, we would have

$$X(DUD) = 96, \quad X(UDD) = 120,$$

and so on.

The multi-period model has the same essential features that we observed in the one-period model.

- The financial market consisting of the stock and the bank account is arbitrage-free if and only if condition (20.1) holds.

- Any contingent claim can be priced uniquely so as to prevent arbitrage. One method is to find a replicating self-financing trading strategy. The price of the claim is then the cost of setting up the initial portfolio for this strategy. A second way is to take the expected discounted value with respect to the *risk-neutral* probability measure Q on Ω, which is simply the measure obtained by applying the appropriate probabilities p or $1 - p$ as given by formula (20.3) to each branch of the binomial tree. That is, if ω has m entries of U and $N - m$ entries of D,

$$Q(\omega) = p^m(1 - p)^{N-m}$$

These facts can be verified by using the results from the one-period model, and working backwards in time. The definition of contingent claim gives us directly its value at time N. We use those to determine the value and strategy applicable to each node at time $N-1$, and then use these to get determine the value and strategy applicable to each node at time $N-2$, and continue to iterate the procedure until we get to time 0.

Example 20.2 Consider a two-period model where the price of a stock evolves as shown by the tree in Figure 18.1 up to time 2, and $r = 0.10$. The contingent claim X is a call option at time 2, with a strike price of 92. So

$$X(\mathrm{UU}) = 52, \quad X(\mathrm{UD}) = X(\mathrm{DU}) = 4, \quad X(\mathrm{DD}) = 0.$$

Find the replicating strategy and the price of the option which will prevent arbitrage,

Solution. Suppose at time 1, the value of the stock is 120. We know that $u = 1.2, d = 0.8$, and we can solve the system (20.4) with $A = 52, B = 4$ to get $\alpha = 1, \beta = -920/11$. This means that if the process is in the upper node at time 1, then in order to replicate the payoff at time 2, we should own 1 unit of stock, and carry a debt of 920/11. The total value $V(1)$ is 400/11.

Similarly, if the value of stock is 80 at time 1, we solve the system (20.4) with $A = 4, B = 0$, and we arrive at a required portfolio of 1/8 units of stock and a debt of 80/11 for a total value $V(1)$ of 30/11.

We now move back to time 0 and again solve the system (20.4) with $A = 400/11$ and $B = 30/11$, to obtain a initial portfolio consisting of 37/44 shares of stock and a debt of 7100/121. The value of this initial portfolio is $V(0) = 3075/121$ which must be the price of the option.

To summarize, one can replicate this contingent claim by the following self-financing trading strategy. At time 0, buy 37/44 units of the stock, using 3075/121 of one's own capital and borrowing the remaining 7100/121 at the risk-free rate. At time 1, if the stock moves up, increase the stock holding to 1 unit, borrowing additional funds to do so. If the stock goes down, sell enough to reduce the stock holding to 1/8 unit, using the proceeds to partially repay the loan.

This replicating strategy gives in addition a *hedging strategy*. Suppose you have just sold such an option. You run the risk that the stock will move up both periods. If you do not actually own the stock, you will be required to buy it at 144 and sell it at 92 (which shows the danger of selling a so-called *naked option* on a stock you do not own). If you follow the trading procedure outlined above, you will be sure to be able to meet your obligation in any event, assuming of course that the given model for the evolution of the stock price is correct.

To calculate only the option price, rather than the complete replicating strategy, the second method can be used. That is

$$\text{option price} = (1+r)^{-N}\left[\sum_{\omega\in\Omega} X(\omega)Q(\omega)\right]. \tag{20.8}$$

For the particular case of a call option with strike K, this takes the form

$$\text{option price} = (1+r)^{-N}\left[\sum_{m=0}^{N}\binom{N}{m}p^m(1-p)^{N-m}\left(S(0)u^m d^{N-m} - K\right)_+\right], \tag{20.9}$$

where p is determined by formula (20.3). In our case $p = 3/4$, and we can verify that, as before, the price is

$$(1.1)^{-2}[52(9/16) + 4(6/16) + 0(1/16)] = 3075/121$$

Example 20.3 In the example given above, suppose the interest rate is 0. Find a price and self-financing trading strategy for the so-called *lookback option*, which pays at time N the maximum value of the stock at time $0, 1, 2$.

Solution. The payoff is 144 for the outcome UU, 120 for the outcome UD, and 100 for each of the outcomes DU and DD. We can find the price exactly as we did for the option above. For $r = 0$ we can calculate $p = 1/2$ and

$$\text{Price} = [144(1/4) + 120(1/4) + 100(1/2)] = 116.$$

To find the trading strategy, we need a more complicated diagram. See Figure 20.1. In the original diagram (Figure 18.1), the paths of UD and DU both led to the same position at time 2. This is fine in cases where the prices of the stock at that point was all we were interested in, since this took the same value of 96 on both paths. However, in this case, the contingent claim is path-dependent and we need two different nodes to distinguish the two paths.

For the upper node at time 1, we need to hold α units of stock and β units of bank account where

$$144\alpha + \beta = 144, \quad 96\alpha + \beta = 120,$$

so that

$$\alpha = 1/2 \quad \beta = 72, \quad V(1) = 132.$$

For the lower node at time 1 we similarly solve

$$96\alpha + \beta = 100, \quad 64\alpha + \beta = 100.$$

so that

$$\alpha = 0, \quad \beta = 100 \quad V(1) = 100.$$

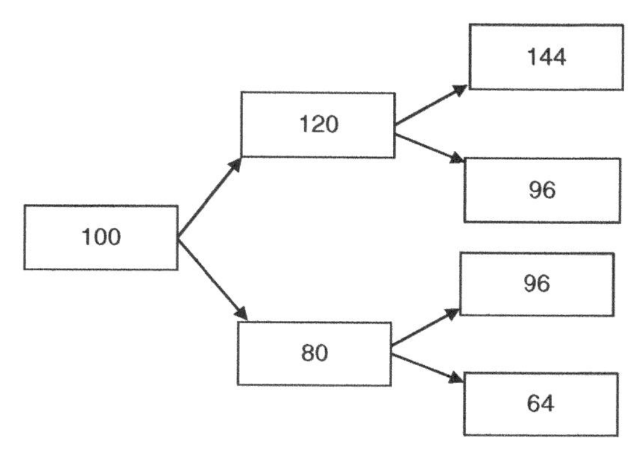

Figure 20.1 Example 20.3

We are then back to a one period model with an asset that has value 132 if the upward movement occurs, or 100 if the downward movement occurs. So the initial portfolio must have α units of stock and β units of the bank account where

$$120\alpha + \beta = 132, \quad 80\alpha + \beta = 100$$

so that

$$\alpha = 0.8, \quad \beta = 36, \quad V(0) = 116.$$

So the trading strategy is to start with 116 (as we knew from the first solution above), buy 0.8 units of stock and put the rest in the bank account. If the upward move occurs, sell 0.3 units of stock, or if the downward move occurs, sell all 0.8 units of stock, in each case putting the proceeds into the bank account.

For contingent claims which depend only on final prices, the first type of diagram, (like Figure 18.1) known as a *recombining* tree, provides a significant reduction in computation. This was not readily apparent in our simple example where $N = 2$, but suppose instead that $N = 10$. The recombining tree would have 11 final nodes, while the more general version would have $2^{10} = 1024$ final nodes.

20.7 American options

We include a brief discussion here on American options, which can have some surprising and initially puzzling features. Recall that such an option can be exercised at any trading date up to and including the final expiration date. One's intuition tells us that the price of this should be greater than the price of the corresponding European option, since there is more choice and therefore a chance for more potential gain. However one's intuition is not completely correct. In fact, given our assumptions, an American call option should never be exercised prior to the expiration date, so in fact the two options are equivalent and should bear the same price. If $r = 0$, the same phenomenon holds for an American put option. It is never correct to exercise early. However, in the more usual case when $r > 0$, it may well be correct to exercise the put option early. We now clarify these rather curious facts.

Consider a particular example. You hold an American call option to buy an asset for a strike price of 100 and on a certain trading date $n < N$, the asset price is 300. Your desire to take advantage of this high price might induce you to exercise the option, making an immediate profit of 200. After all, at a later date the asset price could be lower, with a corresponding reduced gain. However, one should not exercise the option, since there a better way to take advantage of the higher price. You just sell the asset, relying on the option to protect you against further increases, the usual danger with short sales. By doing this and waiting, you would receive 300 immediately. In the worst case scenario, you then have to buy the asset at maturity for 100 and settle your short position. But your gain as of the expiry date would be 200 plus the interest earned on the entire 300 that you received at time n, in addition to an extra gain if the price is below 100 on the expiry date. If you exercised the option early, your gain at expiry would be limited to 200 plus the interest earned on the 200 received at time n. Note that this conclusion depends heavily on our reasonable assumption that $r > 0$ and is not true for negative interest rates.

Our argument also does not apply to dividend-bearing stocks, since it is possible that by exercising early and receiving the stock, the dividends paid will more than compensate for the loss of interest. It does show however that the only possible times when one should exercise are those which coincide with dividend payments. Exercising before such a date will at least incur the loss of interest up to the dividend payment date. We will not go into the complete analysis in this case.

Now consider an American put option. Suppose now that at time n the strike price is 300 and the price of the asset is 100. Should one take advantage of the low price, by buying at 100 (assuming you don not already own the asset), then exercising the put to sell at 300 and making an immediate profit of 200? If the interest rate is 0, the answer is no, because similarly to the call option case there is a better way to take advantage of the current low price. In this reversed situation, we simply borrow 100 and buy the asset, relying on the put to protect us again future drops in the price. At expiry, we sell the asset for a minimum of 300, repay the loan, and have a gain of at least 200. So again with an interest rate of 0, the American and European puts are equivalent. But consider the more realistic case of a positive interest rate. Our previous argument does not hold now, since by waiting, we are paying out interest rather than receiving it as in the case of a call option. Suppose that in any event, we decide to borrow 100 at time n to buy the asset, and the interest charged over the period from n to N is 5%. If we exercise immediately, we receive 300, which increases with interest to 315 by expiry, and after repayment of the loan, our gain at time N is 210. If we wait to exercise, we will be better off if and only the asset price is higher than 315, in which case we keep the asset and tear up the option. So it is not immediately clear whether to exercise or not.

In our discrete model, we can effectively work out the price and trading strategy for an American put option by the same backwards induction process that we illustrated in Examples 20.2 and 20.3. One simply must do an extra comparison at each node. Suppose we have calculated data for all nodes at times greater than n and we are considering a node at time n. One first works out the strategy and a temporary value V exactly as in the European case. One then compares that value with what could be obtained from immediate exercise at time n. This is calculated by buying (or selling) a sufficient quantity of the asset so that you hold one unit and then selling that unit for the strike price. If this exercise value is greater than V, then that replaces V as the value, and the strategy is to exercise at that node.

The following is a simple one-period example, which is sufficient to illustrate the technique, since, in all cases, you just follow the procedure below at each node.

Example 20.4 An asset sells now for 100, and at time 1, will have a price of either 120 or 80, both with positive probability. The risk-free interest rate is 0.10. Find as a function of K the price of an American put option with a strike price of K. Compare this with the price of a corresponding European put if (i) $K = 113$ and (ii) $K = 102$.

Solution. The value at time 1 is $(K - 120)_+$ for an upward move and $(K - 80)_+$ for a downward move. By (20.3), the risk-neutral probability of the downward move is $1/4$. In the extreme case that $K \leq 80$, the option is clearly worth nothing. Take the other extreme where $K \geq 120$. The value at time 0 in the European case would be $[(3/4)(K - 120) + (1/4)(K - 80)](1.1^{-1}) = K(1.1)^{-1} - 100$ which is less than $K - 100$. The price of the option is $K - 100$ and the strategy is to exercise immediately at time 0.

Now consider the case when $80 < K < 120$. The price will be the maximum of

$$\{K - 100, \quad 0.25(K - 80)(1.1)^{-1}\}$$

where the second term is the price of the European put. We can solve to show that immediate exercise is optimal precisely when

$$K \geq 1800/17 = 105.88.$$

So for $K = 113$ the price of the American put is 13 as compared to 7.50 for the European put. When $K = 102$, the price of both options is 5.

20.8 A general financial market

We often wish to model situations which are much more complicated than the ones we considered in the previous section. For one thing, we may have several risky assets rather than one. For another, the evolution of prices may be given by a more involved structure than the binomial tree, and even in the binomial case, it may have a more complicated form than the constant up and down ratios of u and d.

A typical example of such a general market is modelled as a tree like evolution which will apply to all assets. See for example Figure 20.2 At each time n we have a number of nodes, which we can think of as representing a certain 'state of nature', and all the asset prices are determined by this state. This market has three assets and the prices $(S_1(n), S_2(n))$ are shown at each node. The asset 0 prices need not be shown as they are known completely once we specify r.

We will now describe the general discrete-time model. The notation is necessarily somewhat involved, but the market of Figure 20.2 is sufficient to capture the main ideas. We have a finite state stationary Markov chain (see Section 18.2) with the following special structure. The set of states is divided up into subsets $\mathcal{S}_0, \mathcal{S}_1 \ldots \mathcal{S}_N$, where $\mathcal{S}_k$ denotes the set of states at time k. There is a single state $s_0 \in \mathcal{S}_0$ reflecting the fact that there is no uncertainty about prices at time 0. The only transitions of positive probability are those from a state to another one which is one period later. That is, given states i in $\mathcal{S}_k$ and j in $\mathcal{S}_m$ where m is not $k+1$, we must have $p_{ij} = 0$. For example, in our multi-period binomial model, there were $k+1$ states at each time k, and for each state at time k, there were exactly two transitions into a state at time $k+1$. In the market of Figure 20.2, we would have $\mathcal{S}_0 = 0$, $\mathcal{S}_1 = \{1, 2, 3\}$, $\mathcal{S}_2 = \{4, 5, 6, 7, 8, 9, 10\}$.

In the binomial case, we described outcomes by sequences consisting of U and D. In the general case, where we can have more than two branches from a node, we need a somewhat different representation. The evolution of our system up to time k can be described by a sequence of states $\{s_0, s_1, \ldots, s_k\}$ where each $s_k \in \mathcal{S}_k$ and $p_{s_j, s_{j+1}} > 0$ for $j < k$. We will call such a sequence an *admissible k-sequence.* We now take our sample space Ω to consist of all admissible N-sequences, and this is equipped with the probability measure P where for any admissible N sequence ω, we have $P(\omega) = \Pi_{k=0}^{N-1} p_{s_k, s_{k+1}}$.

As an example, in the market of Figure 20.2, Ω consists of the 7 elements $\{\{0, 1, 4\}, \{0, 1, 5\}, \{0, 2, 6\}, \{0, 2, 7\}, \{0, 3, 8\}, \{0, 3, 9\}, \{0, 3, 10\}\}$.

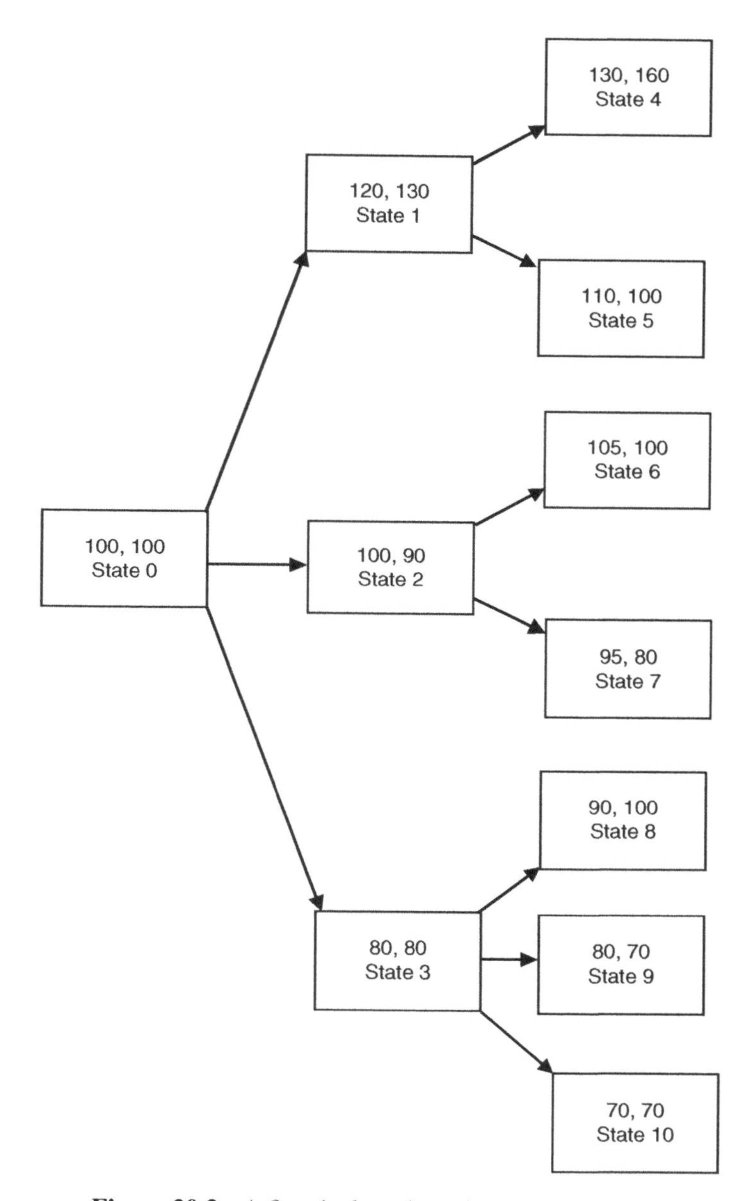

Figure 20.2 A finanical market with two risky assets

It is convenient to adopt the following notational device. For any admissible k-sequence v with $k < N$ we let v° denote the set of all $\omega \in \Omega$ which extend v. That is if $v = s_0, s_1, \dots, s_k$, then

$$v^\circ = \{\omega \in \Omega : \text{ the first } k+1 \text{ entries of } \omega \text{ are } s_0, s_1, \dots s_k, \}. \tag{20.10}$$

For example, in the market of Figure 20.2, $\{0,3\}^\circ$ will denote the subset $\{\{0,3,8\}, \{0,3,9\}, \{0,3,10\}\}$.

We now want to capture formally the nature of quantities like asset prices or asset holdings at a certain time k. These are random before time k, but are then known with certainty at time k or later. For example, in the market of Figure 20.2, the price of asset 1 at time 1 is a random variable which is uncertain at time 0, but then is known precisely at time 1. So it will take the constant value of 120 on the set $\{0, 1\}^\circ = \{\{0, 1, 4\}, \{0, 1, 5\}\}$. Similarly it will take the constant value of 100 on the set $\{0, 2\}^\circ$ and the constant value of 80 on the set $\{0, 3\}^\circ$.

We handle this in general by the following definition.

Definition 20.5 For any integer $k = 1, 2, \ldots, N$, a random variable V defined on Ω is said to be *k-determined* if V is constant on any set v° where v is an admissible $k-$ *sequence*.

For a k-determined random variable V, we will write $V(v)$ to denote the constant value of V on the set v°.

It follows from the definition that a k-determined random variable is also m-determined if $m \geq k$. This just reflects the fact that if we know something at time k, we will know it just as well at some later time.

Note that a 0-determined random variable is just a constant, since the only 0-admissible sequence is that with the single entry of s_0 and s_0° is the entire set Ω.

Definition 20.6 For any random variable W, we define a k-determined random variable $E_k(W)$ as follows. Suppose that the point $\omega \in v^\circ$. That is, v comprises the beginning $k + 1$ entries of ω. We then define

$$E_k(W)(\omega) = E(W|v^\circ)$$

using the notation of (20.10).

So $E_k(W)$ then just gives the expected value of W conditional on the first k-steps of the evolution.

It is clear from the definition that $E_k(W)$ is k-determined. $E_0(W)$, being 0-determined, is a constant and just equal to the usual expected value $E(W)$. Consider the other extreme where $k = N$. Then for any ω, the set v° is the single point ω and $E(W|\omega)$ is just $W(\omega)$, showing that $E_N(W) = W$. So as k increases, $E_k(W)$ gives us more and more information about W until we reach time N and know W exactly.

Example 20.5 Consider Figure 20.2. Suppose that where there are two branches emanating from a node, the probability of an upward move is $2/3$ and that of a downword move is $1/3$, while in the case of three branches emanating from a node, each has probability $1/3$. Describe the random variable $E_1(S_1(2))$.

Solution. Consider the set $B = \{0, 1\}^\circ$, which consists of the two points, namely $\{0, 1, 4\}$ that has probability $4/9$ and $\{0, 1, 5\}$ that has probability $2/9$. We could calculate that the conditional probabilities given B are $(4/9)/(6/9) = 2/3$ for $\{0, 1, 4\}$ and $(2/9)(6/9) = 1/3$ for $\{0, 1, 5\}$. Observe now that we did not have to do all of this calculation, since the tree-like structure makes it possible to read off these conditional probabilities from the future branches of the tree without worrying about the past. In this case, with only one future step, they are immediate. We then have that $E_1((S_1(2))$ takes the value of $2/3(130) + 1/3(110) = 123\ 1/3$ on $\{0, 1, 4\}$ and$\{0, 1, 5\}$. Similarly, it takes the value of $101\ 2/3$ on $\{0, 2, 6\}$ and $\{0, 2, 7\}$ and the value of 80 on $\{0, 3, 8\}$, $\{0, 3, 9\}$ and $\{0, 3, 10\}$.

To summarize, a financial market with $M + 1$ risky assets and of duration N is modelled by a Markov chain with the special structure as noted above, a probability measure on the set Ω of all paths from time 0 to time N, a risk-free interest rate r, and random variables $S_j(n)$, $j = 0, 1, \ldots M$, $n = 0, 1, \ldots, N$, on Ω where each $S_j(k)$ is k-determined. A trading strategy consists of a collection of random variables $\alpha_j(n)$, $j = 0, 1, \ldots M$, $n = 0, 1, \ldots N - 1$ where each $\alpha_j(k)$ is k-determined.

For an important application to follow, we now turn to the concept of a *martingale* introduced in Section 18.3, and look at conditions for this to occur in our present context. Fix a probability measure Q on Ω which is equivalent to P and let $\{W_n\}$, $n = 0, 1, \ldots, N$ be a sequence of random variables such that each W_k is k-determined We claim that this will be a martingale, provided

$$E_k(W_{k+1}) = W_k, \quad k = 0, 1, \ldots, N - 1. \tag{20.11}$$

To see this, suppose that the above holds. Fix any k and a sequence of real numbers $\{w_0, w_1, \ldots, w_k\}$. Consider any set which has positive probability under Q and is of the form

$$A = \{\omega \in \Omega : W_i(\omega) = w_i, i = 0, 1, \ldots, k\}.$$

Now by definition, membership in A is determined by what happens up to time k. If a sequence $\omega \in A$, any sequence which has the same first $k + 1$ entries must also be in A. This implies that A must be the union of subsets of the form ν° for some k-admissible sequence ν. For any such ν, it follows from our hypothesis (20.11) that

$$E(W_{k+1}|\nu^\circ) = E_k(W_{k+1})(\nu^\circ) = W_k(\nu) = w_k$$

and from (A.22), (applied to the sample sample space A with the conditional probability $Q(\cdot|A)$ we can conclude that

$$E(W_{k+1}|A) = w_k,$$

showing that the sequence is a martingale.

To apply this, refer again to Figure 20.2. Let Q be the probability measure which assigns $1/3$ to each transition when there are three transitions out of a state and $1/2$ to each transition when there are two. We can then see that the sequence of prices of asset 1 is a martingale under this measure, by simply verifying the condition (20.11) at each node. For example, at state 3, we have that the value of $S_1(1) = 80$ and the value of $E_1(S_1(2)) = (1/3)90 + (1/3)80 + (1/3)70 = 80$. The same holds at all other states. Similarly, we can show that the same holds for $S_2(n)$, the sequence of prices of asset 2.

20.9 Arbitrage-free condition

To decide when a general financial market is arbitrage-free, directly from Definition 20.2, could be extremely complicated. We would have to consider all possible initial portfolios with value 0 and all possible self-financing trading strategies. Fortunately, there is often a faster way. Suppose we can find a probability measure Q, equivalent to P, such that for each asset i, the sequence $\{S_i(n)\}$ is a *martingale* under Q. Consider any self-financing trading strategy.

At any time n, the portfolio has a value $V(n)$. For each asset i, the expected value at time $n + 1$ will again be $S_i(n)$ and so the expected value of the portfolio before trading will be $V(n)$. Since our trading strategy is self-financing, the expected value after trading will again be $V(n)$. Since this is true for all possible values of the portfolio at time n, we must have that $E_Q[V(n + 1)] = E_Q[V(n)]$. (The subscript indicates that expectations are with respect to the probability measure Q.) Working inductively, we have that $E_Q[V(N)] = V(0) = 0$. It is impossible for $V(N)$ to be nonnegative for all outcomes, have a positive probability of being positive, and still have an expectation of 0, so we cannot have an arbitrage opportunity.

This seems like a nice simple answer but on the face of it there is a major problem. It is not reasonable to expect that our stochastic processes for stock prices are martingales, as we indicated in Section 18.3. In fact, the bank account, by definition, cannot be a martingale unless $r = 0$. So our result above may appear at first to be meaningless, but the following trick saves the day.

We do not have to measure our assets in terms of dollars. They can be expressed relative to some other asset. Define

$$\hat{S}_j(n) = S_j(n)/S_0(n).$$

That is, $\hat{S}_j(n)$ is the value of asset j at time n in terms of the bank account. We can think of $\hat{S}_j(n)$ as a *discounted* or *present value*, since it is what we would have to invest in our risk-free bank account in order to accumulate to $S_j(n)$ at time n. It is a random variable rather than a number since $S_j(n)$ is a random variable. The same argument we gave above clearly goes through if each $\hat{S}_j(n)$ is a martingale. This is now possible since $\hat{S}_0(n)$ takes a constant value of 1. We have therefore proved the 'if' direction of the following major result.

Theorem 20.4 (The fundamental theorem of asset pricing) *A financial market is arbitrage-free if and only if there is a probability measure Q on Ω which is equivalent to P, and for which $\hat{S}_j$ is a martingale for $j = 1, 2, \ldots, M$.*

A major example is the the multi-period binomial model, where the given risk-neutral measure satisfies the conditions of the above theorem, as we verify from Equation (20.11). Indeed, suppose that $\hat{S}(k) = s$, so that $S(k) = s(1 + r)^k$. Referring to Equation (20.3), for any $\omega \in \Omega$,

$$E_k(S_{k+1})(\omega) = s(1 + r)^k \, \frac{u[(1 + r) - d) + d[u - (1 + r)]}{u - d} = s(1 + r)^{k+1}$$

so that

$$E_k[\hat{S}(k + 1)] = s.$$

As another application, we can conclude immediately from our observations in the preceding section that the market of Figure 20.2 is arbitrage-free when $r = 0$.

Note that the risk-neutral probability p that we gave in the one-period binomial market was the only possible value that would make $\hat{S}_1$ a martingale, as shown by Equation (20.2). The terminology is carried over and any probability measure Q satisfying the conditions of Theorem 20.4 is known as a *risk-neutral measure.* The main conclusion of this section then is that usually the best way to show a given a given financial market is arbitrage-free is to show the existence of a risk-neutral measure.

The converse of the fundamental theorem will be proved in the following section.

20.10 Existence and uniqueness of risk-neutral measures

20.10.1 Linear algebra background

To complete our study of financial markets, we require a knowledge of some facts in linear algebra. We assume familiarity with the concept of a linear space (also known as a vector space) and linear subspaces. We also assume familiarity with the concepts of closed and bounded sets. Any basic text on multivariate calculus should contain the necessary details. The following is a brief review, adapted to our ultimate goals.

Consider in particular the vector space W consisting of all real-valued functions defined on some finite set S, with the operations of point-wise addition and scalar multiplication. This is an n-dimensional space where n is the number of points in S. We let 0 denote the function which takes the value 0 at each point of s. (The context should distinguish this from the number 0.) For any f, g in W, we have an inner product

$$f \cdot g = \sum_{s \in S} f(s)g(s).$$

A subset K of W is said to be *convex* if it contains the line segment joining any two of its points. That is, given f and g in K and and a scalar $0 < \gamma < 1$, the function $\gamma f + (1 - \gamma)g$ is in K.

A *hyperspace* in W is a proper linear subspace of maximum dimension, that is one less than the dimension of the space. So for example, a hyperspace can be visualized in two-dimensional space as a line through the origin, or in three-dimensional space as a plane through the origin. We need the following two facts about hyperspaces. The first is a fairly standard result and not difficult to verify. The second is quite a bit more advanced.

1. Any hyperspace H is determined by its so called *orthogonal* vector. That is, there is an element $q \neq 0$ in W such that

   $$H = \{h \in W : q \cdot h = 0\}.$$

   The element q is unique up to a scalar multiple. In two or three dimensions, we can visualize it geometrically as a vector perpendicular to H.

2. Let L be any linear subspace of W and let K be a closed and bounded convex set that does not intersect L. Then there a hyperspace H containing L such that K does not intersect H.

   It is simple enough to visualize this geometrically in three-dimensional space. If a line does not intersect a closed and bounded convex set, we can find a plane containing the line which does not intersect the set. This of course does not hold if the set is not convex. Suppose that K is a doughnut-shaped region, and the line goes through the hole. Then any plane containing the line must intersect K.

20.10.2 The space of contingent claims

We return now to our model as described above and apply our linear algebra concepts. For a given financial market, define the following sets. Let

W be the linear space of all real valued functions on Ω.

We can view this as the space of *contingent claims*, those payments at time N which are determined by the particular path. An important subspace of W is given by

$$L = \{f \in W : \text{ there exists a self-financing trading strategy such that for all } \omega \in \Omega, \\ f(\omega) = V(N)(\omega)\}.$$

So L is the subspace of all *replicable claims* as defined above in Section 20.5. It is a linear subspace since, given f and g in L, we can replicate $f + g$ by just holding at each stage the sum of the holdings in the trading strategies replicating f and g, and we can similarly achieve any scalar multiple of f by multiplying our holdings by that scalar. Let

$$L_0 = \{f \in L : \text{ there is a replicating self-financing trading strategy for } f \text{ with } V(0) = 0$$

This easily seems to be a linear subspace of L. We let

$$K = \left\{ f \in W : f(\omega) \geq 0 \text{ for all } \omega \in \Omega, \sum_{\omega \in \Omega} f(\omega) = 1 \right\}.$$

which is a convex subset of W.

So a nice linear algebra definition for a financial market to be arbitrage-free is to simply say that L_0 does not meet K. (Of course any nonzero, nonnegative function f in L_0 represents an arbitrage opportunity, but an appropriate scalar multiple of such an f will be in K and also in the subspace L_0. We also use the fact that our original probability measure P must take a positive value on each ω.)

To illustrate, Figure 20.3 gives a geometric picture of the one-period binomial market. Any contingent claim is represented by a point $(f(U), f(D))$ in the plane. The set K is the line

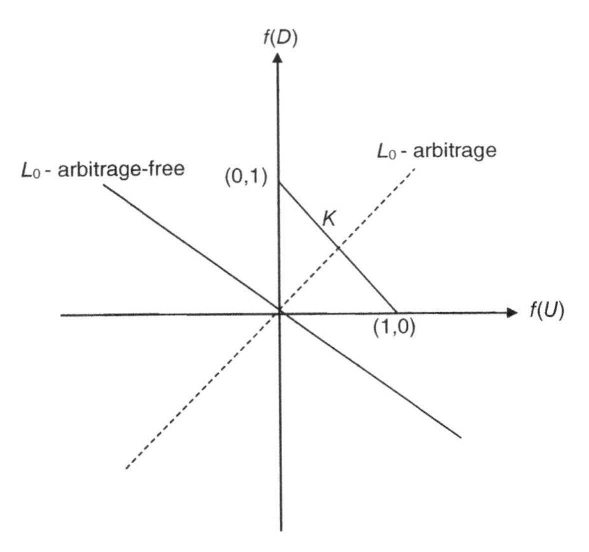

Figure 20.3 A picture of the one period binomial market

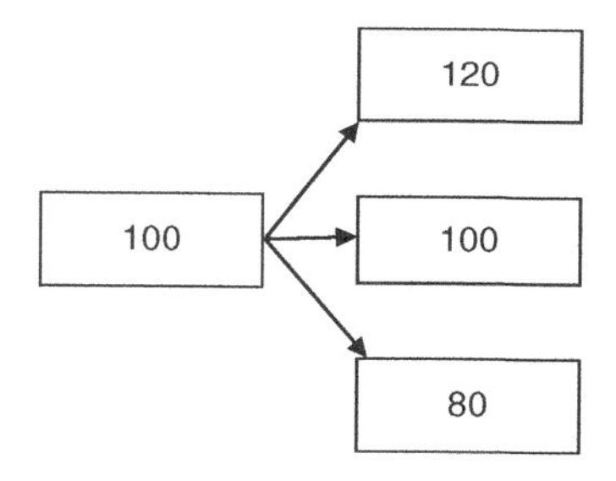

Figure 20.4 A market in which not all contingent claims are replicable

segment joining the points (0, 1) and (1, 0). The subspace L_0 is a proper subspace and therefore must be a line through the origin. In the arbitrage-free case, this line will have negative slope and not meet K. In the case of an arbitrage opportunity, L_0 as represented by the dotted line, has a slope that is either nonnegative, or equal to ∞, and it must intersect K. The picture also makes it clear that L is the entire plane, as we noticed enough, since it a subspace that properly contains L_0.

Further examples are furnished by the markets of Figures 20.4, 20.5 and 20.6. Assume that $r = 0$. Alternatively, we can assume any positive r and interpret the asset values that are given as as $\hat{S}_j(n)$ rather than $S_j(n)$. The conclusions will be the same in either case.

In the single risky asset market of Figure 20.4, the set Ω will have three points U, M, D (for 'up', 'middle', 'down'). If the initial portfolio has α units of stock and β units of the bank account, the time 1 value of portfolio will be $120\alpha + \beta$ for the upward movement, $100\alpha + \beta$ if the price stays the same, or $80\alpha + \beta$ for the downward movement. It follows that

$$L = \{f : f(U) + f(D) - 2f(M) = 0\},$$

a two-dimensional subspace of W, showing that not all contingent claims are replicable in this market. In particular, a call option with strike price 110 will have $f(U) = 10, f(M) = f(D) = 0$ which is not in L.

For initial portfolios of value 0, we must have in addition that $100\alpha + \beta = 0$ leading to

$$L_0 = \{f : f(M) = 0 : f(U) + f(D) = 0\}.$$

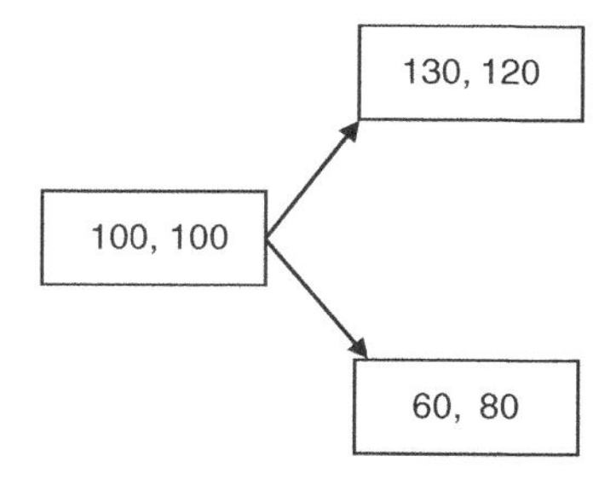

Figure 20.5 A market that is not arbitage-free

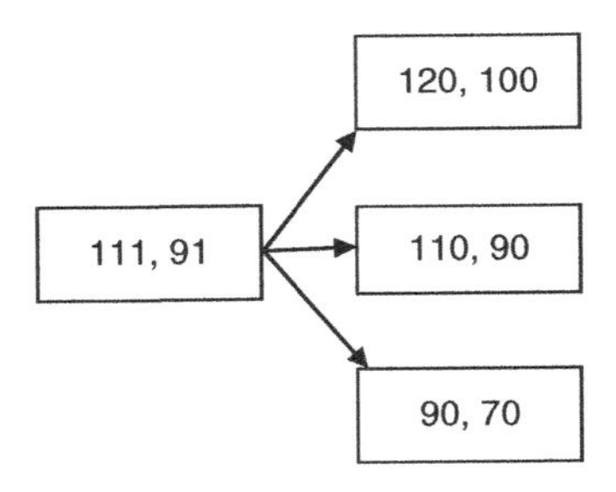

Figure 20.6 See Exercise 20.7

The intersection of L_0 and K is clearly 0, showing that this market is arbitrage-free. Of course we could have immediately deduced this from the Fundamental Theorem, since assigning probabilities of $1/3$ to each branch yields a risk-neutral measure.

Consider the two risky asset market of Figure 20.5. An initial portfolio with value 0 will be given by the vector of the form $100(-(\alpha+\beta), \alpha, \beta)$. Then, a function f will be in L_0 if we can find α, β satisfying

$$30\alpha + 20\beta = f(U), \quad -40\alpha - 20\beta = f(D). \tag{20.12}$$

These equations are readily solved to give $\alpha = -(f(U)+f(D))/10, \beta = (4f(U) + 3f(D)/20$. It follows that $L_0 = L = W$. So this market is about as far from being arbitrage-free as we could possibly get. Any contingent claim can be replicated for an initial cost of 0! Obviously, the prices here as shown could not be maintained by rational investors.

To see the delicacy of situations like this, look at this market again, but make the modification that $S_1(1) = 70$ instead of 60. We leave to the reader to verify that we still have $L = W$, but L_0 is quite different. The coefficient of α in the first equation of system (20.12) is 40 instead of 30, so that now

$$L_0 = \{f \in W : f(U) + f(D) = 0$$

as in the market of Figure 20.3, which shows that the market is arbitrage-free. We can also deduce this immediately from the Fundamental Theorem, since now there is a risk-neutral measure, $Q(U) = Q(D) = 1/2$.

The market of Figure 20.2 that we investigated in Section 20.8 is more complicated, and it would involve a great deal of calculation to try to deduce L_0 exactly, although we do know that it cannot meet K due to the arbitrage-free condition. It is possible to show that $L = W$, but this is far from obvious from the figures as given. As a particular case, consider the following.

Example 20.6 Let X be the contingent claim that that takes the value 60 on UU and 0 elsewhere. Take $r = 0$. Find a self-financing trading strategy to replicate this claim.

Solution. For any such strategy, each of the lower two nodes at time 1 will lead to a claim of 0 at time 2, so by the martingale property, our portfolio must have value 0 at these nodes. Therefore, looking at the pre-trading values at time 1,

$$\alpha_0(0) + 100\alpha_1(0) + 90\alpha_2(0) = 0, \quad \alpha_0(0) + 80\alpha_1(0) + 80\alpha_2(0) = 0,$$

which gives

$$\alpha_1(0) = -0.5\alpha_2(0), \quad \alpha_0(0) = -40\alpha_2(0).$$

Similarly, the value at the upper node at time 1 must be 60 times the probability of an upward move, which is $60(1/2) = 30$. So

$$\alpha_0(0) + 120\alpha_1(0) + 130\alpha_2(0) = 30,$$

and substituting from above we have,

$$\alpha_0(0) = -40, \quad \alpha_1(0) = -0.5, \quad \alpha_2(0) = 1.$$

To summarize, the trading strategy at time 0 is to buy 1 unit of asset 2, financing this by selling 1/2 unit of asset 1, borrowing 40 and putting up the remaining 10. This checks out since the initial cost must be $60Q(UU) = 60(1/3)(1/2)$. At time 1, if the middle or lower branch occurs, sell the unit of asset 2, which is just enough to cover the short position and pay off the loan.

We must now decide what to do at time 1 if the upper branch occurs. In this case, we have

$$\alpha_0(1) + 130\alpha_1(1) + 160\alpha_2(1) = 60, \quad \alpha_0(1) + 110\alpha_1(1) + 100\alpha_2(1) = 0.$$

There are several solutions to these equations, which indicates that a replicating self-financing trading strategy need not be unique. One example is to take

$$\alpha_0(1) = -100, \quad \alpha_1(1) = 0, \quad \alpha_2(1) = 1.$$

This strategy involves borrowing an additional 60 to cover the short position in asset 1. At time 2, we pay off the loan of 100, and have either 60 or 0 left, depending on what happened to asset 2 at time 2.

Similarly, for each of the other six paths, we could find the replicating strategy for a claim which pays off only on that path. We would take a suitable linear combination of these seven strategies to replicate any possible contingent claim. This will show that $L = W$. In Section 20.11, we will prove a result which provides a much easier way to see this.

20.10.3 The Fundamental theorem of asset pricing completed

In this section, we prove the converse to the result established above, and show that in any arbitrage-free financial market, we can find a risk-neutral measure Q. We will proceed in two stages.

Stage 1: Defining Q:

By the arbitrage-free assumption, L_0 does not meet the set K, a convex, closed and bounded set. By our linear algebra results of Section 20.10.1 we can find a hyperspace H containing L_0 and not meeting K. Let q be an element orthogonal to H. That is $H = \{h : q \cdot h = 0\}$.

Now it cannot be that for two distinct points f and g in K, we have $q \cdot f < 0$ and $q \cdot g > 0$, for if so we could find γ such that the function $h = \gamma f + (1 - \gamma)g$ satisfies $q \cdot h = 0$ and so

$h \in H$. But by convexity $h \in K$, and this would contradict the fact that H does not intersect K. So by a change of sign, if necessary, we can assume that $q \cdot f > 0$ for all $f \in K$. Now in particular the functions 1_ω which take the value of 1 on ω and the value 0 elsewhere are in K, and so we can infer that for all $\omega \in \Omega$,

$$q(\omega) = q \cdot 1_\omega > 0,$$

and by multiplying by a suitable scalar we can ensure that

$$\sum_{\omega \in \Omega} q(\omega) = 1,$$

which means that the function q is the probability function for a probability measure Q on Ω.

Stage 2: Showing the martingale condition:

Fix any j. We will show that the stochastic process $\hat{S}_j$ is a martingale under Q. For any time $n < N$ and any possible value s of $S_j(n)$, let A denote the event that $S_j(n) = s$, which means that $\hat{S}_j(n) = s(1+r)^{-n}$.

We must show that

$$E_Q(\hat{S}_j(n+1)|A) = s(1+r)^{-n},$$

or equivalently that

$$E_Q(Y|A) = 0, \tag{20.13}$$

where

$$Y = \hat{S}_j(n+1) - s(1+r)^{-n}.$$

Consider the following trading strategy. Do nothing before time n. If the price of asset j at time n is not equal to s, do nothing at all. If the price at that time is s, buy 1 unit of asset j, borrowing to do so, and sell it at time $n+1$. Apply the proceeds to repaying the loan and let the difference (which could be negative) accumulate in the bank account. Let f be the function in W corresponding to this strategy.

If $S_j(n) = s$, this strategy yields a bank account of $[S_j(n+1) - s(1+r)]$ at time $n+1$. Multiplying by $(1+r)^{N-n-1}$, the accumulated amount at time N is $(1+r)^N Y$ if the purchase is made and 0 if the purchase is not made. Our trading strategy is self-financing and requires an initial investment of 0. Therefore, the function g given by

$$g(\omega) = \begin{cases} (1+r)^n Y(\omega) & \text{if } \omega \in A \\ 0, & \text{if } \omega \notin A \end{cases}$$

is in L_0. This means that

$$0 = q \cdot g = \sum_{\omega \in A} q(\omega)(1+r)^N Y(\omega) = (1+r)^N Q(A) E_Q(Y|A).$$

The fact that s is a possible value of $S_j(n)$ implies that $Q(A) > 0$, and so we must have $E_Q(Y|A) = 0$, establishing Equation (20.13). □

One method of showing that a market is not arbitrage-free is to find the subspace L_0 and show that it intersects K. But as we saw above, this can be computationally infeasible in all but the simplest cases. The converse of the Fundamental Theorem provides an easier way.

Example 20.7 Use the above result to show that the financial market of Figure 20.5 is not arbitrage-free.

Solution. Given a probability measure on Ω, let q be the probability of an upward move. For $\hat{S}_1$ to be a martingale, we need $130q + 60(1 - q) = 100$ so that $q = 4/7$. For $\hat{S}_2$ to be a martingale, we need that $120q + 80(1 - q) = 100$ so that $q = 1/2$. No such measure exists.

20.11 Completeness of markets

In this section, we pose the following questions. Given an arbitrage free market, can we price all contingent claims by the two methods we had in the binomial model? Can we do so uniquely?

The uniqueness question is easily answered for an arbitrage-free market. As we showed in the binomial case, if we can replicate a contingent claim with a self-financing trading strategy, then the cost of that claim should be the cost $V(0)$ of setting up the initial portfolio. What happens, however, if there are several different replicating self-financing trading strategies? This can certainly occur, but in an arbitrage-free market, they necessarily have the same $V(0)$ which means a unique price. Suppose to the contrary that there were two replicating strategies for the same contingent claim, one with an initial cost of 100 and the second with an initial cost of 60. The investor could follow both the second strategy and the reverse of the first strategy for a net gain of 40 at time 0 which would be placed in the bank account, resulting in an overall initial value of 0. At time N, the payments on these two strategies would cancel, leaving a certain positive amount in the bank account, contrary to the fact that there were no arbitrage opportunities.

We turn now to the existence question, beginning with a definition.

Definition 20.7 A financial market is said to be *complete* if, given any contingent claim X, there is a self-financing trading strategy that replicates X. In other words, using the notation of the preceding section, the subspace L is all of W.

We have already shown that the multi-period binomial market is complete. Moreover in Figure 20.4, we gave an example of an incomplete market.

The following theorem gives a characterization of completeness for arbitrage-free markets.

Theorem 20.5 *An arbitrage-free market is complete if and only if there is a unique risk-neutral measure.*

Proof. Suppose that the market is complete. Let Q be any risk-neutral measure. Fix any $\omega \in \Omega$. Let X^{ω} be the contingent claim that pays 1 if ω occurs and pays 0 for all other outcomes, and choose a self-financing trading strategy that replicates X^{ω}. If $V(0)$ is the cost of the initial portfolio, the martingale property ensures that

$$V(0) = (1 + r)^{-N} E_Q(X^{\omega}) = (1 + r)^{-N} Q(\omega),$$

so that

$$Q(\omega) = V(0)(1+r)^N,$$

showing that Q is uniquely determined.

Conversely, suppose that the market is not complete, so that L is not equal to all of W. We can then choose a nonzero function $h \in W$ such that

$$h \cdot f = 0 \text{ for all } f \in L. \tag{20.14}$$

(Since we can do this for a hyperspace, we can clearly do it for any proper subspace which is contained in some hyperspace.) The function which takes the constant value 1 is in L, (achieved by investing $(1+r)^{-N}$ in the bank account at time 0), so we must have that

$$\sum_{\omega \in \Omega} h(\omega) = 0. \tag{20.15}$$

Let Q be the probability measure with the probability function q as constructed in proving the 'only if' part of Theorem 20.4. We will produce a different function q' with the same properties as q, namely

$$q'(\omega) > 0, \text{ for all } \omega \in \Omega, \tag{20.16}$$

$$\sum_{\omega \in \Omega} q'(\omega) = 1, \tag{20.17}$$

$$q' \cdot f = 0, \text{ for all } f \in L. \tag{20.18}$$

Then q' will induce a second martingale measure Q'. To construct q', we note that since $q(\omega) > 0$ for all ω, we can choose a positive number δ sufficiently small so that the function

$$q' = q + \delta h$$

satisfies (20.16), and since $h \neq 0$, q' is different from q. In view of Equations (20.14) and (20.15), it is clear that q' also satisfies Equations (20.17) and (20.18). □

To illustrate the proof of the last part, look again at the market of Figure 20.3. The function $q(\omega) = 1/3$ for all ω gives us a risk-neutral measure. The space L is, as we have seen, the set of all functions such that

$$f(U) - 2f(M) + f(D) = 0$$

and the perpendicular function h can be taken as

$$h(U) = 1, \quad h(M) = -2, \quad h(D) = 1.$$

We can take then δ to be any number strictly between $-1/3$ and $1/6$, which yields an infinite number of risk-neutral measures.

As a consequence of the above theorem, we can see immediately that the market of Figure 20.2 is complete, without going through the somewhat involved calculation of the replications that we did before. The probability assignment which we gave in Section 20.8 is clearly the unique risk-neutral measure.

Incompleteness means that there are not sufficiently many assets to account for all the variations in possible contingent claims. Comparing Figures 20.4 and 20.5, we see that with three branches we need at least three assets in order to achieve completeness. This explains also the fact that with only one risky asset, we need a binomial model to achieve completeness.

For an additional example, consider the two risky asset market of Figure 20.5. We leave it to the reader to decide whether or not it is arbitrage-free, and whether or not it is complete.

At this point, we summarize our conclusions. Suppose we have an arbitrage-free financial market. If the market is complete, then for any contingent claim X, there is a unique price which will prevent arbitrage opportunities. This can be found in one of two ways. First, choose a self-financing trading strategy to replicate X, and the price will be $V(0)$. Second, take the discounted expected value of $(1 + r)^{-N}X$ with respect to the unique risk-neutral measure Q. If the market is incomplete, then we can still find a unique price for the replicable contingent claims. However the non-replicable claims cannot be priced uniquely, since different choices of a risk-neutral measure can give different results. Some other criteria must be used to arrive at prices. This does not mean of course that there are no restrictions on the price of such claims. Often, a range of values can be computed.

Example 20.8 In the market of Figure 20.4, find the possible no-arbitrage prices for a call option with exercise date 1 and strike price 110.

Solution. Add the option as another asset S_2 with a price of π at time 0. For a martingale measure which has probability p of an upward movement and probability q of staying the same, we have $120p + 100q + 80(1 - p - q) = 100$ implying that $2p + q = 1$, so that $p < 1/2$. Applying the martingale condition for the new asset gives $\pi = 10p$, which means that we must have $0 < \pi < 5$. Conversely, all such values are admissible since for any such π we obtain a martingale measure for the enlarged market by taking

$$p = \pi/10, \quad q = 1 - \pi/5.$$

In some cases, specifying the price of certain contingent claims will determine others. To illustrate, having added the additional asset and having specified π in Example 20.8, the resulting martingale measure is unique, so we have in effect completed the market. The price specified for the option will determine unique prices for all other contingent claims.

20.12 The Black–Scholes–Merton formula

The discrete-time model illustrates many of the fundamental principles of pricing contingent claims. There is a limitation, however. To be at all close to a realistic model, we would need an enormously large number of time periods and the computations will quickly become intractable. This is apparent just from looking at the cases for $N = 2$ of the previous section. For actual computation and greater realism, the preferred method is to use continuous-time models, which involve more advanced mathematical machinery. The relatively simple calculations effected by solving linear equations above will be replaced by solving differential equations. Sums of random variables are replaced by integrals of random variables, a concept that is technically very complex. The characterizations of arbitrage-freeness and completeness which we gave can be generalized to continuous-time settings but the proofs are far more difficult. A rigorous development of this is beyond the scope of the book, However we do want to investigate the *Black–Scholes–Merton* formula. This is a pivotal result which in fact initiated much of the modern research into stochastic models in finance.

In this section, it is convenient to use the risk-free force of interest $\delta = \log(1 + r)$ instead of r.

We consider again the case of a financial market consisting of the bank account, and a single stock whose price at time t is $S(t)$. The difference is that we now allow trading at any time, and moreover, we allow asset prices to vary continuously. We therefore must model $S(t)$ as a continuous-time stochastic process and will in fact choose a geometric Brownian motion process, as given in Section 18.7. That is, for some constants μ and σ,

$$S(t) = S(0)e^{\mu t+\sigma B(t)},$$

where $B(t)$ is a standard Brownian motion.

The problem is to find the price of a European call option with exercise date N and strike price K. One approach is to make use of the results that we already know for the discrete-time model. Suppose we can find a probability measure Q under which $\hat{S}(t)$ is a martingale. Then we can approximate our process with our discrete time multiple period binomial model with periods of length $1/m$, $u = e^{\sigma\sqrt{1/m}}, d = e^{-\sigma\sqrt{1/m}}$ and with risk-free force of interest = δ/m, provided we take m sufficiently large. This should seem plausible since the log of our binomial process is a random walk. In this discrete setting, the price of the option is given in Equation (20.9). Putting in all the parameters and taking limits as m goes to ∞, we arrive at the celebrated Black–Scholes–Merton formula:

$$\text{Option price} = S(0)\Phi(\alpha + \sigma\sqrt{N}) - Ke^{-\delta N}\Phi(\alpha), \tag{20.19}$$

where

$$\alpha = \frac{\log(S(0)/K) + (\delta - \sigma^2/2)N}{\sigma\sqrt{N}}, \tag{20.20}$$

and Φ is the c.d.f. of the standard normal distribution. An alternative method to derive the formula is to directly calculate $e^{-\delta N}E_Q[(S(N) - K)_+]$. This is more straightforward, involving basic calculus, although the calculation is somewhat involved. For both of these methods, we will omit the detailed derivations.

Note that the parameters to determine the option price are the risk-free force of interest δ, the strike price K, the duration N, and the quantity σ. The latter is a measure of the tendency for prices of the underlying stock to vary and is known as the *volatility* of the stock. Note however that the formula does not depend on μ. We comment more on this below.

The resulting formula seems somewhat formidable, but we can provide some motivation. The following is not intended as a rigorous exposition, but more to provide a method of remembering and understanding the structure of the formula. First consider the random variable

$$X = \log[(S(N)/S(0)].$$

What is its distribution under Q? From the definition of Brownian motion, we know that under our original probability measure, it is normal with mean μN and variance $\sigma^2 N$. If we make the assumption that it is still normal under Q with the same variance, and calculate the resulting mean M, we know from Equation (A.58) that

$$E_Q[S(N)/S(0)] = e^{M+(\sigma^2/2)N}.$$

But, since $\hat{S}_N$ is a martingale with respect to Q,

$$S(0) = E(\hat{S}_N) = e^{-\delta N}E(S_N),$$

so that

$$E_Q[S(N)/S(0)] = e^{\delta N},$$

leading to

$$M = (\delta - \sigma^2/2)N. \tag{20.21}$$

We conclude that under Q, we have a new drift parameter $(\delta - \sigma^2/2)$ which is completely independent of the original drift parameter μ. This is analogous to the observation in the simple one-period binomial model where the original probability played no role in the option price and it was only the risk-neutral probability that mattered.

Now let us ask, what is the probability under Q that the buyer of this option will exercise it? This will occur if $S(N) > K$. The probability of this is the same as the probability that $X \geq \log(K/S(0)$ which equals the probability that $-X \leq \log(S(0)/K)$, which by the results calculated above is

$$\Phi\left(\frac{\log(S(0)/K) + (\delta - \sigma^2/2)N}{\sigma\sqrt{N}}\right). \tag{20.22}$$

Given our above assumption regarding the distribution of X, we have now identified the term $\Phi(\alpha)$ in the Black-Scholes-Merton formula as the probability under the risk-neutral measure that the option will be exercised.

Now consider the following rather naive reasoning to arrive at the option price. If we exercise the option, we will have an expected gain at time N of $E[S(N) - K)]$ and this will have a present value of $E[S(N) - K]e^{-\delta N}$. Multiply this by the probability of exercise to get

$$S(0)\Phi(\alpha) - Ke^{-\delta(N)}\Phi(\alpha).$$

This looks something like the actual Black–Scholes–Merton formula, but unless $\sigma = 0$, the coefficient of $S(0)$ differs. Our naive approach, however, ignores the nature of the option contract. With positive values of σ, the stock price at time N will vary, and could be above or below the exercise price. Calculating $E(S(N) - K)$ will include negative values when the price is below K, but there is no loss in these cases since the option will not be exercised. The naive formula therefore understates the true price. Another way of expressing this is to note that what we really want is the expected value of $S(N) - K$, *given that the option is exercised.* To correct this understatement, it turns out that the coefficient of S(0) must be increased from $\Phi(\alpha)$ to $\Phi(\alpha + \sigma\sqrt{N})$. This makes sense as we would expect that this correction should increase as the variability in the stock price and length of period increase.

For further insight and verification, we can directly verify the formula for $N = 0$. In this case, $\Phi(\alpha) = \Phi(-\infty) = 0$ if $S(0) < K$, and the formula gives an option price of 0, which it should, while $\Phi(\alpha) = \Phi(\infty) = 1$ if $S(0) > K$ and the formula gives an option price of $S(0) - K$ which it should.

There is still another important feature involving the two terms in the Black–Scholes – Merton formula, which we now present. Start with the following question. For the call option as above, what is its value at any time t, $0 \leq t \leq N$? By the same argument as above, the value will be given by the same formulas as above except that in both (20.19) and (20.20), $S(0)$ is replaced by by $S(t)$ which is the stock price at time t and N is replaced by $N-t$, the remaining duration. It turns out that the two terms of this formula yield a replicating strategy for the option, and therefore a hedging strategy, in terms of holdings of the stock and the bank account. To explain this, we will take new units for the bank account, by adopting a slightly different but equivalent point of view, We can view the bank account as an investment in risk-free zero-coupon bonds maturing at time N. A unit of this asset will be a bond with face amount K, so that its value at time $0 = K(1+r)^{-N}$. Let α_t denote the value of (20.20) with $S(0)$ replaced by $S(t)$ and N replaced by $N-t$. The replicating trading strategy is to hold $\Phi(\alpha_t + \sigma(N-t))$ units of the stock, and carry a short position of $\Phi(\alpha_t)$ units of the bonds we just described, at any time t. At each time t, the value of this portfolio will be precisely the value of the option as given above and it will reach the right amount at time N. In the case where the option is to be exercised the final portfolio will consist of 1 unit of stock, and a short position of 1 bond. In the other case, the portfolio will be reduced to zero units for both assets. It can be observed that the idea behind this strategy is similar to what we observed in the call option of Example 20.2, where the replication is accomplished by buying stock as the price increases, or selling as the price declines.

One must also show that this strategy is self-financing, using an appropriate modification of this definition to apply to continuous time. The idea is that at each instant, the amounts received from selling one of the assets is exactly what is required to buy the other. A precise definition is based on the formulation given in Exercise 20.3, with the differences replaced by differentials. This is not easy, however. The problem is that these are differentials of stochastic processes rather than deterministic functions, and a rigorous presentation involves some knowledge of the subject known as stochastic calculus.

Remark In real life of course this replication by continuous rebalancing is not possible. One could try to get close by very frequent rebalancing but there is no guarantee that this discrete-time approximation is self-financing. It could well require additional amounts of cash or release such. But one can hope that if our model is sufficiently accurate that these extra amounts are relatively small, so that the option could be replicated for something close to the Black–Scholes–Merton price.

To conclude this section, we note that while the Black–Scholes–Merton formula has persisted as the main tool for option pricing, it is based on several simplifying assumptions. One such assumption is that of the log-normal evolution of stock prices. There has been evidence to show that this is not completely realistic, and alternative models have been investigated. Another assumption is that both the risk-free force of interest and the volatility are constant and known, as these must be inserted into the formula to obtain numerical results. More realistic models have been proposed where these quantities are both considered as random.

20.13 Bond markets

20.13.1 Introduction

We now return to the discrete-time case. In this section, we deal with markets where the assets are risk-free zero-coupon bonds. These were already introduced in Section 2.12, where we considered forward prices, but here we want to consider the actual prices which will be random variables. A major result will be to give an appropriate version of 2.1 to ensure no arbitrage in the case of stochastic discount functions.

Our market will consist of $N + 1$ assets where for $n \neq 0$, asset n is a zero-coupon bond maturing for 1 at time n. We will define S_0 later.

We will denote the random variable $S_k(n)$ by $v(k, n)$. For $k \leq n$, $v(k, n)$ is the price you would pay at time k to receive 1 at time n, which ties in well with our original use of this notation in Chapter 2, as well as our notation for forward prices. These are not the forward prices however, but the actual prices. Of course $v(0, n)$ is a real number and the same as the forward price $\tilde{v}(0, n)$, since at time 0 we know the prices of the bonds. But future prices are unknown and therefore modelled as random variables.

A question which may come to mind now is whether the fundamental Equation 2.1 could hold for these quantities only interpreted as a multiplication of random variables. The answer is no. Observe, for example, that with randomness

$$v(0, m) \neq v(0, n)v(n, m),$$

since the left side is a real number while the right side is truly random.

Now in order to avoid a completely trivial situation, we cannot assume a constant deterministic risk-free rate r as we did before. Before discussing how we modify this idea, we want to recall how risk arises from investment in risk-free assets like bonds, (elaborating on the discussion at the end of Section 2.10.3). Note first that unlike a stock, where a value on any future date is unknown, the payoff on a risk-free bond is absolutely certain if held to maturity. The risk arises if one wants to buy or sell before that date. For example, if the risk-free interest today is 5%, then the price of a 3-year bond will be $(1.05)^{-3} = 0.864$. If at time 1, the risk-free interest rate has risen to 0.08, then the value of the bond drops to $(1.08)^{-2} = 0.857$. Buying long-term bonds therefore carries the risk of a rise in interest rates, which can lower the price. However, in our present model, where trading occurs only at integer times, buying bonds which mature in 1 period does not carry any risk. This leads us to define the bank account by

$$S_0(0) = 1, \quad S_0(n) = [v(0, 1)v(1, 2) \cdots v(n-1, n)]^{-1}, n \geq 1. \tag{20.23}$$

Our bank account is formed by starting with 1 at time 0, using that to buy a bond maturing at time 1, then taking the proceeds of $v(0, 1)^{-1}$ to buy a bond maturing for $v(0, 1)^{-1}v(1, 2)^{-1}$ at time 2, and to continue rolling over the account into a new 1-year bond each year. In the case of a constant and deterministic interest rate of r, we would have $v(k, k + 1) = (1 + r)^{-1}$, and this definition of S_0 reduces to the one given before.

Note that under this new definition, $S_0(k)$ is not a definite amount, but strictly random for $k > 1$.

Using the bank account, we can extend the definition of $v(k, n)$ to all ordered pairs (k, n) by defining

$$v(k, n) = S_0(k)/S_0(n), \quad k > n,$$

since this is the amount accumulated at time k, by taking 1 from a bond maturing at time n and placing it in the bank account. Note that with this definition, all prices at the final time N are determined by the previous values of $S_0(n)$ (and of course the fact that $V(N, N) = 1$).

This means that we can model our bond market with a tree going up to only time $N-1$. (See Example 20.9 below.)

20.13.2 Extending the notion of conditional expectation

For our analysis, we will need some additional results about k-determined random variables, We will in fact investigate this idea in more generality and further extend the discussion in Section A.8, since the same ideas are needed in Chapter 24. Suppose we are given a sample space Ω, a probability measure P on Ω and a partition $\Pi = \{B_1, B_2, \dots B_n\}$ of Ω into pairwise disjoint sets with union equal to all of Ω, such that for all i, $P(B_i) > 0$. For any random variable W on Ω, define a random variable $E_\Pi(W)$ as follows. For $\omega \in B_i$,

$$E_\Pi(W)(\omega) = E(W|B_i).$$

To illustrate, in the model of Section 20.8, fix k and take the partition Π which consists of all sets v° where v is an admissible k sequence. (For example, in the market of Figure 20.2 if we take $k = 1$, the partition will consist of the three sets $\{0,1\}^\circ, \{0,2\}^\circ, \{0,3\}^\circ$.) Then $E_\Pi(W)$ is just $E_k(W)$ as defined in that section.

We summarize the facts that we need in the following theorem.

Theorem 20.6 *Take any random variables V, W and a scalar c.*

(a) E_Π is linear. That is

$$E_\Pi(V + W) = E_\Pi(V) + E_\Pi(W), \quad E_\Pi(cV) = cE_\Pi(V).$$

(b) Suppose that W is a random variable which is constant on each subset of Π. Then

$$E_\Pi(WV) = WE_\Pi(V).$$

(c) Suppose that Π' is a finer *partition than Π which means that every set in Π is a union of sets in Π'. Then*

$$E_\Pi[E'_\Pi(W)] = E_{\Pi'}[E_\Pi(W)] = E_\Pi(W),$$

Proof. (a) This follows directly by applying (A.23) and (A.8) to each subset B_i of the partition equipped with the probability measure $P(\cdot|B_i)$

(b) As in (a) apply (A.8) to each set of the partition.

(c) It is clear from the definitions that if a random variable is constant on each subset of Π, then applying E_Π to it will leave it unchanged. Now $E_\Pi(W)$ is constant on each subset of Π therefore constant on each subset of the finer partition Π'. It follows that $E_{\Pi'}[E_\Pi W] = E_\Pi W$. To show the other order of composition, choose any set A of the partition Π and let A be the union of sets $B_1, B_2, \dots B_m$ where each B_i is a set of the partition Π'. To simplify the notation denote $E_\Pi(W)$ by Z. Now by definition, Z takes the constant value of $E(W|B_i)$ on each set B_i so that clearly

$$E(Z|B_i) = E(W|B_i).$$

Now apply (A.29) to the sample space A equipped with the probability measure $P(\cdot|A)$. We have that

$$E(Z|A) = \sum_{i=1}^{m} E(Z|B_i)P(B_i) = \sum_{i=i}^{m} E(W|B_i)P(B_i) = E(W|A)$$

which shows that

$$E_{\Pi}(Z) = E_{\Pi}(W)$$ □

It is of interest to look at the extreme cases. If we take Π to the finest possible partition where the sets are singletons, then $E_{\Pi}(W)$ is just W. If we take Π to be the partition consisting of just one set, namely the whole space, which is the least fine partition, then $E_{\Pi}(W)$ is just the usual expectation $E(W)$. The latter observation leads to some results of interest which we use in Chapter 24. Immediately from part (*c*) of the Theorem 20.6, for any W,

$$E[E_{\Pi}W] = E(W) \tag{20.24}$$

More generally, if Z is any random variable which is constant on the sets of the partition Π, then from part (b) of Theorem 20.6

$$E(ZW) = E[E_{\Pi}(ZW)] = E[ZE_{\Pi}W]. \tag{20.25}$$

20.13.3 The arbitrage-free condition in the bond market

Return to our bond market with the assumption that for all k, the random variable $v(k, n)$ is k-determined. What is the condition on $v(k, n)$ to ensure that the market is arbitrage-free?

Lets go back to the deterministic setting of Chapter 2 for a moment. Note that if a special case of Equation (2.1) holds, namely

$$v(k, n) = v(k, k+1)v(k+1, n) \tag{20.26}$$

for all nonnegative integers $k \leq n$, then a straightforward induction argument shows that (2.1) will hold for all nonnegative integers $k \leq m \leq n$. (Example 20.9 below will make it clear why we begin with Equation (20.26) in place of the more general Equation (2.1))

We saw above that (2.1) does not hold when considered as a statement about random variables, but could it possibly be that this revised version will be valid? The answer is still no, since the left hand side is k-determined, but in general, since $v(k+1, m)$ is only $k+1$-determined, the right side will only be $k+1$-determined. It is plausible, however, that the following natural modification of our statement holds. Namely

$$v(k, n) = v(k, k+1)E_k v(k+1, n). \tag{20.27}$$

It turns out in fact that Equation (20.27) is the correct condition to prevent arbitrage.

Theorem 20.7 *If we can find a probability measure Q on Ω such that Equation (20.27) holds for all nonnegative integers $k < n \leq N$, then the given bond market is arbitrage-free.*

Proof. We will show that under Q, each $\hat{S}_n$ satisfies (20.11) and we can then apply Theorem 20.4. We know from the definition of S_0 that

$$S_0(k+1) = S_0(k)v(k,k+1)^{-1}.$$

Moreover,

$$\hat{S}_n(k+1) = \frac{v(k+1,n)}{S_0(k+1)} = \frac{v(k+1,n)v(k,k+1)}{S_0(k)}.$$

Now $S_0(k)$ is certainly a k-determined random variable, being a product of k-determined random variables, and so therefore is $S_0(k)^{-1}$. Using Theorem 20.6(b) and invoking Equation (20.27)

$$E_k(\hat{S}_n(k+1)) = \frac{E_k[v(k+1,n)]v(k,k+1)}{S_0(k)} = \frac{v(k,n)}{S_0(k)} = \hat{S}_n(k)$$

completing the proof. □

20.13.4 Short-rate modelling

We now deal with a problem that is different from that in previous sections. Instead of being given the asset prices, and asked if the market is arbitrage-fee or not, we are given some prices and want to determine other prices to satisfy the arbitrage-free condition.

For example, refer back to Equation (2.6). This shows in effect that in an deterministic setting, the prices for bonds of one period determine those for all periods. We do in fact have a stochastic version of this formula obtained by taking expectations.

Suppose we have a probability measure Q on Ω and random variables $v(0,1)$, $v(1,2),\dots v(N-1,N)$, where $v(k,k+1)$ is k-determined. We extend this to all our prices by the rule that for $k < n$,

$$v(k,n) = E_k[v(k,k+1)v(k+1,k+2)v(k+2,k+3)\dots v(n-1,n)]. \tag{20.28}$$

This will indeed result in an arbitrage-free market, since

$$\begin{aligned} v(k,k+1)E_k[v(k+1,n)] &= v(k,k+1)E_k[E_{k+1}[v(k+1,k+2)\dots v(n-1,n)]] \\ &= v(k,k+1)E_k[v(k+1,k+2)\dots v(n-1,n)] \\ &= E_k[v(k,k+1)v(k+1,k+2)\dots v(n-1,n)] \\ &= v(k,n). \end{aligned} \tag{20.29}$$

establishing condition (20.27). Here we used the definition (20.28) in the first and last equality, For the second inequality we used Theorem 20.6(c) noting that for any k-admissible sequence v, the set v° is a disjoint union of sets v_i° where each v_i is a $k+1$-admissible sequence – simply add on all possible choices for the last element. So as k increases, the partitioning by k-admissible sequences gets finer. Finally, we use Theorem 20.6(b) for the third inequality.

Example 20.9 Take $N = 3$. As remarked above, we need only consider a two-period model. We will take the binomial model, and let Q be the probability measure that assigns equal probability of $1/4$ to each of the four elements in $\Omega = \{UU, UD, DU, DD\}$. Define the random variables $v(0,1)$, $v(1,2)$ and $v(2,3)$ by $v(0,1) = 0.7$, $v(1,2)(\mathrm{U}) = 0.8$ $v(1,2)(\mathrm{D}) = 0.6$. $v(2,3)(\mathrm{UU}) = 0.9$, $v(2,3)(\mathrm{UD}) = 0.7$, $v(2,3)(\mathrm{DU}) = 0.7$, $v(2,3)(\mathrm{DD}) = 0.5$.

(a) Use (20.28) to find the distribution under Q of the other random variables $v(k,n)$ for $k < n$, which will make the market arbitrage-free.

(b) Show that it is not necessarily true that $v(k,n) = v(k,m)E_k v(m,n)$ if $m \neq k+1$.

Solution. (a) The other random variables are $v(0,2), v(0,3), v(1,3)$which we will calculate directly from Equation (20.28)

$$v(0,2) = E[v(0,1)v(1,2)] = \frac{1}{2}(0.7 \times 0.8 + 0.7 \times 0.6) = 0.49.$$

The 2-determined random variable $v(0,1)v(1,2)v(2,3)$ takes the value of $0.7 \times 0.8 \times 0.9$ on UU, the value of $0.7 \times 0.6 \times 0.7$ on DU, the value of $0.7 \times 0.8 \times 0.7$ on UD, and the value of $0.7 \times 0.6 \times 0.5$ on DD. Each of these paths has probability $1/4$, so

$$v(0,3) = E[v(0,1)v(1,2)v(2,3)] = \frac{1}{4}[0.504 + 0.294 + 0.392 + 0.210] = 0.35$$

Now $v(1,2)v(2,3)$ takes the value of 0.8×0.9 on UU and the value of 0.8×0.7 on UD so

$$v(1,3)(U) = \frac{1}{2}(0.72 + 0.56) = 0.64$$

Also, $v(1,2)v(2,3)$ takes the value of 0.6×0.7 on DU and the value of 0.6×0.5 on DD so

$$v(1,3)(D) = \frac{1}{2}(0.42 + 0.30) = 0.36$$

(b) $v(0,2)E_2[v(2,3)] = (0.49)(0.7) = 0.343 \neq v(0,3)$

To tie this example in with previous material, the reader may find it instructive to reproduce the figures shown in Figures 20.7 and 20.8. Figure 20.7 shows all the asset prices for all four assets, Figure 20.8 shows the values of $\hat{S}_k(n)$ for $k = 1, 2, 3$. The values of $\hat{S}_0(n)$ are of course all equal to 1.

Instead of specifying the one-period bond prices, we could equivalently have specified the interest rates as we did in Chapter 2, except that now $i_k = v(k, k+1)^{-1} - 1$, which is a k-determined random variable.

We have shown that we can model stochastic interest rates in much the same way as we did in the deterministic model, that is, by choosing the one period rates first. For this reason, this procedure is sometimes referred to as short-rate modelling. What may be surprising at first, is that in the stochastic case the result in far from unique. We can indeed specify any probability measure we want for each i_k, and achieve an arbitrage-free bond market. The procedure does not tell us how to choose the random variables i_k and the probability measure

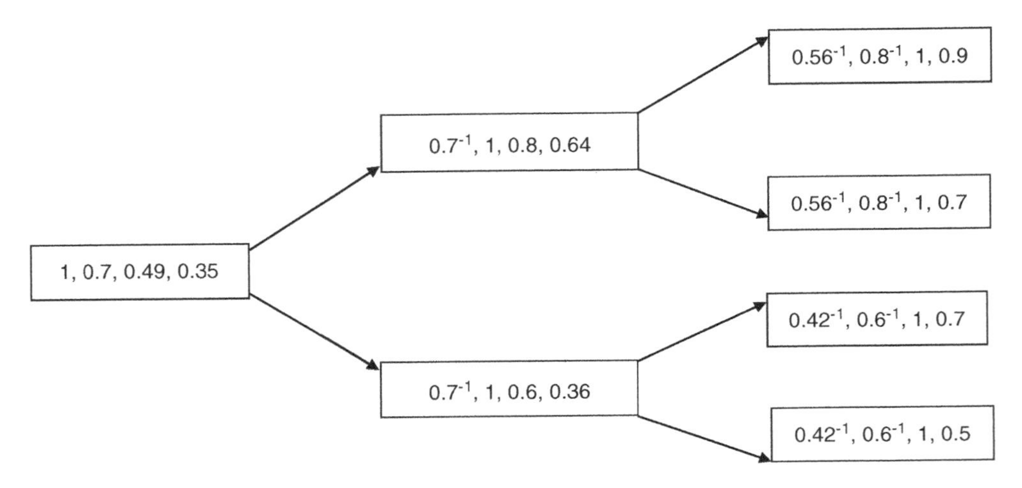

Figure 20.7 Example 20.9. Values of $S_k(n)$, $k = 0, 1, 2, 3$

Q. In practice, this can be motivated by trying to satisfy other conditions that one wants to impose. For example, at time 0, one knows the values $v(0, n)$. But as we saw in Example 22.7, the values $v(0, n)$ are determined from the one-period rates. A problem of some interest, which we will not deal with here, is to choose the distributions of the i_k so as to recover specified values of $v(0, n)$.

20.13.5 Forward prices and rates

We now expand on our definition of forward bond prices, which we introduced in Section 2.12

Suppose we are given a financial bond market as above. For $j \leq k \leq n$, we let

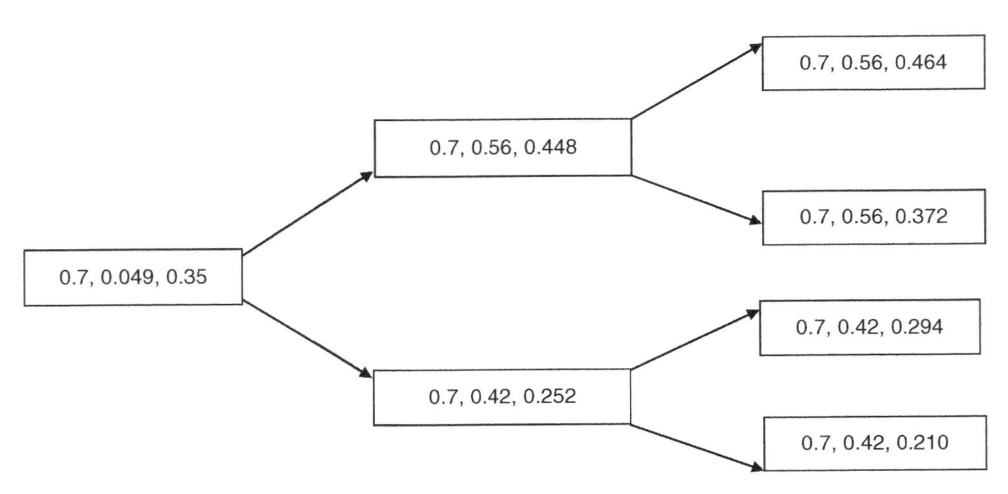

Figure 20.8 Example 20.9. Values of $\hat{S}_k(n)$, $k = 1, 2, 3$

$\tilde{v}_j(k,n)$ = the forward price to be paid at time j for a 1-unit zero-coupon bond issued at time k and maturing at time n.

This will be a j-determined random variable, since it is random depending on the state of nature up to time j. The quantity $\tilde{v}(k,n)$ defined in Section 2.12 could be labeled as $\tilde{v}_0(k,n)$.

The same argument as used in Section 2.12 shows that to prevent arbitrage, we must have

$$\tilde{v}_j(k,n) = \frac{v(j,n)}{v(j,k)}.$$

Indeed, if there is any sample point at which there is a discrepancy between these two random variables, an individual could wait and see if this particular sample point materialized at time j, which would be known by the fact that the random variables are j-determined. One could then follow one of two strategies, depending on which way the inequality went, to achieve a sure profit at time n.

In particular $\tilde{v}_j(j,n) = v(j,n)$ which is clear from the definition.

In place of the forward prices, we can describe the information available from the bond prices at time j by interest rates as we did before. For $j \leq k$, we define a j-determined random variable $i_j(k)$, the *forward interest rate* for contracts at time time j applicable to time period k to $k+1$, by

$$\tilde{i}_j(k) = (\tilde{v}_j(k,k+1))^{-1} - 1$$

Note that $i_k(k) = i_k$. In the deterministic case $i_j(k) = i_k$ for all k.

It follows immediately from the definitions that

$$v(k,n) = \tilde{v}_k(k,k+1)\tilde{v}_k(k+1,k+2)\ldots\tilde{v}_k(n-1,n) = [(1+\tilde{i}_{k,k})(1+\tilde{i}_{k,k+1})\ldots(1+i_{k,n-1})]^{-1} \tag{20.30}$$

so that bond prices are determined by the forward interest rates. This suggests an alternative procedure to short-rate modelling for bond prices. The method is to model all the forward interest rates in place of the short rate i_k. In this case, the forward rates at time 0 are all known, so the model will clearly reproduce the observed prices at time 0. One cannot, however, choose the distributions of the forward rates arbitrarily as for the short rates. Certain conditions must be imposed to ensure that the model is arbitrage-free. (See Exercise 20.12). This is another topic that is beyond our scope for further elaboration.

20.13.6 Observations on the continuous time bond market

We have concentrated on the discrete time setting for bond market, but many applications involve a continuous time framework. In this section, we give a very quick overview. It is intended mainly to bridge the gap and prepare those who wish to look at other sources and compare the material there with what we have done above.

Our market now will consist of zero-coupon bonds maturing at any time t, with a price at time s, of $v(s,t)$. This will be a s-determined random variable under an appropriate extension of this definition to cover continuous time. As well we can define an extension of the conditional expectation E_s. We will not go into details.

In place of the forward and spot rates of interest, one often wants to deal with the corresponding forces of interest, generalizing the quantity $\delta(t)$ to a stochastic setting. The forward force of interest, for contracts at time s applicable to time t is given by

$$\tilde{\delta}_s(t) = -\frac{\partial}{\partial t} \log v(s,t),$$

an s-determined random variable.

In the deterministic case, when we can assume that our basic Equation (2.1) holds, then $\log(v(s,t) = \log v(t) - \log v(s)$ and all forward forces will be equal to $\delta(t)$ as defined in (8.3).

Knowing the forward forces determines bond prices uniquely, since by the definition

$$v(s,t) = e^{-\int_s^t \tilde{\delta}_s(r)dr},$$

which we can view as a stochastic version of Equation (8.4) and as a continuous version of Equation (20.30).

We can also define a continuous and stochastic version of the short rate. Let $\delta(t)$ denote $\tilde{\delta}_t(t)$. This perhaps requires some care, since the partial derivative applies only to the second variable. Precisely

$$\delta(t) = -\lim_{h \to 0} \frac{\log v(t, t+h)}{h},$$

which will be a t determined random variable, and one which agrees with our Chapter 8 definition in the deterministic case. As in the discrete time case, this will not uniquely determine bond prices in a stochastic setting. Given any stochastic process for $\delta(t)$, we can produce an arbitrage-free bond market by taking

$$v(s,t) = E_s[e^{-\int_s^t \delta(r)dr}],$$

under a suitable interpretation of an integral of a random integrand.

Notes and references

This chapter constitutes a basic introduction to financial markets. For more comprehensive coverage of the basic concepts, see Hull (2014) or McDonald (2012). Readers particularly interested in the mathematics of continuous-time models can consult Björk (2009) or Etheridge (2002) for sources that are not overly advanced.

For many years, formula (20.19) was referred to as the Black–Scholes formula, recognizing the authors of the original paper that was published. Recently, many writers have added the name of Robert Merton, who made important contributions to developing and extending the ideas behind this result.

A proof of the extension result on hyperspaces can be found in Steland (2012), Theorem 2.4.5.

Those reading the financial literature should be aware that the terminology can differ from the actuarial conventions that we have adopted. Financial economists often use the word *rate* of interest to mean the continuously compounded rate, which we have called the force of interest.

Exercises

20.1 A one-period financial market has in addition to the bank account, one risky asset. The price at time 0 is 50, and at time 1, the price will either be 52 or 55 each with positive probability. For what value of the risk-free interest rate r will this market be arbitrage-free?

20.2 For the option contract introduced in Section 20.5, find an arbitrage opportunity if the option price is 11.

20.3 For the option contract introduced in Section 20.5, find the price if r is changed to (a) 11%, (b) 9 %. Can you explain why the prices change in this way?

20.4 Show that a trading strategy is self-financing if and only if, for $0 \leq n < N$,

$$V(n+1) - V(n) = \sum_{j=0}^{M} \alpha_j(n)(S_j(n+1) - S_j(n)).$$

20.5 Assume the data of Example 20.2. Find the self-financing trading strategy for a 2-year American put option with a strike price of 106. Compare the price with the corresponding European put option.

20.6 Consider the financial market with $N = 1$, $M = 1$, $r = 0$, $S_1(0) = 100$, $S_1(1) =$ 130, 110, or 80, all with positive probability.

(a) Find all possible risk-neutral probability measures.

(b) Consider a 1-year call option with a strike price of 100. (This is known as an 'at-the-money' option). Find the range of possible prices for this option that will avoid arbitrage.

(c) If the price of the option in part (b) is 12, find an arbitrage opportunity.

20.7 Consider the two risky asset market of Figure 20.6 with $r = 0$;

(a) Use Theorems 20.4 and 20.5 to decide whether or not this market is (i) arbitrage-free, (ii) complete.

(b) Describe the subspaces L and L_0.

20.8 In Figure 20.3, find the slopes of the two lines representing L_0 in terms of u, d and r.

20.9 When pricing options with the Black–Scholes–Merton formula, describe how option prices change, as changes occurs in volatility, strike price, duration, and risk-free force of interest. Use the formula and put-call parity to compute put and call prices for 3-month European options, on a stock selling now for 100 with a strike price of 110, 100 and 90 respectively, assuming σ is 0.2 and the force of interest for a 1-year period is 0.06. Verify your conclusions above by repeating the calculations for other values of σ, δ and duration.

20.10 For a certain stock, the price of a call option is 7.80 and the price of the corresponding put option with the same strike price and expiration date is 1.50. In each of the following scenarios, find the new price of the put option.

(a) A change in volatility raises the call price to 11.30.

(b) The stock price at time 0 is 100. A 10% increase in the strike price lowers the call price to 5.00.

(c) The risk-free force of interest is doubled and the time to expiration is cut in half, which combine to lower the call price to 7.00.

20.11 Consider a 6-month European call option on a stock now selling for 100, with a strike price of 97. You are given that $\sigma = 0.25$ and the force of interest for a 1-year period is 0.10. You would like to replicate this option by a trading strategy involving holdings in the stock and risk-free zero-coupon bonds of face amount 97, maturing in 6 months. You plan to use the Black–Scholes–Merton formula.

(a) What is the initial portfolio at time 0?

(b) At the end of 2 months, the stock price has risen to 105. What is your portfolio now?

(c) At the end of 4 months, the stock price has fallen to 95. What is your portfolio now?

20.12 Give a direct verification for the Black–Scholes–Merton formula in the case that $\sigma = 0$.

20.13 Verify directly from Figure 20.8 that the market of Example 20.9 is arbitrage-free and complete.

20.14 Redo Example 20.9 only assuming that the probability of an upward move is $3/4$ and that of a downward move is $1/4$.

20.15 A bond market has two transitions, U and D from time 0 to time 1. You are given the following forward prices:

(i) $\tilde{v}(0,1) = \tilde{v}(1,2) = \tilde{v}(2,3) = 0.7$;

(ii) $\tilde{v}_1(1,2)$ and $\tilde{v}_1(2,3)$ both take the value 0.8 on U and 0.6 on D.

(a) If $Q(U) = Q(D) = 0.5$, show that Equation (20.27) is not satisfied.

(b) Show that there is no probability measure Q for which Equation (20.27) is satisfied.

20.16 Consider the bond market with prices as given in Figure 20.7. Consider a call option on a bond maturing for 1000 at time 3, with expiration date time 2. Find the price of the option if the strike price is (a) 900, (b) 650.

*20.17 A universal life contract provides that your account will be credited with a minimum of 4% interest, up to a maximum of 10% interest. Describe the options present in this arrangement.

Part IV

RISK THEORY

21

Compound distributions

21.1 Introduction

In earlier parts of this book, we concentrated on the present value of the benefits paid on a *single* insurance or annuity contract. The insurer, of course, is interested in the total benefits paid on an entire portfolio of policies. An obvious way to handle this is simply to obtain the present value of the total amount paid on all policies in the portfolio, as the sum of the individual random variables. This is known as the *individual* risk model. There is another method for estimating the total amount paid on a group of policies, known as the *collective risk* model, which has advantages in certain cases. In this chapter, we deal with a static version of this model, covering a 1-year time period. Chapter 23, concerned with ruin theory, will involve a dynamic multi-period version of the collective risk model. The combined subject matter of these chapters has traditionally been referred to as *risk theory* in the actuarial literature. The collective risk model is particularly useful for casualty insurance such as automobile, home, or health policies. The following are three main ways in which such contracts differ from life insurance:

1. In a given period, there can be several claims under a single policy. Clearly, you can have several accidents or several visits to the doctor, even in a relatively short period. However, no matter how long the period, you can only die once.

2. The amount of each claim can vary substantially. A collision claim under an automobile policy can range from a small amount for a dented fender, to the complete cost of the vehicle. A health claim may involve a single visit to a doctor, or it may involve prolonged treatment, drugs, and hospital care costing a large amount of money. By contrast, although the amount paid on a life insurance policy can vary by time of death, we do not have variation of the amount for different 'kinds' of dying.

Fundamentals of Actuarial Mathematics, Third Edition. S. David Promislow.

Companion website: http://www.wiley.com/go/actuarial

3. Such contracts are usually written for a relatively short period such as a year, and are then renewed if the insured wishes to continue. They do not have the long-term nature of the life contracts we have discussed. Consequently, the effect of interest is not so important. To simplify the mathematics, the effects of interest will be ignored in the models of the subsequent chapters.

The collective risk model views total claims as a compound distribution, which we will now examine. To motivate the idea, consider the following game. Toss two coins, and for each head that comes up, throw a die. What is the distribution of the total? First, we identify the range. The possible totals can range from 0, which occurs if you toss two tails, to 12, which occurs if you toss two heads, and get a 6 on each of the two throws of the die. After an elementary but somewhat tedious calculation, we can arrive at the following distribution, which the reader should verify before proceeding any further. The probabilities of 0 to 12 respectively, in multiples of $1/144$, are 36, 12, 13, 14, 15, 16, 17, 6, 5, 4, 3, 2, 1.

We could complicate this problem tremendously. Instead of two coins, toss 1000. Instead of a simple die throw, choose a much more complicated random variable, possibly one with a continuous distribution. It may become impossible to actually calculate the exact distribution as we did above, but we still may want to say something. At the very least we want to compute the mean and variance of the resulting distribution, or possibly higher moments. We may want to calculate the moment generating function. We may be able to find a known distribution that closely approximates the one we are interested in.

What is the relation of this game to insurance? The collective risk model identifies two main factors that influence the total claims. One is the claim *frequency*, that is, the number of claims that will occur over a certain period. This will be a discrete random variable taking nonnegative integers as values. The second factor is the amount that will be paid, given that a claim has occurred. This is known as the *severity* of the claim. We have observed that in any fixed period under a life insurance policy, the claim severity is normally just a constant, but under other types of insurance, it will vary substantially. Even though the insurer is ultimately interested in the total payout, it has been found advantageous to first model frequency and severity separately and to then combine the results to determine total claims. One reason for this is that changed conditions can affect these factors in different ways. For example, requiring automobile passengers to wear seat belts has little effect on the frequency of car accidents, but it certainly tends to reduce the claim payments for personal injuries. On the other hand, the introduction of daytime headlights is likely to have little effect on the severity of claims, but it might well reduce the number of accidents. Another example involves the effect of seasonal differences. It may be that people drive faster during summer months, when the weather is better, so a typical summer accident is more serious than one in the winter. By contrast, one might well expect more accidents in the winter.

We will now describe the formal model. We have a fixed period, a collection of policies, and we want to predict S, the total claims from all policies over that period. We let N denote the frequency of claims and X the severity, both of which we model as random variables. Throughout the discussion, we make some standard assumptions, which are reasonable in most insurance situations, although they may not always hold exactly. We assume that the severity of claims is independent of the frequency and that the severity of any one claim is independent of that of others. We also postulate that the severity follows the same distribution over the period. This will normally hold for a sufficiently small period, although it may not for longer periods due to seasonal differences such as that alluded to above.

Let X_i denote the amount of the ith claim. Our last assumption says that there is a single severity distribution given by a random variable X, and we assume that each X_i is distributed as X. The total amount of claims is then simply given by the independent sum

$$S = X_1 + X_2 + \cdots + X_N. \tag{21.1}$$

We have encountered sums of random variables before, but the above is quite different, since the number of summands is *random*, rather than a fixed integer.

We can now observe that the simple coin–dice problem we mentioned at the beginning is really an example of this type, where N takes the values $0, 1, 2$ with respective probabilities $1/4$, $1/2$, $1/4$, and X takes the value $1, 2, 3, \ldots, 6$ with equal probabilities.

The distribution of S is known as a *compound distribution*, which is usually prefaced by referring to the distribution of N. For example, if N is Poisson, we call S a *compound Poisson* distribution. This comes from the fact that when X is a random variable that takes the value 1 with certainty, then the resulting compound distribution is just N itself.

To summarize the goals in this chapter, we will be given N and X, and our object is to investigate the compound distribution S as given by (21.1). We will denote this distribution by the symbol $\langle N, X\rangle$.

21.2 The mean and variance of S

As mentioned, although it may be difficult to calculate the distribution of S exactly, it is quite simple to find its mean and variance, given the corresponding quantities for N and X. Throughout this chapter, we will let p denote the probability function of N. From the law of total expectation (A.29)

$$E(S) = \sum_n E(S|N = n)p(n).$$

When $N = n$, S is just the sum of n independent copies of a random variable with the distribution of X. By using the fact that N and X are independent, we can write

$$E(S|N = n) = nE(X|N = n) = nE(X),$$

leading to

$$E(S) = E(X)E(N), \tag{21.2}$$

an intuitively obvious result. Similarly,

$$E(S^2) = \sum_n E(S^2|N = n)p(n).$$

The second moment of any random variable is the sum of the variance plus the square of the mean, so that

$$E(S^2|N = n) = \mathrm{Var}(S|N = n) + [E(S|N = n)]^2 = n\mathrm{Var}(X) + n^2(E(X))^2.$$

We use here the fact that the variance of a sum of independent random variables is the sum of the variances. The last two formulas yield

$$E(S^2) = E(N)\mathrm{Var}(X) + E(N^2)E(X)^2,$$

and subtracting the term $E(S)^2 = E(N)^2E(X)^2$, we obtain

$$\mathrm{Var}(S) = E(N)\mathrm{Var}(X) + E(X)^2\mathrm{Var}(N). \qquad (21.3)\ddagger$$

There is an informative explanation of the above formula. Variance represents uncertainty, and this decomposes the uncertainty in the value of S into two parts. The first term gives the uncertainty resulting from the severity, and second term gives the uncertainty arising from the frequency.

21.3 Generating functions

The same conditioning technique as used above can be employed to deduce the moment generating function (m.g.f) $M_S(t)$ and the probability generating function (p.g.f.) $P_S(t)$ (see Sections A.9 and A.10). When $N = n$, we have that $S = X_1 + X_2 + \cdots + X_n$, and so

$$E(e^{tS}|N = n) = E[e^{t(X_1+X_2+\cdots X_n)}] = E[e^{tX_1}e^{tX_2}\ldots e^{tX_n}] = E(e^{tX_1})E(e^{tX_2})\ldots E(e^{tX_n}),$$

where we invoke independence in order to write the expectation of a product as a product of expectations. Since each X_i is distributed as the random variable X,

$$E(e^{tS}|N = n) = M_X(t)^n,$$

and

$$M_S(t) = \sum_{n=0}^{\infty} E(e^{tS}|N = n)p(n) = \sum_{n=0}^{\infty} M_X(t)^n p(n),$$

from which we conclude that

$$M_S(t) = P_N(M_X(t)). \qquad (21.4)\ddagger$$

When X takes nonnegative integer values, we can use A.34 and A.35 to conclude that

$$P_S(t) = M_S(\log t) = (M_X(\log t)) = P_N(P_X(e^{\log t})),$$

giving us the nice result that

$$P_S(t) = P_N\big(P_X(t)\big). \qquad (21.5)\ddagger$$

21.4 Exact distribution of S

In the previous two sections, we considered the problem of getting partial information about S through moments and generating functions, but it is natural to ask if we can find the *exact* distribution of this random variable. The answer is that it is easy enough to write down a formula for this, but in all but some very simple cases, it is not at all easy to actually use the formula to calculate numbers.

What is the probability that S takes a value less than or equal to s? Once again we use the conditioning technique. If $N = n$, then the answer is just the probability that $X_1 + X_2 + \cdots + X_n \leq s$, which is just the n-fold convolution of F_X at the point s, which we have denoted by $F^{*n}(s)$ (see Section A.12). It then follows that

$$F_S(s) = \sum_{n=0}^{\infty} p(n) F_X^{*n}(s), \tag{21.6}$$

or similarly, by using density/probability functions,

$$f_S(s) = \sum_{n=0}^{\infty} p(n) f_X^{*n}(s), \tag{21.7}$$

where $f^{*0}(k)$ takes a value of 1 for $k = 0$ and zero elsewhere, and f^{*1} is just f.

Example 21.1 Suppose that N takes the values $0, 1, 2$ with probabilities 0.5, 0.3, 0.2, respectively, and X takes the values $1, 2, 3$ with probabilities 0.4, 0.2, 0.4, respectively. Find the distribution of S.

Solution. We form the following table, in which for each row, the entries are multiplied by the weights in the bottom row to get the totals in the far right hand column:

k	$f^{*(0)}(k)$	$f^{*(1)}(k)$	$f^{*(2)}(k)$	$f_S(k)$
0	1	0	0	0.500
1	0	0.4	0	0.120
2	0	0.2	0.16	0.092
3	0	0.4	0.16	0.152
4	0	0	0.36	0.072
5	0	0	0.16	0.032
6	0	0	0.16	0.032
Weights	0.5	0.3	0.2	

21.5 Choosing a frequency distribution

Given a certain portfolio of insurance policies, how does the insurer select appropriate distributions in order to model the aggregate claims S? In this section, we focus on the claim frequency N. We could conceivably do this by strictly empirical means. We might use past data from similar policies to try to estimate a distribution. This is a statistical problem that we do not concentrate on in this book. There are, however, many advantages to choosing a distribution from one of several well-known families of discrete distributions. We then have nice mathematical expressions for the distribution. These families are based on one or more parameters. The estimation procedure is confined to choosing just these parameters from the observed data rather than the entire distribution. Three families that play major roles in modelling claim frequency are the binomial, Poisson, and negative binomial distributions. Details and notation are given in Sections A.11.1, A.11.2 and A.11.3 respectively.

Is the binomial a suitable distribution for claim frequency? To illustrate, suppose that our period of time is 30 days, and our observed data show that on average we can expect 10 claims over each 30-day period. Suppose also that we now assume a *time homogeneity* for frequency, which is analogous to the assumption for severity that we made as part of our general postulates. That is, we assume that the rate of claims remains constant over the period. (This assumption may not be completely realistic in certain cases. For automobile insurance, for example, there are more chances of an accident occurring during the rush hour than in the middle of the night.) As a final condition, suppose we assume that we will get at most one claim per day. We then can look upon this as 30 repeated trials. Each day we either get a claim, which constitutes 'success', or no claim which constitutes 'failure'. In order that the expected number of claims equals our estimated value of 10, we must take the probability of a claim each day to be $1/3$. So indeed, under our assumptions, we can model N by $\text{Bin}(30, 1/3)$.

However, what if we decide that our limit of one claim a day is not really an accurate assumption, and that we may well experience more on some days? We could get a more accurate model by assuming that there would be no more than one claim each half-day period. We would still get a binomial distribution, but now with $m = 60$, and we have to change p to $1/6$ to preserve an expectation of 10. Perhaps, however, even this half-day limitation is not quite accurate, and we should replace it by an hour, or perhaps a minute, or even a second. In fact, why not allow complete freedom and take the limiting distribution? This leads immediately to the the *Poisson distribution* which is one of the most common distributions used for modelling claim frequency. It arises in a natural way from our independence and time-homogeneous assumptions, by taking the limit of binomials.

It is worthwhile to note that the variance of a compound Poisson distribution has a particularly simple form. If $S \sim \langle N, X\rangle$, where $N \sim \text{Poisson}(\lambda)$, then from A.42 , $E(N) = \text{Var}(N) = \lambda$, so that

$$\text{Var}(S) = \lambda E(X^2). \tag{21.8}$$

The *negative binomial* is also a popular choice for modelling insurance claims, but the reason is not immediate. It stems from the following idea. Suppose we assume that each insured individual produces claims according to a Poisson distribution, but that the parameter of this distribution can differ according to this individual. For a simple example, suppose that automobile drivers are classified as either good or bad. Assume that the claims of the good drivers are distributed as Poisson(1), while the claims of the bad drivers are distributed as Poisson(2). Assume, furthermore, that 60% of drivers are good and the rest are bad, but that the insurer has no way of distinguishing between the classes. It would then be reasonable to model the claims frequency N as a mixture of the two distributions Poisson(1) and Poisson(2), with respective weights 0.6 and 0.4 (see Section A.13). Of course, this is oversimplified and we could strive for a more sophisticated model by considering a mixing distribution involving several different values. We may even consider letting the parameter vary continuously over an entire interval of positive numbers and take a continuous mixing distribution. Remarkably, it turns out that if we take the *gamma distribution* (see Section A.11.6) for this purpose, our mixed Poisson is a negative binomial. Precisely, if we want a mixture of random variables $N_\Lambda \sim \text{Poisson}(\Lambda)$, in which Λ has a Gamma(α, β) distribution, then the resulting mixed distribution N satisfies

$$
\begin{aligned}
f_N(k) &= \frac{\beta^\alpha}{\Gamma(\alpha)} \int_0^\infty \lambda^{\alpha-1}\, \mathrm{e}^{-\beta\lambda}\, \mathrm{e}^{-\lambda} \frac{\lambda^k}{k!} \mathrm{d}\lambda \\
&= \frac{\beta^\alpha}{\Gamma(\alpha)k!} \int_0^\infty \lambda^{\alpha+k-1}\, \mathrm{e}^{-\lambda(\beta+1)} \mathrm{d}\lambda = \frac{\beta^\alpha}{\Gamma(\alpha)k!} \frac{\Gamma(\alpha+k)}{(\beta+1)^{\alpha+k}}.
\end{aligned}
$$

Applying (A.53) repeatedly gives

$$
\begin{aligned}
\Gamma(\alpha+k) &= (\alpha+k-1)\Gamma(\alpha+k-1) = (\alpha+k-1)(\alpha+k-2)\Gamma(\alpha+k-2) = \ldots \\
&= (\alpha+k-1)(\alpha+k-2)\ldots\alpha\Gamma(\alpha).
\end{aligned}
$$

By comparing with (A.45) we see that $N \sim \text{Negbin}(\alpha, (\beta+1)^{-1})$.

21.6 Choosing a severity distribution

What distributions are suitable for measuring claim amounts? For many types of insurance, claims can assume a large number of values, and it is usually convenient to model claims by a continuous distribution. We will discuss some possibilities. See Sections A11.5–A11.8 for details and notation.

Is a *normal* distribution a suitable one for modelling severity? A possible drawback is that the claim distribution will almost always be positive-valued (it is not usual for the policyholder to pay the insurer), and the normal of course takes values over the entire real line. This by itself is not a major concern. If the mean of a normal is sufficiently high relative to its standard deviation, there will be so little chance of a negative value that for all practical purposes, we may as well consider it as positive-valued.

A more major difficulty is that the normal density is not the right shape for most applications, as it does not give sufficiently high weight to lower valued claims. We usually want a distribution that has a greater concentration of mass on the left. (This could be described as a distribution with the mean greater than the median.) The family of *gamma* distributions does have this general shape that we want, and provide a popular choice for severity modelling.

Another important criterion for selecting a severity distribution is *tail behaviour*. For any distribution X, the function $s_X(t) = P(X > t)$ approaches 0 as t approaches ∞. However, the rate at which convergence to 0 occurs will differ. We say that the distribution X has *heavier right tails* than the distribution of Y if $s_Y(t)/s_X(t)$ approaches 0 (or equivalently, by L'Hôpital's rule, if $f_Y(t)/f_X(t)$ approaches 0). A heavier-tailed distribution therefore gives more weight to large values. If one desires a heavier-tailed distribution than the gamma, the *Pareto* distribution (see Section A.11.8.) is a possible choice. Using L'Hôpital's rule to find the limit of the ratio of density functions, we can verify that this is heavier-tailed than any gamma distribution. The heavy-tailed feature of this distribution is further revealed by the fact that for large enough k, the kth moment becomes infinite.

21.7 Handling the point mass at 0

When X is discrete, then S is clearly discrete, but what happens when X is continuous? Provided that N takes the value 0 with positive probability, we will have a distribution of *mixed type*. Since $S = 0$ whenever $N = 0$, the distribution for S will have what is known as a *point mass* at the point 0. It is often convenient to split off the continuous part. Let

$$S^+ \text{ denote the random variable } S|S > 0,$$

(see Section A.8) which will be continuous when X is.(Note that this is different from $S_+ = S|S \geq 0$). Then S can be considered as a mixture of S^+ and 0 (the random variable that always takes the value 0), with respective weights $1 - p(0)$ and $p(0)$. It follows from (A.67) that

$$F_S(s) = p(0)P(0 \leq s) + (1 - p(0))F_{S^+}(s), \qquad s \geq 0.$$

Since the zero random variable is always less than or equal to s,

$$F_{S^+}(s) = \frac{F_S(s) - p(0)}{1 - p(0)}.$$

Similarly, since the m.g.f. of the zero random variable is identically equal to 1, the m.g.f. of S^+ is given by

$$M_{S^+}(t) = \frac{M_S(t) - p(0)}{1 - p(0)}.$$

Example 21.2 Suppose that N has a Geom(p) distribution and X has an Exp(λ) distribution. What is the distribution of S?

Solution. For a general severity distribution X,

$$M_S(t) = P_N(M_X(t)) = (1 - p)(1 - pM_X(t))^{-1},$$

and since $p(0) = 1 - p$,

$$M_{S^+}(t) = \frac{(1 - p)(1 - pM_X(t))^{-1} - (1 - p)}{p} = (1 - p)\left(\frac{M_X(t)}{1 - pM_X(t)}\right).$$

By substituting $M_X(t) = \lambda/(\lambda - t)$ from (A.56), we have

$$M_{S^+}(t) = \frac{\lambda(1 - p)}{\lambda(1 - p) - t},$$

which from (A.56) again is the m.g.f. of an Exp($\lambda(1 - p)$) distribution. Invoking Theorem A.2, we know that the distribution of S is a mixture of 0 and an Exp($\lambda(1 - p)$) distribution with weights $1 - p$ and p, respectively.

21.8 Counting claims of a particular type

21.8.1 One special class

Suppose we divide our claims into two groups, 'special' and 'non-special,' and we are interested in knowing the number of special claims, as well as the total. Denote the special claim frequency by N_1. Our problem is to deduce the distribution of N_1 given the distribution of N. The special claims can be anything at all – those of high amount, those of low amount, those divisible by 79. It does not really matter. All we need to know is the probability of a special claim and we can write down a formula for the distribution of N_1. What is the probability that $N_1 = k$, given that the probability of a special claim is π? In order to get k special claims, there must be at least k claims in total. That is, N must take the value of $k + r$ for some nonnegative integer r. From the law of total probability (A.30),

$$P(N_1 = k) = \sum_{r=0}^{\infty} P(N_1 = k|N = k + r)p(k + r).$$

Given that we have $k + r$ claims in total, the number of special claims out of these is distributed as $\text{Bin}(k + r, \pi)$, so we can substitute in the above to get

$$P(N_1 = k) = \sum_{r=0}^{\infty} \frac{(k + r)!}{k!r!}\pi^k(1 - \pi)^r p(k + r). \tag{21.9}$$

Example 21.3 N takes the values 0, 1, 2, 3, 4 with probabilities 0.3, 0.1, 0.3, 0.2, 0.1 respectively, and X takes the values 1 to 100 with equal probabilities. What is the probability that we have exactly 2 claims for an amount less than or equal to 60?

Solution. The special claims are those for an amount less than or equal to 60, so $\pi = 0.6$.

$$P(N_1 = 2) = 0.6^2(0.3 + 3 \times 0.4 \times 0.2 + 6 \times 0.4^2 \times 0.1) = 0.228\,96.$$

While formula (21.9) may be good for calculating individual probabilities, it can be tedious to calculate the entire distribution. It is often better to proceed by calculating the p.g.f. of N_1. We do this by making the ingenious observation that N_1 is itself a compound distribution. In fact, $N_1 \sim \langle N, \delta_\pi \rangle$, where δ_π takes the value 1 with probability π and 0 with probability $1 - \pi$. This is clear, since we can count special claims by simply assigning a value of 0 whenever we get a non-special claim. We note that δ_π is a Bernoulli random variable, that is, it has a binomial distribution with $m = 1$ and, therefore, its p.g.f. is $1 - \pi + \pi t$. From (21.5),

$$P_{N_1}(t) = P_N(1 - \pi + \pi t).$$

This formula allows us to show that for our three major counting distributions, Poisson, binomial, and negative binomial, N_1 is of the same type as N, but with a changed parameter.

If $N \sim \text{Poisson}(\lambda)$, then

$$P_{N_1}(t) = e^{\lambda(1-\pi+\pi t-1)} = e^{\lambda\pi(t-1)},$$

showing that $N_1 \sim \text{Poisson}(\lambda\pi)$.

If $N \sim \text{Bin}(m, p)$, then

$$P_{N_1}(t) = (1 - p + p(1 - \pi + \pi t))^m = (1 - p\pi + p\pi t)^m,$$

showing that $N_1 \sim \text{Bin}(m, p\pi)$.

The last case is a bit tricker. Suppose $N \sim \text{Negbin}(r, p)$. Can we expect N_1 to be a negative binomial with changed parameters? Motivated by the binomial case, we might try leaving r the same and modifying p. We cannot, however, take πp for the new value of p since that would not give us the correct value of $\pi p/(1 - p)$ for $E(N_1)$ that we must have by formula (A.46). We will, however, at least get the right mean if we multiply α by π, where $\alpha = p/(1 - p)$ is the alternate parameter to p mentioned in Section 21.5. This indeed is the right answer. To verify this, it is convenient to express the probability function of N in terms of this new parameter. We can write

$$P_N(t) = \left(\frac{1 - pt}{1 - p}\right)^{-r} = (1 + \alpha - \alpha t)^{-r} = [1 - \alpha(t - 1)]^{-r}.$$

So if M is the negative binomial with the first parameter r and the second (modified) parameter $\alpha\pi$, then

$$P_{N_1}(t) = P_N(1 - \pi + \pi t) = [1 - \alpha(\pi t - \pi)]^{-r} = P_M(t).$$

Therefore, reverting to our original parameter,

$$N_1 \sim \text{Negbin}\left(r, \frac{p\pi}{1 - p + p\pi}\right).$$

21.8.2 Special classes in the Poisson case

Suppose now that we have two special classes, with the number of special claims in the two classes denoted by N_1 and N_2, respectively. We can write down a formula similar to (21.9) for the joint distribution of N_1 and N_2. If the probability of a claim in the first class is π_1 and that of a claim in the second class is π_2, then

$$P(N_1 = k \text{ and } N_2 = m) = \sum_{r=0}^{\infty} \frac{(k + m + r)!}{k!m!r!} \pi_1^k \pi_2^m (1 - \pi_1 - \pi_2)^r p(k + m + r).$$

What do you suppose the covariance of N_1 and N_2 will be? We would naturally expect this to be negative. A high value for one type of claim would seem to indicate that there are fewer of the other type. Indeed, this will be the case for most distributions, but remarkably, for the particular case where N is Poisson, the two distributions are independent. If $N \sim \text{Poisson}(\lambda)$, we obtain from the above

$$\begin{aligned}
P(N_1 = k \text{ and } N_2 = m) &= \sum_{r=0}^{\infty} \frac{(k+m+r)!}{k!m!r!} \pi_1^k \pi_2^m (1-\pi_1-\pi_2)^r \mathrm{e}^{-\lambda} \frac{\lambda^{k+m+r}}{(k+m+r)!} \\
&= \frac{\mathrm{e}^{-\lambda\pi_1}(\lambda\pi_1)^k}{k!} \frac{\mathrm{e}^{-\lambda\pi_2}(\lambda\pi_2)^m}{m!} \sum_{r=0}^{\infty} \mathrm{e}^{-\lambda(1-\pi_1-\pi_2)} \frac{\lambda(1-\pi_1-\pi_2)^r}{r!}.
\end{aligned}$$

The third term above equals 1, since it is the sum over all nonnegative integers of the probability function for a Poisson($\lambda(1-\pi_1-\pi_2)$) distribution. We know from above that $N_i \sim \text{Poisson}(\lambda\pi_i)$ for $i = 1, 2$, so

$$P(N_1 = k \text{ and } N_2 = m) = P(N_1 = k)P(N_2 = m),$$

and the independence result is proved. We can similarly show that if we have r special classes, with N_r denoting the number of claims in class r, then the collection $(N_1, N_2, \ldots, N_r)$ will be independent.

This allows us to write certain compound Poisson distributions in an alternate form, which is sometimes useful. Suppose that X takes finitely many values, say $x_1, x_2, \ldots, x_n$. Let π_i denote the probability that $X = x_i$. Let N_i be the number of claims for amount x_i. We can then write

$$S = \sum_{i=1}^{n} x_i N_i.$$

This does not use the fact that N is Poisson and is true for any frequency distribution. The problem is that, in general, this formulation is of little use. We may not be able to easily identify the distribution of the various N_i and, even if we can, such as in the binomial or negative binomial cases, they will not be independent. In general, it can be very difficult to deal with dependent sums. In the Poisson case, however, we know that $N_i \sim \text{Poisson}(\lambda\pi_i)$ and that the N_i are independent. The above expression is often easier to deal with than formula (21.7) if we want to compute the exact distribution for the compound Poisson distribution where X is finite-valued. We need only compute a single n-fold convolution.

21.9 The sum of two compound Poisson distributions

The sum of two independent Poisson distributions is itself Poisson distributed as shown by Example A.1 in Section A.11.2. We now show that the same statement is true for *compound* Poisson distributions. Given two independent compound Poisson distributions, $S_1 = \langle N_1, X_1 \rangle$ and $S_2 = \langle N_2, X_2 \rangle$, where $N_1 \sim \text{Poisson}(\lambda_1)$ and $N_2 \sim \text{Poisson}(\lambda_2)$, let $S = S_1 + S_2$. Then

$$\begin{aligned}
M_S(t) = M_{S_1}(t) M_{S_2}(t) &= \mathrm{e}^{\lambda_1[M_{X_1}(t)-1]} \mathrm{e}^{\lambda_2[M_{X_2}(t)-1]} \\
&= \mathrm{e}^{(\lambda_1+\lambda_2)[(\lambda_1/(\lambda_1+\lambda_2))M_{X_1}(t)+(\lambda_2/(\lambda_1+\lambda_2))M_{X_2}(t)-1]}.
\end{aligned}$$

It follows that $S \sim \langle N, X \rangle$, where $N \sim \text{Poisson}(\lambda_1 + \lambda_2)$ and X is a mixture of X_1 and X_2 with respective weights $\lambda_1/(\lambda_1+\lambda_2)$ and $\lambda_2/(\lambda_1+\lambda_2)$.

21.10 Deductibles and other modifications

Up to now we have assumed that the insurer pays the totality of any loss as given by the random variable X. In practice, the insurer often only covers part of the loss, leaving the insured to pay the remainder. The prime motivation is to make the insured party partially responsible, so that they have an interest in taking steps to avoid loss.

This has major implications when we look at the statistical problem of inferring details about loss distributions from the data furnished by insurers on their claims experience. In practice, this data will give the amounts actually paid on claims, rather than the actual losses. In order to estimate loss distributions, one needs to understand clearly the relationship between the amount of the loss and the amount actually paid under the common types of modifications.

21.10.1 The nature of a deductible

One of the most common modification devices is a *deductible*. Under this arrangement, the insurer only pays the losses that are above some amount d fixed in advance. The purchaser of the insurance pays for the first d units of loss, and of course if the loss is less than d, the insurer pays nothing, and the insured is fully responsible. This has an added advantage to the insurer of preventing an undue expense involved in processing small claims.

If the original severity distribution is given by the random variable X, and there is a deductible of d, the amount actually paid by the insurer is

$$(X-d)_+ = \begin{cases} X-d, & \text{if } X \geq d, \\ 0 & \text{if } X < d. \end{cases}$$

It is not always easy to describe the exact distribution of this random variable given the distribution of X, but it is relatively simple to compute its expectation. This is given for continuous X by

$$E(X-d)_+ = \int_d^{\infty} (x-d) f_X(x)\, dx = \int_d^{\infty} s_X(x)\, dx, \tag{21.10}$$

where the second expression follows from the first by integrating by parts (or directly from (A.15).

Another random variable of interest, which is associated with the above, is

$$X \wedge d = \min(X,d) = \begin{cases} d, & \text{if } X \geq d, \\ X, & \text{if } X < d. \end{cases}$$

By looking separately at the case where X is less than or greater than d, it easily follows that, in general,

$$X = (X-d)_+ + (X \wedge d),$$

so that

$$E(X-d)_+ = E(X) - E(X \wedge d). \tag{21.11}$$

For continuous X, it follows from (21.10) by calculating directly that

$$E(X \wedge d) = \int_0^d x f_X(x)\,\mathrm{d}x + d s_X(d) = \int_0^d s_X(x)\,\mathrm{d}x. \tag{21.12}$$

Example 21.4 X takes the values 100, 200, 300, 400 with probabilities 0.4, 0.3, 0.2, 0.1, respectively. Describe the distributions of $(X-d)_+$ and $(X \wedge d)$, for $d = 230$. Find $E(X - 230)_+$ and $E(X \wedge 230)$.

Solution. $(X-230)_+$ takes the value 0 with probability 0.7, 70 with probability 0.2, and 170 with probability 0.1, while $X \wedge 230$ takes the value 100 with probability 0.4, 200 with probability 0.3, and 230 with probability 0.3. Calculating directly,

$$E(X \wedge 230) = 100 \times 0.4 + 200 \times 0.3 + 230 \times 0.3 = 169.$$

$$E(X - 230)_+ = 70 \times 0.2 + 170 \times 0.1 = 31.$$

It is often convenient to compute $E(X-d)_+$ from formula (21.12). This is particularly true when X is infinite and discrete. We can compute $E(X \wedge d)$ by a finite sum, as opposed to the infinite series involved in computing $E(X-d)_+$ directly. Here is a typical example.

Example 21.5 Suppose $X \sim \text{Geom}(p)$. Find $E(X-1/2)_+$.

Solution. Since X is always greater than 1/2 unless it is equal to 0, we know that $X \wedge 1/2$ takes the value 0 with probability $1-p$ and the value 1/2 with probability p. So $E(X \wedge 1/2) = p/2$. Since $E(X) = p/1-p$, we conclude that

$$E(X-1/2)_+ = \frac{p}{1-p} - \frac{p}{2}.$$

21.10.2 Some calculations in the discrete case

Suppose that X takes integer values. The integrals involving f_X in (21.10) and (21.12) must be replaced by summations with the same limits, and with f_X now equal to the probability function. They apply only to integer values of d. The expressions involving s_X, however, remain valid as is and apply to any value of d. Of course, in this case, s_X is a step function and the integral may be rewritten as a sum. For example, if $k \leq d < k+1$ for some integer k, then

$$E(X-d)_+ = \int_d^\infty s_X(x)\mathrm{d}x = (k+1-d)s_X(k) + s_X(k+1) + s_X(k+2) + \cdots. \tag{21.13}$$

To illustrate, look again at Example 21.4, except we take a unit to be 100, so now X takes the values 1, 2, 3, 4. We have $s(2) = 0.3$, $s(3) = 0.1$, $s(4) = 0$, and

$$E(X-2.3)_+ = 0.7 \times 0.3 + 0.1 = 0.31,$$

as above.

It is quite simple to compute all values of $E(X-d)_+$ in the case of an integer-valued X. When d is an integer, (21.13) just says that

$$E(X-d)_+ = \sum_{k=d}^{\infty} s_X(k),$$

and we get the recursion formula

$$E(X-(d+1))_+ = E(X-d)_+ - s_X(d), \tag{21.14}$$

where we start the recursion with $E(X-0)_+ = E(X)$. Noninteger values of d are computed exactly by linear interpolation, since (21.13) immediately implies that, for $d = k + r$ where k is an integer and $0 < r < 1$,

$$E(X-d)_+ = (1-r)E(X-k)_+ + rE(X-(k+1))_+.$$

Example 21.6 Suppose that the probability function of X takes the values $f(0) = 0.2$, $f(1) = 0.2$, $f(2) = 0.3$, $f(3) = 0.1$, $f(4) = 0.2$. Find $E(X-d)_+$ for $d = 0, 1, 2, 3, 4$, and 2.6.

Solution. We first calculate $s(0) = 0.8, s(1) = 0.6, s(2) = 0.3, s(3) = 0.2, s(4) = 0$, and

$$E(X) = \sum_{k=0}^{4} s(k) = 1.9,$$

so that $E(X-0)_+ = 1.9$, $E(X-1)_+ = 1.9 - 0.8 = 1.1$, $E(X-2)_+ = 1.1 - 0.6 = 0.5$, $E(X-3)_+ = 0.5 - 0.3 = 0.2$, $E(X-4)_+ = 0.2 - 0.2 = 0$. This final value of 0 serves as a check that we have done the recursion correctly. Finally, we have

$$E(X-2.6)_+ = 0.4E(X-2)_+ + 0.6E(X-3)_+ = 0.32.$$

21.10.3 Some calculations in the continuous case

In general, $(X-d)_+$ will have a point mass at 0, as it will take a value of 0 with probability $F_X(d)$. Therefore, if X is continuous, $(X-d)_+$ will be a mixed distribution. In such cases, we may want to proceed as in Section 21.7 and consider the random variable $(X-d)^+ = (X-d)|X > d$. We then have that $(X-d)_+$ is a mixture of 0 and $(X-d)^+)$ with weights $F_X(d)$ and $s_X(d)$, respectively. It follows that

$$E(X-d)_+ = E((X-d)^+ s_X(d). \tag{21.15}$$

Example 21.7 If $X \sim \mathrm{Exp}(\lambda)$, what is the distribution of $(X-d)^+$?

Solution. If Y denotes $X - d^+$, then

$$s_Y(y) = \frac{s_X(y+d)}{s_X(d)} = \frac{e^{-\lambda(y+d)}}{e^{-\lambda d}} = e^{-\lambda y},$$

so that Y has exactly the same distribution as X. This only happens with an exponential distribution. The result at first glance seems surprising. It says that no matter how high the deductible is, the excess of the loss over the deductible is distributed as the original loss. The point to keep in mind is that Y is conditioned on the loss being above the deductible, which of course will have a very small chance of occurring for high values of d.

Example 21.8 If $X \sim$ Pareto(θ, α), what is the distribution of $(X-d)^+$?

Solution. If Y denotes $(X-d)^+$,

$$s_Y(y) = \frac{s_X(y+d)}{s_X(d)} = \left(\frac{\theta}{y+d+\theta}\right)^{\alpha} \Big/ \left(\frac{\theta}{d+\theta}\right)^{\alpha} = \left(\frac{\theta+d}{y+d+\theta}\right)^{\alpha},$$

and we see that

$$Y \sim \text{Pareto}\ (\theta+d, \alpha).$$

In general, we will not be able to easily identify the distributions associated with the deductible d as we did in the above two examples, although in some cases, we may be able to compute expectations. This is true for the gamma distribution with first parameter 2.

Example 21.9 If $X \sim$ Gamma $(2, \beta)$, find $E[(X-d)^+]$ and $E[(X-d)_+)]$.

Solution. Integrating by parts,

$$\int x\mathrm{e}^{-\beta x}\,\mathrm{d}x = -\frac{(1+\beta x)}{\beta^2}\mathrm{e}^{-\beta x},$$

so that

$$s_X(x) = \beta^2 \int_x^{\infty} y\mathrm{e}^{-\beta y}\,\mathrm{d}y = (1+\beta x)\mathrm{e}^{-\beta x}.$$

From (21.13),

$$E[(X-d)_+] = \int_d^{\infty} (1+\beta x)\,\mathrm{e}^{-\beta x}\,\mathrm{d}x = \frac{\mathrm{e}^{-\beta d}}{\beta} + \frac{(1+\beta d)}{\beta}\mathrm{e}^{-\beta d} = \left(\frac{2}{\beta}+d\right)\mathrm{e}^{-\beta d},$$

and from (21.15),

$$E[(X-d)^+] = \frac{E[(X-d)_+]}{s_X(d)} = \frac{1}{\beta}\left(\frac{2+\beta d}{1+\beta d}\right).$$

As a check, both these expectations reduce to $E(X)$ when $d=0$.

21.10.4 The effect on aggregate claims

Up to now we have focused on the severity distribution X. We now turn our attention to the effect of a deductible on the aggregate claim distribution S. It is important to distinguish two situations.

In one case, the deductible is applied directly to aggregate claims. This could arise from a reinsurance arrangement. It is often the case that a party known as a *reinsurer* agrees to cover part of the losses of the original insurer in return for a premium. In one common type of arrangement, the reinsurer would impose a deductible on the aggregate claims for a certain portfolio. This is known as *stop-loss reinsurance*. In such a case, we are interested in the random variable $(S-d)_+$, which will be the amount paid by the reinsurer.

In the second situation, the deductible d is applied to each individual claim. There are two ways to proceed here. The obvious way is to simply note that in place of the distribution $S \sim \langle N, X \rangle$, which applies without the deductible, we are now interested in the distribution

$$S' \sim \langle N, (X-d)_+ \rangle. \tag{21.16}$$

There is an alternate representation for S' that is useful with certain distributions. Let us motivate this by asking the following deep philosophical question. Is a claim for an amount of 0 really a claim? The simple answer is that it either is or is not, depending on which way you want it. In the first representation of S, there will be claims for zero amount, namely those that are less than the deductible and for which nothing is reimbursed. Suppose we decide not to count these as claims. Our severity will then be distributed as $(X-d)^+$ rather than $(X-d)_+$, since we now only consider a claim to have occurred if it is over the deductible. We must then, however, also change the distribution N to count only those claims for an amount above the deductible. We know how to do this from Section 21.8. The special claims in this case are those for an amount greater than d. Under this approach, we have

$$S' \sim \langle \langle N, \delta_\pi \rangle, (X-d)^+ \rangle, \qquad \pi = s_X(d). \tag{21.17}$$

To use (21.17) effectively, we have to know that claim frequency and severity remain independent when we make these transformations. A general proof of this fact becomes somewhat involved in notation, and we will not present it, but the idea is straightforward as the following example illustrates.

Example 21.10 Suppose that N takes values of 0,1,2, and and let $p(i)$ be the probability that $N = i$. Suppose that X takes the values x_1, x_2, x_3, where $x_1 \le d$ and x_2 and x_3 are greater than d. Let N_1 denotes the number of claims for an amount greater than d. Show that

$$P[N_1 = 1 \text{ and } (X-d)^+ = x_2 - d] = P[N_1 = 1]P[(X-d)^+ = x_2 - d]$$

Solution. Let a, b, c denote respectively the probability that X takes the values x_1,x_2, x_3.

For the event on the left to occur we need either one claim for an amount of x_2 or two claims, where one is of amount x_2 and the other is of amount x_1. The required probability then is $p(1)b + p(2)2ab$. Now, for $N_1 = 1$ we need either one claim for an amount of either x_2 or x_3, or two claims where one is for x_1 and the other is for either x_2 or x_3.

So $P[N_1 = 1] = p(1)(b+c) + p(2)2a(b+c)]$. Moreover $P[(X-2)^+ = x_2 - d] = b/b+c$. Multiplying the last two quantities gives the first.

Use of (21.17) in place of (21.16) works particularly well when N is one of three basic cases of Poisson, binomial, or negative binomial, where we know what $\langle N, \delta_\pi \rangle$ is, and when we know what $(X-d)^+$ is, as in the exponential or Pareto distributions for severity.

Example 21.11 Suppose that $N \sim$ Poisson(2) and $X \sim$ Exp(3). A deductible of 2 is applied to each claim. Find the variance of the resulting distribution of aggregate claims.

Solution. We know that $(X-2)^+$ has the same distribution as X and therefore a second moment of $\frac{2}{9}$. We replace the original N by a Poisson($2e^{-6}$) distribution, and by using (21.8), the resulting variance is $\frac{4}{9}e^{-6}$.

21.10.5 Other modifications

Another method of modifying the original claim amount is to set a maximum value m. The insurer will pay at most m regardless of the actual value of the loss. If X is the original severity distribution, the amount paid on a claim would then be $X \wedge m$. There could be both a deductible d and and a maximum m imposed. The amount paid on a claim in this situation would be $X \wedge (m+d) - X \wedge d$. Yet another modification is for the insurer to pay only a certain percentage of the loss. The amount paid on a claim will now be αX, for some $0 \le \alpha \le 1$. In doing calculations where all these modifications are present, it is useful to keep in mind the fact that

$$\alpha(X \wedge d) = \alpha X \wedge \alpha d.$$

Example 21.12 A policy will cover 80% of all losses in excess of 100, with the further provision that a maximum payment of 900 will be made regardless of the amount of the loss. Express the amount paid on a claim with terms of the form $X \wedge d$.

Solution. The amount paid is $0.8(X-100)_+$ provided that $0.8(X-100) \le 900$, which will occur for $X \le 1225$. We can express this as

$$0.8(X \wedge 1225) - 0.8(X \wedge 100).$$

21.11 A recursion formula for S

21.11.1 The positive-valued case

We suppose that X takes *positive* integers as values, and we seek a recursion formula to compute the probability function of S. It turns out that this is possible provided that the probability function of N satisfies a certain recursion. The required property is that for some constants a and b,

$$p(k) = \left(a + \frac{b}{k}\right) p(k-1), \qquad k = 1, 2, 3, \ldots. \tag{21.18}$$

To develop our recursion formula, we first need some identities for convolutions of $f = f_X$. By definition, we know that for all n and all positive integers x,

$$f^{*n}(x) = \sum_{i=1}^{x-1} f(i)f^{*(n-1)}(x-i). \tag{21.19}$$

There is, however, another curious identity relating convolutions.

Proposition 21.1 *For all n and all positive integer values of x,*

$$f^{*n}(x) = \frac{n}{x}\sum_{i=1}^{x-1} if(i)f^{*(n-1)}(x-i).$$

Proof. Let $X_1, X_2, \ldots, X_n$ be independent and each distributed as X, and let $A = \sum_{i=1}^{n} X_i$. Then, for any positive integer $i \le x$,

$$\begin{aligned} P(X_1 = i|A = x) &= \frac{P(X_1 = i \text{ and } A = x)}{P(A = x)} \\ &= \frac{f(i)f^{*(n-1)}(x-i)}{f^{*n}(x)}, \end{aligned}$$

since in order for the event in the numerator to occur, $X_1 = i$ and the other $n-1$ random variables add up to $x - i$. It follows that

$$E(X_1|A = x) = \sum_{i=1}^{x-1} i\frac{f(i)f^{*(n-1)}(x-i)}{f^{*n}(x)}.$$

There is, however, nothing special about X_1, and, by symmetry, we get exactly the same equality, with X_1 replaced by X_j for $j = 2, 3, \ldots, n$. Add up this last equality for all n values of j. The left hand side is just $E(A|A = x)$, which is simply x. The right hand side of each equation is a constant that gets multiplied by n in the sum. Equating and rearranging gives the stated identity. □

In the remainder of this section, we will let g denote the probability function of S.

Theorem 21.1 (The recursion formula) *Suppose that p satisfies (21.18). Then, for all positive integers k,*

$$g(k) = \sum_{i=1}^{k}\left(a + \frac{bi}{k}\right)f(i)g(k-i).$$

Proof. For any positive integer k, we have

$$g(k) = \sum_{n=1}^{k} p(n)f^{*n}(k) = a\sum_{n=1}^{k} p(n-1)f^{*n}(k) + b\sum_{n=1}^{k} p(n-1)\frac{f^{*n}(k)}{n}. \tag{21.20}$$

Consider the term multiplying a in the above, which is

$$p(0)f(k) + p(1)f^{*2}(k) + p(2)f^{*3}(k) + p(3)f^{*4}(k) + \cdots + p(k-1)f^{*k}(k).$$

By applying (21.19) to each summand (after the first), we can write this as

$$\begin{array}{llll} p(0)f(k)+ & p(1)f(1)f(k-1)+ & p(2)f(1)f^{*2}(k-1)+ & p(3)f(1)f^{*3}(k-1)+\cdots \\ & p(1)f(2)f(k-2)+ & p(2)f(2)f^{*2}(k-2)+ & p(3)f(2)f^{*3}(k-2)+\cdots \\ & p(1)f(3)f(k-3)+ & p(2)f(3)f^{*2}(k-3)+ & p(3)f(3)f^{*3}(k-3)+\cdots \\ & \vdots & \vdots & \vdots \end{array}$$

The sum of the first row, excluding the leading term $p(0)f(k)$, is

$$\begin{aligned} &f(1)[p(1)f(k-1) + p(2)f^{*2}(k-1) + p(3)f^{*3}(k-1) + \cdots + p(k-1)f^{*(k-1)}(k-1)] \\ &\quad = f(1)g(k-1). \end{aligned}$$

Similarly, the sum of the ith row is just $f(i)g(k-i)$. The leading term $f(k)p(0)$ equals $f(k)g(0)$, since, given the restriction of strictly positive values for X, the only way for S to be equal to 0 is if $N = 0$. The sum of the entire array is then simply

$$\sum_{i=1}^{k} f(i)g(k-i). \tag{21.21}$$

Next, consider the term multiplying b in (21.20). We do exactly what we did above except we use the identity in Proposition 21.1 in place of (21.19). In this case, the term from (21.20) introduces a coefficient of $1/n$ in the column of the array involving $p(n-1)$. Had we used (21.19), we could not have conveniently summed along rows. The beauty of the other identity is that it introduces a term n/k in the column involving $p(n-1)$ that conveniently cancels with the $1/n$. The sum in this case is the same as (21.21), except that it is multiplied by $1/k$ and $if(i)$ replaces $f(i)$, giving

$$\frac{1}{k}\sum_{i=1}^{k} if(i)g(k-i).$$

Substituting in (21.20) gives the recursion formula. □

The starting value for the recursion is given by $g(0) = p(0)$, as indicated above.

Of course, in order that this formula be useful, we need to know that there are distributions of N that satisfy the given condition on p. Fortunately, this occurs for our three main families.

If $N \sim \text{Poisson}(\lambda)$, then

$$\frac{p(k)}{p(k-1)} = \frac{\lambda}{k},$$

so (21.18) holds with $a = 0, b = \lambda$.

If $N \sim \text{Bin}(m, p)$, then

$$\frac{p(k)}{p(k-1)} = \frac{m-k+1}{k}\frac{p}{1-p},$$

so that (21.18) holds with $a = -p/(1-p), b = (m+1)p/(1-p)$.

If $N \sim \text{Negbin}(r, p)$, then

$$\frac{p(k)}{p(k-1)} = \frac{r+k-1}{k}p,$$

so that (21.18) holds with $a = p, b = (r-1)p$.

What are the other possibilities? It turns out that there are none, and that, remarkably, only these three families satisfy the required recurrence relation.

Theorem 21.2 *Suppose that N is a nonnegative-valued random variable satisfying (21.18). Then:*

1. *if $a = 0, N \sim$ Poisson(b);*
2. *if $a > 0, N \sim$ Negbin($b/a + 1, a$);*
3. *If $a < 0, N \sim$* Bin$(m, -a/(1-a))$ *for some positive integer m.*

Proof.

(i) If $a = 0$, then clearly $p(k) = (b^k/k!)p(0)$. This gives

$$1 = \sum_{0}^{\infty} p(k) = p(0)\sum_{k=0}^{\infty}\frac{b^k}{k!} = e^b p(0),$$

so that $p(0) = e^{-b}$. Substituting this into the expression for $p(k)$, we see that $N \sim$ Poisson(b).

(ii) Suppose $a > 0$. Note first that $p(0)$ cannot be 0. For, if so, then all $p(k)$ would be 0, and we would not have a probability distribution. Let $r = (b/a) + 1$. Then $r \geq 0$, for, otherwise we have $p(1) = (a+b)p(0) < 0$. Writing $b = (r-1)a$, we calculate inductively $p(1) = rap(0)$, $p(2) = [r(r+1)/2]a^2p(0), \ldots, p(k) = [r(r+1)\ldots(r+k-1)/k!]a^kp(0), \ldots$. This gives

$$1 = \sum_{k=0}^{\infty} p(k) = p(0)\sum_{k=0}^{\infty}\frac{r(r+1)\ldots(r+k-1)}{k!}a^k = p(0)(1-a)^{-r},$$

so that $p(0) = (1-a)^r$, showing that $a < 1$. By substituting this into the expression for $p(k)$, we see that $N \sim \text{Negbin}(r, a)$.

(iii) Suppose that $a < 0$. Since $a + (b/k)$ will become negative for sufficiently large k, it must necessarily become zero for some value of k. If not, we would eventually get negative probabilities. Therefore, we must have that $b > 0$ and $a = -b/(m+1)$ for some positive integer m. This means that $p(k) = 0$ for $k > m$. We note that $a + b/k = -a(m-k+1)/k$, so that $p(1) = -amp(0)$, $p(2) = (-a)^2[m(m-1)/2!]p(0), \ldots, p(k) = (-a)^k[(m(m-1)\ldots m-k+1)/k!]p(0), \ldots$. Then

$$1 = \sum_{k=0}^{m} p(k) = p(0)(1-a)^m.$$

Let $p = -a/(1-a)$. Then, $-a = p/(1-p)$, so that $p(0) = (1-p)^m$, and by substituting this into the expression for $p(k)$, we see that $N \sim \text{Bin}(m, p)$. □

Example 21.13 Go back to the coin–dice problem that started this chapter and illustrate that you can find these probabilities by recursion.

Solution. In this case $N \sim \text{Bin}(2, 0.5)$, so we have $a = -1, b = 3$. Moreover, $f(i) = 1/6$ for $i = 1, 2, \ldots, 6$. We will do the first few calculations here to illustrate the procedure.

$$g(0) = p(0) = \frac{1}{4}, \qquad g(1) = \frac{1}{6}[2g(0)] = \frac{1}{12},$$

$$g(2) = \frac{1}{6}\left[\frac{1}{2}g(1) + 2g(0)\right] = \frac{13}{144}, \qquad g(3) = \frac{1}{6}[g(1) + 2g(0)] = \frac{7}{72}.$$

21.11.2 The case with claims of zero amount

There is a more general recursion formula that allows for the possibility that X can take a value of 0. This is

$$g(k) = \frac{1}{1 - af(0)} \sum_{i=1}^{k} \left(a + \frac{bi}{k}\right) f(i) g(k-i), \tag{21.22}$$

which reduces to that given above when $f(0) = 0$. This can be derived by a suitable modification of the proof of Theorem 21.1. We note that there will be two extra rows in the first array, which we used to compute the sum of the term multiplying a. One row at the beginning will have terms of the form $f(0)f^{*r}(k)$, and one at the end will have terms of the form $f(k)f^{*r}(0)$. The beginning row will just sum to $f(0)g(k)$. The ending row will combine with the leading term to sum to $f(k)g(0)$. In this case, $g(0)$ is not equal to $p(0)$. In the second calculation, when we compute the summation multiplying b, we only get this row at the end, in view of the extra coefficient of i, which makes entries in the new first row equal to 0. The final conclusion is that

$$g(k) = af(0)g(k) + \sum_{i=1}^{k} \left(a + \frac{bi}{k}\right) f(i) g(x-i),$$

which leads to (21.22).

The disadvantage of using (21.22) is that we have a more complicated calculation for the initial value than before:

$$g(0) = \sum_{k=0}^{\infty} p(k)f(0)^k.$$

An easier procedure is to follow the alternate method we mentioned for handling per-claim deductibles in the previous section. That is, we simply get rid of the zero claims by counting only the positive ones. From Section 21.8, we know now to modify N in all the relevant cases. We must also replace X by X^+, but that is done simply by multiplying each probability by $1/(1-f(0))$.

Example 21.14 Suppose $N \sim \text{Negbin}(0.5, 0.4)$. X takes the values 0,1,2 with probabilities 0.25, 0.35, 0.40, respectively. Write down a recursion formula for computing g.

Solution. We follow the second procedure. The probability of a nonzero claim is 0.75. Recall that in the negative binomial, the ratio $p/(1-p)$ gets multiplied by this probability and changes from $2/3$ to $1/2$. So the new value of p is $1/3$, and we have $a = 1/3, b = -1/6$. The random variable X^+ takes the values $1, 2$ with probabilities $7/15, 8/15$, respectively. The recursion becomes

$$g(0) = \left(\frac{2}{3}\right)^{0.5}, \qquad g(1) = \frac{1}{6} \times \frac{7}{15}\, g(0),$$

$$g(k) = \left(\frac{1}{3} - \frac{1}{6k}\right)\frac{7}{15}\, g(k-1) + \left(\frac{1}{3} - \frac{1}{3k}\right)\frac{8}{15}\, g(k-2), \qquad k > 1.$$

Notes and references

The reader is cautioned that some authors use an alternative definition of the negative binomial random variable. In the formulation in terms of repeated trials, they would count the total number, rather than just the successes. Their random variable would then be equal to $N + r$, where N is the definition that we have adopted. Moreover, parameters chosen for the different distributions are not standardized and different choices are made by various authors. We indicated an alternate choice for the negative binomial. Another example occurs with the exponential distribution. While we chose to parametrize this by the hazard rate, some may use the mean and take the parameter to be the reciprocal of ours. This is carried forward to the gamma distributions. So for example, what we call a Gamma(α, β) distribution could be termed a Gamma(α, β^{-1}) distribution by others.

Exercises

21.1 If $N \sim \text{Binomial}(9, 1/3)$ and $X \sim \text{Gamma}(2, 0.5)$, find $E(S)$ and $\text{Var}(S)$.

21.2 If N takes the values 0, 1, 2, 3 with probabilities 0.3, 0.4, 0.2, and 0.1 respectively, and X takes values 1, 2, 3, 4, 5 each with probability 0.2, find the probability that $S = 4$.

21.3 If N takes values 0, 1, 2, 3 with probabilities 0.5, 0.3, 0.1, 0.1 respectively, and X takes values 1, 2, 3, 4, 5, 6 with probabilities 0.2, 0.2, 0.2, 0.2, 0.1, 0.1 respectively, find the probability that $S = 6$.

21.4 Suppose that N is negative binomial with mean 4 and variance 12, and X is exponentially distributed. Let N_1 denote the number of claims that are less than the average claim amount. Find the variance of N_1.

21.5 The frequency of accidents for automobile drivers over a certain period follows a Poisson distribution. Good drivers can expect to have on average one accident over that period, while bad drivers can expect to have two accidents. It is estimated that 80% of drivers are good and 20% are bad. If an accident occurs, the claim amount is exponentially distributed with a mean of 100. Calculate the expected value and variance of the aggregate claims over this period.

21.6 Suppose that $N \sim$ Negbin(2, 0.8) and $X \sim$ Gamma(4, 3). Find $E(S)$ and Var(S).

21.7 Suppose that N takes the values 0, 1, 2 with probabilities 0.5, 0.3, 0.2 respectively, and X takes values 1, 2, 3, with probabilities 0.3, 0.6, 0.1 respectively.

(a) Find the probability that $S = 3$.

(b) Let N_i = the number of claims of size i, for $i = 1, 2, 3$. Are N_1 and N_2 independent?

21.8 Suppose that N is a continuous mixture of Poisson(λ) distributions where $\lambda \sim$ *Gamma* (3, 2). Find $E(N)$ and Var(N).

21.9 Suppose that N takes the values 0, 1, 2, 3 with probabilities 0.4, 0.3, 0.2, 0.1 respectively, and that X takes the values 10 with a probability of 0.5, 20 with a probability of 0.3, 30 with a probability of 0.1, and various other values, all higher than 30, with a total probability of 0.1. (You are not given these values.)

(a) Find the probability that $S = 30$.

(b) Find $E[(S - 15)_+]$, given that $E(X) = 20$.

21.10 Suppose $N \sim$ Poisson(2) and $X \sim$ Exp(3).

(a) Find $E(S)$ and Var(S).

(b) Find $M_S(1)$, where M_S is the m.g.f. of S.

21.11 The number of customers arriving at a restaurant is Poisson distributed with a mean of 15 per hour. The amount that each customer spends is exponentially distributed with an average of 20. The restaurant is open 16 hours each day and the daily expenses are 4500. Using a normal approximation, estimate the probability that on a given day, revenue will cover expenses.

21.12 Suppose that N is a continuous mixture of Poisson distributions where the mean is itself a random variable. Find $P(N = 0)$ in each of the following cases.

(a) The mean is uniformly distributed on the interval [1, 3].

(b) The mean has a Gamma (2, 3) distribution.

21.13 Suppose we have two mutually exclusive special classes of claims N_1 and N_2. Show, by an example, that if N has a binomial distribution, then N_1 and N_2 need *not* be independent.

21.14 Suppose that the probability of n claims is $1/2^{n+1}$ and the probability is e^{-2x} that a given claim will be greater than x. What is the probability that the aggregate claims will be less than or equal to log (10)?

21.15 A policy will cover 75% of all losses in excess of 240 with the further provision that a maximum payment of 2700 will be made regardless of the loss. Express the amount paid on a claim with terms of the form $X \wedge d$, where X is the actual loss.

21.16 Each hour, vehicles pass a certain point on a highway in accordance with a Poisson distribution. The expected number of vehicles that pass during the hour is four. Assume that one half of all passing vehicles are trucks, and one quarter are sports cars. Find the probability that, in a given hour, the passing vehicles include exactly two trucks and exactly one sports car. There is no restriction on the number of vehicles other than trucks and sports cars (so, for example, the event of two trucks, one sport car and seven other vehicles would satisfy the given condition).

21.17 Suppose that N is geometric with mean = 1, and X takes the values 10 or 20 with equal probability. Find $E(S - 30)_+$.

21.18 For a certain insurer, N has a Poisson distribution, and X is exponentially distributed. Each claim is subject to a deductible of d. If $d = 2$, the expected amount paid by the insurer is equal to 100. If $d = 3$, this expected payout reduces to 50. What is the expected payout if $d = 1$?

21.19 The density function of X is given by

$$f(x) = \frac{10 - x}{50}, \qquad 0 \leq x \leq 10.$$

In order to reduce the expected amount paid, the insurer is considering two possibilities. One is to introduce a deductible of two per claim. The other is to pay only a maximum of 7 per claim. Which scheme should they adopt if they want to minimize the expected payout?

21.20 The manufacturer of a television set costing 1000, offers a guarantee to repair or replace the set for free for the first year. The number of defective sets follows a Poisson distribution with mean 4. Half the defective sets require replacement and half require a repair costing 500. The manufacturer purchases an insurance policy that will cover the total cost of this guarantee above 1500. (So, for example, if there were four defective sets and each required replacement the insurer would pay 2500 to the manufacturer.) Find the expected amount that the insurer will pay.

21.21 For a certain collection of contracts, $N \sim$ Poisson(4), while $X \sim$ Exp(3). Suppose that each individual claim is subject to a deductible of 2. If S' is the total amount actually paid on all claims, find the variance of S'.

21.22 For a certain firm, the number of losses of a certain type has a Poisson(2) distribution. The amount of a loss takes a value of 100, 200 or 300, with probabilities 0.5, 0.3, 0.2, respectively. The firm purchases an insurance policy that will cover all losses above an aggregate deductible of 200. (So, for example, if there was one loss of 100 and three losses of 300, the insurer would pay 800.) What is the expected reimbursement by the insurer?

21.23 A manager of a certain office is offered a bonus each month if the total expenses are under 1000. The bonus is half the difference between 1000 and the expenses. So, for example, if expenses were 800, the bonus would be 100. Find the expected value of the bonus in each of the following cases.

(a) Expenses are exponentially distributed with a mean of 2000.

(b) Expenses are uniformly distributed on the interval $[0, 4000]$.

21.24 The distribution of N is a mixture of Poisson(λ) distributions, where λ follows a Gamma distribution with $\alpha = 10$ and $\beta = 2$. Moreover, X has a Pareto distribution with $\theta = 4, \alpha = 3$. A per-claim deductible of 2 is applied. Find the expectation and variance of the aggregate payments made on all claims.

21.25 You are the manufacturer of a product that gives guarantees against failure. Each month there is a 50% chance that there will be exactly one failure, a 30% chance that there will be exactly two failures, and a 20% chance that there will be exactly three failures. Moreover, 20% of failures will be complete, requiring a full reimbursement of 800, while 80% will require only partial reimbursement of 400. Each month you purchase insurance that will provide all reimbursements for that period above a total of 1000. (So, for example, if there were three complete failures, the insurer would pay 1400.) What is the expected amount of reimbursement that the insurer will pay each month?

21.26 For a certain insurer, the frequency of claims has a negative binomial distribution with an expected value of 16 and the claim severity distribution is Pareto (600, 2). The insurer is planning to introduce a per-claim deductible of 200. If this is done, what would be the reduction in the expected value of aggregate claims?

21.27 $E(X \wedge d)/E(X)$ is known as the *loss elimination ratio* (LER), since it gives the proportion of the risk to the insurer that is eliminated by a deductible of d. For each of the following distributions, find the LER in terms of d and the parameters. (a) $X \sim \text{Exp}(\lambda)$, (b) $X \sim \text{Pareto}(\Theta, \alpha)$, (c) $X \sim \text{Gamma}(2, \beta)$. In each case verify that your answers have the correct limits as d approaches 0 or ∞.

21.28 Suppose that the severity distribution changes from X to $(1 + r)X$ due to inflation.

(a) Show that if the deductible d is increased to $(1 + r)d$, the LER is unchanged. What happens to the LER if d is unchanged?

(b) Suppose that X is exponentially distributed and for a certain value of d the LER is 0.3. If $r = 0.10$ and d is unchanged, what is the new LER?

21.29 Suppose that N is a continuous mixture of Poisson(λ) distributions, where $\lambda \sim$ Gamma $(2, 1)$ and X takes the value 1 with probability 0.6 and 2 with probability 0.4. If $f_S(10) = c$ and $f_S(11) = d$, find a formula for $f_S(12)$ in terms of c and d.

21.30 For a certain collection of contracts, X takes values 1, 2, 3, each with equal probabilities. You know that the insurer uses a recursion formula to calculate $g(s) = P[S = s]$ and you are trying to determine what distribution is being used for N. All you have to go on is a scrap of paper on which the following appears:

$$g(4) = \frac{1}{3}[xg(3) + yg(2) + zg(1)],$$

where x, y, and z are numbers that have been smudged and are unreadable. You can, however, read enough of the numbers to definitely conclude that y is strictly less than $2x$. What is the distribution of N and why? (Just identify the basic type. You do not have enough information to determine the parameters.)

21.31 Compute $g_S(x)$ for $x = 0, 1, \ldots, 5$ for the following three compound distributions, each with claim amount distribution given by $f_X(1) = 0.7$ and $f_X(2) = 0.3$: (a) Poisson with $\lambda = 4.5$; (b) negative binomial with $r = 4.5$ and $p = 0.5$; (c) binomial with $m = 9$ and $p = 0.5$.

21.32 Suppose you have 11 boxes, numbered 1 to 11, and each box contains three balls, of which two are numbered 1 and one is numbered 2. A coin has a probability of 0.6 of coming up heads. You are going to toss the coin 11 times, and if a head occurs on the ith toss, you are going to take box number i and randomly select a ball. Let g denote the probability function of the random variable S, the total of all the selected balls. Given that $g(10) = 0.1386$ and $g(11) = 0.1055$, find $g(12)$.

21.33 Aggregate claims S follow a compound Poisson distribution with $\lambda = \log(4)$ and with the probability function of X given by $f_X(k) = 2^{-k}/(k\log(2))$, $k = 1, 2, \ldots$. What is the distribution of S?

21.34 Suppose that (N_i), $i = 1, 2, 3$, is an independent family of random variables where $N_i \sim \text{Poisson}(i)$. If $S = 3N_1 + 2N_2 + 5N_3$, find distributions N and X so that $S \sim \langle N, X \rangle$.

22

Risk assessment

22.1 Introduction

The previous chapter was largely devoted to computing or approximating the distribution of aggregate claims for the losses on an insurance portfolio. The next problem that arises is to effectively use this information to assess and manage the risk associated with the insurer's commitment to pay these losses. Similarly, a consumer is interested in assessing the extent to which their risk is transferred by the purchase of insurance. We alluded to this theme somewhat in Part II of the book but we now wish to investigate some of the issues in more detail. Our concentration will be on a more general basic question that has application in many areas. Given two or more uncertain alternatives, how do we compare or measure the amount of risk associated with each? This is a large topic and we confine ourselves here to a survey of some of the main ideas. It is important in what follows to distinguish between two cases. The quantities in question may involve losses in which case less is better, or gains, in which case more is better. We could conventionally fix one or the other, by introducing minus signs, but that complicates the notation, so we rely on the context to clarify what is intended. We start in the next section by talking about gains.

22.2 Utility theory

One method which may seem natural for deciding between two random payouts is to compare the expected amounts that you will receive. However, this does not always give reasonable answers and does not always conform to choices that rational people actually make, as the following example indicates. Imagine that you are offered the following two risky alternatives. In alternative 1, you gain 200 with probability 0.99, or lose 10 000 with probability 0.01. In alternative 2, you gain 200 with probability 0.5 or lose 10 with probability 0.5. Alternative 1 has an expectation of 98, which is greater than 95, the expectation of alternative 2. However,

Fundamentals of Actuarial Mathematics, Third Edition. S. David Promislow.

Companion website: http://www.wiley.com/go/actuarial

nearly everybody would reject alternative 1 with the possibility, although small, of a very large loss. Expectation does not take into account the amount of the risk involved.

Perhaps the most famous example of the drawbacks inherent in using the expected value as a decision tool is the *St. Petersburg Paradox*, formulated by Daniel Bernoulli in the eighteenth century. His explanation marked the beginning of the concept of utility theory. A game consists of tossing a coin until a head appears, with a payout of 2^n if this occurs on the nth toss. The expectation of the amount to be won is then $\sum_{n=1}^{\infty} 2^n(1/2^n) = \infty$, but it is not reasonable to expect that someone would pay an arbitrarily large amount to play this game. Bernouli introduced the idea that one should not consider the actual amounts paid but rather the 'satisfaction' or 'utility' that comes with possessing a certain level of wealth. In other words, he postulated that each individual has a so-called *utility function* u, where $u(x)$ denotes the utility that the person derives from having x units of wealth. It is expected that u is an increasing function of x, as more wealth gives additional utility, but that the rate of increase diminishes with increasing x. A person who is already a multimillionaire will derive little satisfaction from an additional 1 unit of wealth, while somebody who is destitute would welcome it greatly. In mathematical terms, it is expected that for most people u is a concave function. (We give a precise definition in Section 22.3.) If u is differentiable, the above features simply mean that the first derivative of u is nonnegative and the second derivative is nonpositive. A typical example of such a function is $\log x$ which was chosen by Bernouli in his explanation of the St. Petersburg Paradox. He argued that one should consider the *expected utility* rather than the expected value of the actual amounts, and indeed $\sum_{n=1}^{\infty} \log(2^n)(1/2^n)$ is finite.

People with concave utility functions are called *risk-averse*, since they prefer certainty to uncertainty, and therefore derive utility from insuring, as the following example illustrates.

Example 22.1 People are faced with a potential loss of 100, which will occur with probability 0.1. Their goal is to maximize the expected utility of their resulting wealth. How large a single premium P would they pay to insure against such a loss if their utility function is given by $u(x) = \log(x)$, and their initial wealth is (a) 1000? (b) 500?

Solution. In part (a), if they do insure, they will have utility of $u(1000 - P) = \log(1000 - P)$, which is certain; while if they do not insure, they will have an expected utility of $0.9u(1000) + 0.1u(900) = \log[(1000)^{0.9}(900)^{0.1}]$. Equating the resulting expected utility, the largest P they would pay is given by

$$1000 - P = 1000^{0.9}\ 900^{0.1},$$

which is solved to give $P = 10.48$. So such individuals would be willing to pay more than the expected loss of 10 in order to acquire the utility they derive from the extra security.

In part (b), the equation changes to $500 - P = 500^{0.9} 400^{0.1}$ which is solved to give $P = 11.03$. The premium is higher, reflecting the fact that people with the lower wealth are less prepared to suffer a loss and will pay more for insurance. This indicates the important fact that in general one must take into account initial wealth and not just the particular transaction when comparing alternatives as to expected utility. (See Exercise 22.1 for an exception to this statement.)

Economists use many other choices of utility functions in their desire to model people's preferences. One of the most popular choices is the family of so-called *power utility* functions. This is a parametrized family, which includes the log function, and is defined by

$$u_\gamma(x) = \frac{x^{1-\gamma} - 1}{1 - \gamma},$$

for some parameter $\gamma \geq 0$. It is clear that the first two derivatives have the property stated above.

Readers should not be dismayed by the fact that $u_\gamma(x)$ can take negative values, for as long as we use these for comparative purposes there is no difficulty. Indeed, if we replace any utility function u by the function $au + b$, where a and b are constants with $a > 0$, it follows from the linearity of expectation that we will always obtain the same results when comparing two alternatives as to expected utility.

From this point of view we could have described the above family by the functions $x^{1-\gamma}$ for $\gamma < 1$ or $-x^{1-\gamma}$ for $\gamma > 1$. However, an advantage of the given form is that we can define the value at $\gamma = 1$ by taking a limit. Applying L'Hopitals rule, we get

$$u_1(x) = \lim_{\gamma \to 1} u_\gamma(x) = \log x.$$

As γ increases, individuals become more risk-averse, as indicated by the fact that they will pay more to reduce risk. For example if we redo part (a) of Example 22.1 with $\gamma = 2$, we get the equation

$$-(1000 - P)^{-1} = -0.9(1000)^{-1} - 0.1(900)^{-1},$$

which is solved to give $P = 10.99$, an amount greater than the premium of 10.48 for $\gamma = 1$.

Note that $\gamma_0(x) = x - 1$ which indicates there is no risk aversion and comparison is done simply by expected values. A person with such a utility function is the *risk-neutral* individual described in Section 20.5.

People with a convex (defined in Section 22.3) utility function would be termed *risk-seekers* as such people will pay to gamble, even at unfavourable odds, (typical of the usual casino). The following example illustrates the effect of such a utility function.

Example 22.2 People with an initial wealth of 10 are offered a chance to play a game in which they win either 2 or 0, each with probability 1/2. If their utility function is given by the convex function $u(x) = x^2$, what is the most they will pay to play this game.

Solution. Let P be the amount paid to play the game. We equate the expected utility of not playing versus playing which gives the equation

$$100 = 0.5[(12 - P)^2 + (10 - P)^2].$$

Solving, $P = 1.0501$, which is more than the expected winnings of 1.

The examples of this section should provide further clarification of the difference between insurance and gambling that we alluded to in Section 1.1. In the next section, we provide some more precise definitions and mathematical verification of our conclusions.

22.3 Convex and concave functions: Jensen's inequality

22.3.1 Basic definitions

Definition 22.1 A real-valued function g defined on an interval I of the real line is said to be *convex* if for all x and y in I and $0 \le \alpha \le 1$:

$$g(\alpha x + (1-\alpha)y) \le \alpha g(x) + (1-\alpha)g(y). \tag{22.1}$$

Geometrically, this says that a line segment joining any two points of the graph of g will lie above the graph.

A feature of convex functions that is often used is the *increasing slope* condition which states that for three points $x < y < z$ in I,

$$\frac{g(y)-g(x)}{y-x} \le \frac{g(z)-g(y)}{z-y}. \tag{22.2}$$

To derive (22.2) note that

$$y = \frac{z-y}{z-x}\,x + \frac{y-x}{z-x}\,z,$$

so that

$$g(y) \le \frac{z-y}{z-x}\,g(x) + \frac{y-x}{z-x}\,g(z).$$

Now multiply this equation by $(z-x)/(z-y)(y-x) = (y-x)^{-1} + (z-y)^{-1}$ to get

$$g(y)\left[\frac{1}{y-x} + \frac{1}{z-y}\right] \le \frac{g(x)}{y-x} + \frac{g(z)}{z-y},$$

and rearrange to get (22.2).

Many readers will be familiar with the result from basic calculus that for twice differentiable functions, convexity is characterized by the fact the second derivative is nonnegative. The advantage of the general definition above is that we can apply it to the functions that have points of nondifferentiability, such as the two families of functions (using the notation of Section 21.10.1):

$$u_d(x) = (x-d)_+, \quad v_d(x) = d - x \wedge d,$$

whose graphs are shown in Figure 22.1. Any piecewise linear function (one whose graph consists of a finite number of straight line segments) can be written as a linear combination of

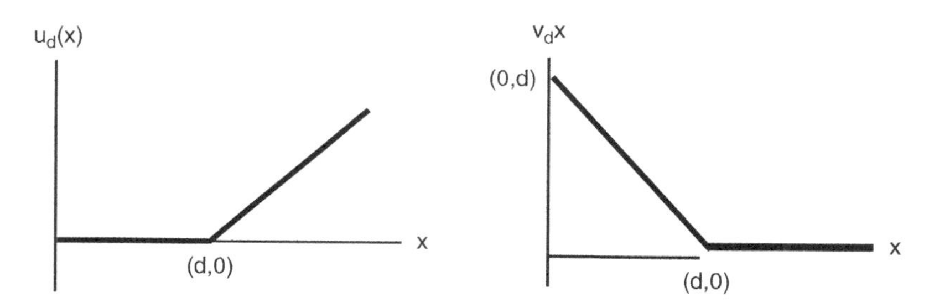

Figure 22.1 Graphs of $u_d(x)$ and $v_d(x)$

the $u_d{'}s$ and $v_d{'}s$. In view of the increasing slope condition, a *convex* piecewise linear function can be written as a linear combination of these with *positive* coefficients. For example, the function defined on the real line by

$$f(x) = \begin{cases} |x|, & \text{if } x < 2, \\ 3x - 4, & \text{if } 2 \le x \end{cases}$$

can be written as $v_0 + u_0 + 2u_2$.

Definition 22.2 A function defined on an interval I of the real line is said to be *concave* if the inequality reverses in (22.1) and therefore also in (22.2).

For concave functions, straight line segments between any two points of the graph are now *above* the graph. For twice differentiable functions the second derivative is nonpositive. Clearly, a function g is concave if and only if $-g$ is convex.

22.3.2 Jensen's inequality

We introduce a basic inequality for risk assessment. To motivate, imagine that you are presented with a bag with two numbered balls. You draw one at random and receive the square of the number. If both the balls had number 5, you get 25 for sure. What if they were numbered 4 and 6, which average 5? You now get an expected payoff of $0.5(36 + 16) = 26 > 25$, so the randomness has produced an extra expected return over the certain case. We can see exactly why this happens by writing

$$(5 + 1)^2 = 5^2 + 2(5) + 1^2$$

and

$$(5 - 1)^2 = 5^2 - 2(5) + (-1)^2.$$

The key to this result then is the fact that $(-1)^2 = 1$. When you average, the middle terms cancel but the squared term is the same for the positive and negative deviations, giving the extra amount. This calculation shows that you will get the same conclusion for any two numbered balls. What about for three balls? Say you have numbers 1, 2, 6 which average 3.

The average of the squares is 13 2/3 which is greater than 3^2. Indeed, the principle involved holds in great generality since the inequality

$$0 \leq E(X-\mu)^2 = E(X^2) - \mu^2,$$

where $\mu = E(X)$, shows that for any random variable X, the expectation of X^2 must be greater than or equal to the square of the expectation of X.

What happens for other functions other than the square? For the square root function, the inequality goes the other way. If you have two balls numbered 16 and 4 with an average of 10, the average of the square roots is 3 which is less than $\sqrt{10}$. It turns out that we get the extra return with any *convex* function, and therefore a reduced return from the average with any concave function. The formal statement is as follows.

Theorem 22.1 (Jensen's inequality) *For any convex function g,*

$$E[g(X)] \geq g\big(E(X)\big).$$

Proof. We will give the proof in the case where g has a continuous second derivative. The idea is that by our remarks above, it is true for any polynomial of degree two which has a *positive* coefficient of x^2. Taylor's theorem from basic calculus tells us that g can be approximated by such a polynomial. Precisely, if $E(X) = \mu$, then

$$g(x) = g(\mu) + (x-\mu)g'(\mu) + \frac{(x-\mu)^2}{2}g''(\xi),$$

for some point ξ between μ and x. Of course we do not know ξ but we do not need it, since if g is convex, then g'' is nonnegative, and the left-hand side is greater than or equal to the sum of the first two terms. Taking expectations,

$$E[g(X)] \geq g(\mu) + g'(\mu)E(X-\mu) = g(\mu),$$

completing the proof. □

Jensen's inequality obviously reverses for a concave function, as seen by applying the statement above to the function $-g$. This shows that in general we have the conclusions shown by the particular examples of the last section. That is, as we would expect, risk-averse individuals prefer certainty to risk. Such individuals with initial wealth w who pay $E(X)$ to insure against a loss X will have resulting utility of $u(w - E(X))$ which is greater than $E[u(w - X)]$. Therefore, they are willing to pay somewhat more than the expected value of the benefits. This makes the business of insurance economically feasible, since as we indicated in previous chapters, the insurer must necessarily charge more than the expected value of the loss in the form of loadings for expenses, profits and risk.

22.4 A general comparison method

In this section, we consider the general problem of comparing two random variables as to their degree of risk. There are many ways of defining such an order relationship. We will concentrate on one of the early definitions, introduced in Rothschild and Stiglitz (1970) which ties in with the utility theory concept. For simplification, we concentrate on the equal mean case.

Definition 22.3 For nonnegative random variables X and Y with $E(X) = E(Y)$, we say that X is *less risky* than Y, if for all d,

$$E(X - d)_+ \leq E(Y - d)_+. \tag{22.3}$$

For example, if our random variables represent losses, then the net premium for deductible insurance that covers losses above a certain amount is always less for the less risky option. It is clear that this relation does not depend on the actual random variables but only on their distribution.

In view of the equal mean hypothesis, formula (21.11) shows that an equivalent formulation is that for all $d \geq 0$,

$$E(X \wedge d) \geq E(Y \wedge d). \tag{22.4}$$

We can apply the above inequalities to the functions introduced in the previous section. The result is that if X is less risky than Y, then for all $d \geq 0$,

$$E[u_d(X)] \leq E[u_d(Y)], \quad E[v_d(X)] \leq E[v_d(Y)], \tag{22.5}$$

and therefore, for any piecewise linear convex function g,

$$E(g(X)) \leq E(g(Y)), \tag{22.6}$$

since the positive coefficients, when we express g as a linear combination of the u_d and v_d for various values of d, will preserve the order. Now given an arbitrary convex function, we can approximate it by a piecewise linear one as follows. Choose a large number of points on the graph and join them with straight line segments. The increasing slope condition shows that this approximating function will be convex. The more points that we choose, the better the approximation will be. By standard approximation techniques in analysis it follows that (22.6) holds for all convex functions (we do not give the exact details here), and of course the reverse inequality holds for concave functions. We are therefore led to the conclusion that if risk-averse individuals have to choose between two distributions of their final wealth, both of the same mean, they will choose the less risky one according to our definition, in order to maximize expected utility. This justifies the definition given by Rothschild–Stiglitz as one which truly captures the concept of riskiness.

It is not always easy to decide whether one random variable is less risky than another, but there are certain instances when we can verify this. It is assumed in the following two theorems that $E(X) = E(Y)$.

Theorem 22.2 *Suppose that X takes the values $a_1 \leq a_2 \leq \ldots \leq a_{N-1} \leq a_N$ each with probability $1/N$ and Y takes the values $b_1 \leq b_2 \ldots \leq b_{N-1} \leq b_N$ each with probability $1/N$. Then X will be less risky than Y if and only if*

$$\sum_{i=1}^{k} a_i \geq \sum_{i=1}^{k} b_i \tag{22.7}$$

for $1 \leq k \leq N$.

Proof. Suppose the condition holds. Then, given d where $a_k \leq d < a_{k+1}$, we have

$$E(X \wedge d) = \frac{1}{N}\left[\sum_{k=1}^{k} a_i + (N-k)d\right] \geq \frac{1}{N}\left[\sum_{k=1}^{k} b_i + (N-k)d\right] \geq E(Y \wedge d).$$

The last inequality follows from the fact that $b_i \wedge d$ is less than or equal to both b_i and d for all i.

Conversely, suppose X is less risky than Y. Fix any index $k < N$, and let $d = \max\{a_k, b_k\}$. Invoking (22.5), we see that, if $d = b_k$, then

$$\frac{1}{N}\sum_{i=1}^{k}(d - a_i) \leq E[v_d(X)] \leq E[v_d(Y)] = \frac{1}{N}\sum_{i=1}^{k}(d - b_i),$$

while if $d = a_k$, then

$$\frac{1}{N}\sum_{i=k+1}^{N}(a_i - d) = E[u_d(X)] \leq E[u_d(Y)] \leq \frac{1}{N}\sum_{i=k+1}^{N}(b_i - d),$$

In either case, condition (22.7) follows in view of the fact that the equal mean hypothesis implies that $\sum_{i=1}^{N} a_i = \sum_{i=1}^{N} b_i$. □

Remark Since repetition of values is allowed, the above theorem can be applied to all finite discrete distributions where the probabilities are rational numbers.

Here is another condition which applies to all distributions. We can view this geometrically as saying that if two distribution functions intersect at one point, then the steeper curve gives the the less risky distribution (see Figure 22.2).

Theorem 22.3 (The cut condition) *Suppose that for some point c,*

$$F_X(t) \leq F_Y(t), \text{ for } t < c \text{ while } F_X(t) \geq F_Y(t), \text{ for } t > c.$$

Then X is less risky than Y.

Proof. We use formulas (21.10) and (21.11), and note that the stated inequalities in the premise reverse with s in place of F. If $d < c$,

$$E(X \wedge d) = \int_0^d s_X(x)dx \geq \int_0^d s_Y(x)dx = E(Y \wedge d),$$

while if $d > c$,

$$E(X - d)_+ = \int_d^\infty s_X(x)dx \leq \int_d^\infty s_Y(x)dx = E(Y - d)_+.$$

By continuity of the function $g(d) = E(X - d)_+$, we must obtain the same result at $d = c$. In all cases we have by definition that X is less risky than Y. □

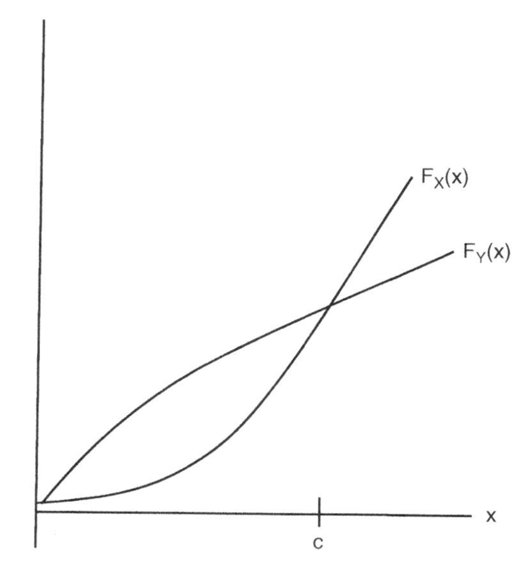

Figure 22.2 The cut condition: X is less risky than Y

Remark The definition of being less risky and the cut condition apply equally well to random variables that take negative values. The complication in the proof is that formula (21.12) no longer holds as given and a suitable modification is required. We leave this to the interested reader.

Readers should be aware that our ordering in this case is what mathematicians call a *partial order*, meaning that certain pairs are incomparable. Given a choice of X and Y, it may happen that some risk -averse people would prefer a final wealth of X and others would prefer Y, so neither is less risky than the other according to our definition. Here is a typical example.

The random variable X, representing a gain, takes the value 1 with probability 0.1 and 10 with probability 0.9, while an alternative Y takes the value 2 with probability 0.9 and 73 with probability 0.1. Both have a mean of 9.1. One might expect a risk averter to choose X, ensuring themselves of a return of 10 in most cases, rather than gamble on the higher return which most of the time will lead to a return of only 2. However, by Theorem 22.2, since $1 < 2$, while $11 > 4$, they are incomparable. This happens since there could be someone so risk averse that they could not tolerate even the small chance of a return of only 1. Indeed, consider the following scenario. Suppose individuals absolutely need 2 units of wealth or something dreadful will happen to them. They might well want to ensure that this terrible fate cannot occur by choosing Y.

Of course, when we are able to make a comparison, the conclusion is that much stronger. As a typical example we will prove a well-known result on optimal choices of insurance, which was proved in Arrow (1963).

The idea is as follows. Suppose that individuals want to ensure against a loss by paying some fixed premium, which is however insufficient to provide for full coverage, and so they must arrange for partial reimbursement.

Let X denote a random loss (assumed to be nonnegative) and P denote the premium they want to pay. They must choose some $I(X)$, a function of X, which will be the amount paid to them when loss occurs. We make the natural assumption that $0 \leq I(X) \leq X$, so that the amount paid cannot be negative and cannot exceed the amount of the loss. We assume that to provide the coverage, the insurer charges a premium that is a function π of $E[I(X)]$. Any of the modifications discussed in Section 21.10.1, as well as others, could be used. As an example, suppose the insurer charges a 20% loading above the expected value of the loss and $E(X) = 100$, while $P = 60$. The insured then is only going to pay one-half of the premium necessary for full coverage. One way of doing so is to choose $I(X) = 0.5X$ so whatever the loss is, the insured will receive one-half of this amount as reimbursement. Another alternative is to set a maximum on the amount paid, and of course there are many other possibilities. Which choice is best?

Theorem 22.4 *Given the assumptions above, any risk-averse individual should choose deductible insurance, in order to maximize the utility of resulting wealth.*

Proof. Suppose the individuals start with an initial wealth of a. Their resulting wealth after paying the premium, incurring the loss and being reimbursed, will be

$$W = a - P - X + I(X).$$

In particular let $I_0(X) = (X - d)_+$, where d is chosen to satisfy $\pi[E(I_0(X))] = P$, and let W_0 be the resulting wealth for this case. Apply the cut condition with $c = a - P - d$. Since W_0 is never less than c, then $F_{W_0}(t) = 0 \leq F_W(t)$, for any $t < c$. Consider now the case when $t > c$. If $W_0 > t$, then $X - I_0(X) < d$, so the loss must have been under the deductible. From our assumption that reimbursement cannot exceed the loss, we must have $I(X) \leq d$, showing that $W \geq W_0 > t$. It follows that $F_W(t) \geq F_{W_0}(t)$ completing the proof. □

The conclusion seems intuitively clear since deductible insurance avoids large catastrophic losses, which should appeal to the risk averter.

Remark We leave to the interested reader the following extension of the above problem. The individual truly interested in maximizing utility would normally not specify the desired premium in advance, but would let that vary as well. The problem now is not only to choose the form of insurance but also to choose P, which will be a function of the initial wealth.

22.5 Risk measures for capital adequacy

22.5.1 The general notion of a risk measure

In some cases we want to do more than just compare the riskiness of two random variables. We actually want to assign a number that in some sense quantifies or measures the risk, and of course that will then allow us to compare two or in fact any number of random variables. A function that assigns a number to a certain class of random variables is known appropriately enough as a *risk measure*. We already encountered this concept in Section 15.6.4. Assigning a premium to a random variable representing the benefits paid on an insurance contract is a form of risk measure.

Another important use of this concept is to arrive at the amount of capital that should be held to cover possible losses. Now we have already introduced the concept of a reserve to provide for future obligations. Recall, however, from Section 15.5 that the reserve is the *expected* amount that one needs, and it is provided for by the excess premiums that are collected in early years. The reserve is not intended to provide for unexpected large losses. If these occur, the company will have to draw on surplus in order to maintain the required reserves, and the question is, how much capital should be on hand for this purpose. This idea is not just confined to insurance. Measures for this purpose are extensively used by the banking industry. The method that has been almost universally adopted by banks is a quantile-based approach, which we have already discussed in Section 15.7. For present purposes, we now introduce some more precise definitions and terminologies.

22.5.2 Value-at-risk

Suppose we have a random variable X and a number α between 0 and 1. A number x, such that $F_X(x) = \alpha$, is known as an *α-quantile* of X. (An alternate terminology is to multiply α by 100 and speak of a *percentile*. For example, a 0.7 quantile can be referred to as 70th percentile.) The 0.5 quantile is commonly referred to as the *median* of the distribution.

When the values of F_X vary continuously from 0 to 1 over some interval, then $F_X^{-1}(\alpha)$ is the unique *α-quantile*. We will denote this number by q_α. In other cases there might not exist any such number, or there might exist infinitely many. The former arises when the value of F_X jumps at a point x from a number less than α to one more than α. In that case we take q_α to be the point of the jump. Such a point will of course be equal to q_β for several different values of β. The case of infinitely many values arises when for some x, $F_X(x) = \alpha$, but X does not take any values in some open interval with a left endpoint of x. For example, if $F_X(3) = 0.7$ and the probability of X taking a value in the interval (3,3.1) is zero, then $F_X(x)$ is also equal to 0.7 for any x in the interval (3, 3.1). For our purposes we will want to single out the smallest such number. We can then cover all cases by the following.

Definition 22.4

$$q_\alpha = \min\{x : F_X(x) \geq \alpha\}.$$

When X represents losses, or an amount that to be paid out, q_α can be viewed as a type of risk measure, with higher values signifying more risk. The percentile premium was a risk measure of this type. In the banking industry, this risk measure has been termed *value-at-risk* and abbreviated as VaR. (The capital R at the end distinguishes it from the common notation for 'variance'. It is pronounced to rhyme with 'far'.) VaR is expressed with a certain time horizon and a confidence level α, often taken to be a number reasonably close to 1 such as 0.95 or 0.99. For example, to say that a certain investment portfolio has a 1 day VaR of 100 000 means that 100 000 will be sufficient to ensure that the losses over the next day will be covered *most of the time* where 'most' is measured by the specified confidence level.

22.5.3 Tail value-at-risk

There are problems with VaR, as we have already noted in connection with percentile premiums. It does not take into account how bad the losses can be when they exceed the chosen quantile. In the above example, if there were a possibility of a loss of several million, which occurred with probability less than $(1-\alpha)$, we still would have a VaR of only 100 000. The possibility of this large loss would be completely ignored in our risk measure. For this, and other reasons, many people have advocated a modification of VaR known as *tail value-at-risk* (abbreviated as TailVar or just TVar). The same measure is also known as *conditional tail expectation*, abbreviated as CTE or sometimes TCE.

For a given confidence level α, TVaR_α is essentially defined as the expected loss given that the loss is in excess of the quantile q_α. Problems can arise in interpreting the words 'in excess of'. Do these words mean 'strictly greater than' or 'greater than or equal to'? The distinction is irrelevant for continuous distributions, but in the discrete case they can give different results, and, strangely enough, *neither* may be the one that you want. For example, suppose that somebody tosses a penny and a dime and you must pay them 1 for each head. Set $\alpha = 0.5$. The median loss is 1, and $\text{TVaR}_{0.5}$ should be the expected value in the worst one-half of the distribution. There are four possibilities of equal probability giving respective losses of 0, 1, 1, 2, so we want to select the two worst cases. Of course there is a tie, but a logical way to handle this is to arbitrarily pick any one of the two coins, say the penny, and we then say that the the two worst outcomes will be when both coins come up heads, or the penny only comes up heads. With this reasoning $\text{TVaR}_{0.5}$ should be $0.5(1+2) = 1.5$. However, the expected loss, given that the loss is strictly greater than 1, will be 2, and the expected loss, given that the loss is greater than or equal to 1, will be $4/3$. Readers are cautioned that there are examples in the literature where the 'strictly greater than' or 'greater than equal to' methods are used in defining TVaR and similar risk measures. However, both of these can lead to inconsistencies. See, for example, Exercise 22.6. These are avoided by the definition below, which follows from the idea presented in this simple example.

For a continuous distribution we define our desired risk measure by

$$\frac{1}{1-\alpha}\int_{q_\alpha}^{\infty} x f_X(x)dx. \tag{22.8}$$

Now by definition

$$\int_{q_\alpha}^{\infty} f_X(x)dx = P(X > q_\alpha) = (1-\alpha),$$

so we can add and subtract q_α on the right-hand side to get the following.

Definition 22.5

$$\text{TVaR}_\alpha(X) = q_\alpha + \frac{1}{1-\alpha}\int_{q_\alpha}^{\infty}(x-q_\alpha)f(x)dx = q_\alpha + \frac{1}{1-\alpha}E(X-q_\alpha)_+. \tag{22.9}$$

The above formula is the best form of the definition to use as it applies to any distribution, not just a continuous one.

As a check, we verify that it gives the results we expect in the discrete case. (In particular we retrieve the answer of 1.5 obtained in the coin-flip example.) We calculate TVaR for the particular type of discrete random variables considered in Theorem 22.2. For a definite example, let $N = 10$ and suppose that our confidence level $\alpha = 7/10$. Then $\text{VaR}_{0.7}(X)$ clearly equals a_7 which will cover the loss whenever the outcome is a_i, $i \leq 7$ which occurs $7/10$ of the time. What about $\text{TVaR}_{0.7}$?

From Definition 22.5, this is just

$$a_7 + (10/3)(1/10)[(a_8 - a_7) + (a_9 - a_7) + (a_{10} - a_7)] = (1/3)[a_8 + a_9 + a_{10}], \quad (22.10)$$

as we would expect. In general, if $\alpha = r/10$ for an integer r, then $\text{VaR}_\alpha(X) = r/10$ and $\text{TVarR}_\alpha(X) = (a_{r+1} + \cdots + a_{10})/(10 - r)$.

Things are a bit more complicated when for some integer r, $(r-1)/10 < \alpha < r/10$. Suppose in the above example that $\alpha = 13/20$ which is between $6/10$ and $7/10$. We still get $\text{VaR}_{\alpha}(X) = a_7$, but now $(1-\alpha)^{-1} = 20/7$, so Equation (22.9) yields

$$\text{TVaR}_{13/20}(X) = \frac{a_7 + 2a_8 + 2a_9 + 2a_{10}}{7} = \frac{4}{7}\left(\frac{a_7 + a_8 + a_9 + a_{10}}{4}\right) + \frac{3}{7}\left(\frac{a_8 + a_9 + a_{10}}{3}\right).$$

The general formula is as follows. If $\alpha = (r - p)/N$ where r is an integer and $0 < p < 1$, then

$$\text{TVaR}_\alpha = \beta\,\text{TVaR}_{(r-1)/N} + (1-\beta)\,\text{TVaR}_{r/N}, \quad (22.11)$$

where $\beta = p(N - r + 1)/(N - r + p)$. We leave this for the reader to verify. To actually compute TVaR in this discrete case, it is usually more efficient to use Definition 22.5 directly, but (22.11) is useful for demonstrating properties of TVaR, as we will illustrate later.

Many people believe that a risk measure H used for premiums or capital adequacy should be subadditive. Namely they want that

$$H(X + Y) \leq H(X) + H(Y).$$

After all, if we have to set aside $H(X)$ of capital to provide for a risk X and a further $H(Y)$ of capital to provide for a risk Y, then it is reasonable to suppose that we will not need more than this sum if we take on both risks. (This viewpoint is not completely universal and some argue that such activities as mergers can produce inefficiencies and cause other problems that actually increase the total risk.) We already saw in Example 15.6 that a quantile risk measure does not satisfy subadditivity, and this has been a major criticism levelled against the use of VaR. On the other hand, TVaR is subadditive. A complete general proof is somewhat advanced and we will confine attention here to show this for the discrete random variables that we introduced above, where the proof is straightforward and clearly indicates the reason for the result.

Suppose we have a sample space $\{\omega_1, \omega_2, \ldots, \omega_N\}$ each with probability $1/N$ and X takes the value a_i on ω_i. where $a_i \leq a_j$ for $i \leq j$. Suppose that the random variable Y takes the value b_i on ω_i.

Assume first that $b_i \leq b_j$ for $i < j$. The random variables X and Y in the this case are said to be *comonotonic*. Precisely this means that for two sample points, if X takes a higher value on one of them, then Y will also take a higher value on that point. In other words they move together. (This concept can been generalized to arbitrary distributions and plays a major role in risk assessment.) Take $\alpha = r/N$. In this case, the $N - r$ highest values of $X + Y$ will be of the form $a_{r+1} + b_{r+1}, a_{r+2} + b_{r+2}, \ldots, a_N + b_N$ and from (22.10) it is clear that

$$\text{TVaR}(X + Y) = \text{TVaR}(X) + \text{TVaR}(Y). \tag{22.12}$$

This is a reasonable conclusion. In the case of comonotonicity, we cannot hope to have a high value in one random variable offset by a low value in the other. There is no diversification effect and combining the random variables does not lead to a reduction in the total risk measure. Now suppose that we remove the comonotonicity by rearranging the $b's$. Obviously, the $N - r$ highest values of $a_i + b_i$ cannot get any larger than what we had in the previous case, where we included the $N - r$ highest values of *both* the a's and b's. The left side of (22.12) must stay the same or decrease, and this implies the required subadditivity when α is as given. Now, we can invoke (22.11) to see that it holds for any α.

Besides subadditivity, there are other desirable features of risk measures that are satisfied by TVaR but not VaR. See, for example, Exercise 18.13.

Following are two examples which compute TVaR for familiar distributions.

Example 22.3 Compute TVaR for a normal distribution.

Solution. In the case of a standard normal Z, the density function satisfies $xf_Z(x) = -f'_Z(x)$. From (22.8) and the fundamental theorem of calculus,

$$\text{TVaR}_\alpha(Z) = \frac{1}{1-\alpha} f_Z\big(\Phi^{-1}(\alpha)\big),$$

where Φ is the c.d.f. of Z.

It is not difficult to show that for any X and constants a and b, with $b > 0$, $\text{TVaR}_\alpha(a + bX) = a + b\text{TVaR}_\alpha(X)$. Therefore, if X is a normal distribution with mean μ and variance σ^2, we have

$$\text{TVaR}_\alpha(X) = \mu + \frac{\sigma}{1-\alpha} f_Z\big(\Phi^{-1}(\alpha)\big).$$

Example 22.4 Compute TVaR for a exponential distribution.

Solution. First we note that if $X \sim \text{Exp}(\lambda)$ then $q_\alpha = s_X^{-1}(1-\alpha) = -\log(1-\alpha)/\lambda$. From Equation (21.15), Example 21.7, and the fact that $s_X(q_\alpha) = 1 - \alpha$ by definition, it follows that $E(x - q_\alpha)_+ = (1-\alpha)/\lambda$. Now from Definition 22.5,

$$\text{TVaR}_\alpha(X) = \frac{1 - \log(1-\alpha)}{\lambda}.$$

We conclude this section by providing an equivalent formulation of TVaR for continuous distributions, which serves to illustrate further how the information in the tail which is ignored in VaR gets incorporated into TVaR. In the integral (22.8), make the substitution $\beta = F_X(x)$. We have $x = q_\beta(X)$, $d\beta = f_X(x)dx$ so that

$$\text{TVaR}_\alpha(X) = \frac{1}{1-\alpha}\int_\alpha^1 q_\beta(X)d\beta.$$

This says that we can view TVaR_α as an average of all the VaRs at confidence levels greater than α.

22.5.4 Distortion risk measures

Here is another equivalent formulation of TVaR. For any α in the interval (0, 1), let g_α denote the function on [0, 1] defined by

$$g_\alpha(x) = \begin{cases} \dfrac{x}{1-\alpha}, & \text{if } 0 \le x < 1-\alpha, \\ 1, & \text{if } 1-\alpha \le x \le 1. \end{cases}$$

Then from (21.10) and Definition (22.5),

$$\text{TVaR}_\alpha(X) = \int_0^\infty g_\alpha(s_X(x))dx.$$

A whole family of other risk measures arises if, in the above formula, we replace g_α by any continuous function g that increases from 0 to 1. These are known as *distortion risk measures.* Note that when $g(x) = x$ we just get $E(X)$ as the risk measure. The concept has been termed by some as a sort of dual approach to that of taking expected utility. In the latter case we distort the amounts paid by converting them into utilities. In this case we distort the probabilities. Smaller values of the survival function, corresponding to right tail events, are increased in value, in an attempt to reflect the risk. It can be shown that any concave function g will give a subadditive risk measure.

Notes and references

The order relation we introduced in Section 22.4 is sometimes referred to as the *convex* order in view of (22.6). The same definition with the equal mean hypothesis eliminated is known as the *stop loss order*.

More detailed information on ordering risks can be found in Kass *et al.* (2008). This same reference contains additional material on utility theory.

Readers particularly interested is risk measures may consult Artzner *et al.* (1999), which deals with 'coherency', a much discussed topic in recent years.

Exercises

22.1 For any $\alpha > 0$ define a utility function by $u_\alpha(x) = -e^{-\alpha x}$. (This is known as *exponential utility*.)

(a) Show that a person with this utility function is risk averse.

(b) Show that when comparing risky alternatives to maximize utility by using u_α, the result is independent of initial wealth.

(c) In Example (22.1) let P_α denote the premium that would be paid when using u_α as a utility function. Calculate $P_{0.02}$, $P_{0.01}$ and $\lim_{\alpha \to 0} P_\alpha$. What happens to risk adversity as α decreases?

22.2 Use Jensen's inequality to prove the well-known arithmetic–geometric mean inequality. For positive numbers a_i, $1 \le i \le n$,

$$(a_1 a_2 \cdots a_n)^{1/n} \le \frac{a_1 + a_2 + \ldots + a_n}{n}.$$

Hint: Take an appropriate distribution and let $g(x) = \log x$.

22.3 Consider the two random variables $X \sim \exp(2)$ and $Y \sim \exp(3) + 1/6$. Compare as to riskiness according to the definition given in Section 22.4. That is, is X less risky than Y, or is Y less risky than X or are they incomparable?

22.4 Compare the following three distributions as to riskiness:

X takes the value 1 with probability 1/6, 2 with probability 1/3, 3 with probability 1/3 and 6 with probability 1/6.

Y takes the value 2 with probability 1/2, 3 with probability 1/3 and 5 with probability 1/6.

Z takes the value 1 with probability 1/6, 2 with probability 1/6, 3 with probability 1/3, and 4 with probability 1/3.

22.5 (a) Show that X less risky than Y implies that the variance of X is less than the variance of Y.

(b) Show that the converse is true in the normal case. That is, for normal random variables X and Y with the same mean, the one with the smaller variance will be less risky.

22.6 (a) On a sample space of three points, ω_1, ω_2, ω_3, a random variable X takes the values (1,1,3) and a random variable Y takes the values (1,2,3). Consider the risk measure $H(Z) = E[Z|Z > q_{0.3}]$. Show that despite the fact that Y takes values at all sample points that are greater than or equal to those of X, we have $H(X) > H(Y)$.

(b) Find an example that works as the above only now with $H(Z) = E[Z|Z \ge q_\alpha]$ for some α.

22.7 Suppose $X \sim \text{Gamma}(2, 3)$. For a certain value of α, $\text{VaR}(X) = 4$. Find $\text{TVaR}(X)$.

22.8 The random variable X has a density function given by

$$f(x) = \begin{cases} x, & \text{if } 0 \le x < 1, \\ 2 - x, & \text{if } 1 \le x \le 2. \end{cases}$$

Find $\text{VaR}_\alpha(X)$ and $\text{TVAR}_\alpha(X)$ for $\alpha =$ (a) 1/2, (b) 31/32.

22.9 Find $\text{VaR}_\alpha(X)$ and $\text{TVaR}_\alpha(X)$ when X is uniform on [0,N].

22.10 A random variable X takes the values 1 with probability $4/7$, 3 with probability $2/7$ and 6 with probability $1/7$. Find $\text{VaR}_{0.8}(X)$ and $\text{TVaR}_{0.8}(X)$.

22.11 A random variable takes on the values $x_1, x_2, \ldots, x_{40}$, each with probability $1/40$. If $\text{TVaR}_{0.85}(X) = 100$ and $\text{TVaR}_{0.875}(X) = 150$, find $\text{TVaR}_{0.86}(X)$.

22.12 On a sample space consisting of four points that have equal probability, the random variables X and Y take, respectively, the values (1,2,3,4) and (4,1,2,3). Consider the distortion risk measure H given by the function $g(x) = x^{1/2}$.

(a) Calculate $H(X), H(Y), H(X + Y)$ and verify that the subadditivity holds.

(b) Suppose that Z is an other random variable on this space taking the values (a, b, c, d), where $a \leq b \leq c \leq d$. Verify that $H(X + Z) = H(X) + H(Z)$.

22.13 Suppose that X and Y both take N values each with probability $1/N$, $E(X) = E(Y)$ and that X is less risky than Y. Show that $\text{TVaR}_\alpha(X) \leq \text{TVaR}_\alpha(Y)$, but it is not necessarily true that $\text{VaR}_\alpha(X) \leq \text{VaR}_\alpha(Y)$.

22.14 (a) If X is a distribution such that $\text{TVaR}_\alpha(X) - \text{VaR}_\alpha(X)$ is independent of α, what must this constant difference be?

(b) Show that an exponential distribution has this property.

22.15 (This question refers back to material introduced in Part I of the book.) Assume that interest rates are positive, that is, the investment discount function $v(t)$ is decreasing with t.

(a) Show that $\bar{a}(1_t; v)$ is a concave function of t.

(b) Use Jensen's inequality to show that $\bar{a}(1_{\mathring{e}_x}) \geq \bar{a}_x$

(c) Now assume constant interest. A 1-unit whole life policy on (x) has premiums payable continuously at the annual rate of P. Show that

$$P\, E[\text{Val}_{T(x)}(1_{T(x)}; v)] \geq 1.$$

That is, the expected amount of premiums accumulated at death by a policyholder is greater than or equal to the benefit payment at that time.

(d) Explain the inequality in (c) by general reasoning. (You may want to refer to Section 8.4.3.)

(e) Show that the inequalities in (b) and (c) are equalities at 0 interest.

23

Ruin models

23.1 Introduction

This chapter involves extending some aspects of the collective risk model to a multi-period setting. It will require a sound knowledge of the material in Chapter 18. We begin with the discrete-time case and consider another interpretation of Equation 18.3.

Consider an insurer who each period collects total premiums of c and experiences aggregate claims of $\langle N, X \rangle$ as defined at the end of Section 21.1. Then the gain of the insurer in the nth period is given by a random variable G_n where

$$G_n \sim c - \langle N, X \rangle.$$

If we assume that claims each period are independent of those in other periods, we can interpret Equation (18.3) as representing a surplus process of the insurer, where U_n is the surplus at time n resulting from an initial surplus of u at time 0. This will be a major application for the theory in this chapter, although it applies as well to the original gambling formulation.

Let T be the first time the surplus becomes negative. We call this the *time of ruin*. In the discrete-time case, we define this formally as

$$T = \min\{n : U_n < 0\}.$$

The random variable T is different from the other random variables we have encountered since it is not necessarily real valued. For any realization for which the surplus is nonnegative at all times, the value of T will be ∞. The set of all such realizations can have positive probability, in which case ruin is not certain. We are interested in the probability that ruin will eventually occur. This will, of course, depend on the initial surplus u, so we denote this by $\psi(u)$. That is,

$$\psi(u) = P(T < \infty | U_0 = u).$$

Fundamentals of Actuarial Mathematics, Third Edition. S. David Promislow.

Companion website: http://www.wiley.com/go/actuarial

Note that $\psi(u)$ is the probability of eventual ruin, which may seem to be of little interest since nobody is planning to gamble or to run an insurance company forever. From a practical point of view, one may want to compute the probability of ruin over some finite time horizon. That is, one may want the probability that ruin will occur before a fixed time t. We denote this by $\psi(u, t)$. So,

$$\psi(u,t) = P(T \le t | U_0 = u).$$

It is difficult to find general methods for calculating this quantity, and normally each case must be treated individually. The following example exhibits some of the possible techniques in a simple discrete time example.

Example 23.1 A special insurance company has a *single contract*. There can be at most one claim, and the probability that a claim does not occur by time t is $1/(1+t)$. If a claim occurs, the amount is 100 with probability 0.6 or 200 with probability 0.4. Premiums are paid continuously at the rate of 20 per year. The company begins with an initial surplus of 60. What is the probability of eventual ruin?

Solution. This is really a finite-time question, since after time 7 ruin cannot occur, for the insurer will have collected the maximum claim amount of 200. We break this interval up into the relevant time periods.

(i) From time 0 to time 2, a claim will occur with probability $2/3$, and the insurer will necessarily be ruined since the initial surplus and premiums collected will be under the minimum claim of 100.

(ii) From time 2 to time 7, a claim will occur with probability $\frac{7}{8} - \frac{2}{3} = \frac{5}{24}$, and ruin will occur only if the claim is for 200. So the probability of ruin in this interval is $0.4 \times \frac{5}{24} = \frac{1}{12}$.

The total probability of ruin is therefore $\frac{2}{3} + \frac{1}{12} = \frac{3}{4}$.

When G is finite valued, the quantity $\psi(u, t)$ can be both childishly simple and fiendishly difficult to compute. To illustrate this paradoxical statement, consider an example. You flip a coin with probability of a head equal to p, and you win 1 for a head and lose 1 for a tail. What is $\psi(1, 2)$? This is answered immediately since the only way you can be ruined by time 2 is to get two tails in a row, and we conclude that $\psi(1, 2) = (1 - p)^2$. In fact, whenever G is finite valued, we can always, in theory, compute $\psi(u, t)$ by simply looking at all possible paths up to time t and seeing which ones lead to ruin. The problem is that if the range of G and the time t are large enough, the number of such paths could be enormous, rendering any computation infeasible. We therefore need other ways to get information. One such method is to compute $\psi(u)$, which gives an upper bound since

$$\psi(u,t) \le \psi(u), \qquad \text{for all } t.$$

The previous discussion then serves as a motivation for the main theme of this chapter, which is to derive methods for calculating the infinite time ruin probability $\psi(u)$. There are several ways to either compute this or estimate it, and we will discuss them in turn. Each is useful for certain cases.

Remark Some authors define ruin as the first time the surplus reaches zero, rather than the first time it becomes negative. Let $\hat{\psi}(u)$ denote the probability of ruin in this case. Then $\hat{\psi}(u) = \psi(u)$ if either G is continuous or u is not an integer. In the discrete case, if u is a positive integer, then $\hat{\psi}(u) = \psi(u-1)$.

23.2 A functional equation approach

Suppose you sit down to gamble with 300 units of capital. You divide this initial stake into two piles, one with 200 and the other with the remaining 100. Now, to be ruined you first have to lose the 200 pile, and following that you have be ruined all over again starting with the 100 pile. So, it seems reasonable to conclude that $\psi(300) = \psi(200)\psi(100)$, or, more generally, since there is nothing special about these particular amounts, that $\psi(u+v) = \psi(u)\psi(v)$. Assuming that our reasoning here is correct, we would know already with no calculation at all (and ruling out certain degenerate cases) that $\psi(u)$ is an exponential function. Unfortunately, our reasoning is not quite accurate. The problem is that the ruining bet, or claim, by definition, will leave us with a deficit. In the original problem, we have to draw on our second pile in order to pay for this, and we would not have the full 100 to continue the procedure. Suppose, we knew that our deficit at the time of ruin was some $d > 0$. That is, we would have a surplus of $-d$ at ruin, and the appropriate equation would be

$$\psi(u+v) = \psi(u)\psi(v-d).$$

The following trick allows us to convert this into the form above. Let $\rho(u) = \psi(u-d)$; then,

$$\rho(u+v) = \psi(u+v-d) = \psi(u-d+v) = \psi(u-d)\psi(v-d) = \rho(u)\rho(v).$$

This is the same functional equation we encountered in Section 2.6. Assuming that we know that ruin probabilities are positive, and assuming some minimal regularity condition, such as continuity at one point, we know that $\rho(u) = z^u$ for some z between 0 and 1. Therefore,

$$\psi(u) = z^{u+d}. \tag{23.1}$$

Once again, however, we have to question our assumption. Is it ever possible that we could know that the deficit at ruin had to be some fixed number d? The answer is 'not very often', but it does happen in one particular case. Take the discrete-time model, for which the values of G are all nonnegative integers except for a single negative value of -1 – as for a simple coin flip where $G = 1$ or -1 – and for which the initial surplus u is a positive integer. In this case, the only way to be ruined is to reach a position where your surplus is 0 and then to lose 1 in the next period. The deficit at ruin can only be 1. Formula (23.1) would give us the ruin probabilities if we could only determine z. We will attempt to do so by using a recursive technique that is a basic tool in ruin theory. Let p denote the probability function of G. Suppose you start with an initial surplus of 0. If your gain is -1 in the first period, you are immediately ruined. If your gain is k in the first period, your subsequent probability of ruin is $\psi(k)$. Considering all possibilities, we have

$$\psi(0) = p(-1) + \sum_{k=0}^{\infty} p(k)\psi(k). \tag{23.2}$$

From (23.1), with $d = 1$,

$$z = p(-1) + \sum_{k=0}^{\infty} p(k)z^{k+1},$$

and dividing this equation by z, we can write

$$P_G(z) = 1, \tag{23.3}$$

where P_G is the p.g.f. of G.

Example 23.2 Consider a game where you win 1 with probability $p \geq 1/2$, and lose 1 with probability $1 - p$. What is $\psi(u)$?

Solution. From (23.3), we have

$$pz + \frac{1-p}{z} = 1.$$

There are unfortunately two possible solutions to this, either $z = (1 - p)/p$ or $z = 1$. We do not have a definite answer at this point but can only conclude that either

$$\psi(u) = \left(\frac{1-p}{p}\right)^{u+1} \tag{23.4}$$

or

$$\psi(u) = 1, \qquad \text{for all } u.$$

Note that $z = 1$ is a solution of (23.3) for all G, so our method would seem to have accomplished little, leaving us in all cases with a possible conclusion that ruin is certain. We will show however in the following sections that we can often rule out this possibility. This will be in fact be true in the present example for $p > 1/2$, and we will then know that the probability of ruin is given by (23.4).

We can already deduce an interesting result in the case that $p = 1/2$. In that case the only root is 1, and we know definitely that ruin is certain, regardless of the initial surplus. (This is of course also true if $p < 1/2$.) This is one of the well-known results in ruin theory. It says that even if you are playing a perfectly fair game, if you play it long enough, you will eventually lose all your money. This may been somewhat strange since you would seem to be on equal grounds with your opponent (the casino, for example). You are not, however, since there is an implicit assumption that the opponent has unlimited resources at its disposal, while you only have the u units you started with. We will return to a variation on this problem in Example 23.4 below.

To summarize, this section has achieved only limited success in deducing ruin probabilities, but it is mainly intended as a motivation for methods to follow. One point that should be emphasized is that it illustrates the importance of considering the *deficit at time of ruin* (given that ruin occurs) when trying to deduce ruin probabilities. In the following sections, we will refer frequently to this random variable and denote it by $D(u)$ (D for deficit). That is,

$$D(u) = -U_T|(T < \infty, U_0 = u).$$

In the example above $D(u)$ was always 1, but in general it is random and can depend on the initial surplus u.

23.3 The martingale approach to ruin theory

23.3.1 Stopping times

We will motivate the concept discussed here by looking at the gambling situation. Some gamblers claim that they can overcome unfavourable odds by a clever 'system'. This often takes the form of planning to stop at a certain point. For example, they will continue gambling until they have won \$100 and then quit. That way, they claim, they are always a winner. Or, they will continue to bet on black until the wheel comes up black four times in a row, and then quit. They are using what is called a *stopping time*. Intuitively, a stopping time is a rule that tells you when to stop, and it must be such that you know about it when the time occurs. In other words, it depends only on the past and not the future. Stopping after four blacks in a row is a legitimate stopping time. A rule which says that whenever there are four blacks in a row, then you stop after the third one, is *not* a stopping time since you clearly will not be aware of that time when it occurs. (The concept is similar to that we encountered in Section 20.5 when discussing trading strategies).

Here is a more formal definition. A stopping time for a discrete-time stochastic process is a rule that assigns to each realization (x_n) of the process an integer k, the stopping time, in such a way that if we assign k to a realization (x_n), and (y_n) is a realization such that $x_n = y_n$ for $n = 1, 2, \dots, k$, then we must assign the stopping time k to (y_n) as well.

To illustrate this definition, take the coin flipping example, starting with an initial surplus of 3, winning 1 for a head, and consider a rule which tells you to stop after the first head whenever you get two consecutive heads. This should not be a stopping time, and we can see that it does not satisfy the definition. A realization of the form $(3, 4, 5, \dots)$ would be assigned 1, but a realization of the form $(3, 4, 3, \dots)$, which agrees with the first one up to time 1, would not be assigned 1.

A stopping time will often be denoted by a letter such as S. A major example that we have already encountered is when S equals the time of ruin, which certainly satisfies the requirement for a stopping time. A particularly simple example of a stopping time is $S = k$, for some fixed k. That is, one stops at time k regardless of what has happened.

A fundamental fact about the fixed stopping time is that for any martingale (X_k), as defined in Section 18.3,

$$E(X_k) = E(X_0), \qquad \text{for all } k. \tag{23.5}$$

This is intuitively clear. In the case of a Markov chain, it is derived easily from

$$E(X_{k+1}) = \sum_x E(X_{k+1}|X_k = x)P(X_k = x) = \sum_x xP(X_k = x) = E(X_k),$$

and by induction we derive (23.5).

Example 23.3 Consider again the game of flipping a fair coin, winning 1 for heads, starting with an initial surplus of 3. The gambler decides to stop at time 4 or whenever two consecutive heads come up, if earlier. What is the expected surplus at the end of the game?

Solution. This is complicated by the fact that we no longer have a Markov chain. There is, however, a useful general technique that allows us to recover the Markov property by adding states. In this example, instead of having a single state for each integer w, we insert a state wu to signify that there was a win on the previous play, following a loss on the play before; and we insert a state wd to signify a loss on the previous play. See Figure 23.1, where the shaded boxes indicate points where play stops. By counting paths, we calculate the expected surplus at stopping as

$$E(X_S) = 5 \times \frac{1}{4} + 4 \times \frac{1}{8} + 5 \times \frac{1}{16} + 3 \times \frac{4}{16} + \frac{4}{16} - \frac{1}{16} = 3.$$

The stopping rule has not helped to raise the expectation above the initial stake.

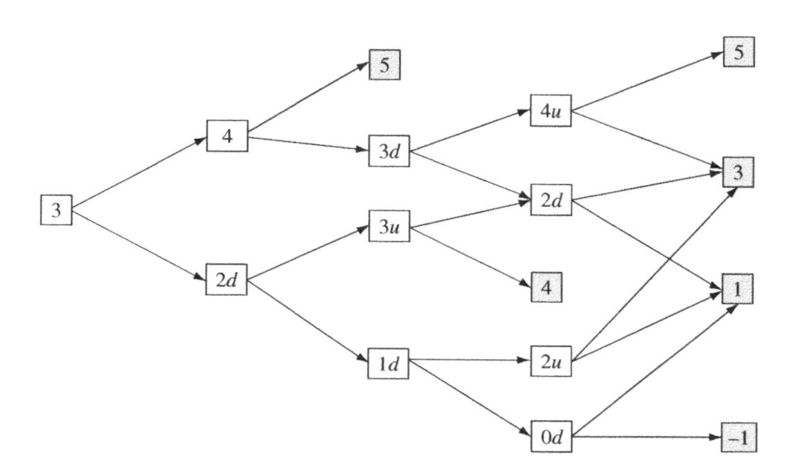

Figure 23.1 Tree for Example 23.3

23.3.2 The optional stopping theorem and its consequences

This motivates the question of whether it is always true in the case of a martingale that $E(X_S) = E(X_0)$? That is, does (23.5) hold when the fixed time k is replaced by an arbitrary stopping time? The answer is no. Starting with a positive initial surplus of u, we flip a fair coin repeatedly, wagering any amount b that we choose, and receiving back $2b$ for a head and nothing for a tail. There are two situations we want to present. In the first, we bet 1 unit each time and continue until we lose all of our initial stake. The stopping time S is the first time the surplus reaches 0. Trivially, $E(X_S) = 0 \neq u$. The second case is the familiar doubling strategy. Bet 1 unit on the first toss, and double the bet each successive play. S is the first time a head comes up, and it is not hard to see that $E(X_S)$ is $u + 1$, so we are sure to gain 1. The problem is that neither of these strategies are feasible in practice. (Of course the first is irrational as well.) In both cases, the amount of time we need is unbounded, and in the second case the amount of capital we need is unbounded as well. We will, however, get an affirmative answer to our question if there exists a suitable bound on a combination of the stopping time and values. The following theorem gives a precise condition. Observe first that in place of X_S, which is not defined if $S = \infty$, we want in general to consider $X_S | S < \infty$.

Theorem 23.1 (Optional stopping theorem) *Suppose that $\{X_n\}$ is a martingale and S is stopping time such that*

$$\lim_{n\to\infty} E(X_n | S > n)P(S > n) = 0. \tag{23.6}$$

Then

$$E(X_0) = E(X_S | S < \infty)P(S < \infty).$$

Proof. For any n,

$$E(X_0) = E(X_n) = E(X_n | S \leq n)P(S \leq n) + E(X_n | S > n)P(S > n). \tag{23.7}$$

Using a modification of Equation (A.29), the third term above can be written as

$$\sum_{k=0}^{n} E(X_n | S = k)P(S = k).$$

In view of the martingale property,

$$E(X_n | S = k) = E(X_k | S = k) = E(X_S | S = k).$$

The entire third term therefore reduces to $E(X_S | S \leq n)P(S \leq n)$. We now simply take limits as $n \to \infty$ to reach the conclusion. □

Corollary *In the case where S is finite valued, we have $E(X_0) = E(X_S)$ in either of the following cases:*

- *(i) S is bounded. That is, for some $N > 0$, we have $S \leq N$.*
- *(ii) The values of X_n are bounded in absolute value prior to stopping. That is, there is a constant C such that, for all n, if $S > n$ then $|X_n| \leq C$.*

Proof. In (i), the second factor in (23.6) is 0 for $n \geq N$, so (23.6) necessarily holds. In (ii), the second factor in (23.6) approaches 0 by the fact that S is finite, and the first factor is bounded in absolute value by C, so the product converges to 0. □

Example 23.4 A gambler starting with an initial fortune of a units repeatedly plays an even money game against an adversary with an initial fortune of b. Each has an equal chance of winning each game. They continue the play until one is broke. What is the probability that the gambler with a units will eventually lose his initial stake before the other does? (As an equivalent formulation, we can remove any restriction from the opponents' initial stake and instead postulate that the gambler decides to quit upon losing a or winning b.)

Solution. In Example 23.2 above, we essentially considered the case that b was infinite, that is there was no restriction on how much the opponent might lose, and we saw that ruin was certain. Here, we deal with the more realistic case that b is finite. Even a casino has some upper bound on its available wealth. Let U_k be the fortune of our gambler at time k. In view of the fairness of the game, this is indeed a martingale. Let S be the time that the gambler either loses the initial stake of a or wins b from the opponent. This is a stopping time that certainly satisfies condition (ii) of the Corollary to Theorem 23.1 since the values of U_k range from 0 to $a + b$. We will show later that S must take a finite value, which allows us to use the Corollary. Let π be the probability that our gambler loses his initial stake. We can invoke the Corollary to conclude that $E(U_S) = a$. But also, considering the two possibilities for S, we have

$$E(U_S) = (a + b)(1 - \pi) + 0\pi,$$

and by equating, we obtain

$$\pi = \frac{b}{a + b}.$$

Note that π approaches 1 as b approaches ∞, verifying our conclusion following Example 23.2.

Example 23.5 Redo Example 23.4 assuming now that the probability of a win is $p \neq 1/2$. (This is a classical problem, often referred to as *gambler's ruin.*)

Solution. The difficulty now is that we no longer have a martingale. However, the following ingenious trick allows us to transform the process into one. To simplify notation, let $q = 1 - p$. Consider the process

$$X_n = \left(\frac{q}{p}\right)^{U_n}.$$

For any n, $U_{n+1} = U_n + G_{n+1}$, where G_n takes the value 1 with probability p and -1 with probability q. Therefore,

$$E\left[\left(\frac{q}{p}\right)^G\right] = p\left(\frac{q}{p}\right) + q\left(\frac{q}{p}\right)^{-1} = 1,$$

leading to

$$E(X_{n+1}) = E\left[\left(\frac{q}{p}\right)^{U_{n+1}}\right] = E\left[\left(\frac{q}{p}\right)^{U_n+G_{n+1}}\right] = E\left[\left(\frac{q}{p}\right)^{U_n}\left(\frac{q}{p}\right)^{G_{n+1}}\right]$$
$$= E\left[X_n\left(\frac{q}{p}\right)^{G_{n+1}}\right]. \qquad (23.8)‡$$

Now, invoking the independence of U_n and G_{n+1},

$$E(X_{n+1}|X_n = x) = xE\left[\left(\frac{q}{p}\right)^{G_{n+1}}\right] = x,$$

showing that $\{X_n\}$ is a martingale. We can now duplicate the calculations in Example 23.4, applied to X_n. We have that

$$E(X_S) = E(X_0) = \left(\frac{q}{p}\right)^a,$$

and

$$E(X_S) = (1-\pi)\left(\frac{q}{p}\right)^{a+b} + \pi\left(\frac{q}{p}\right)^0.$$

Solving,

$$\pi = \frac{(q/p)^a - (q/p)^{a+b}}{1-(q/p)^{a+b}}.$$

We return to the unfinished business of showing that the stopping time S of the last two examples must assume a finite value. We have a finite-state Markov Chain with states $0, 1, 2, \dots, a+b$, corresponding to the amount held by the person who started with a. We can see, similarly to Example 18.3, that the states $1, 2, \dots, a+b-1$ are transient and the states 0 and $a+b$ are both absorbing and therefore recurrent. By Theorem 18.2 the process must reach one of these two recurrent states, implying that S cannot take a value of ∞.

Here is another striking application of this idea.

Example 23.6 A gambler plays a game in which she will either win 1000 with probability p, where $p < 1$, or lose 1 with probability $1-p$. Suppose, however, that whenever she accumulates more than 10 000 in winnings, a companion takes everything in excess of 10 000 to spend in the casino gift shop. What is the probability that the gambler will eventually go broke?

Solution. The probability is 1. Regardless of the initial fortune or the value of p, the gambler in this case is sure to lose everything if she plays for a sufficiently long time. Once again we have a finite Markov chain with states taking values from 0 to 10 000. All states except 0 are transient, as we showed in the previous example, and we are certain to reach the one recurrent state of 0.

Remark The key to the above example is the phrase 'a sufficiently long time'. As a practical matter, most people would be happy to play this game for a high value of p and would expect to eventually emerge a winner.

23.3.3 The adjustment coefficient

We now wish to apply Theorem 23.1 to the discrete surplus process (18.3) as interpreted in Section 23.1. We cannot, however, expect this to be a martingale. It will only be one if $E(G) = 0$, that is, if $c = E(N)E(X)$. This is unrealistic, since, as we noted in previous chapters, insurers will invariably charge an amount above this expected value to guard against the possibility that the aggregate claims will be higher than expected. That is, they take $c = (1 + \theta)E(N)E(X)$ for some $\theta > 0$. However, we can transform this process to be a martingale by the same type of procedure as we used in Example 23.4. The following definition gives the basic tool for doing this.

Definition 23.1 An *adjustment coefficient* of a random variable G is a positive number R satisfying

$$M_G(-R) = 1. \tag{23.9}$$

If R is an adjustment coefficient of a discrete random variable, then for $z = e^{-R}$ it follows that $P_G(z) = 1$, so we have already essentially seen this idea in (23.3).

To justify the word 'the' in the title of this subsection, we will show that there cannot be two positive numbers satisfying (23.9). Let γ be the supremum of all points for which $M_G(-r)$ is defined. (For example, if $G = c - W$ where $W \sim \text{Exp}(\beta)$, then γ will simply be equal to β.) In many cases $\gamma = \infty$. Define a function $\phi(r) = M_G(-r) - 1 = E(e^{-rG}) - 1$ on the interval $[0, \gamma)$. Then

$$\phi'(r) = -E(Ge^{-rG}), \qquad \phi''(r) = E(G^2e^{-rG}) > 0,$$

so ϕ is a convex function that takes the value 0 at the point 0 and, therefore, cannot have more than one positive root (see Figure 23.2).

An equally pertinent problem is to decide if a positive root of ϕ exists. We can show that this will almost always happen in view of the following two properties that G will satisfy in any realistic insurance context:

(i) $E(G) > 0$;

(ii) $P(G < 0) > 0$.

As we mentioned above, (i) will hold due to the relative risk loading. Moreover, there must be the possibility of paying out more in claims than the premiums collected, or nobody would ever buy insurance. This will imply (ii).

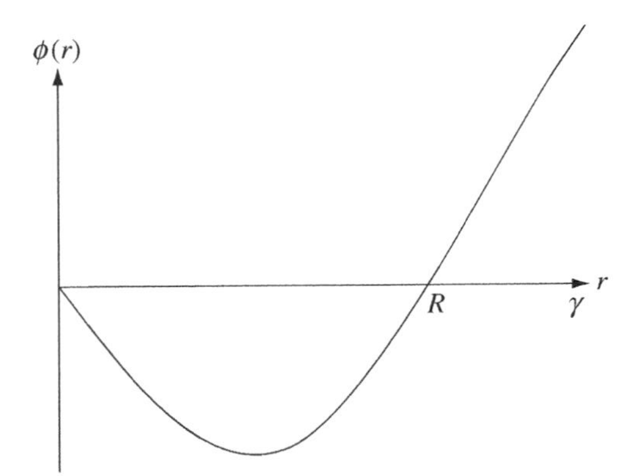

Figure 23.2 Graph of the function ϕ

Now (i) implies that $\phi'(0) = -E(G) < 0$, so ϕ will start out negative. We must, therefore, have a positive root as long as $\lim_{r\to\gamma} M_G(-r) > 1$. From condition (ii), we can find a positive numbers s and δ such that $P(G < -s) > \delta$. This means that $E(\mathrm{e}^{-rG}) > \delta \mathrm{e}^{rs} > 1$ for sufficiently large r. This guarantees a positive root whenever $\gamma = \infty$ (for example, when G is finite).

There are a few rare examples, which we will not give here, where γ is finite but $\lim_{r\to\gamma} M_G(-r) \le 1$, so the adjustment coefficient will not exist. (This is aside from those cases where the m.g.f. does not exist in the first place, so we cannot even define the adjustment coefficient.)

It is sometimes convenient to extend the definition of the adjustment coefficient to the extreme cases. If $\phi(t) \le 0$ for all t, we say that $R = \infty$; while if $\phi(t) \ge 0$ for all t, we say that $R = 0$.

The adjustment coefficient can be taken as a measure of safety. The higher R is, the less risk there is for the insurer. This is indicated by Theorem 23.4 below, which shows that as R increases, our upper bound for the probability of ruin decreases. This principle is also seen in Examples 23.2 and 23.5. These examples show that for the case where G takes the value 1 with probability p and -1 with probability $1 - p$, the adjustment coefficient is $\log(p) - \log(1 - p)$, which increases as p does. The following is yet another example.

Example 23.7 Find the adjustment coefficient for a normal distribution with mean μ and variance σ^2.

Solution. The m.g.f. of the normal is given in (A.52). We have $\mathrm{e}^{-R\mu+\sigma^2 R^2/2} = 1$, so that $-R\mu + \sigma^2 R^2/2 = 0$, and since $R > 0$,

$$R = \frac{2\mu}{\sigma^2}. \tag{23.10}$$

We see that for two normals with the same mean, as the variance gets smaller, which should signify less variation and therefore less risk, the adjustment coefficient goes up.

Equation (23.10) is often used as an approximation to the adjustment coefficient for other distributions. This is justified by the following argument. The function $\log(M_G(t))$ has first derivative equal to $M'_G(t)/M_G(t)$, which at $t = 0$ equals $E(G)$. Differentiating this latter expression yields that the second derivative at 0 is equal to Var(G). We can then write (23.9) in the form

$$0 = \log M_G(-R) = -R\mu + \frac{R^2}{2!}\sigma^2 + \cdots,$$

and ignoring powers of R higher than 2 gives (23.10).

The following example will be extensively used in the continuous-time models.

Example 23.8 Suppose that $G = c - \langle N, X\rangle$, where $N \sim$ Poisson (λ). Find an expression for R in terms of X and λ.

Solution. Using (A.50) and (A.51), $M_G(-r) = \mathrm{e}^{-cr}M_{\langle N,X\rangle}(r)$, and then using (A.42) and (21.4), this is equal to $\mathrm{e}^{-cr}\mathrm{e}^{\lambda(M_X(r)-1)} = \mathrm{e}^{-cr+\lambda(M_X(r)-1)}$. For this to equal 1, the exponent must equal 0, and dividing by λ, we get

$$M_X(R) = 1 + \left(\frac{c}{\lambda}\right)R. \tag{23.11}$$

Since $c = (1+\theta)\lambda E(X)$, where θ is the relative risk loading, we can also write

$$M_X(R) = 1 + (1+\theta)E(X)R. \tag{23.12}$$

23.3.4 The main conclusions

A key fact about the adjustment coefficient is that the process $X_n = \mathrm{e}^{-RU_n}$ is a martingale. This follows by exactly the same calculation as given in Example 23.5 where we used the fact that $E(\mathrm{e}^{-RG}) = 1$. We now obtain a major conclusion by applying Theorem 23.1 to this martingale. We must first verify that condition (23.6) holds.

Theorem 23.2 *For the process given in (18.3), suppose that $E(G) > 0$ and that Var(G) exists. Then, for any positive r,*

$$E(\mathrm{e}^{-rU_n}|T > n)P(T > n) \to 0 \qquad \text{as } n \to \infty.$$

Proof. Let μ be the mean and let σ be standard deviation of G, so that

$$E(U_n) = u + n\mu, \qquad \mathrm{Var}(U_n) = n\sigma^2.$$

Consider the following events referring to the situation at time n. Let A_n be the event that $T > n$ and $U_n \geq E(U_n)/2$, and let B_n be the event that $T > n$ and $0 \leq U_n < E(U_n/2)$. Since $T > n$ means that U_n must be nonnegative, we see that the event $T > n$ is a disjoint union of A_n and B_n. □

We also know from Chebyshev's inequality (A.11) that

$$P(B_n) \leq \frac{4\mathrm{Var}(U_n)}{(EU_n)^2} = \frac{4n\sigma^2}{(u+n\mu)^2},$$

so that $P(B_n) \to 0$ as $n \to \infty$. Then,

$$E(\mathrm{e}^{-rU_n}|T > n)P(T > n) = E(\mathrm{e}^{-rU_n}|A_n)P(A_n) + E(\mathrm{e}^{-rU_n}|B_n)P(B_n).$$

As n goes to ∞, the second term approaches zero because the first factor is bounded by 1, and the first term is less than $\mathrm{e}^{-r(u+n\mu)/2}$ and also approaches 0.

We can now apply Theorem 23.1 to the martingale e^{-RU_n} with the stopping time T to conclude that

$$E(\mathrm{e}^{-RU_T}|T < \infty)\psi(u) = \mathrm{e}^{-Ru}.$$

While stopping theorems are normally used to derive the expectation at time of stopping, in this instance we get an expression for the probability of ruin. Note that $-U_T|T < \infty$ is just the deficit at ruin, which we termed $D(u)$ before, so we have the following theorem.

Theorem 23.3 *If the adjustment coefficient R exists, then*

$$\psi(u) = \frac{\mathrm{e}^{-Ru}}{E[\mathrm{e}^{RD(u)}]}.$$

Note that in the case that $D(u)$ has a constant value of 1, we recover (23.1) with $z = \mathrm{e}^{-R}$.

In general, this theorem does not give us the exact value of $\psi(u)$ since we will not know the denominator. We can say, however, that since R is positive, the denominator is greater than 1, and we conclude the following.

Theorem 23.4 *If the adjustment coefficient R exists, we have*

$$\psi(u) \leq \mathrm{e}^{-Ru},$$

and therefore

$$\lim_{u\to\infty} \psi(u) = 0.$$

This theorem tells us that we can make the probability of ruin as small as we like by taking the initial surplus sufficiently high, which is certainly a reasonable conclusion. It also tells us that the probability of ruin reduces exponentially as a function of initial surplus. For example, if the upper bound to the probability of ruin given by Theorem 23.4 is less than 0.05, and we double the initial surplus, we then know that the probability of ruin is less than 0.0025.

It is also sometimes useful to have a lower bound for the ruin probability. This is possible if G is bounded below by $-M$ for some $M > 0$. We then know that $D(u) \leq M$, and we can conclude from Theorem 23.3 that

$$e^{-R(u+M)} \leq \psi(u).$$

We conclude this section with some remarks on what happens in the rare case that the adjustment coefficient does not exist. We can still reach the second conclusion in Theorem 23.4. As long as $E(G) > 0$ and M_G is defined on some interval of positive length about 0, we can find $\beta > 0$ such that $M_G(-\beta) < 1$. We then have that $e^{-\beta U_n}$ is a supermartingale. We obtain a version of (23.7) with $\geq$ replacing the first equality sign, and this is enough to derive the conclusion of Theorem 23.4 with β replacing R.

23.4 Distribution of the deficit at ruin

What can we say about the denominator in the expression for the ruin probability given in Theorem 23.3, other than that it is greater than 1? Can we ever evaluate it exactly in cases other than that mentioned in Section 23.1, where the values of G are nonnegative integers except for a single negative value of -1? In this section, we will try to provide some insight into these questions. To do so, we introduce some additional random variables.

Let $Y = -G|G < 0$. For example, if G takes values $\{4, 3, -1, -2, -3\}$ with respective probabilities $\{0.4, 0.1, 0.1, 0.25, 0.15\}$, then Y will take the values $\{1, 2, 3\}$ with respective probabilities $\{0.2, 0.5, 0.3\}$. The significance of Y is that only the negative values of G can bring about ruin.

Let $J(u)$ denote the value of the surplus in the period prior to ruin, assuming an initial surplus of u (the J stands for 'just before'). The connection between all these is that $D(u)$ will be equal to some value of $Y - J(u)$. (In what follows, we will just write D and J, suppressing u, which will be fixed.)

To illustrate, take G as given above, and suppose that $u = 2$. Given a realization $G_1 = 1$, $G_2 = -1$, $G_3 = -3, \ldots$, ruin will occur at time 3, J will equal 2, and D will equal 1. For a realization $G_1 = 1$, $G_2 = -1$, $G_3 = -2$, $G_4 = -3, \ldots$, ruin will take place at time 4, $J = 0$, and $D = 3$.

We can make the following observations about this example. J can take possible values of 0, 1, or 2. When J takes the value 2, then D will take the value 1 with certainty. When J takes the value of 0, then any of the three negative amounts will cause ruin, so D will have the same distribution as Y. When J takes a value of 1, then the ruining claim must be either 2 or 3, so D will take the value 1 with probability $5/8$, and 2 with probability $3/8$, as these are the conditional probabilities of values of Y given that $Y > 1$.

In the general case, if $J = j$, then we know that the ruining claim Y must take a value greater than j, and in particular for D to take a value of d, we need that $Y = j + d$. This gives

$$P(D = d) = \sum_j P(J = j) \frac{f_Y(d + j)}{s_Y(j)}, \qquad d = 1, 2, \ldots.$$

If only we knew the distribution of J, this would give us the distribution of D. Unfortunately, deducing the distribution of J is just as difficult as getting that of D, so we would appear to have simply gone around in circles, with no gain of information.

This is not quite true, however. For one thing, our analysis indicates some features about the distribution of D. It is somewhat related to that of Y. It takes exactly the same values, and it will in fact have the same distribution as Y whenever $J = 0$. (In our simple example of Section 23.2 we knew in fact that J was always 0.) In general, however, it will involve more mass at the lower values than Y, for when J takes values higher than 0, there is less chance for D to assume higher values.

A second observation is that there is one case where we can indeed use the formula above. Suppose that $Y \sim \text{Geom}(p)^+$, using the notation we introduced at the beginning of Section 21.7. For this distribution, $f_Y(d+j)/s_Y(j) = (1-p)p^{d+j-1}/p^j = f_Y(d)$. The second factor in the summation is independent of j, so it can be factored out, and we conclude that

$$P(D = d) = f_Y(d)\sum_j P(J = j) = f_Y(d).$$

In other words, we have the following:

$$\text{If } Y \sim \text{Geom}(p)^+, \text{ then for all } u,\ D(u) \sim \text{Geom}(p)^+. \tag{23.13}$$

We will use this example later to motivate a result in the continuous-time case.

23.5 Recursion formulas

In some cases, we can use recursion to calculate exact ruin probabilities as well as the exact distribution of surplus at time of ruin.

23.5.1 Calculating ruin probabilities

Suppose we have the insurance claims model

$$G = c - \langle N, X\rangle,$$

and that the following restrictions hold:

(i) $c = 1$.

(ii) N takes the values 0 or 1 with probabilities $1-q$ or q, respectively.

(iii) X takes positive integer values $1, 2, \ldots, K$, with probabilities $f(1), f(2), \ldots, f(K)$, respectively.

(iv) $1 > qE(X)$, so that $E(G) > 0$.

(v) The initial surplus u is a positive integer.

Note that (i) and (iii) simply say that all claim amounts are integer multiples of the premium, since we can always take the amount of the premium as the unit of capital.

To simplify the notation, we will first illustrate with $K = 3$. Clearly, the following calculations will work for any value of K. Note that G takes the values $1, 0, -1, -2$ with probabilities $1 - q, qf(1), qf(2)$, and $qf(3)$, respectively.

Suppose we start with u units at time 0. In the first period, the four possible outcomes for G lead to four possible values of surplus at time 1. This gives us the following set of equations, one for each value of u:

$$\begin{aligned}
\psi(0) &= (1-q)\psi(1) + qf(1)\psi(0) + qf(2)\psi(-1) + qf(3)\psi(-2)\\
\psi(1) &= (1-q)\psi(2) + qf(1)\psi(1) + qf(2)\psi(0) + qf(3)\psi(-1)\\
&\vdots\\
\psi(u) &= (1-q)\psi(u+1) + qf(1)\psi(u) + qf(2)\psi(u-1) + qf(3)\psi(u-2)\\
&\vdots
\end{aligned}$$

Note that for convenience we have included terms of the form $\psi(i)$ for $i < 0$. These, of course, are just equal to 1. If we start with a negative amount, we are already ruined. Next, rearrange the equations above to give

$$\begin{aligned}
\psi(0) - \psi(1) &= q[-\psi(1) + f(1)\psi(0) + f(2)\psi(-1) + f(3)\psi(-2)]\\
\psi(1) - \psi(2) &= q[-\psi(2) + f(1)\psi(1) + f(2)\psi(0) + f(3)\psi(-1)]\\
\psi(2) - \psi(3) &= q[-\psi(3) + f(1)\psi(2) + f(2)\psi(1) + f(3)\psi(0)]\\
\psi(3) - \psi(4) &= q[-\psi(4) + f(1)\psi(3) + f(2)\psi(2) + f(3)\psi(1)]\\
&\vdots\\
\psi(n) - \psi(n+1) &= q[-\psi(n+1) + f(1)\psi(n) + f(2)\psi(n-1) + f(3)\psi(n-2)]\\
&\vdots
\end{aligned}$$

Sum the first $n + 1$ of these equations. The left-hand side adds up to $\psi(0) - \psi(n + 1)$. To add the right-hand side, it is convenient to add by the diagonals (running northwest to southeast) because they involve the term $\psi(k)$ for the same value of k. Since $f(1) + f(2) + f(3) = 1$, all of the diagonals will sum to zero except for the three on the upper right and the three on the lower left. The third diagonal on the upper right sums to $q\psi(0)$. The first two diagonals on the upper right sum to

$$q\sum_{i=1}^{3}(i-1)f(i) = qE(X-1).$$

The three diagonals on the lower left involve terms in $\psi(n + 1), \psi(n)$, and $\psi(n - 1)$. These will all converge to 0 as n approaches ∞ by Theorem 23.4. Taking limits gives

$$\psi(0) = q[\psi(0) + E(X-1)],$$

and solving

$$\psi(0) = \frac{q}{1-q}E(X-1). \tag{23.14}$$

It is instructive to write this in terms of θ, the relative risk loading. Since

$$1+\theta=\frac{1}{qE(X)}, \tag{23.15}$$

we can substitute for $E(X)$ in (23.14) to obtain

$$\psi(0)=\frac{1/(1+\theta)-q}{1-q},$$

which tells us that

$$\lim_{q\to 0}\psi(0)=\frac{1}{1+\theta}. \tag{23.16}$$

Now consider the general case with G taking the K values $1,2,\ldots,K$. Formula (23.14) gives us a starting value, and all of the ruin probabilities can be calculated recursively by rearranging our first set of equations to get

$$\psi(n)=\frac{1}{1-q}[\psi(n-1)-q[f(1)\psi(n-1)+f(2)\psi(n-2)+\cdots+f(K)\psi(n-K)]], \tag{23.17}$$

which we can write more compactly, using the notation of Section A.12.3, as

$$\psi(n)=\frac{1}{1-q}[\psi(n-1)-q(f*\psi)(n)].$$

Remark The quantity $\psi(0)$ is of importance since it gives us our starting value. One may think, however, that it is intrinsically of little interest, since we are not likely in practice to have a situation where the initial surplus is 0. It is, however, of great significance since given any starting value u, we can interpret $\psi(0)$ as giving the probability that we will eventually reach some point where our surplus is less than u. Similarly, $D(0)$ is the amount by which we are less than the initial surplus, if this occurs. We will exploit this idea to great advantage in Section 23.7.

23.5.2 The distribution of $D(u)$

The same approach lets us deduce the distribution of $D(u)$. For $k=1,2,\ldots,K-1$, let $\psi_k(u)$ be the probability that, starting with initial surplus u, ruin eventually occurs and the value of $D(u)=k$. It will be convenient again to consider negative values of the argument, and we note that

$$\psi_k(-r)=\begin{cases}0, & \text{if } r\neq k,\\ 1, & \text{if } r=k.\end{cases}$$

Given values of $\psi_k(u)$, we can immediately find $\psi(u)$, as well as the distribution of $D(u)$, since

$$\psi(u) = \sum_{k=1}^{K-1} \psi_k(u) \tag{23.18}$$

and

$$P[D(u) = k] = \frac{\psi_k(u)}{\psi(u)}, \qquad k = 1, 2, \ldots, K-1. \tag{23.19}$$

We calculate $\psi_k(u)$ by following exactly the procedure in the previous subsection. We get the same systems of equations except with ψ replaced by ψ_k. When we sum the second set of equations and take limits, we again will have everything on the right-hand side vanishing, except possibly for a finite number of diagonals on the upper right. Since these all involve negative values of the argument of ψ, these sums will also be zero except for the single diagonal involving the terms $\psi_k(-k)$, and the sum of that will be $f(k+1) + f(k+2) + \cdots + f(K) = P(X > k)$. So, in place of (23.14), we get

$$\psi_k(0) = \frac{q}{1-q} P(X > k). \tag{23.20}$$

From (23.14), (23.19), and (23.20), we immediately have a nice simple formula for the distribution of $D(0)$:

$$P[D(0) = k] = \frac{s_X(k)}{E(X-1)}, \qquad k = 1, 2, \ldots, K-1. \tag{23.21}$$

This verifies our intuition of the previous section. In our present case, we have $Y = X - 1$, and we see precisely how $D(0)$ is a type of 'shifted to the left' version of Y. For example, if Y takes values 1, 2, 3, with probabilities 0.2, 0.5, 0.3 respectively, then $D(0)$ takes the values $1, 2, 3$ with probabilities that are in the ratio, $1 : 0.8 : 0.3$, so they are $10/21, 8/21, 3/21$.

We now obtain the same recursion formula as in (23.17), except with ψ replaced by ψ_k, that is,

$$\begin{aligned}
\psi_k(n) &= \frac{1}{1-q}[\psi_k(n-1) - q[f(1)\psi_k(n-1) + f(2)\psi_k(n-2) + \cdots + f(K)\psi_k(n-K)]] \\
&= \frac{1}{1-q}[\psi_k(n-1) - q(f * \psi_k)(n)].
\end{aligned}$$

Example 23.9 $q = 1/15$, X has a constant value of 4. Calculate $\psi_1(1), \psi_2(1), \psi_3(1)$, and $\psi(1)$. Verify that your answer to the last agrees with that given by Theorem 23.3.

Solution. We note first that $f(1) = f(2) = f(3) = 0, f(4) = 1$. So

$$\psi_1(0) = \frac{1}{14}[P(X > 1)] = \frac{1}{14}, \qquad \psi_1(1) = \frac{15}{14}\left(\frac{1}{14} - 0\right) = \frac{15}{196}.$$

Similarly,

$$\psi_2(0) = \frac{1}{14}, \qquad \psi_2(1) = \frac{15}{196},$$
$$\psi_3(0) = \frac{1}{14}, \qquad \psi_3(1) = \frac{15}{14}\left(\frac{1}{14} - \frac{1}{15}\right) = \frac{1}{196},$$

so that $\psi(1) = 31/196$, and $D(1)$ takes the values 1,2,3 with probabilities $15/31, 15/31, 1/31$, respectively. (Note that $D(1)$ differs from $D(0)$, which has a uniform distribution on 1, 2, 3.)

To check this by Theorem 23.3, we first solve for R, or equivalently $z = \mathrm{e}^{-R}$. Since G takes the value 1 with probability $14/15$ or -3 with probability $1/15$, we have

$$14z + z^{-3} = 15,$$

and we can verify that $z = 1/2$. So,

$$E\left[z^{-D(1)}\right] = \frac{15}{31}z^{-1} + \frac{15}{31}z^{-2} + \frac{1}{31}z^{-3} = \frac{98}{31}.$$

From Theorem 23.3,

$$\psi(1) = \frac{1/2}{98/31} = \frac{31}{196},$$

as above.

23.6 The compound Poisson surplus process

23.6.1 Description of the process

We now turn to ruin calculations in the continuous case. This will be based on a *compound Poisson process* which is simply a process corresponding to the compound Poisson distribution, which we considered in Chapter 21. That is, instead of merely counting 1 every time an event occurs, we take an observation from some given distribution. So we have a distribution X, which we can think of as a severity distribution, and we have a Poisson process, $N(t)$. The resulting compound Poisson process is given by

$$S(t) = \sum_{k=1}^{N(t)} X_k,$$

where $\{X_k\}$ are independent, each has the same distribution as X, and they are independent of $N(t)$. In other words, we can simply write

$$S(t) \sim \langle N(t), X \rangle.$$

We can now provide a model for a surplus process. Suppose that an insurer's aggregate claims up to time t are given by a compound Poisson process $S(t)$, as above. In the discrete case, we postulated a premium of c per period. We now assume that the insurer collects premiums at a *continuous rate* of c per period, so in any period of length h total premiums of ch will be collected. We assume also that the insurer begins with an initial surplus of u. The *compound Poisson surplus process* is the process given by

$$U(t) = u + ct - S(t), \tag{23.22}$$

where $U(t)$ is the surplus at time t. Our relative risk loading is given as in the discrete case by

$$1 + \theta = \frac{c}{\lambda E(X)}.$$

The probability of ruin is defined similarly to the definition in the discrete case, although we need an infimum to replace the minimum. That is,

$$T = \inf\{t : U(t) < 0\}, \qquad \psi(u) = P(T < \infty | U_0 = u).$$

See Figure 23.3 for a typical realization of this process. The diagonal lines have slope c, and show the increase in surplus arising from the premium payments. Downward jumps then occur whenever there is a claim.

We can approach this case by considering an approximating discrete model. Suppose that we divide up our time into periods of length h for some small h. Then, if we only view our surplus at the end of each period, we just have a discrete surplus model with the gain in each period given by

$$G = ch - \langle N, X \rangle,$$

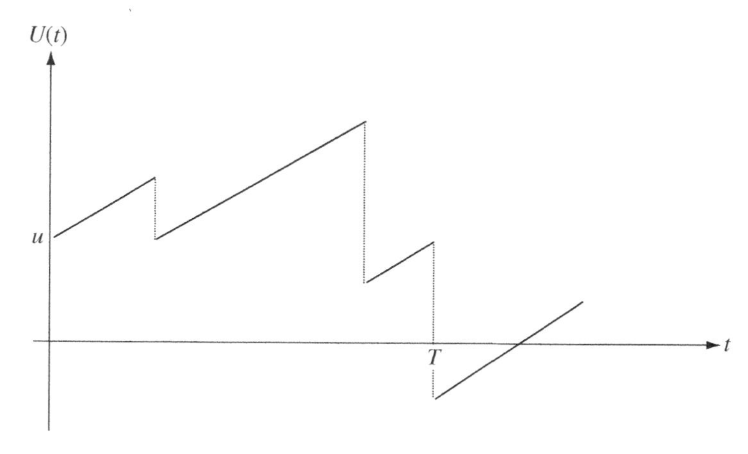

Figure 23.3 A realization of the continuous-time surplus process

where $N \sim \text{Poisson}(\lambda h)$. We will now go over the various results we obtained in the last two sections to see how they are modified for the continuous model. In many instances, we do not give rigorous proofs. However, all the results are motivated by those of the previous sections.

23.6.2 The probability of eventual ruin

We first would like to calculate the adjustment coefficient in the continuous model. It is natural to do this by calculating the adjustment coefficient of G for the approximating discrete model as given above, and then taking the limit as h approaches 0. This turns out to be extremely simple in the Poisson case. The term c/λ that we get in (23.11) is independent of h, so we get exactly the same answer as before. The adjustment coefficient is given by (23.11) or (23.12), as it was in the discrete model. The probability of ruin is now given by Theorems 23.3 and 23.4, which carry over unaltered to the surplus model with compound Poisson aggregate claims.

23.6.3 The value of $\psi(0)$

In light of the Poisson assumption, and Theorem 18.3, we can view our continuous-time model as a limiting case of the particular model discussed in Section 23.5. As h goes to 0, so will q and we deduce, directly from (23.16), that

$$\psi(0) = \frac{1}{1+\theta}.$$

This is somewhat surprising, since $\psi(0)$ does not depend on the particular distribution of X, except through the expectation $E(X)$, which affects θ. In other words, given any two distributions for X, as long as they have the same mean, they will produce the same value of $\psi(0)$.

23.6.4 The distribution of $D(0)$

$D(0)$ is necessarily continuous here in view of the fact that premiums are collected continuously. We can deduce, analogously to (23.21), that if $f_{D(0)}$ is the density function of $D(0)$, then

$$f_{D(0)}(d) = \frac{s_X(d)}{E(X)}. \tag{23.23}$$

For later purposes, we will need the m.g.f. of this distribution. Integrating by parts,

$$\int_0^\infty s_X(t)e^{rt}dt = s_X(t)\frac{e^{rt}}{r}\bigg|_0^\infty + \frac{1}{r}\int_0^\infty f_X(t)e^{rt}dt = \frac{1}{r}(M_X(r)-1),$$

so that

$$M_{D(0)}(r) = \frac{M_X(r)-1}{rE(X)}. \tag{23.24}$$

23.6.5 The case when X is exponentially distributed

It is easy to deduce from the distribution of $D(0)$ given above that when $X \sim \text{Exp}(\beta)$, then also $D(0) \sim \text{Exp}(\beta)$. However, much more than this is true. As a continuous analogue to (23.13), we obtain the following:

When $X \sim \text{Exp}(\beta)$, then $D(u) \sim \text{Exp}(\beta)$ for all $u \geq 0$.

The last observation tells us that we can get an exact calculation of the ruin probability when X is exponential. We first must calculate the adjustment coefficient. From (23.12), we solve

$$\frac{(1+\theta)R}{\beta} = \frac{\beta}{\beta - R} - 1 = \frac{R}{\beta - R}$$

and it is immediate that

$$R = \frac{\beta\theta}{1+\theta}$$

Substituting in (23.12) gives

$$M_X(R) = 1 + \theta. \tag{23.25}$$

Let us verify that our formula for R makes sense intuitively, recalling that our adjustment coefficient is a measure of safety. It increases as β increases, which it should since the mean of the severity distribution is decreasing. It must be 0 in the extreme case that $\theta = 0$, and it must increase with θ.

Note now that the denominator in Theorem 23.3 is just $M_{D(u)}(R)$. When $X \sim \text{Exp}(\beta)$, $D(u)$ has the same distribution, and from (23.25) we know the denominator is $1 + \theta$. We substitute in the statement of Theorem 20.3 to obtain a key result.

Theorem 23.5 *For the surplus model, with compound Poisson aggregate claims, where the severity distribution* $X \sim \text{Exp}(\beta)$,

$$\psi(u) = \frac{1}{1+\theta} \mathrm{e}^{-u[\beta\theta/(1+\theta)]}.$$

23.7 The maximal aggregate loss

We continue with the same continuous model as in the previous section and give an alternate approach to ruin probabilities. This will provide another proof of Theorem 23.5, as well as enabling us to deduce the ruin probability when X is a mixture of gamma distributions.

Definition 23.2 The *maximal aggregate loss*, denoted by $\mathcal{L}$, is the largest amount by which our surplus will be less than the beginning surplus. In other words,

$$\mathcal{L} = \max\{u - U(t) : t > 0\}.$$

For example, suppose we start out with an initial surplus of 10. Then, given a realization in which the smallest value of surplus ever attained is −7, the maximal aggregate loss will be 17. It is clear that $\mathcal{L}$ is independent of u. In fact, we could write it as

$$\mathcal{L} = \max\{S(t) - ct : t > 0\}.$$

The significance of this random variable for ruin models is that

$$\psi(u) = P(\mathcal{L} > u). \tag{23.26}$$

Therefore, if we know the distribution of $\mathcal{L}$, we can immediately write down the probability of ruin. We will show in this section that for our compound Poisson model, we can find the m.g.f. of $\mathcal{L}$, which makes a significant step towards this goal.

To do this, we consider what is often termed as the *record low process*. Imagine we are observing the surplus process and each time that we reach a new record low point in surplus, we record the amount by which we 'beat the record', that is, how much we are below the previous record low. Let $\mathcal{L}_n$ be the amount we record on the nth occasion the record is broken, should this occur.

As an example, suppose $u = 10$ and $c = 1$. If we have a claim of size 5 at time 3, we have a new record low surplus of 8, so $\mathcal{L}_1 = 2$. Suppose the next claim is 3 at time 7. Our surplus is 9 at that time, so we do not have a record low. Then, suppose the next claim is 5 at time 8. This gives us a new record low of 5, so we have $\mathcal{L}_2 = 3$, and so on. We may of course have many more record lows, but suppose that we do not and a surplus of 5 is as low as we get. Then the maximal aggregate loss is just 5, which is the sum of the two record low increments, that is, $\mathcal{L} = \mathcal{L}_1 + \mathcal{L}_2$.

Observe that the random variable $\mathcal{L}_1$ is just $D(0)$, as is clear from our remarks in the previous section. Similarly, it follows from the stationarity of the process that each $\mathcal{L}_n$ will have the same distribution. How many new record lows will we encounter? Well, after any new record low has occurred, the probability of another occurrence is just $\psi(0) = 1/(1+\theta)$. If N is the number of records lows that occur, then $N \sim \text{Geom}(1/(1+\theta))$. Now $\mathcal{L}$, the maximal aggregate loss, is simply the sum of the L_n. So, we have

$$\mathcal{L} = \mathcal{L}_1 + \mathcal{L}_2 + \cdots + \mathcal{L}_N.$$

By the independence in the process, this shows that $\mathcal{L}$ can be expressed as a compound distribution $\langle N, D(0)\rangle$, where N is geometric. We have already worked out the m.g.f. for a compound geometric distribution in Example 21.2. As we noted in that example, in order to have a chance of recognizing a compound distribution from its m.g.f. we will want to get rid of the point mass at 0 and look at $\mathcal{L}^+ = \mathcal{L} | \mathcal{L} > 0$. From this example,

$$M_{\mathcal{L}^+}(r) = (1-p)\left(\frac{M_{D(0)}(r)}{1 - pM_{D(0)}(r)}\right),$$

where $p = 1/(1+\theta)$. Substituting from (23.24) for $M_{D(0)}$ and simplifying gives us

$$M_{\mathcal{L}^+}(r) = \frac{\theta[M_X(r) - 1]}{(1+\theta)E(X)r - [M_X(r) - 1]}. \tag{23.27}$$

Since $\mathcal{L}$ is a mixture of 0 and $\mathcal{L}^+$ with weights $\theta/(1+\theta)$ and $1/(1+\theta)$, respectively, we know that

$$P(\mathcal{L} > u) = \frac{\theta}{1+\theta} P(0 > u) + \frac{1}{1+\theta} P(\mathcal{L}^+ > u).$$

The first term in (23.27) is trivially equal to 0 for nonnegative u, yielding the key result

$$\psi(u) = \frac{1}{1+\theta} P(\mathcal{L}^+ > u). \tag{23.28}$$

We summarize the whole procedure of finding ruin probabilities by the maximal aggregate loss procedure:

1. Calculate $M_X(r)$ and use (23.27) to find $M_{\mathcal{L}^+}(r)$.
2. Try to deduce from this the survival function of $\mathcal{L}^+$.
3. Calculate $\psi(u)$ directly from (23.28).

Step 2 is of course the difficult one. Will we actually be able to recognize the distribution of $\mathcal{L}^+$ from its m.g.f.? Here is one particularly easy example.

Example 23.10 Use the method described above to find $\psi(u)$ when $X \sim \text{Exp}(\beta)$.

Solution. Example 17.2 showed that $\mathcal{L}^+ \sim \text{Exp}(\beta\theta/(1+\theta))$, and from (23.28),

$$\psi(u) = \frac{1}{1+\theta} e^{-u[\beta\theta/(1+\theta)]},$$

giving us an alternate proof of Theorem 18.3.

Are there any other cases where we can recognize the distribution of $\mathcal{L}^+$ from its m.g.f.? We make two key observations. First, from (23.27), if $M_X(r)$ is a rational function (a quotient of two polynomials), then so is $M_{\mathcal{L}^+}(r)$. Second, if X is a mixture of gamma distributions that have an integer for the first parameter, its m.g.f. will be a linear combination of rational functions, therefore rational, and hence $M_{\mathcal{L}^+}(r)$ will be rational. So, the steps to handle such X are as follows. For such a distribution X, calculate $M_{\mathcal{L}^+}(r)$ and decompose it into partial fractions. Suppose

$$M_{\mathcal{L}^+}(r) = \sum_{i=i}^{n} w_i \left(\frac{r_i}{r_i - r} \right)^{\alpha_i}.$$

This shows that $\mathcal{L}^+$ is a mixture of $(Y_1, Y_2, \ldots, Y_n)$ with weights $(w_1, w_2, \ldots, w_n)$, where for each i, $Y_i \sim$ Gamma (α_i, r_i). From (23.28), we deduce that

$$\psi(u) = \left(\frac{1}{1+\theta} \right) \sum_{i=1}^{n} w_i [P(Y_i > u)].$$

Example 23.11 Find $\psi(u)$, given that X has the density function

$$f(x) = \frac{1}{2}\mathrm{e}^{-x} + \mathrm{e}^{-2x}$$

and $\theta = 7/9$.

Solution. X is a mixture of Exp(1) and Exp(2) distributions with equal weights, so

$$M_X(r) - 1 = \frac{1}{2}\left(\frac{1}{1-r}\right) + \frac{1}{2}\left(\frac{2}{2-r}\right) - 1 = \frac{3r - 2r^2}{4 - 6r + 2r^2}.$$

As a check, this quantity must take the value 0 for $r = 0$. Substituting $M_X(r) - 1$ into (23.27) gives

$$M_{\mathcal{L}^+}(r) = \frac{7}{9}\left(\frac{9 - 6r}{7 - 18r + 8r^2}\right) = \frac{7}{9}\frac{9 - 6r}{(1 - 2r)(7 - 4r)}.$$

The partial fraction expression for this is

$$\frac{14}{15}\left(\frac{1}{1-2r}\right) + \frac{1}{15}\left(\frac{7}{7-4r}\right).$$

So $\mathcal{L}^+$ is a mixture of Exp(1/2) and Exp(7/4) distributions with weights 14/15 and 1/15, respectively. It follows that

$$\psi(u) = \frac{9}{16}\left[\frac{14}{15}\mathrm{e}^{-0.5u} + \frac{1}{15}\mathrm{e}^{-1.75u}\right].$$

What happened to our adjustment coefficient? It played a prominent role in our previous method but seems to be absent here. It is not, however. The denominator of (23.27) is precisely the quantity we set equal to 0 to find R. To find the partial fraction decomposition, we must find the roots of this equation, and the adjustment coefficient appears as the *smallest* such root. In Example 20.11, we found the two roots $r = 1/2$ and $r = 7/4$. We then know that $R = 1/2$.

How do we reconcile the other root of 7/4 with our previous statement that the adjustment coefficient was uniquely defined as the *only* positive root of the defining equation? The point is that $M_X(r) = E(\mathrm{e}^{rX})$ may only be defined over a certain region. In this example, it is defined only for $0 \leq r < 1$, and if $r > 1$, then the expectation will not exist. However, the algebraic expression that gives the m.g.f. over this range does make sense for values greater than 1. It just no longer represents an expectation of a function of X. The resulting algebraic expression can have other roots, as it does in this example.

The procedure followed in the previous example can be used to write a general closed-form formula for $\psi(u)$ when X is a mixture of two exponential distributions, as given by the following theorem. We leave the details of the proof to the reader.

Theorem 23.6 *Suppose that in the compound Poisson surplus process, X is a mixture of* Exp(α) *and* Exp($k\alpha$), *for some $k > 1$, with respective weights w and $1 - w$.*

(a) Let $\varphi = 1 - w + wk$. The adjustment coefficient is $r\alpha$, where r is in the interval (0,1) and is a solution to the quadratic equation

$$\frac{\varphi - r}{(1-r)(k-r)} = (1+\theta)\frac{\varphi}{k}.$$

(b) Let

$$s = \frac{k}{r}\frac{\theta}{(1+\theta)}.$$

Then $\mathcal{L}^+$ is a mixture of Exp($r\alpha$) *and* Exp($s\alpha$) *with respective weights*

$$\left(\frac{\varphi - r}{s - r}\right)\frac{s}{\varphi} \text{ and } \left(\frac{s-\varphi}{s-r}\right)\frac{r}{\varphi}.$$

(c)

$$\psi(u) = \frac{1}{1+\theta}\left[\left(\frac{\varphi - r}{s - r}\right)\frac{s}{\varphi}\mathrm{e}^{-r\alpha u} + \left(\frac{s-\varphi}{s-r}\right)\frac{r}{\varphi}\mathrm{e}^{-s\alpha u}\right].$$

Notes and references

The inequality of Theorem 23.4 is an early result of ruin theory known as Lundberg's inequality. See Bowers *et al.* (1997, Example 13.4.3) for a case where the adjustment coefficient does not exist. An alternate method of handling the case of a mixture of gamma distributions is by developing an integrodifferenial equation for $\psi(u)$ (see Klugman *et al.*, 2012). There is an extensive literature on ruin theory which goes beyond the treatment here. See Grandel (1991) for some of the extensions. Several results concerning the distribution of the surplus just before ruin and just after ruin were developed in Gerber and Shiu (1998). Similar results along these lines are found in Powers (1995).

Exercises

23.1 Consider the discrete-time surplus process $U_n = u + G_1 + G_2 + \cdots + G_n$, where the G_is are independent and each distributed as a random variable G which takes the values $1, 0, -1, -2$ with probabilities 0.5, 0.2, 0.2, 0.1, respectively. Let $\psi(u)$ be the probability of eventual ruin, starting with an initial surplus of u.

(a) If you start with an initial surplus of 1, what is the probability that you will be ruined by time 2 or before?

(b) Given $\psi(6) = a, \psi(5) = b, \psi(4) = c$, express $\psi(7)$ in terms of a, b, c.

23.2 You are repeatedly flipping coins and will win 1 for a head and lose 1 for a tail. You plan to stop playing when you get four consecutive wins, or after 100 tosses at the very latest. You start with 10 units. Let U_S denote the amount of units you will have upon stopping.

(a) Suppose that the probability of a head is $1/2$. What is $E(U_S)$?

(b) Suppose again that the probability of head is $1/2$, but now you are given the additional information that you began with three wins, followed by two losses. What is $E(U_S)$ now?

(c) Suppose that the probability of a head is $1/3$. What is $E(2^{U_S})$?

23.3 An insurer's portfolio consists of a single possible claim. You are given the following information. The claim amount is uniformly distributed over (100, 500). The probability that the claim occurs after time t is $e^{-0.1t}$ for $t > 0$. The claim time and amount are independent. The insurer's initial surplus is 20. Premium income is received continuously at the rate of 40 per year. Determine the probability of ruin.

23.4 (a) A random variable G takes the value 1 with probability $6/7$ and -1 with probability $1/7$. Show that the adjustment coefficient of G is $\log(6)$.

(b) A random variable G takes the value 1 with probability $1/2$ and 2 with probability $1/2$. Show that the adjustment coefficient is ∞.

23.5 You are repeatedly playing a game in which at each stage you win 1 with probability $15/19$ or lose 2 with probability $4/19$.

(a) Show that the adjustment coefficient is $\log(3/2)$.

(b) If U_n denotes the amount you have at time n, show that U_n is not a martingale.

(c) At time 20 you have a total of 4 units. What is $E((2/3)^{U_{50}})$?

(d) Suppose you start with 3 units at time 0. You decide to stop play when you have a total of 20 units, or after 100 plays if that occurs earlier. If U_S is the amount that you will have after stopping play, what is $E((2/3)^{U_S})$?

23.6 You are playing a game repeatedly in which at each turn you either win 1 with probability $12/13$ or lose 2 with probability $1/13$.

(a) Write an equation for the adjustment coefficient R, and verify that $R = \log(3)$.

(b) Show that for an initial stake of u, the probability of eventual ruin is between $(1/3)^{u+2}$ and $(1/3)^{u+1}$.

23.7 Show that π in Example 23.5 approaches $b/(a+b)$ as p approaches $1/2$, verifying the conclusion of Example 23.4.

23.8 Refer to Example 23.4.

(a) Show that $X_n = (U_n - a)^2 - n$ is a martingale.

(b) Use your result in part (a) to show that $E[S] = ab$.

*23.9 This extends Exercise 23.8 to the case for which the probability of winning 1 is $p \neq 1/2$, as in Example 23.5.

(a) Show that $g(U_n) - n$ is a martingale, where

$$g(u) = \frac{r^u - ur + u - 1}{p(r-1)^2}, \qquad r = \frac{q}{p}.$$

(b) Use your result in part (a) to show that

$$E[S] = \frac{b}{p-q} - \frac{a+b}{p-q} \cdot \frac{1-(p/q)^b}{1-(p/q)^{a+b}}.$$

23.10 Redo Example 20.9, now assuming that $q = 1/7$ and that X has a constant value of 3.

23.11 In the model of Section 23.5, let q equal $1/3$, and let X take the values 1, 2, with equal probability. Show that $\psi(k) = 1/4^{k+1}$ for all k.

The remaining exercises deal with the continuous-time surplus process:

$$U(t) = u + ct - \langle N(t), X \rangle, \qquad \text{where } \{N(t)\} \text{ is a Poisson process at rate } \lambda.$$

23.12 Answer the following parts separately:

(a) Suppose that the initial surplus is 10. How large should the adjustment coefficient R be, to ensure that the probability of ruin will be less than 0.10?

(b) Suppose that the adjustment coefficient $R = 2/9$, $\lambda = 5$, and X is exponentially distributed with mean 3. Find c.

(c) You are given that $c = 20$, $\lambda = 5$. All you know about X is that $E(X) = 3$. What is the probability that the surplus will eventually drop below its initial value?

(d) Suppose that X takes some fixed value with certainty. Show that if the initial surplus does drop below its initial value, the amount of deficit the first time this occurs will be uniformly distributed.

23.13 You are given that $X \sim \text{Exp}(3)$, $c = 2$, $\lambda = 3$. You want the probability of eventual ruin to be no more than 0.05. How much initial surplus should you begin with?

23.14 You are given that X is uniformly distributed on the interval [0,2], the initial surplus $u = 10$, and the adjustment coefficient $R = \log(3)$.

(a) What is the probability that the surplus will eventually drop below 10?

(b) Given that the surplus does eventually drop below 10, what is the probability that the first time this happens, the surplus will be between 8.5 and 9?

23.15 Suppose that X is exponentially distributed. Is $\psi(2u)$ equal to, strictly less than, or strictly greater than $\psi(u)^2$?

23.16 For any random variable A, let R_A be the adjustment coefficient. That is, R_A is the positive solution of $\phi_A(r) = 0$, where $\phi_A(r) = M_A(-r) - 1$. Suppose that G and H are random variables such that both adjustment coefficients exist and $R_G \leq R_H$. If K is a mixture of G and H, show that

$$R_G \leq R_K \leq R_H$$

by using properties of the graph of the function ϕ_A.

23.17 (a) If the adjustment coefficient $R = 2$, find u so that $\psi(u) \leq 0.05$.

(b) Suppose that X has a Gamma(2,3) distribution and that the adjustment coefficient $R = 1$. Find θ, the relative risk loading.

(c) Suppose that X is exponential with mean 2. If $\psi(0) = 0.8$, what is $\psi(1)$?

(d) Suppose that X is uniformly distributed on [0, 4] and that $u = 5$. Given that the surplus eventually drops below 5, what is the probability that, the first time this happens, the surplus is between 3 and 5?

23.18 Use the m.g.f. of the maximal aggregate loss $\mathcal{L}$ to show that $E(\mathcal{L}) = E(X^2)/2\theta E(X)$.

23.19 For a surplus process with a compound Poisson claim process, you are given that the adjustment coefficient is 0.25, the claim amount has a density function $f(x) = \mathrm{e}^{-2x} + 2.5\mathrm{e}^{-5x}$ for $x > 0$. If $\mathcal{L}$ is the maximal aggregate loss, determine $P(\mathcal{L} = 0)$.

23.20 If X is a mixture of Exp(α) and Exp(β) with weights w and $1 - w$, respectively, find a formula for $\psi(u)$ in each of the following cases:

(a) $\alpha = 3,\ \beta = 7,\ \omega = 1/2,\ \theta = 2/5$;

(b) $\alpha = 1,\ \beta = 2,\ \omega = 1/3,\ \theta = 4/11$;

(c) $\alpha = 3,\ \beta = 6,\ \omega = 1/9,\ \theta = 4/5$.

23.21 Suppose that $\lambda = 3, c = 3$ and X has a density function $f(x) = 36x\mathrm{e}^{-6x}$. Find a formula for $\psi(u)$.

24

Credibility theory

24.1 Introductory material

24.1.1 The nature of credibility theory

Suppose that X is the random variable denoting claims on an insurance policy during a specific period, and an insurer charges $E(X)$ as a net premium for each individual. Of course, a typical group of purchasers is not homogenous and there will be both bad and good risks within the group, but these cannot normally be identified as such at the outset of the contract. This means that some are paying more than they should be and some less. Suppose, however, that after issuance of the policy, the insurer is presented with additional information on a policyholder, usually as the result of claim experience, which provides some indication as to the degree of risk. For example, a purchaser of automobile insurance incurs a few costly claims. Does this indicate they are a poor driver, and their subsequent premiums should increase, or could they really be good drivers with the costly claims simply a result of bad luck? Credibility theory deals with the problem of analyzing this additional information and deciding on how it should be used to modify future premiums. The key question then is, how 'credible' is the additional data, providing the source for the name of this concept.

24.1.2 Information assessment

We first explore the basic idea of utilizing information to reassess probabilities, beginning with a few simple examples.

Example 24.1 A bag contains three dies. Two of these are standard, having sides marked with one to six dots. A third is special, in that the side with one dot has been replaced with six dots, so there are two sides with six dots and none with one dot. A game consists of drawing

Fundamentals of Actuarial Mathematics, Third Edition. S. David Promislow.

Companion website: http://www.wiley.com/go/actuarial

a die at random from the bag and then tossing it. An amount of money equal to the resulting face is paid to the participant.

(a) What is the fair price that a person should pay for one play of this game?

(b) Suppose now that after selecting the die, you are allowed to throw it once, observe the result, and then decide if you want to play the game or not with this selected die. How does this additional information change the fair price you would pay to play?

Solution.

(a) If X_1 is the number that comes up after throwing a standard die, and X_2 the number that comes up after throwing the special die, the resulting payoff X is a mixture of X_1 and X_2 with respective weights of $2/3$ and $1/3$. It is clear that $E(X_1) = 21/6$ and $E(X_2) = 26/6$, so the fair price is

$$E(X) = (2/3)(21/6) + (1/3)(26/6) = 34/9.$$

(b) The fair price will of course depend on the number we observe, for this will alter the respective probabilities that we assign to the two types of dice. If, for example, 1 is thrown, then we know for certain that the selected die is standard, and the fair price will change to $21/6$. Suppose that a 6 is thrown. We now cannot be certain which type of die was selected, but the observation clearly tends to provide evidence that it was the special one with two 6 dot sides rather than 1. To find the fair price we must reassess the respective probabilities of the selected die being either standard or special. The tool for doing so is a well-known result in probability and statistics known as *Bayes theorem*, which formally amounts to a simple calculation. Given two events A and B you can sometimes easily infer $P(A|B)$, and you want to use that to compute the reverse conditional probability, namely $P(B|A)$. (Refer to Section A.2 for notation.) This is easily done via

$$P(B|A) = \frac{P(A \cap B)}{P(A)} = \frac{P(A \cap B)}{P(B)} \frac{P(B)}{P(A)} = \frac{P(B)P(A|B)}{P(A)}.$$

Bayes Theorem therefore says the following. Suppose we observe that event A has occurred. Our new assessment of the probability of any event B is proportional to the number obtained by taking the original probability of B and weighting it with the probability that event B would have caused us to observe the event A. The constant of proportionality is $P(A)^{-1}$. In the present example, the probability of a 6 on a standard die is $1/6$ as opposed to a probability of $2/6$ for the special die. The original probabilities of $(2/3,\ 1/3)$ for the selection of a standard or special die respectively, will under the observation of a 6 being thrown, change to something proportional to $\big((2/3)(1/6), (1/3)(2/6)\big) = (2/18, 2/18)$. We could compute the probability of the observation to be $2/9$, in order to derive the constant of proportionality as $9/2$, but there is no need to do so. Since we know that the two probabilities add to 1, they must each be $1/2$. Our fair price is now

$$E(Y) = (1/2)(21/6) + (1/2)(26/6) = 47/12.$$

Finally, suppose a 2, 3, 4, or 5 is thrown. Since each of these has the same chance of being thrown by either type of die, the weightings are equal. The original probabilities and therefore the original fair price will remain unchanged.

To summarize, the fair price is equal to $g(y)$, where y is the number thrown and g is the function given by

$$g(1) = 7/2, \quad g(2) = g(3) = g(4) = g(5) = 34/9, \quad g(6) = 47/12.$$

Example 24.2 (**The Monte Hall Problem**) This example has nothing to do with insurance, but we include it since it is perhaps one of the most famous examples on the theme of using information to alter probabilities, and it indicates just how subtle the transfer of information can be. It keeps on reappearing and never fails to baffle several people (including mathematicians and others well versed in technical matters) who frequently get an incorrect answer. It is named after the original host of a popular T.V. game show, as it is typical of the type of decision making that the quests on this show sometimes had to make. A valuable prize is randomly hidden behind one of the three doors, and you will receive the prize if you can guess the correct one. Suppose that after you choose a door, say number 1, the host opens another door, say number 2, revealing that there is nothing behind it, and then allows you to alter your choice if you so desire. Should you remain with your original guess of door 1, or instead change your selection to the remaining door 3?

Solution. This is the way the problem is usually formulated, but there is possibly some room for ambiguity (which may explain some of the wrong answers). It is a crucial condition that the host is obligated to open a blank door on every guess. The situation changes drastically if there is an option of doing so or not. For example, if the strategy of the host was to show you a blank door only when your guess was right, in an attempt to talk you out of a prize, then obviously you should never switch.

A great number of people think that you should not switch in any event. They reason that since there is always one blank door among the two remaining ones, the revelation gives you no information at all and you may just as well stick with your original choice. This is incorrect however under our stated condition. We can apply Bayes Theorem as we did above. Let B_i denote the event that the prize is behind door i, and A the observed event that the host opens door 2. We know that precisely one of the events B_1 or B_3 occurred and we have to decide which is more likely. Now the original probabilities are $1/3$ for each event. Note however that $P(A|B_3) = 1$ since if the prize was behind door 3, the host had no choice but to open door 2, in order to give you a guess. Without knowing the strategy the host uses to pick between two doors that have nothing behind them, we can only say that $P(A|B_1)$ is some number p. The new probabilities given the event A are then proportional to p for A_1 and 1 for A_3. Since p cannot be bigger than 1 it is clear that you should switch.

As an example, in the case that the host picks randomly so that $p = 1/2$, the probability that the prize is behind door 3 would now be $2/3$ as opposed to $1/3$ for door 1. Suppose instead that the host always picks the lowest numbered door when there are two choices. There is no gain by switching in the given scenario, but if the host had opened door 3 in response to your selection of door 1, you would be certain to win if you switched. In all cases then you cannot lose and will normally gain by switching.

People go wrong on this since they fail to evaluate the extent of the information correctly. The host is telling you much more than just the fact that one of the two remaining doors is the losing one. They are indeed naming *one particular such door*, and this lends evidence to the fact that the remaining door has the prize.

Another fallacious line of reasoning is that the information tells you only that either door 1 or door 3 is the winning one, and therefore you have no reason to switch. This would be correct had the host told you that door 2 had nothing behind it, *before* you made your choice. But the fact that you are told this after the choice, alters the respective probabilities, as Bayes Theorem shows.

As an aside, there is a simple way to see that you should switch without the use of Bayes theorem. The effect of switching is to convert an original wrong guess into a correct one and an original correct guess into a wrong one. But since you will be wrong on the average $2/3$ of the time, it is definitely an advantage to switch.

We next present a situation, which is similar in nature to Example 24.1, but which ties in directly to our insurance application discussed in Section 24.1.1.

Example 24.3 A population consists of 60% good drivers and 40% bad drivers. During any period, the claims for a good driver will be 10 with probability 0.2 and 0 with probability 0.8. The claims for a bad driver will be 10 with probability 0.5 and 0 with probability 0.5.

(a) Calculate the net premium per period, assuming we cannot distinguish between good and bad drivers.

(b) Suppose we observe that a particular driver had a claim of 10 in each of the first two periods. What is the expected claim amount for this driver in period 3?

Solution.

(a) The expected claims are 2 for a good driver and 5 for a bad driver. Since we have a mixture of good and bad drivers, the net premium is the overall expectation of

$$2(0.6) + 5(0.4) = 3.2.$$

(b) We argue exactly as in the Example 24.1. The probability that a good driver will produce the observation of a claim of 10 in each of the first two periods is $0.2^2 = 0.04$. The probability that a bad driver will produce the observation is $0.5^2 = 0.25$. Therefore, the original probabilities of (0.6, 0.4) for good and bad drivers respectively will be revised for this driver to something proportional to $\big((0.6)(0.04), (0.4)(0.25)\big) = (0.024, 0.1)$ and in order to add to 1, the two probabilities must be $(6/31, 25/31)$. Our revised expectation of third period claims is

$$2(6/31) + 5(25/31) = 137/31 = 4.42.$$

Let us summarize the reasoning of a typical insurer in this situation, which will be familiar to many readers who have had insurance premiums raised due to their claims experience. In this instance, a possible conclusion for the insurer to make is that they are dealing with a bad driver, and raise this individual's premium for the next period to 5, if this is allowed by the contract. This could be unfair however, since there is some possibility that the person is a good driver with the two high claims occurring only by chance. Nonetheless there is enough evidence to raise the premiums in the next period from 3.2 to 4.42.

To complete the analysis as in the dice example, we would want to know the expected claims for the other three possibilities for claims of either 10 or 0 in the first two periods. We leave this as an exercise for the reader.

24.2 Conditional expectation and variance with respect to another random variable

24.2.1 The random variable $E(X|Y)$

We refer back to Section 20.13.2.

Definition 24.1 Given discrete variables X and Y, defined on a sample space, we define the random variable $E(X|Y)$ to be $E_\Pi(X)$ where the set is partitioned according to the values of Y. In other words, for a point ω such that $Y(\omega) = y$,

$$E(X|Y)(\omega) = E(X|Y = y).$$

The reader is invited to show that $E(X|I_B)$, where I_B is the indicator random variable as defined in Section A.5, takes the value $E(X|B)$ (as defined in Section A.7) on the set B and 0 on the complement of B.

Example 24.4 Suppose that Ω consists of four points $\omega_1, \omega_2, \omega_3, \omega_4$ with probabilities 0.4, 0.3, 0.2, 0.1, respectively, and that

$$Y(\omega_1) = Y(\omega_2) = 1, \qquad Y(\omega_3) = Y(\omega_4) = 2, \qquad X(\omega_i) = i, \quad i = 1, 2, 3, 4.$$

Describe the random variable $E(X|Y)$.

Solution. The set $\{Y = 1\}$ consists of the two points ω_1 and ω_2 with conditional probabilities equal to $4/7$ and $3/7$, respectively. Therefore $E(X|Y)$ takes the value of $4/7 + 2(3/7) = 10/7$ on ω_1 and ω_2. Similarly it takes the value of $3(2/3) + 4(1/3) = 10/3$ on ω_3 and ω_4.

For another example, which will indicate how this concept ties in with credibility theory, we leave the reader to verify that in Example (24.1) where Y is the observed number on the first throw and X is the final outcome, the random variable $E(X|Y)$ is equal to $g(Y)$ where g is the function constructed in that example.

We motivate the use of $E(X|Y)$ by looking at a problem of prediction. Suppose we want to predict the outcome of a random experiment which is modeled by the random variable X. Our method of doing so must obviously depend on the objective that we have in mind. For a simple example, suppose that X takes the value 0 with probability 0.4 and 100 with probability 0.6. If our objective is to maximize the probability of an exact answer, we would clearly predict

100, but of course that means that we will be far off 40% of the time. Suppose our objective is to be 'close' on the average. This criteria demands that we clarify what we mean by 'close'. Assume we select the time honoured way of measuring closeness in statistics, which is via squared deviation. In this context we should choose as a prediction that number b such that $E[(X-b)^2]$ is minimal. Let μ denote $E(X)$ and consider the identity

$$E[(X-b)^2] = E[(X-\mu)^2] + (b-\mu)^2, \tag{24.1}$$

which is easily derived by expanding the right-hand side. From this, it is evident that we choose $b = \mu$, as we might well have guessed without any calculation.

Now suppose that before making our prediction we can observe the outcome of another random variable Y and then base our prediction on the value of Y. Our prediction Z then will be a random variable, since it depends on the random value of Y. We will show that the least squares minimizing choice for Z is the random variable $E(X|Y)$. That is, $E[X|Y]$ can be viewed as that random variable which is a function of Y, and which gives, in some sense, the best estimate of a value of X after being given a value of Y.

In the rest of this section, when Y is fixed, we will write $E[X|Y]$ as $\hat{X}$ for convenience in notation. In addition to the general results given in Theorem 20.6 we need some additional facts dealing with the relationship between the two random variables.

Theorem 24.1

(a) *If X is a function of Y, then $\hat{X} = X$.*

(b) *If X and Y are independent, then $\hat{X}$ takes the constant value of $E(X)$.*

(c) *$\hat{X}$ is a function of Y that minimizes the expected squared deviation from X.*

Proof.

(a) If X is a function of Y, then its value is constant on the sets $Y = y$. Part (a) follows immediately from Theorem 20.6(b), taking V to the random variable with constant value 1.

(b) We make use of probability functions. On any set $\{Y = y\}$, $\hat{X}$ will have the constant value

$$\sum_x x\frac{f_{X,Y}(x,y)}{f_Y(y)} = \sum_x x\frac{f_X(x)f_Y(y))}{f_Y(y)} = \sum_x xf_X(x) = E(X).$$

(c) We will derive the following which will imply the desired result. If Z is a function of Y, then

$$E(X-Z)^2 = E(X-\hat{X})^2 + E(\hat{X}-Z)^2.$$

Using linearity, the right-hand side expands to

$$E(X^2) - 2E(X\hat{X}) + E(\hat{X}^2) + E(\hat{X}^2) - 2E(Z\hat{X}) + E(Z)^2.$$

From Equation (20.25)

$$E(\hat{X}\hat{X}) = E(\hat{X}X), \quad E(Z\hat{X}) = E(ZX),$$

and substitution into the expression above gives the left -hand side.

□

Remark The results above remain true in much more generality than for discrete random variables and will be so applied in the material that follows. The main problem is that when Y is continuous, a set B of the form $Y = y$ will have probability zero, and we cannot directly define the conditional probability $P(\cdot|B)$. In the case where a joint density function $f_{X,Y}$ exists, this problem can be handled, since we can define the conditional density function on such a set B as $f_{X,Y}(x, y)/f_Y(y)$ and things will work out as above. A full treatment, however, applicable to general random variables requires knowledge of some advanced topics, such as measure theory, and we will not do this here.

We conclude this section with a generalization. In what follows we will deal with cases where instead of making an observation of just one random variable before making our prediction of X we may observe n random variables. In fact we have already encountered this situation in Example 24.3 where we had two observations. This means that our observation is not just a random variable but a *random vector*:

$$\mathbf{Y} = (Y_1, Y_2, \ldots, Y_n),$$

where each Y_i is a random variable. We can define the random variable $E(X|\mathbf{Y})$ as $E_\Pi(X)$ where Π is just the partition of the sample space into the sets on which the vector $\mathbf{Y}$ is constant. As in the case where $n = 1$, $E(X|\mathbf{Y})$ will be that random variable which is equal to $g(\mathbf{Y})$ for some function g of n variables and which minimizes $E[(X - Z)^2]$ over all random variables Z which are functions of $\mathbf{Y}$.

24.2.2 Conditional variance

We now introduce another important random variable.

Definition 24.2 We define the *conditional variance of* X given Y, denoted by $\mathrm{Var}(X|Y)$, as the random variable that takes the value $\mathrm{Var}(X|Y = y)$ on a point ω of the sample space such that $Y(\omega) = y$.

An equivalent formulation is to write

$$\mathrm{Var}(X|Y) = E[(X - \hat{X})^2|Y]. \tag{24.2}$$

A major use of this concept is in the following decomposition of variance formula.

Theorem 24.2

$$\mathrm{Var}(X) = E[\mathrm{Var}(X|Y)] + \mathrm{Var}[E(X|Y)] \tag{24.3}$$

Proof. Let

$$w = \text{Var}\,(\hat{X}), \quad v = E[\text{Var}\,(X|Y)], \quad \mu = E(X) = E(\hat{X}.)$$

Use linearity to expand the right-hand side of (24.2) to get

$$\text{Var}\,(X|Y) = E(X^2|Y) - 2E(X\hat{X}|Y) + E(\hat{X}^2|Y).$$

Now $\hat{X}^2$ being constant on the sets $\{Y = y\}$ is clearly equal to $E[\hat{X}^2|Y]$ and Theorem 20.6(b) shows that $\hat{X}^2$ is equal to $E[X\hat{X}|Y]$. Making these substitutions and taking expectations gives

$$v = E(X^2) - E(\hat{X}^2).$$

We also have

$$w = E(\hat{X}^2) - \mu^2.$$

Adding the last two equations gives

$$\text{Var}\,(X) = w + v$$

to complete the proof. □

Remark We have encountered particular cases of this decomposition previously. The first was for the variance of deferred annuity values given in formula (15.16). In that case we used Y in place of X and the conditioning random variable was the indicator random variable for survival to time m. The second was the frequency-severity decomposition given in formula (21.3). In that case we had S in place of X and N in place of Y.

Example 24.5 Verify the decomposition of variance formula for Example 24.3 with Y as the random variable that takes the value of 1 for a good driver or 2 for a bad driver.

Solution. In this example X takes the value 10 with probability $(0.2)(0.6) + (0.5)(0.4) = 0.32$, and the value 0 with probability 0.68. Therefore

$$\text{Var}(X) = 100(0.32)(0.68) = 21.76.$$

(A variance of a two-valued random variable is simply the square of the difference, multiplied by the product of the two probabilities).

$E(X|Y)$ takes the value 2 when $Y = 1$ or 5 when $Y = 2$. Therefore

$$\text{Var}(X|Y) = 3^2(0.4)(0.6) = 2.16.$$

Now $(X|Y = 1)$ takes the value 10 and 0 with probabilities 0.2 and 0.8, respectively, so $\text{Var}(X|Y = 1) = 100(0.2)(0.8) = 16$, and similarly $\text{Var}(X|Y = 2) = 100(0.5)(0.5) = 25$. So

$$E[\text{Var}(X|Y) = 16(0.6) + 25(0.4) = 19.6.$$

We return again to the theme of transfer of *information*. The quantities w and v are useful tools in measuring the information about X that we get from observing Y. Before the observation, we start out with some uncertainty as measured by Var (X). We may view w as representing that part of the uncertainty that is removed by the information we get from being told the value of Y, leaving the uncertainly of v. A high value of w tends to increase the value of the information. On the other hand when v is high, it means that on average $\text{Var}(X|Y = y)$ is high. This suggests that even knowing the value of Y, there is a great deal of uncertainty in the value of X. A high value of v then tends to diminish the value of the information.

Both w and v depend on the units used in measuring X and Y. We can make use of a similar idea as we used in Definition 15.3 to get a quantity independent of units by defining

$$k = \frac{v}{w},$$

which we will refer to as the *information coefficient* of Y with respect to X. It will play a major role in the sequel. The lower the value of k, the more useful the information is to us. Consider the extreme cases.

(i) When X and Y are independent, then $\hat{X}$ has the constant value of μ, so $w = 0$. We get absolutely no information from knowing Y as reflected by the value of $k = \infty$.

(ii) X is a function of Y. Now $\hat{X} = X$ so $w = \text{Var}(X)$ and $v = 0$. The information removes all our uncertainty. This is reflected by the value of $k = 0$.

24.3 General framework for Bayesian credibility

Now, having given some basic examples and theory, we will go back and look in detail at the original insurance situation. We start with a random variable X that represents total claims in a period from an insured individual. We make the assumption that individuals can be distinguished by a risk parameter Θ which varies over some subset of the real numbers, but the value of Θ is unknown for each particular individual, Θ then constitutes another random variable. We assume however that we have the following information:

$f(x;\theta) = \frac{f_{X,\Theta}(x,\theta)}{f_\Theta(\theta)}$, the probability or density function of $(X|\Theta)$,
$\pi(\theta)$, a probability or density function for Θ.

So for example where X is discrete, $f(x;\theta)$ would give the probability that for an individual with risk parameter θ, the value of X will be x.

The probability or density function of X then can be written as

$$f(x) = \sum_{\theta} f(x;\theta)\pi(\theta)$$

when Θ is discrete or

$$f(x) = \int_{-\infty}^{\infty} f(x;\theta)\pi(\theta)d\theta$$

when Θ is continuous.

Let us show how this notation fits into Example 24.3. The two values of Θ were 1 and 2, denoting good and bad drivers, respectively, and the distribution of Θ was given by

$$\pi(1) = 0.6, \quad \pi(2) = 0.4.$$

We then needed two density functions which were

$$f(0;1) = 0.8, \quad f(10;1) = 0.2,$$

and

$$f(0;2) = 0.5, \quad f(10;2) = 0.5.$$

Suppose we observe the claims of a certain individual over n periods. We then have a random vector $\mathbf{X} = (X_1, X_2, \ldots, X_n)$ where X_i denotes the claims in the ith period. Our assumptions are that each X_j is distributed as X and $(X_1, X_2, \ldots, X_n)$ are independent, *conditional* on Θ.

Remark It is important to note that the random variables X_i themselves will *not* in general be independent. Referring again to Example 24.3, if some X_i has a value of 10, this is evidence that we have a bad driver, and this will affect the likelihood that the other values X_j will be high. In general (when say X is discrete), the probability that $X_1 = x_1$ and $X_2 = x_2$, given that $\Theta = \theta$, is equal to $f(x_1;\theta)f(x_2;\theta)$ but the probability that $X_1 = 1$ and $X_2 = 2$ is not necessarily equal to $f_X(x_1)f_X(x_2)$.

Our overall goal is to estimate the claims in the next period. That is, we want to determine the number $E(X_{n+1}|\mathbf{X} = \mathbf{x})$ which, as we have seen, will give us the least squares minimizing prediction. We usually will write this as $E(X_{n+1}|\mathbf{x})$, with $\mathbf{X}$ being understood. This number is known as the *Bayesian credibility premium.* We will not calculate it directly in the form given, however, but rather use the procedure that we have illustrated in Examples 24.1 and 24.3, where we make use of the random variable Θ as an intermediary, which usually simplifies the computation. Formally we are calculating $E[E(X|\Theta)|\mathbf{x}]$.

To summarize, the procedure is as follows.

Step 1: Calculate a revised distribution of Θ. (This is known as a *posterior* distribution as opposed to the original distribution,which is known as the *prior.*) Following Bayes Theorem, we do this by multiplying original probabilities or densities by the probability or density that a parameter of θ would cause the observed sample.

For the latter, we invoke the assumption of independence of the operations given the value of Θ to write the new probability or density function as

$$\pi^*(\theta) \propto \pi(\theta) \prod_{i=1}^{n} f(x_i; \theta).$$

(The symbol $\propto$ stand for 'proportional to' and means that there is some constant (i.e. a value independent of the variable(s)) multiplying the right-hand side. Since this constant is determined uniquely by the fact that the function on the left side sums or integrates to 1, we normally do not need to write it down exactly.)

Step 2: Identify the random variable $X^* = E^*(X|\Theta)$, where * indicates that we are using the posterior distribution found in step 1.

Step 3: Calculate the credibility premium as $E(X^*)$.

24.4 Classical examples

To illustrate the procedure above we will give two classical examples.

Example 24.6 (**Normal–Normal**) Suppose that for each θ, X is normal with mean θ and variance v, while Θ is itself normal with mean μ and variance w. Find the credibility premium.

Solution. Consider first the case when $v = 1$. We then have

$$f(x; \theta) \propto e^{-\frac{1}{2}(x-\theta)^2},$$

and

$$\pi(\theta) \propto e^{-\frac{1}{2w}(\theta-\mu)^2}.$$

To simplify the calculation, we will use the following fact. Given any n numbers $x_1, x_2, \dots, x_n$, let $\bar{x} = (1/n)\sum_{i=1}^{n} x_i$. Then for any number c

$$\sum_{i=1}^{n}(x_i - c)^2 = \sum_{i=1}^{n}(x_i - \bar{x})^2 + n(\bar{x} - c)^2. \tag{24.4}$$

This follows by applying (A.9) with X as a random variable that takes the value $x_i - c$ for $i = 1, 2, \dots n$, each with probability $1/n$. Rearrange and multiply by n.

Step 1:
$$\pi^*(\theta) \propto e^{-\frac{1}{2}\sum_{i=1}^{n}(x_i-\theta)^2} e^{-\frac{1}{2w}(\theta-\mu)^2} \propto e^{-A/2}, \tag{24.5}$$
where

$$\begin{aligned} A &= \textstyle\sum_{i=1}^{n}(x_i - \theta)^2 + w^{-1}(\theta - \mu)^2 = \sum_{i=1}^{n}(x_i - \bar{x})^2 + n(\bar{x} - \theta)^2 + w^{-1}(\theta - \mu)^2 \\ &= \textstyle\sum_{i=1}^{n}(x_i - \bar{x})^2 + n\bar{x}^2 + w^{-1}\mu^2 + \theta^2(n + w^{-1}) - 2\theta(n\bar{x} + \mu w^{-1}). \end{aligned}$$

We use (24.4) for the second equality above. Let

$$w^* = (n + w^{-1})^{-1}$$

and

$$\mu^* = (n\bar{x} + \mu w^{-1})w^* = \left(\frac{w^{-1}}{n + w^{-1}}\right)\mu + \left(\frac{n}{n + w^{-1}}\right)\bar{x}. \tag{24.6}$$

In the above expression for A, simplify by completing the square for the terms in Θ, then lump together everything else as a constant, to obtain

$$A = \frac{(\theta - \mu^*)^2}{w^*} + K,$$

where K is independent of θ. From (24.5),

$$\pi^*(\theta) \propto \mathrm{e}^{-\frac{(\theta-\mu^*)^2}{2w^*}},$$

so we have succeeded in identifying the posterior distribution of Θ as again being normal but now with mean μ^* and variance w^*.

Step 2: Since Θ is the mean of X, this means that $X^* = E^*(X|\Theta) = \Theta$ equipped with the posterior distribution.

Step 3: The credibility premium is $E^*(\Theta) = \mu^*$.

Consider now the general case where X has variance v instead of 1. Then the random variable $\tilde{X} = X/\sqrt{v}$ has variance 1. It will have mean $\Theta/\sqrt{v}$ which is normally distributed with mean $\mu/\sqrt{v}$ and variance w/v. From (24.6)

$$E(X_{n+1}|\mathbf{x}) = \left(\frac{vw^{-1}}{n + vw^{-1}}\right)\frac{\mu}{\sqrt{v}} + \left(\frac{n}{n + vw^{-1}}\right)\frac{\bar{x}}{\sqrt{v}}.$$

To verify the last term, keep in mind that we must divide our observed values by $\sqrt{v}$ to get values of $\tilde{X}$ rather than X. We then multiply by $\sqrt{v}$ to get

$$E(X_{n+1}|\mathbf{X} = \mathbf{x}) = \left(\frac{vw^{-1}}{n + vw^{-1}}\right)\mu + \left(\frac{n}{n + vw^{-1}}\right)\bar{x}.$$

Let

$$k = vw^{-1}, \quad Z = \frac{n}{n + k}.$$

This gives the convenient form

$$E(X_{n+1}|\mathbf{x}) = (1 - Z)\mu + Z\bar{x}. \tag{24.7}$$

Note also that $E[X|\Theta] = \Theta$ so that Var $E[X|\Theta] = w$. Moreover Var $(X|\Theta)$ is a constant v, so that $E[\text{Var}\,(X|\Theta)] = v$. The constant k then is the information coefficient of Θ relative to X as we defined in Section 24.2.2.

In this case we obtain a very satisfactory prediction of the next value of X as a weighted average of the over-all mean μ and the mean of our observed values. The quantity Z is known as the *credibility factor*. The higher the value of Z, the more weight is given to the observations, as opposed to our prior estimate. In this case, it behaves just as we would expect. As k decreases, which means that more information is obtained from an observation, Z increases. Of course Z also increases as n increases. This is natural enough, since with a large number of observations, we can expect the data to be a reliable source of information.

Example 24.7 (Poisson–Gamma) Suppose that

$$f(x;\theta) = \frac{\theta^x e^{-\theta}}{x!}, \quad x = 1, 2, \ldots,$$

a Poisson distribution with mean θ, and

$$\pi(\theta) \propto \theta^{\alpha-1} e^{-\beta\theta},$$

a Gamma distribution with parameters α and β. Find the credibility premium.

Solution.

Step 1: $\pi(\theta|\mathbf{x}) \propto f(x_1)f(x_2)\ldots f(x_n)\pi(\theta) \propto \theta^{x_1}\theta^{x_2}\ldots\theta^{x_n}e^{-n\theta}\theta^{\alpha-1}e^{-\beta\theta} \propto \theta^{\alpha+n\bar{x}-1}$ $e^{-(\beta+n)\theta}$. The posterior distribution of Θ is again a Gamma distribution but now with parameters $\alpha^* = \alpha + n\bar{x}$ and $\beta^* = \beta + n$.

Step 2: $X^* = E(X|\Theta) = \Theta$ so X^* has a Gamma distribution with parameters α^* and β^*.

Step 3: From (A.55),

$$E(X_{n+1}|\mathbf{x}) = \frac{\alpha^*}{\beta^*} = \left(\frac{\beta}{\beta+n}\right)\frac{\alpha}{\beta} + \left(\frac{n}{\beta+n}\right)\bar{x}$$

which is exactly the form of (24.7) with credibility factor $Z = n/(n+\beta)$.

Note also that

$$w = \text{Var}\,E(X|\Theta) = \frac{\alpha}{\beta^2},$$

and since the variance of a Poisson is the same as its mean

$$v = E[\text{Var}\,(X|\Theta)] = E(\Theta) = \frac{\alpha}{\beta}.$$

It follows that the information coefficient of Θ with respect to X is $v/w = \beta$, so the credibility factor is the same form as found in Example 24.6.

24.5 Approximations

In many cases, we cannot apply the exact methods shown above since we may only have partial information about the distributions. Even in cases where we do know the distributions exactly, the computation may be too complex to be feasibly carried out. We seek approximate methods.

Considering Examples 24.6 and 24.7, we might ask if we can approximate the revised expected value of X_{n+1} as a linear combination of the observations and the original expected value. We look therefore for constants $\alpha_0, \alpha_1, \ldots, \alpha_n$ such that we can approximate $E(X_{n+1}|\mathbf{X})$ by the random variable

$$\tilde{X}_{n+1} = \alpha_0 + \sum_{i=1}^{n} \alpha_i X_i \tag{24.8}$$

24.5.1 A general case

In view of our least squares minimization result of the section above, it is natural to choose the coefficients α_i so as minimize the squared error between X_{n+1} and $\tilde{X}_{n+1}$. In other words rather than choosing our minimizing function g from all possible functions of n variables, we confine ourselves to functions of the form

$$g(x_1, x_2, \ldots, x_n) = \alpha_0 + \sum_{i=1}^{n} \alpha_i x_i$$

In cases such as Examples 24.6 or 24.7 above when the actual credibility premium is already of this form, our approximation will necessarily be exact.

As a simplification we demonstrate the procedure with $n = 2$. We want to minimize

$$Q = E[X_3 - \alpha_0 - \alpha_1 X_1 - \alpha_2 X_2]^2.$$

By the convexity of the square function, we can minimize by setting the partial derivatives of Q equal to 0. For the derivative of Q with respect to α_0 this gives

$$E(X_3) = \alpha_0 + \alpha_1 E(X_1) + \alpha_2 E(X_2). \tag{24.9}$$

Taking the derivative of Q with respect to α_1 and setting this equal to 0 gives

$$E(X_3 X_1) = \alpha_0 E(X_1) + \alpha_1 E(X_1 X_1) + \alpha_2 E(X_2 X_1). \tag{24.10}$$

Multiply (24.9) by $E(X_1)$ and subtract from (24.10) to get

$$\mathrm{Cov}(X_3, X_1) = \alpha_1 \mathrm{Cov}(X_1, X_1) + \alpha_2 \mathrm{Cov}(X_2, X_1).$$

Similarly, setting the derivative with respect to α_2 equal to 0 leads

$$\mathrm{Cov}(X_3, X_2) = \alpha_1 \mathrm{Cov}(X_1, X_2) + \alpha_2 \mathrm{Cov}(X_2, X_2).$$

For general n, the procedure is similar and we get a system of equations

$$\mathrm{Cov}(\mathrm{X_j}, \mathrm{X_{n+1}}) = \alpha_j \mathrm{Var}\,(X_j) + \sum_{i \neq j} \alpha_i \mathrm{Cov}(\mathrm{X_j}, \mathrm{X_i}), \tag{24.11}$$

$$\alpha_0 = (1 - s)\mu, \tag{24.12}$$

where $\mu = E(X)$ and $s = \sum_{i=1}^{n} \alpha_i$. If we know the covariances, these equations are easily solved to yield the desired coefficients.

24.5.2 The Bühlman model

The formulas in the last section apply to a quite general situation, but it could be difficult to actually find the covariances. Using our framework, with the intermediate random variable of Θ, we can determine these from our basic three parameters:

$$\mu = E(X) = E(X|\Theta)], \quad v = E[\mathrm{Var}\,(X|\Theta)], \quad w = \mathrm{Var}\,[E(X|\Theta)].$$

Terminology v is sometimes called the *variance of the hypothetical means* while w is called the the *expected process variance*.

Applying Equation (20.24)

$$E(X_iX_j) = E[E(X_iX_j|\Theta) = E[E(X_i|\Theta)E(X_j|\Theta)] = E[E(X|\Theta)^2],$$

where we use the fact that the X_i's are independent conditional on Θ. This shows that

$$\mathrm{Cov}(\mathrm{X_i}, \mathrm{X_j}) = \mathrm{E}[\mathrm{X_iX_j}] - \mu^2 = \mathrm{E}[(\mathrm{EX}|\Theta)^2] - \mu^2 = \mathrm{w}. \tag{24.13}$$

From Theorem 24.2, $v + w = Var(X_j)$ for all j and we substitute into (24.11) to get

$$w = \alpha_j v + sw. \tag{24.14}$$

This verifies that the α_j's are all equal (a fact that seems clear from the outset since all X_j have the same distribution). Summing the above equations for $j = 1, 2, \ldots, n$ we get $nw = sv + sna$ so that

$$s = \frac{n}{n+k}, \tag{24.15}$$

where $k = v/w$. Now, using (24.12) we can write Equation (24.8) as

$$\tilde{X}_{n+1} = (1 - s)\mu + s\bar{X}, \tag{24.16}$$

which is precisely of the form we noticed in the examples of the last section, with s as the credibility factor. This estimate is known as the *Bühlman credibility premium*, named after Hans Bühlman. The general form of the credibility factor (24.15) had been suggested from the early days of the subject, but the choice and determination of the constant k remained somewhat mysterious. Bülhman's contribution was to identify this constant under least squares minimization. The Bühlman credibility premium will be equal to the Bayesian credibility theorem, if and only if the latter is in fact a linear combination of the prior mean and the observed sample mean, as observed in Examples 24.6 and 24.7.

24.5.3 Bühlman–Straub Model

The assumption that the X_i's are identically distributed does not always hold in applications. Credibility theory is often applied to group insurance where an observation may be the average of several individual claims, and the number of such individuals may vary by period. In addition, there may be differences in the length of the observation period at different times. In many such cases we can handle the situation by a modification of the Bühlman model, which involves introducing a weight p_j for period j such that

$$\text{Var}\,(X_j|\Theta = \theta) = \tilde{v}(\theta)/p_j.$$

for some function $\tilde{v}$. We now let $v = E[\text{Var}\,(X|\Theta)] = E[\tilde{v}(\Theta)]$.

This will result in a generalization of our least squares estimate, known as the *Bühlman–Straub credibility premium*. The original model covered the case where each $p_i = 1$.

The decomposition of variance formula now becomes $\text{Var}(X_j) = w + v/p_j$ so that when we substitute into Equation (24.11), we must modify Equation (24.14) to

$$p_j w = \alpha_j v + p_j s w,$$

for each j. Letting $p = \sum_{j=1}^{n} p_j$ we sum this over all j to get

$$pw = sv + psw = s(v + pw).$$

So that

$$s = \frac{pw}{pw + v} = \frac{p}{p + k}.$$

Then $(1 - s) = v/(pw + v)$ and from each of the starting equations we can write

$$\alpha_i = \frac{p_i(1 - s)w}{v} = \frac{p_i w}{pw + v}.$$

So we obtain the same formula as before only now with the credibility factor applied to a *weighted* average of the observations namely, for

$$\bar{x} = \frac{1}{p}\sum_{j=1}^{n} p_j x_j,$$

the Bühlman–Straub credibility premium is

$$(1 - s)\mu + s\bar{x}.$$

Example 24.8 Let Y denote the aggregate claims per period for an individual, and suppose that $E(Y) = 100$, $E[\text{Var}(Y|\Theta] = 125$, $\text{Var}[E(Y|\Theta)] = 5$. The last three years of data on a group insurance plan show that in the first year 40 employees had aggregate claims of 4000, in the second year 25 employees had aggregate claims of 2000 and in the third year 35 employees had aggregate claims of 2500. In the fourth year the group will consist of 45 employees. Estimate the aggregate claims for the fourth year.

Solution. Let X_i denote the average claim in period i. The Bühlman–Straub model applies with weights $p_1 = 40, p_2 = 25, p_3 = 35$. (To verify this, note, for example, that $(X_1|\Theta)$ is the sum of 40 independent copies of $(Y|\Theta)$ divided by 40. So $\text{Var}(X_1|\Theta) = 40\,\text{Var}(Y|\Theta)/40^2$.) The credibility factor is given by

$$Z = \frac{100}{100 + 125/5} = 0.8$$

and

$$\bar{X} = \frac{4000 + 2000 + 2500}{100} = 85$$

so that the estimate of the average claims in the fourth year is

$$0.8(85) + 0.2(100) = 88$$

and our estimated aggregate claims for year 4 is $45(88) = 3960$.

24.6 Conditions for exactness

Looking at Examples 24.6 and 24.7, one might wonder whether the Bayesian credibility premium will always be a linear combination of the prior mean and the observed sample mean, so that the Bühlman credibility estimate is exact. The answer is no.

Example 24.9 Show that the Bühlman credibility estimate in Example 24.3 is not exact.

Solution. In this case $\mu = 3.2$ and from the solution to Example 24.5,

$$Z = \frac{2}{2 + 19.6/2.16} = 0.18060,$$

and so the Bühlman credibility premium for an observation of (10,10) will be

$$3.2(1 - 0.1806) + 10(0.1806) = 4.43.$$

The approximation is close to but not exactly equal to the true answer of 137/31.

We can however identify some sufficient conditions for exactness.

Theorem 24.3 *Suppose that the following conditions hold.*

(i) There exist functions b and c, with c continuously differentiable, such that the density or probability functions of $(X|\Theta)$ are given by

$$f(x;\theta) = \frac{b(x)e^{-\theta x}}{c(\theta)}, \tag{24.17}$$

for all x in (m, n) where (m, n) is an interval (possibly infinite) that does not depend on θ.

(ii) There exists parameters j, k, such that the density or probability function of Θ is given by

$$\pi(\theta) \propto c(\theta)^{-k}\mathrm{e}^{-j\theta}, p < \theta < q \tag{24.18}$$

for the same function c as in (i).

(iii) $\lim_{\theta\to p+} \pi(\theta)$ and $\lim_{\theta\to q-} \pi(\theta)$ exist and are equal.

Then the Bühlman credibility premium is exact.

Proof. We give the proof in the case that f is a density function. The discrete case can be verified in the same manner with summation replacing the integration. The argument will be broken up into various steps.

Step 1: We note first that conditions (i) and (ii) imply the property observed in the Section 24.4 examples, namely that π^* will have the same form as π but with changed parameters. Indeed,

$$\pi^*(\theta) \propto f(x_1)f(x_2)\ldots f(x_n)\pi(\theta) \propto c(\theta)^{-(n+k)}\mathrm{e}^{-\theta(n\bar{x}+j)},$$

which is the same form as π but with parameters

$$k^* = n + k, \quad j^* = n\bar{x} + j.$$

Step 2: We show that

$$E(X|\Theta = \theta) = -\frac{c'(\theta)}{c(\theta)}. \tag{24.19}$$

For any θ in (p, q) we know that $f(x; \theta)$ integrates to 1 over the interval (m, n) so that

$$c(\theta) = \int_m^n \mathrm{e}^{-\theta x}b(x)dx.$$

Differentiating with respect to θ

$$c'(\theta) = \int_m^n -xe^{-\theta x} b(x)dx = -E(X|\Theta = \theta)c(\theta),$$

establishing (24.19).

Step 3: We show that

$$E(X) = j/k. \tag{24.20}$$

Taking logarithms in (24.18) gives

$$\log \pi(\theta) = \log(K) - k \log c(\theta) - j\theta,$$

where K is the constant of proportionality. Differentiating with respect to θ and substituting from (24.19),

$$\frac{\pi'(\theta)}{\pi(\theta)} = -k\frac{c'(\theta)}{c(\theta)} - j = kE(X|\theta) - j.$$

Now multiplying by $\pi(\theta)$, integrating, substituting from (24.19) and using Equation 20.24.

$$\int_p^q \pi'(\theta) = k \int_p^q E(X|\theta)\pi(\theta)d\theta - j \int_p^q \pi(\theta)d\theta = kE(X|\Theta) - j = kE(X) - j.$$

The left-hand side is $\lim_{\theta \to q-} \pi(\theta) - \lim_{\theta \to p+} \pi(\theta)$, which equals 0 by condition (c) and establishes (24.20)

Step 4: In conclusion, we apply Step 3 to the new distribution of X and note that the credibility premium is

$$E(X^*) = \frac{j^*}{k^*} = \frac{n\bar{x} + j}{k + n} = \left(\frac{n}{k+n}\right)\bar{x} + \left(\frac{k}{k+n}\right)E(X)$$

Since the credibility premium is a linear combination of $\bar{x}$ and $E(X)$, we know from our previous remarks that the Bühlman estimate is exact. □

Remark The final step shows also that the information coefficient of Θ with respect to X is given by the parameter k in the density of $\pi(\theta)$. This can be verified by direct calculation.

Remark Distributions satisfying (24.17) are said to be from a *linear exponential family*. The name reflects the fact that X and θ are related only through an exponential term. Distributions of the form (24.18) are what are called *conjugate priors* for $f(x; \theta)$. This means that, as we noted in Step 1, the altered distribution has the same form as the original, but with different parameters.

Condition (iii) is automatically satisfied when the interval (p, q) is the whole real line, since the fact that π integrates to the finite number 1, means that the limits in questions must be both zero. In another common case where the interval is (p, ∞), we need the limit at p to be zero, which will hold if and only if $\lim_{\theta \to p+} c(\theta) = 0$.

The form of the density in (i) of Theorem 24.3 is not as restrictive as might appear, and many cases can be put into this form with a change of parameter. Several of these are covered by the following generalization.

Theorem 24.4 *Suppose that (iii) of Theorem 24.7.3 holds while (i) and (ii) are modified to*

(i) There exist functions b, c, a, where c and a are continuously differentiable, such that the density or probability functions of $(X|\Theta)$ *is given by*

$$f(x; \theta) = \frac{b(x)e^{a(\theta)x}}{c(\theta)} \tag{24.21}$$

for all x in an interval (m, n) *that does depend on* θ,

(ii) There exist parameters j, k, such that the density or probability function of Θ *is given by*

$$\pi(\theta) \propto c(\theta)^{-k} e^{ja(\theta)}, p < \theta < q \tag{24.22}$$

for the same functions a and c as in (i).

Then the Bühlman premium is exact.

Proof. Follow the proof of the previous theorem, but simply note that in Step 2 we now get

$$c'(\theta) = E(X|\theta)c(\theta)a'(\theta).$$

The $a'(\theta)$ is cancelled out in Step 3 when we multiply by $\pi(\theta)$, and the proof is completed as above. □

Note that this generalizes Theorem 24.3, which is the particular case with $a(\theta) = -\theta$.

We now show that the two examples given in Section 24.4 are covered by the above theorem and therefore we could have written down the Bayesian credibility premium immediately without going through the calculations.

In Example 24.6, with $v = 1$, we can expand the square and write $f(x, \theta)$ in the form of (24.17) with $c(\theta) = e^{\theta^2/2}$. By again expanding the square in the expression for $\pi(\theta)$, we see that it is of the required form (24.18) for the same $c(\theta)$ with $k = 1/w$ and $j = \mu k$.

In Example 24.7 we can apply Theorem 24.4 with $a(\theta) = \log(\theta), c(\theta) = e^{\theta}, k = \beta$, $j = \alpha$.

24.7 Estimation

In many cases we do not have any information regarding the distribution of X given Θ or the distribution of Θ, and we must rely strictly on the observed data alone. The data now performs double duty, giving us not only the credibility factor, but also an estimate of the parameters of the distributions. We begin with the Bühlman model and provide one possible method for estimating μ, v and w.

24.7.1 Unbiased estimators

This is the only time in the book where we delve into the field of statistical inference. Many readers will be quite familiar with the material in this section, but for completeness we give a quick discussion of the concept of an unbiased estimator.

Suppose we want to estimate a parameter v of a random variable X on the basis of observations $X_1, X_2, \ldots, X_n$. To do so we choose a function $\hat{v}$ of n-variables and if the observed values are $(x_1, x_2, \ldots, x_n)$, then we estimate the value of v as $\hat{v}(x_1, x_2, \ldots, x_n)$. So $\hat{v}$ is a random variable, since it depends on the particular observations. It is known as an *estimator* of v.

For an easy example, suppose we want to estimate the mean μ of a distribution. A natural way of doing this is to take the sample mean as an estimate. That is

$$\hat{\mu} = \bar{X} = \frac{X_1 + X_2 + \cdots + X_n}{n}. \tag{24.23}$$

So, for example, if $n = 3$ and we observe the values $10, 5, 9$, we would estimate the mean to be 8.

One desirable quantity of an estimator is that the values it gives will average to the true value. That is, we would like that

$$E(\hat{v}) = v.$$

Such an estimator is said to be *unbiased*. For example, $\hat{\mu}$ as given above is unbiased since

$$E(\hat{\mu}) = (1/n)[E(X_1) + E(X_2) + \ldots + E(X_n)] = (1/n)nE(X) = \mu.$$

Another quantity which we often want to estimate is the variance. Obtaining an unbiased estimator is not quite as obvious in this case. A possible guess might be the estimator

$$\frac{1}{n}\sum_{i=1}^{n}(X_i - \bar{X})^2,$$

but this turns out to be biased. We will derive a natural unbiased estimator in the case that the observations are independent.

First note that

$$\text{Var}(\bar{X}) = \frac{1}{n^2}\, n\text{Var}(X) = \frac{1}{n}\text{Var}(X). \tag{24.24}$$

This is a natural result. When we average over independent observations, the high and low values will tend to cancel each another. We can expect to get a result closer to the mean than with just one observation, so that the variance reduces.

Now consider (24.4) with $c = \mu$, applied to random variables X_i instead of numbers, and take expectations. For the term on the left of (24.4) we get $n\text{Var}(X)$ and for the term on the far right we get $n\text{Var}(\bar{X}) = \text{Var}(X)$ by (24.24). Rearranging,

$$E\left[\sum_{i=1}^{n}(X_i - \bar{X})^2\right] = n\text{Var}(X) - v(X) = (n-1)\text{Var}(X).$$

This shows that an unbiased estimator of Var (X) is given by

$$\frac{1}{n-1}\sum_{i=1}^{n}(X_i - \bar{X})^2. \tag{24.25}$$

Note what happens for the case that $n = 1$. Clearly one observation gives us absolutely no information at all regarding the variance, and that is reflected in the fact that our estimator is not defined.

24.7.2 Calculating Var($\bar{X}$) in the credibility model

Section 24.7.1 is perfectly general. We now want to consider the particular case of the variance of the sample mean in the basic setup of credibility theory. A major problem is that the observations are *not* independent. so that (24.24) does not hold.

Consider, however, our basic decomposition from Theorem 24.2,

$$\text{Var}(\bar{X}) = \text{Var}[E(\bar{X}|\Theta)] + E[\text{Var}(\bar{X}|\Theta)].$$

Substituting for $\bar{X}$ and using the fact that the observations are independent, conditional on Θ, we see that the first term will be equal to Var $[E(X|\Theta)] = w$, while the second term will be $E[X|\Theta]/n = v/n$. So we can write

$$\text{Var}(\bar{X}) = w + v/n. \tag{24.26}$$

This result can be explained intuitively. Consider that part of the uncertainty that is due to the lack of homogeneity, that is, the variation due to differences in the risk parameter. That does *not* get reduced by taking several observations from a single policyholder, and averaging. Only the part that is due to the variation for a given value of Θ is divided by n and therefore reduced.

24.7.3 Estimation of the Bülhman parameters

In our previous models we used the observations only to obtain $\bar{X}$. In this case where we as well want to estimate parameters from the data, we must make some more refined observations. In this instance, it will not be sufficient to look at claims from a single individual only. We need to have data from r individuals, where $r \geq 2$. We suppose, therefore, that we are going to take n observations from each of r individuals. We distinguish between observations from different individuals, which are independent, and observations from the same individual, which are only independent conditional on Θ. Let X_{ij} be the value of the jth observation from individual i. We now show how to derive unbiased estimators of each of the three parameters from this data.

Estimation of $\bar{X}$. In this case there is no need to consider the independence structure and we can just take the mean of the entire sample as an estimate as we did in the case of known distributions. That is we take

$$\hat{\mu} = \bar{X} = \frac{1}{rn}\sum_{i,j} X_{ij}. \tag{24.27}$$

This can also be calculated as

$$\hat{\mu} = \frac{1}{r}\sum_{i} \bar{X}_i,$$

where $\bar{X}_i = (1/n)\sum_{j=1}^{n} X_{i,j}$, the mean of the observations for individual i.

Estimation of v. Let θ_i be the (unknown) risk parameter of individual i. By the independence conditional on θ we can take

$$\hat{v}_i = \frac{1}{n-1}\sum_{j=1}^{n}(X_{ij} - \bar{X}_i)^2$$

as an unbiased estimator of $\mathrm{Var}\,(X|\Theta = \theta_i)$. Therefore, an unbiased estimator of $v = E[\mathrm{Var}\,(X|\Theta)]$ is given by averaging the above over all r individuals. The estimator is given by

$$\hat{v} = \frac{1}{r}\sum_{i=1}^{r} \hat{v}_i.$$

Estimation of w. From (24.26)

$$w = \mathrm{Var}\,(\bar{X}) - v/n.$$

Now $\bar{X}_1, \bar{X}_2, \ldots, \bar{X}_r$ constitute r independent observations of $\bar{X}$, so from (24.25)

$$\frac{1}{r-1}\sum(\bar{X}_i - \bar{X})^2$$

is an unbiased estimator of $\mathrm{Var}\,(\bar{X})$. Then, from the estimate of v, invoking (A.8) and (A.22) we can take

$$\hat{w} = \frac{1}{r-1}\sum(\bar{X}_i - \bar{X})^2 - \hat{v}/n$$

as an unbiased estimator of w.

One problem with this approach is that the estimate $\hat{w}$ often turns out be negative, while we know w is the expectation of a nonnegative random variable. Note that this does not contradict unbiasedness. It simply means that in this particular case the estimate of w is too low and indicates that there will be other sets of observations that will give an estimate of w which is too high. It points out the fact that the property of being unbiased is not always by itself sufficient to ensure a reasonable estimator. The method of handling this problem is normally to set $w = 0$. This means that $Z = 0$, and the credibility premium is simply the mean of the data. The conclusion is in effect that the data does not show sufficient variability between individuals to indicate any departure in our estimates from the observed overall mean.

Example 24.10 Policyholder 1 has aggregate claims in the first three periods of (4,2,6). Policyholder 2 has aggregate claims in the first three period of (6,5,10). Estimate the credibility premium for each policyholder.

Solution. We have that

$$\bar{X}_i = 4, \quad \bar{X}_2 = 7, \quad \bar{X} = 5.5$$

$$\hat{v}_1 = (0+4+4)/2 = 4, \quad \hat{v}_2 = (1+4+9)/2 = 7, \quad \hat{v} = 11/2$$

$$\hat{w} = (1.5^2 + 1.5^2) - 5.5/3 = 8/3$$

Our estimate of the information coefficient is then $(11/2)/\ (8/3) = (33/16)$, so the credibility factor is $3/(\ 3+ (33/16)) = 0.593$.

The estimated credibility premium for policyholder 1 is $0.593(4) + 0.407(5.5) = 4.61$.

The estimated credibility premium for policyholder 2 is $0.593(7) + 0.408(5.5) = 6.39$.

24.7.4 Estimation in the Bülhman–Straub model

The procedure described in this section can be modified to handle estimation in the Bülhman–Straub model. The results are similar, but one must be careful with the handling of the weights. We will list the final result here without going through the details of the derivation. As a further generalization we will allow the number of observations to vary with the individual, which is useful for some applications. As above, we suppose we have observations from r individuals where $r > 1$. Let

n_i be the number of observations for individual i.

p_{ij} be the weight attached to the jth observation of individual i.

$p_i = \sum_j p_{ij}, \quad p = \sum_i p_i$.

Then, if X_{ij} is the value of the jth observation of individual i we can take for unbiased estimators of the parameters,

$$\hat{u} = \bar{X} = \frac{\sum_{ij} p_{ij} X_{ij}}{p},$$

$$\hat{v} = \frac{\sum_{ij}(X_{ij} - \bar{X}_i)^2}{\sum_i (p_i - 1)},$$

where

$$\bar{X}_i = \frac{\sum_j p_{ij} X_{ij}}{p_i}$$

and

$$\hat{w} = \frac{\sum_i p_i(\bar{X}_i - \bar{X})^2 - (r-1)\hat{v}}{p - p^{-1}\sum_i p_i^2}.$$

In the case that each $p_{ij} = 1$ and $n_i = n$ for all i, we have $p_i = n$, $p = rn$ and it is easily verified that the estimators reduce to those given in the previous section.

Notes and References

This chapter constitutes a basic introduction to credibility theory, and certain topics are omitted. We have not covered *limited fluctuation* credibility theory, an older approach to the subject, that is still in use, despite some theoretical drawbacks. There are also many aspects of the estimation of parameters that we have not covered. There are alternate methods which avoid the problem of a negative estimate of w. Another frequent application is where some, but not all, of the features of the underlying distributions are known, so different estimation procedures are employed. A common occurrence of this type is one where $f(x;\theta)$ can be specified, but there is no reasonable way to determine $\pi(\theta)$. Readers can consult Klugman *et al.* (2012) or Herzog (1999) for more information on these topics. For a more advanced and complete treatment of credibility theory, see Bülhman and Gisler (2005).

Exercises

24.1 A die is selected at random from an urn that contains two six-sided dice. Die number 1 has three faces with the number 2, while the other three faces are numbered 1, 3, 4. Die number 2 has three faces with the number 4, while the other three faces are numbered 1, 2, 3. The first five rolls of the die yielded the numbers 2, 3, 4, 1 and 4, in that order. Determine the expected number for the sixth role of the same die.

24.2 Urn A has four balls numbered 1–4. Urn B has six balls numbered 1–6. An urn is selected at random, and then a random draw produces a ball with number 4, which is replaced. A ball is then drawn randomly from the same urn. Find the expected number.

24.3 Three urns contain balls marked either 0 or 1. In urn A, 10% are marked 0; in urn B, 60% are marked 0; and in urn C, 80% are marked 0. An urn is selected at random and three balls selected with replacement. The total of the values is 1. Three more balls are selected with replacement from the same urn. Find the expected total on the three balls.

24.4 One spinner is selected at random from a group of three spinners. Each spinner is divided into six equally likely sectors. The number of sectors marked 0, 12 and 48, respectively on each spinner is as follows: Spinner A: 2, 2, 2; Spinner B: 3, 2, 1; and

Spinner C: 4, 1, 1. A spinner is selected at random and a 0 is obtained on the first spin. What is the expected value for a second spin of the same spinner.

24.5 Complete Example 24.3 by finding the expected claim amounts given observed claims of (a) (0,0), (b) (0,10), (c) (10,0).

24.6 The number of claims in 1 year has a Poisson distribution with parameter Θ. The parameter Θ has a gamma distribution with mean 2 and variance 2. A particular insured had one claim in 1 year. What is the expected number of claims for this policyholder for the next year.

24.7 Suppose that $X \sim \text{Bin}(m, \Theta)$, where θ has a Beta(α, β) distribution. That is

$$\pi(\theta) = \frac{\Gamma(\alpha + \beta)}{\Gamma(\alpha)\Gamma(\beta)}\theta^{\alpha-1}(1-\theta)^{\beta-1}, 0 \leq \theta \leq 1.$$

Given n independent observations of X, find the expected value of the next observation in terms of the parameters $n, m, \bar{x}, \alpha, \beta$.

24.8 Repeat Exercise 24.6 only now assuming that $X \sim \text{Negbin}(r, \Theta)$.

24.9 In Exercise 24.2, find the Bühlman estimate of the expected number on the next ball.

24.10 In Exercise 24.3, find the Bühlman estimate of the expected total on the next three balls.

24.11 In the Bayesian credibility model, $X|\Theta \sim \text{Exp}(\Theta)$, where $\Theta \sim$ Gamma(2, 1). An individual has a claim of 5 in the first period. Find the expected claim in the second period for the same individual.

24.12 X has a Poisson distribution with parameter Θ, where $\pi(\theta) = 3\theta^{-4}$, for $\theta > 1$. A particular insured experienced a total of 20 claims in the previous 2 years.

(a) Determine the Bühlmann credibility estimate.

(b) Determine the exact Bayesian credibility premium in terms of an integral.

24.13 Suppose that X and Θ satisfy the conditions of Theorem 24.2. You are given that $E(X) = 1, E(X|X_1 = 4) = 2$, where X_1 is the value of a single observation. If $E[\text{Var}(X|\Theta)] = 3$, find $\text{Var}[E(X|\Theta)]$.

24.14 Redo Example 24.8, only now assuming that the aggregate claims for the first 3 years are (4500, 2200, 3000) and that $E[\text{Var}\ (Y|\Theta)] = 125$.

24.15 A taxi-cab company keeps a small fleet of cars, the number of which can vary from year to year. For each car, the number of accidents in a year is Poisson distributed where the Poisson parameter is uniform on [0,1]. The data for the past 3 years show one accident from four cars, two accidents from five cars and zero accidents from two cars. Next year the company will have three cars. What is the Bühlmann–Straub estimate for the number of accidents next year.

24.16 (a) Redo Example 24.10 assuming a third policyholder is observed, with claims in the first three periods of (7,5,6).

(b) Now, assume a fourth policy holder is observed with claims of (1,0,2), and calculate the credibility premiums.

(c) Explain briefly why the value of the credibility coefficient Z is lower in (a) than in the original example,despite the fact that there is additional data. Why does it increase in (c).

24.17 Show that exact credibility holds for the distributions of Exercises 24.7 and 24.8. Identify the value of k in each case.

24.18 Suppose that in the Monte Hall problem, the strategy of the host is as follows. If you pick the correct door, the host will either open the lowest numbered blank door, the highest numbered blank door or not open any door, each with probability $1/3$, and if you pick an incorrect door, the host will either open the remaining blank door, or not open any door, each with probability $1/2$. If the host opens a blank door, should you switch or not?

24.19 Show that $E_k(W)$, as defined in Section 20.8 is equal to $E(W|\mathbf{Y})$ as defined in Section 24.2 for a suitable random vector $\mathbf{Y}$.

Answers to exercises

Chapter 2

2.1 (a) 6 (b) 150

2.2 1

2.3 4.1

2.4 (a) $(1, 2/3, 4/9, 8/27, 2/9, 1/6)$ (b) $20\frac{2}{3}$ (c) $46\frac{1}{2}$

2.5 (a) 1.5 (b) 10.75

2.6 (a) $-19/12$ (b) $40/3$

2.7 (a) 360 (b) 440

2.8 For last part, $17 = 4.4 + 0.7 \times 18$

2.11 The first.

2.14 $d_7 = 12.2$

2.15 0.8

2.16 (a) 9980.89 (b) 10 117.40

2.17 (a) better off by 133.72 (b) worse off by 61.38

2.28 $v(0, 1) = 0.7, v(1, 2) = 0.4375.$

2.20 (b) $i_r = 2r/(1 + r)$

2.23 746.92

2.24 0.045

2.25 Initial payment is 8177.15. As examples, outstanding balance at time 5 is 49 560.15, and at time 15 is 19 132.59.

Fundamentals of Actuarial Mathematics, Third Edition. S. David Promislow.

Companion website: http://www.wiley.com/go/actuarial

Chapter 3

3.1 (a) 1000, 800, 600, 450, 315, 189.

(b) (i) 285/800 (ii) 189/600

(c) 2.5052, 2.1315, 1.842, 1.456, 1.08

3.2 0.288

3.3 1/10, 1/6, 4/15, 11/20

3.4 (a) $1 - n/(100 - x)$ (b) $n/(100 - x)$ (c) $k/(100 - x)$, for suitable n, k

3.5 22.8

3.7 (a) $(1 - q)^n$ (b) $(1 - q)/q$

3.9 $e_0 = 79.83$ for original table. New values are 77.72, 78.41.

Chapter 4

4.1 3.64

4.2 3316.62

4.3 (a) 609.01 (b) 755.95

4.4 16.2

4.5 11 111

4.6 $y_{50} \circ 2(2) = 0.384$, $y_{52}(2) = 0.48$

4.7 $1/(1 - v(1 - q))$

4.9 19 900

4.10 3.29

4.11 153.85

4.13 947.83

4.14 274 754

4.15 (a) 20 (b) 13

4.16 $\ddot{a}_{\{40\}+20} = 11.54$, $\ddot{a}_{\{50\}+10} = 10.77$ which is lower since is covers more of the high interest years.

Chapter 5

5.1 0.0760

5.2 135.04

5.3 55.06

5.4 (0.05, 0.1, 0.15)

5.5 (a) $S = 1000\ddot{a}(0_{25}, 1_{\infty})/(1 - A_{40}(1_{25}))$ (b) 2222.22

5.6 9.57%

5.7 ${}_{n}q_x$

5.8 q/(i+q)

5.9 Lower

5.10 625

5.11 0.3015

5.12 0.4

5.16 (a)

$$P = \frac{1000[{}_{20}p_{40}\ddot{a}(0_{20}, 1_{10}; v) + \ddot{a}_{40}(0_{30}, 1_{\infty})]}{\ddot{a}_{40}(1_{10}) - A_{40}(\mathbf{j})}, \quad \mathbf{j} = (1, 2, 3, \dots, 9, \underbrace{10, 10, \dots, 10}_{11 \text{ times}})$$

(b)

$$P = \frac{1000[\ddot{a}(1_{10}; v) + \ddot{a}_{[40]+20}(0_{10}, 1_{\infty})]}{\ddot{a}(1_{10}; v)v(20, 0)}$$

5.19 Change the vector $\mathbf{j}$ so that $j_k = 0.5 \, \text{Val}_{k+1}(1_{k+1}; v)$ if $0 \leq k < 5$

5.20 194.08

5.23 1928.27

5.24 469.67

5.25 (a) 19.53 (b) 12.49

Chapter 6

6.1 444.58, 506.36, 985.84, 4000

6.2 467.74, 451.61, 865.59, 2000

6.3 (a) (1, 0.75, 0.51) (b) (−1005.79, −760.88, 853.19) (c) −1341.05, −853.19
(d) Benefits decrease

6.4 420, 580, 900

6.5 80

6.6 (a) $91.40 + 376.34, 222.88 + 240.86, 0 + 467.74$

(b) Interest gain = −45.97, mortality gain = 56.73

6.7 $39.49 + 405.09, 162.27 + 282.31, -125 + 1014.16$

6.8 (a) 160 (b) $106\frac{2}{3} - 66\frac{2}{3}$ (c) Interest gain = 12, mortality gain = 32

6.10 (a) 311.87 (b) 398.50 (c) 3200

6.11 650

6.13 (a) 3.4 (b) −0.048

6.14 (a) 10 (b) 2

6.17 (a) 200 (b) 40

6.18 (a) 125 (b) ${}_1V = 50,\ {}_kV = 150, k = 2, 3, \ldots$

6.19 $j > i$

6.20 2307.69

6.22 (a) Premium = 368.85. For example, ${}_{25}V = 14.473$.
(b) Decrease for for first 15 years. (c) Increase for first 15 years.

Chapter 7

7.1 (a) 153 002 (b) 152 113

7.2 (a) 3017 (b) 2665

7.3 (a) 96 670 (b) 96 071

7.4 1/3

7.5 0.56

7.6 0.144

7.7 2847

7.8 200

7.9 0.64

7.11 Overstate by 0.0582

7.12 103.19, 502.90

Chapter 8

8.1 0.96

8.2 232.37

8.3 679.50

8.4 (a) 0.2083 (b) 0.179 26

8.5 (a) 1.744 (b) 1.733

8.6 0.292 69

8.7 (a) 0.4541 (b) 0.093 02

8.8 0.777

8.9 (a) 50 (b) 312.50

8.10 (a) 0.488 (b) 0.421

8.11 0.3

8.14 $\pi = \mu(\mu + \delta)/(\mu + \delta - \gamma)$

8.15 0.81

8.16 $2/(140 - n)$

8.17

$$\bar{a}_x = \frac{1 - e^{-n(\mu_1+\delta)}}{\mu_1 + \delta} + \frac{e^{-n(\mu_1+\delta)}}{\mu_2 + \delta}, \qquad \bar{A}_x = \frac{\mu_1\left(1 - e^{-n(\mu_1+\delta)}\right)}{\mu_1 + \delta} + \frac{\mu_2 e^{-n(\mu_1+\delta)}}{\mu_2 + \delta}$$

8.23 20

8.24 (a) 0.00047877 (b) 0.003560, 0.003564

8.25 (a) 0.89944 (b) 0.90281 (c) 0.89673 (d) 0.89674

8.27 (a) $(3/4)(e^{0.05t} - 1)$

Chapter 9

9.1 0.274 68

9.2 (a) 137.27 (b) 51.63

9.3 The probability that a person age 40, first observed age 30, will die between the ages of 46 and 54

9.4 $\ell_{[60]} = 6313$, $\ell_{[61]} = 5553$, $\ell_{[62]} = 4717$

9.5 0.03163

9.6 (a) 2957.31 (b) 35 664.74

9.7 (a) 23.108 (b) 23.089

Chapter 10

10.1 (a) 0.4, 0.1 (b) 0.9, 0.55 (c) 1480 (d) 864 (e) 2392 (f) 528

10.2

	both	at least one	exactly one	neither
live 5 years	0.48	0.92	0.44	0.08
die within 5 years	0.08	0.52	0.44	0.48
die between time 5 and 6	0.02	0.28	0.26	0.72

10.3 403.70

10.4 9/110

10.5 (a) 0.3352 (b) 0.1648 (c) 1.4

10.6 (a) 9/32 (b) 3/32

10.7 0.1757

10.8 0.7294

10.9 1.2

10.10 (a) 0.1547 (b) 0.0859.

10.11 (a) 409.72 (b) 647.87 (c) 374.92 (d) 350.85

10.12 $\ddot{a}_{40}(1_{20}) + \ddot{a}_{50}(0_{10}, 1_{\infty}) - \ddot{a}_{40:50}(0_{10}, 1_{10})$

10.13 $\ddot{a}_{40}(1_{20}) + \ddot{a}_{50}(0_{10}, 1_{20}) - \ddot{a}_{40:50}(0_{10}, 1_{10})$

10.14

$$\frac{3\bar{A}_x + 3\bar{A}_y - \bar{A}_{xy}}{(2/3)\ddot{a}_x + (2/3)\ddot{a}_y - (1/3)\ddot{a}_{xy}}$$

10.15 $\ddot{a}_x(12_{10}, 6_{10}) + \ddot{a}_y(12_{10}, 8_{10}) - \ddot{a}_{xy}(12_{10}, 2_{10})$

10.16 $4\ddot{a}_{40}(1_{30}) + 5\ddot{a}_{50}(1_{20}) + \ddot{a}_{40:50}(-3_{20}\ 2_{10})$

10.17 For all values of t, ${}_tV = 0$ if both are alive, 0.1765 if (x) only is alive, 0.8824 if y only is alive.

10.19 0.6720

10.20 Either $P(D) \leq P(B) \leq P(A) \leq P(C)$ or $P(D) \leq P(A) \leq P(B) \leq P(C)$

10.21 1.0041

10.22 (a) ${}_nq^1_{xy} + {}_nq^1_{xz} - 2{}_nq^1_{xyz}$

(b) ${}_nq_x -{}_n q^1_{xy} -{}_n q^1_{xy} +{}_n q^1_{xyz}$

10.23 (a) $\bar{A}^2_{yx}(\mathbf{b}) = \bar{A}_y(\mathbf{b}) + \bar{A}^1_{xy}(\mathbf{b}) - \bar{A}_{xy}(\mathbf{b})$ (b) $\bar{A}^2_{yx}(\mathbf{b}) = \bar{A}_{\overline{xy}} - \bar{A}^2_{xy}(\mathbf{b})$

10.24 0.14.

10.25 $\bar{A}_y(1_n) + v^n{}_np_yA^1_{y+n:x}$

10.26 $\bar{a}(1_n) + v^n{}_np_x\bar{a}_{x+n:y} + v^n{}_np_y\bar{a}_{y+n:x} - v^n{}_np_{xy}\bar{a}_{x+n:y+n}$

10.27 (a) $-1/\gamma$ (b) $(-1/\gamma) - 1/(\mu - \gamma)$

10.29 20.50

10.30 $v^n{}_np_x\bar{a}_{y+n} - \bar{a}_{y+n:x}$

10.31 (a) 211.44 (b) 158.20 (c) 0.749

Chapter 11

11.1 29/110

11.2 (a) 0.6852 (b) 0.3245, 0.4445

11.3 1/30

11.4 Method 1 gives 0.219 and 0.438. Method 2 gives 0.2235 and 0.4335.

11.5 0.04595

11.6 0.0552, 0.0869, 0.1074

11.7 Method 1: 0.0786, 0.1666, 0.2148 Method 2: 0.0792, 0.1667, 0.2142

11.9 5/9

11.10 (a) 0.162 67, 0.325 33 (b) 0.164, 0.324 (c) 0.182, 0.306

11.11 Letting a denote $q_x^{(1)}$ and b denote $a^2b^2)/4$; for (11.18), $(a^2b + b^2a)/(4 - 2a - 2b + ab)$.

11.12 False. The left hand side equals $\ell_x^{(j)}/\ell_x^{(\tau)}$.

11.13 0.384

11.14 0.14, 0.18

11.16 0.063 07, 0.132 07, 0.208 40, 0.294 07

11.17 $q_x^{(d)} = 0.111$, $q_x^{(w)} = 0.185$

11.18 Using the notation of Example 11.5

$$a = a' - \frac{a'b'}{2} - \frac{14}{30}a'c' + \frac{7}{30}a'b'c'$$
$$b = b' - \frac{a'b'}{2} - \frac{b'c'}{2} + \frac{3}{8}a'b'c'$$
$$c = c' - \frac{16}{30}a'c' - \frac{b'c'}{2} + \frac{47}{120}a'b'c'$$

Chapter 12

12.1

$$\frac{G = 100\,550\,\bar{A}_x(1_{20}) + \ddot{a}_x(1020, 260_{19})}{\ddot{a}_x(0.4, 0.95_9)}$$

12.2 30

12.3 1175.86

12.4 97.46, 24.14, 260.77

12.5

$$\frac{A_{x+2}(\mathbf{b} \circ 2)}{\ddot{a}_{x+2}(1_{n-2})}$$

12.8 (77.65, 4.76, 175.56, −31,20, 0.25, −0.16)

12.9 (a) (−23925.26, −1948,40, 2698.97, 2527.36, 2945.05), 0.0103 (b) 0.0141 (c) 0.0141

12.10 22085

Chapter 13

13.1 29340, 9.85

13.2 (a) −111.08, (b) 612.74

13.4 51285, 1022.36

13.5 New coefficients are (69051, 76751, 85128)

13.7 0.88

13.8 1.085

Chapter 14

14.1 $f(1) = 1/4, f(2) = 3/8, f(3) = 9/32, f(4) = 3/32$. Time 2 is the most likely cause of failure.

14.2 $f(k) = 2k3^k/(k+3)!$

14.3 e^{-1}

14.4 (a) 1.27 (b) 0.8357

14.5 $133\frac{1}{3}$

14.6 Machine 1, which has an expected output of 400 000 copies, as opposed to 360 000 for Machine 2.

14.7 (a) 0.1215 (b) An overstatement of 0.0008

14.8 (b) $\mu_T(t) = \beta^2 t/(1 + \beta t)$ (c) This is a mixture of an exponential and a gamma.

14.9 (a) $0.003(10 - t)^2,\ 0 \leq t \leq 10$ (b) $t/25 - 3t^2/1000,\ 0 \leq t \leq 10$

14.13 (a) Yes (b) No

Chapter 15

15.1 1.1070, 0.0041

15.2 (a) 1.86, 0.7204 (b) 0.70

15.3 $5, 8\frac{1}{3}$

15.4 7.5, 3.75

15.5 16.67, 222.22

15.6 The exact distribution of ${}_1L$ is given by

k	$P(\tilde{T} = k)$	L
1	0.20	−100
2	0.20	−180
2	0.18	−244
≥3	0.42	268

Therefore ${}_1V = 12.64$.

15.7 (a) −13.76, 1840.34, 1344 (b) 0.49

15.8 $L = 50$ with probability 0.2, 10 with probability 0.32, and −2.8 with probability 0.48. So ${}_0V = 11.856$.

15.9 (a) 271 (b) 0.933

15.10 0.45, 0.675

15.11 0.0248

15.12 (a) 17, 28.5 (b)14.78, 0.30 (c) $8\frac{1}{3}$

15.13 (a) $L = 160$ with probability 0.2; 40 with probability 0.32; −20 with probability 0.24; −45 with probability 0.24. ${}_1L = 160$ with probability 0.4; 40 with probability 0.3; −10 with probability 0.3. So ${}_1V = 73$.

(b) $P = 66\frac{2}{3}$

15.15 (a) 483.39 (b) 458.02

15.16 (a) 13 (b) 0.07

15.17 $E(Z) = 54.45$ $\text{Var}(Z) = 341.80$

15.18 $\mu/(\mu + \delta - \gamma)$, $\mu/(\mu + 2\delta - 2\gamma) - (\mu/(\mu + \delta - \gamma))^2$

15.19 (a) $900 + 25c^2$ (b) $900 - 300c + 25c^2$, which equals 0 for $c = 6$

15.20 (a) $\bar{Y} = T$ for $0 \leq T \leq 6$ and $\bar{Y} = 6$ when $T \geq 6$ (b) 3.5, 4.25 (c) 0.5775

15.21 676

15.23 (a) $f_{\overline{xy}}(t) = t$ when $0 \leq t < 1$ and $1/2$ when $1 < t \leq 2$ (b) 2.075, 4.891

15.24 26 870.

Chapter 16

16.1 0.196

16.2 0.32

16.3 20

16.4 0.04902

16.5 (a) $\beta^2/(\beta+\delta)^2$ (b) $(2\beta+\delta)/(\beta+\delta)^2$ (c) $\beta^2/(2\beta+\delta)$
(d) $\beta^2 k/(1+\beta k)(2\beta+\delta)$

16.6 0.24

16.7 (a) 0.5931 (b) 0

16.8 (a) 0.6376 (b) 0.8187

16.9 $F_{\bar{Z}}(z) = 0.76$ for $0 \le z \le 0.1296$, $0.4 + z^{1/2}$ for $0.1296 \le z \le 0.36$, 1 for $0.36 \le z$.

16.11 (a) 0.25 (b) 0.882 (c) 1

16.12 (a) 0.8521 (b) 0.2219

16.13 (b) Not true. Subtract ${}^2\ddot{a}_{\bar{T}}$ from right hand side.

16.14 Change both occurrences of F_T to $\hat{F}_T$.

Chapter 17

17.1 $F(t,1) = F(t,2) = \frac{3}{2}t^2 - t^3, 0 \le t \le 1$, and $\frac{1}{2}$ when $t > 1$.

17.2 $F(t,1) = \frac{3}{2}t^2 - t^3, 0 \le t \le 1$ and $\frac{1}{2}$ when $t > 1$.

17.3 (a) $F(t,1) = \frac{2}{5}(1 - e^{-5t})$, $F(t,2) = \frac{3}{5}(1 - e^{-5t})$ (b) $f_J(1) = 2/5$, $f_J(2) = 3/5$.

17.4 For $0 \le t \le a$, $F(t,2) = (1 - e^{-\mu t})/\mu a$. For $t > a$, $F(t,2) = P(J=2) = (1 - e^{-\mu a})/\mu a$.

17.5 15/16

17.6 (a) 21/40 (b) 4/10

17.7 (a) For $0 \le t < 1$, $F(t,1) = \frac{1}{3}[1 - (1-t)^4]$, $F(t,2) = \frac{2}{3}[1 - (1-t)^3]$.
For $t > 1$, $F(t,1) = 1/3$, $F(t,2) = 2/3$.

(b) $f_J(1) = 1/3, f_J(2) = 2/3$

(c) $\mu(t,1) = 4/(3-t)$, $\mu(t,2) = 6/(1-t)(3-t)$

17.8 (a) For $0 \le t < 1$, $F(t,1) = (8t - 2t^4)/9$, $F(t,2) = (2t^2 - t^4)/3$. For $t > 1, F(t,1) = 2/3$, $F(t,2) = 1/3$.

(b) $f_J(1) = 2/3, f_J(2) = 1/3$

(c) $\mu_1(t) = (8(1-t^3)/(9 - 8t - 6t^2 + 5t^4)$ $\mu_2(t) = 12(t - t^3)/(9 - 8t - 6t^2 + 5t^4)$

17.9 $\mu(t,j) = 2/(1-t)$, $\mu_j(t) = (2 - 6t + 6t^2)/(1 - 2t + 3t^2 - 2t^3)$, $0 \le t < 1$

17.10 $\mu(t,j) = (3t - 3t^2)/(1 - 3t^2 + 2t^3)$, $\mu_j(t) = (6t - 3t^2)/(2 - 3t^2 + t^3)$

17.11 (a) $\alpha/(\alpha+\beta)[1-(1-t)^{\alpha+\beta}]$, $0 \le t \le 1$ (b) $\beta/(\alpha+\beta)$ (c) $\alpha/(\alpha+\beta+1)$

17.12 (a) $0.2884 + 0.0473 + 0.1119 = 0.4476$ (b) 0.2346

17.13 (a) 0.7520 (b) 0.1654 (c) 0.0827

17.14 In all cases $f(0,1) = f(1,0) = 1/2 - f(0,0), f(1,1) = f(0,0)$.
(a) $f(0,0) = 1/2$ (b) $f(0,0) = 0$ (c) $f(0,0) = 1/4$ (d) $f(0,0) = 0.2497, 0.0139, 0.4861$

Chapter 18

18.1 100. None apply.

18.2 For $i \neq 0, r$, $p_{i:i-1} = i^2/r^2$, $p_{ii} = 2i(r-i))/r^2$, $p_{i:i+1} = (r-1)^2/r^2$, $p(0,1) = p(r, r-1) = 1$

18.3 (a) $\begin{pmatrix} 1/2 & 1/2 & 0 \\ 1/3 & 1/3 & 1/3 \\ 0 & 1/2 & 1/2 \end{pmatrix}$ (b) 55/216 (c) (2/7, 3/7, 2/7)

18.4 0,1,3,4, are recurrent, 2 is transient.

18.5 All are recurrent.

18.6 Only 3 is recurrent.

18.7 $q/(p+q)$, $p/(p+q)$

18.8 Limiting distribution is uniform for 5 chairs, does not exist for 4.

18.9 4/5, 6/5, 7/5

18.11 (a) 5 feet, 21/76 (b) 5 feet, 28/93

18.12 (a) $0.5e^{-1}$ (b) $1 - e^{-1/2}$ (c) 5/3, 5/9.

18.13 (a) $4.5e^{-3}$ (b) Exp(3), Gamma(n,3) (c) $54e^{-9}$.

18.14 (a) 0.012 74 (b) 0.264 24 (c) 9,7

18.15 (a) Increments not stationary (ii) If increments stationary, not independent
(c) Given the first number, the second must be $-\log(0.5)/2$

18.16 (a) $2.5e^{-1}$ (b) $1 - e^{-0.4}$ (c) 3.3, 3/4.

18.17 (a) $2e^{-2}$ (b) $2e^{-3}$ (c) 4.8 (d) 18/125 (e) $(7/2)e^{-5/2}$

18.18 7/3, 9

18.19 0.030 904

18.20 (a) 72 (b) 0.2875

18.21 0.434

18.22 0.121

18.23 0.363

Chapter 19

19.1 1.2808

19.2 (a) 82.69, (b) 2036.80

19.3 0.333, 0.249

19.4 (a) 45.24 (b) (i) 71.11 (ii) 14.39 (iii) 21.88

19.7 (a) 0.116, (b) 0.950, 0.1658

19.10 (a) 69.60 (b) 4.177

19.11 (c) Eigenvectors are $-(a+b), -a, -b$ with respective eigenvectors of $(1, 1, 1,), (0, 1, 0, (0, 0, 1))$.

19.13 $5/81, 20/91$

19.14 $9, 24, 94$

19.15 $-15420, 5350$

19.16 (a) Reserves for states 0, 1, 2 (resp.) are $\frac{7}{12}e^{0.05t} - \frac{7}{12}$, $\frac{2}{3}e^{0.05t} - \frac{35}{108}$, $\frac{3}{4}e^{0.05t} - \frac{28}{108}$

Chapter 20

20.1 $0.04 < r < 0.10$.

20.2 For example, take initial portfolio of (64.6, −0.7, 1).

20.3 (a) 12.54 (b) 11.45

20.5 Cost is 7.46 compared with 6.375 for European.

20.6 (a) Probabilities of (U, M, D) resp. are $(p, (2-5p)/3, (2p+1)/3), 0 < p < 4$.
(b) $62/3 < \pi < 12$, (c) For example take initial portfolio of $(-48, 0.6, -1)$.

20.7 (a) Arbitrage-free, not complete. (b) $L = \{f : 2f(U) - 3f(M) + f(D) = 0\}$,$L_0 = \{f : f(U) = -9a, f(M) = a, f(D) = 21a\}$ for some a.

20.8 $(d - (1 + r))/(u - (1 + r))$ in both cases.

20.9 Calls: (1.24, 4.75. 11.86), Puts: (9.61, 3.26, 0.52). Calls decrease with strike, increase with interest, duration and volatility. Puts increase with duration, strike, volatility, decrease with interest.

20.10 (a) 5 (b) 8.07 (c) 0.70

20.11 (a) (0.707, −0.643, (b) (0.803, −0.761), (c) (0.504, −0.463).

20.14 $v(0, 2) = 0.525, v(0, 3 = 0.42525, v(1, 3)(U) = 0.68, v(1, 3)(D) = 0.39$.

20.16 (a) 14 (b) 47.25

Chapter 21

21.1 56

21.2 0.1064

21.3 0.054

21.4 5.7251

21.5 120, 25 600

21.6 32/3, 224/3

21.7 (a) 0.102 (b) No

21.8 3/2, 9/4

21.9 (a) 0.1025 (b)11.75

21.10 (a) 2/3, 4/9 (b) e

21.11 0.753

21.12 (a) 0.1590 (b) 9/16

21.14 0.95

21.15 $0.75(X \wedge 3840) - 0.75(X \wedge 240)$

21.16 0.0996

21.17 4.0625

21.18 200

21.19 The deductible

21.20 1600.74

21.21 0.0022

21.22 180.60

21.23 (a) 106.53 (b) 62.5

21.24 4.44, 55.31

21.25 114.40

21.26 2400

21.27 (a) $1 - e^{-d\lambda}$ (b) $1 - [\theta/(\theta + d)]^{\alpha-1}$ (c) $1 - ((2 + \beta d)/2)e^{-\beta d}$

21.28 (b) 0.2769

21.29 $f_S(12) = \frac{13}{40}d + \frac{7}{30}c$

21.30 Negative binomial

21.31

x	0	1	2	3	4	5
Poisson	0.0111	0.0350	0.0701	0.1051	0.1301	0.1387
Negative binomial	0.0442	0.0696	0.0968	0.1082	0.1110	0.1050
Binomial	0.0020	0.0123	0.0397	0.0858	0.1378	0.1737

21.32 0.0693

21.33 Negbin(2, 0.5)

21.34 $N \sim$ Poisson(6). X takes values 2, 3, 5 with probabilities 1/3, 1/6, 1/2, respectively.

Chapter 22

22.1 (c) $P_{0.02} = 24.70$, $P_{0.01} = 15.86$, $\lim_{\alpha\to\infty} P_\alpha = 10$

22.3 Y is less risky than X

22.4 Y and Z are both less risky than X, but Y and Z are incomparable.

22.6 (a) $H(X) = 3$, $H(Y) = 2.5$

22.7 TVaR(X) = 4 14/49.

22.8 (a) 1, 4/3 (b) 7/4, 11/6

22.9 $N\alpha$, $N(1+\alpha)/2$

22.10 3, 5 1/7

22.11 117.86

22.12 (a) 3.07 3.07, 5.73

22.14 (a) The mean

Chapter 23

23.1 (a) 0.18 (b) $1.6a - 0.4b - 0.2c$

23.2 (a) 10 (b) 11 (c) 1024

23.3 0.4825

23.5 (c) $(2/3)^4$ (d) $(2/3)^3$

23.10 $\psi_1(1) = 7/36$, $\psi_2(1) = 1/36$

23.12 (a) $R > 0.23026$ (b) 45 (c) 3/4

23.13 0.1535

23.14 (a) 0.4160 (b) 3/16

23.15 Strictly greater than, provided $\theta > 0$.

23.17 (a) $u \geq 1.4979$ (b) $7/8$ (c) 0.7239 (d) $3/4$

23.19 0.105

23.20 (a) $\frac{24}{35}e^{-u} + \frac{1}{35}e^{-6u}$

(b) $\frac{32}{45}e^{-u/3} + \frac{1}{45}e^{-4u/3}$

(c) $\frac{4}{9}e^{-2u} + \frac{1}{9}e^{-4u}$

23.21 $\frac{2}{5}e^{-3u} - \frac{1}{15}e^{-8u}$

Chapter 24

24.1 $17/6$

24.2 2.9

24.3 0.9747

24.4 12.89

24.5 (a) 2.61 (b) 3.53 (c) 3.53

24.6 1.5

24.7 $m\left(\frac{\alpha+n\bar{x}}{\alpha+\beta+nm}\right)$

24.8 $r\left(\frac{\alpha+n\bar{x}}{\beta+nr-1}\right)$

24.9 3.107

24.10 1.1929

24.11 3

24.12 (a) 5.75

(b) $\frac{\int_1^\infty \theta^{17}e^{-2\theta}}{\int_1^\infty \theta^{16}e^{-2\theta}}$

24.13 $3/2$

24.14 4391

24.15 1.06

24.16 (a) (4.95, 6.24, 5.81)

(b) (4.08, 6.61, 5.77, 1.54)

24.18 Switch

Appendix

A review of probability theory

This appendix provides a review of the basic probability theory used in this book. In addition, more specialized topics in probability theory, will appear in the relevant chapters as they are needed. It is expected that most readers will be at least somewhat familiar with this material, so the pace is fairly rapid, with few examples or derivations.

A.1 Sample spaces and probability measures

We model the results of a random experiment by a set Ω, known as a *sample space*, where each point of Ω corresponds to a possible outcome. For example, if we throw a pair of dice, the sample space could consist of the 36 ordered pairs (a, b) where a and b take values from 1 to 6. A combination of outcomes, known as an *event*, is represented by a subset of Ω. Such an event occurs if any of the outcomes in the subset occur. In the above example, the event that the total on the dice is 10, would be represented by the set $\{(4, 6), (5, 5), (6, 4)\}$. From a given collection of events, familiar set operations can be applied to build other events. The union of events, denoted with the symbol $\cup$, gives us the event that occurs if any one of the given events occur. The intersection of events, denoted with the symbol $\cap$, gives us the event that occurs if all of the given events occurs. The complement of an event A, denoted by A^c, gives us the event that occurs if A does *not* occur. Two events are said to be *mutually exclusive* if the occurrence of one means that the other cannot occur. These are represented by subsets A and B that are *disjoint*, that is, $A \cap B = \emptyset$ (the empty set).

To each event A we assign a number $P(A)$ in the interval [0,1], known as the *probability of* A. This measures how likely it is for the event A to occur in a single trial of the experiment. The assignment $P(A)$ must satisfy the following fundamental rule: Given a finite or countably

Fundamentals of Actuarial Mathematics, Third Edition. S. David Promislow.

Companion website: http://www.wiley.com/go/actuarial

infinite sequence $A_1, A_2, \ldots, A_n, \ldots$ of pairwise disjoint events – that is, $A_i \cap A_j = \emptyset$ for $i \neq j$ – then

$$P\left[\bigcup_i A_i\right] = \sum_i P(A_i). \tag{A.1}$$

Moreover, we require that $P(\Omega) = 1$.

These requirements can be motivated by adopting the relative frequency interpretation of P, whereby $P(A)$ gives the expected proportion of times that the event A will occur if the experiment is repeated a sufficiently large number of times.

We often refer to P as a *probability measure* on Ω.

The following are some simple, frequently used consequences of (A.1),

$$P(\emptyset) = 0, \quad A \subseteq B \text{ implies } P(A) \leq P(B), \qquad P(A^c) = 1 - P(A), \tag{A.2}$$

and

$$P(A \cup B) = P(A) + P(B) - P(A \cap B). \tag{A.3}$$

We now discuss a technical difficulty that arises when the sample space is not finite or countably infinite. In this case, postulating that $P(A)$ should be defined for all subsets A is overly restrictive, and would prevent one from finding suitable probability measures in many cases. Accordingly we only require that $P(A)$ be defined when A belongs to a certain specified collection of subsets, which we denote by $\mathcal{S}$. We do, however, require certain restrictions on $\mathcal{S}$ to ensure that countable families of events can be combined in the familiar ways we described above. These requirements are as follows:

(a) Given any finite or countably infinite sequence $A_1, A_2, \ldots, A_n, \ldots$ of sets in $\mathcal{S}$,

  (i) their union is in $\mathcal{S}$,

  (ii) their intersection is in $\mathcal{S}$.

(b) For any $A \in \mathcal{S}$, the complement $A^c \in \mathcal{S}$.

(c) The whole set $\Omega \in \mathcal{S}$.

A collection of subsets with these properties is known as a σ-*field*. (Since (a)(ii) follows from (a)(i) and (b), while (a)(i) follows from (a)(ii) and (b), many authors omit one of (a)(i) or (a)(ii) in the definition of a σ-field.) As a consequence of (a)(i) we know that the left hand side of (A.1) makes sense.

To summarize, in modelling a random experiment, we choose a sample space Ω, a probability measure P and a σ-field $\mathcal{S}$. When Ω is finite or countably infinite, the collection $\mathcal{S}$ is invariably taken to be all subsets. In the remaining sections of this appendix, we assume a fixed Ω, P and $\mathcal{S}$.

A.2 Conditioning and independence

If we are told that a certain event B has occurred, we need to reassess probabilities. Assuming $P(B) > 0$, this is done by defining a new probability measure, denoted by $P(\cdot|B)$, as follows:

$$P(A|B) = \frac{P(A \cap B)}{P(B)}, \quad A \in \mathcal{S}.$$

We often refer to the left hand side as 'the probability of A *given* B'. This new measure assigns probability 0 to events disjoint from B, and for $A \subseteq B$ it assigns the probability $P(A)/P(B)$. In a sense, we can view this as changing the sample space from Ω to B, and then dividing by a constant to ensure that the probability of the entire space is 1.

The symbol $P(A|B_1, B_2, \ldots, B_n)$ denotes the probability of A given that all of the events B_i have occurred. That is, it equals $P(A|B)$ where $B = \cap_{i=1}^{n} B_i$.

We say that A and B are *independent* if

$$P(A|B) = P(A).$$

In other words, knowing that B occurred does not affect our original assessment of the likelihood of A. It is convenient to write this condition in the symmetric form

$$P(A \cap B) = P(A)P(B),$$

which shows immediately that we also have $P(B|A) = P(B)$. Moreover, this form makes sense if either set has probability 0, in which case we automatically have independence.

More generally, we define a collection of events (possibly infinite) to be independent if, given any finite subcollection $A_1, A_2, \ldots, A_n$,

$$P\left[\bigcap_{1=i}^{n} A_i\right] = P(A_1)P(A_2)\cdots P(A_n). \tag{A.4}$$

A.3 Random variables

Random variables are, intuitively, numerical quantities associated with a random experiment. For example, we toss 100 coins and count the number of heads. Formally, we represent the random variable by a real-valued function X defined on Ω, where for $\omega \in \Omega$, $X(\omega)$ is the number obtained when the outcome of the experiment is ω. The technical restriction we referred to above when Ω is not finite or countably infinite also puts restrictions on the possible functions we consider. Given a random variable, traditionally denoted by a capital letter like X, and a subset A of real numbers, we would like to find the probability that X takes a value in the set A. This is not possible in our model if the event $\{\omega \in \Omega : X(\omega) \in A\}$ is not in the collection of events $\mathcal{S}$ that we are able to compute probabilities for. Accordingly, in the definition of a random variable we postulate that at least for A of the form $(-\infty, r]$, the corresponding event is in $\mathcal{S}$, so we can always compute the probability that X is less than or equal to r. From the properties that we imposed on $\mathcal{S}$ it then follows that the event $\{X \in A\}$ is in $\mathcal{S}$ for a large collection of subsets A of real numbers, known as the *Borel* sets. This class contains all intervals, and, roughly speaking, all sets that can be formed from intervals, by repeatedly taking unions and intersections over countable index sets.

A.4 Distributions

In many cases, given a random variable, we do not need the details of Ω or the particular function X, but wish only to know the *distribution* of X, which is basically a description of how likely it is that X will take on certain values. We will concentrate on two main types of random variables. There are *discrete* random variables, which in almost all applications in this book will take nonnegative integers as values. Second, there are *continuous* random variables, which take values that vary continuously over some interval of the real line. In almost all of our applications, this interval will be $[0, \infty)$ or $[0, N]$ for some finite N. We will therefore simplify the following discussion by assuming (unless otherwise mentioned) that all our random variables take nonnegative values.

We will occasionally encounter *mixed* random variables that have both a discrete and continuous part. They will be dealt with in turn as they arise.

For a discrete random variable X there are two main functions for describing the distribution. First, the *probability (mass) function* f_X is a function defined on the nonnegative integers by

$$f_X(k) = P(X = k), \quad k = 0, 1, 2, \ldots .$$

Second, the *(cumulative) distribution function* (c.d.f.) F is defined on $[0, \infty)$ by

$$F_X(x) = P(X \le x).$$

The two functions are related by

$$f_X(k) = F_X(k) - F_X(k-1), \qquad \text{for all integers } k \le 0.$$

$$F_X(x) = \sum_{i=0}^{k} f_X(i),$$

where k is the greatest integer $\le x$.

We can define the distribution function F_X exactly as above for any random variable. The precise definition of a continuous random variable is one for which the function F is continuous. In this case we get no information from looking at the probability that X will take on a specific value, as this will always be zero: ($P(X = r) \le P(X \in (r - h, r]) = F_X(r) - F_X(r - h)$, for all $h > 0$, and this goes to 0 as $h \to 0$). In place of the probability function, we define a *probability density function* (p.d.f.) f_X as the function satisfying

$$P(a < X \le b) = \int_a^b f_X(x)\mathrm{d}x. \tag{A.5}$$

The two functions are related in the continuous case by

$$f_X(x) = F'_X(x), \qquad F_X(t) = \int_0^x f_X(y)\mathrm{d}y. \tag{A.6}$$

(Primes denote differentiation when clear from the context.)

The lower limit of 0 on the integral is a consequence of our assumption of nonnegative values the general case we would need a lower limit of $-\infty$.

When there is no confusion, we will sometimes omit subscripts and simply write f or F.

The reader is cautioned that values of the density function are *not* probabilities and can take values greater than 1. We can interpret these probabilistically by the intuitive statement that for a 'small' value of Δx, the probability that X will take a value between x and $x + \Delta x$ is 'approximately' $f(x)\Delta x$. This can verified from (A.5).

Density functions need not exist for a given continuous distribution, but throughout the book it will be assumed that they do exist, unless otherwise indicated.

A.5 Expectations and moments

The *expectation* (also known as the *mean*) of a random variable X, represents in some sense the *average value* that X will take. It is given by

$$E(X) = \sum_{k=1}^{\infty} kf(k) \quad \text{or} \quad \int_0^{\infty} xf(x)\mathrm{d}x,$$

depending on whether X is discrete or continuous.

Of course the above series or integral may diverge, in which case we have $E(X) = \infty$. (In the general case where X is not necessarily nonnegative-valued, the expectation may not exist at all.)

If g is a function defined on a set that includes the range of X, we can define another random variable $g(X)$ that takes the value $g(x)$ when X takes the value x. We refer to such a random variable as a *function of* X. It can be shown that the expectation of a function of X is given by

$$E[g(X)] = \sum_{k=0}^{\infty} g(k)f(k) \quad \text{or} \quad \int_0^{\infty} g(x)f(x)\mathrm{d}x, \tag{A.7}$$

depending on whether X is discrete or continuous.

In particular, taking $g(x) = cx$ for some constant c leads to the formula

$$E(cX) = cE(X). \tag{A.8}$$

Of particular importance are the functions $g(x) = x^n$. For such a function $E[g(X)]$ is known as the nth *moment* of X.

We define the *variance* of X by

$$\mathrm{Var}(X) = E[X - E(X)]^2 = E(X^2) - E(X)^2. \tag{A.9}$$

It is clear from (A.8) that for any constant c

$$\mathrm{Var}(cX) = c^2\mathrm{Var}(X). \tag{A.10}$$

The positive square root of Var(X) is known as the *standard deviation* of X.

The smaller the variance, the more likely it is that values of X are close to the mean. A formal statement along these lines is given by *Chebyshev's inequality*, which states that for $k > 0$,

$$P(|X - E(X)|) \geq k \leq \frac{\mathrm{Var}(X)}{k^2}. \tag{A.11}$$

Up to this point, we have talked first about events and probabilities, and second about random variables and expectations. It is interesting to note that, if we wish, we can subsume events and probabilities under the latter category, and speak only of random variables and expectations. For any event A we define a random variable I_A that takes the value 1 on points in A and the value 0 on points not in A. These are known as *indicator random variables*. In this way the event A can be viewed as a random variable and $P(A)$ is equal to $E[I_A]$, as can be easily verified.

A.6 Expectation in terms of the distribution function

It is often desirable to express expectations in terms of the distribution function rather than the density or probability function. In place of F, it is more convenient to work with the function

$$s(t) = 1 - F(t) = P(T > t).$$

For a simple example, consider a discrete random variable X:

$$\begin{aligned} E(X) &= f(1) + 2f(2) + 3f(3) + \cdots \\ &= f(1) + f(2) + f(3) + \cdots \\ &\quad\quad\quad\; + f(2) + f(3) + \cdots \\ &\quad\quad\quad\quad\quad\quad\;\; + f(3) + \cdots \end{aligned}$$

Summing by rows, we get

$$E(X) = \sum_{k=0}^{\infty} s(k). \tag{A.12}$$

In the continuous case we can similarly derive

$$E(X) = \int_0^{\infty} s(t)\mathrm{d}t. \tag{A.13}$$

We will in fact develop a more general formula. Suppose g is a differentiable function such that $E[g(X)]$ exists and is finite. Noting that $f(t) = -s'(t)$, we integrate by parts to obtain

$$\int_0^N g(t)f(t)\mathrm{d}t = -g(N)s(N) + g(0) + \int_0^N g'(t)s(t)\mathrm{d}t.$$

In the case where

$$g(N)s(N) \to 0, \quad \text{as } N \to \infty \tag{A.14}$$

we can conclude that

$$E[g(X)] = g(0) + \int_0^\infty g'(t)s(t)\mathrm{d}t. \tag{A.15}$$

A similar derivation in the discrete case shows that, for any function g satisfying (A.14) for integer values of N,

$$E[g(X)] = g(0) + \sum_{k=0}^{\infty}[g(k+1) - g(k)]s(k). \tag{A.16}$$

Condition (A.14) is automatically satisfied if g is bounded, since $s(N)$ tends to 0 as N tends to ∞, or if X is bounded, since then $s(N) = 0$ for sufficiently high N. It is also satisfied for *monotone* g. Suppose, for example, g is increasing and nonnegative. Then (in the continuous case),

$$g(N)s(N) = g(N)\int_N^\infty f(t)\mathrm{d}t \le \int_N^\infty g(t)f(t)\mathrm{d}t,$$

and the last term must approach 0, by the hypothesis that $E[g(X)] = \int_0^\infty g(t)f(t)\mathrm{d}t$ is finite.

A.7 Joint distributions

Suppose we have two random variables X and Y defined on the same sample space. For many applications, we are interested not only in the distribution of each random variable, but also in the *joint distribution,* which gives information on how the two are related. When both are discrete, we can describe this by a *joint probability function*. This is a two-variable function $f_{X,Y}$ defined by

$$f_{X,Y}(k,n) = P(X = k \text{ and } Y = n).$$

(As in the single-variable case, we sometimes omit the subscripts on f when no confusion results.) When X and Y are both continuous we define a *joint density function* $f_{X,Y}$. This is a two-variable function, such that for suitable regions A in the plane, the probability that the point (X, Y) lies in A is given by

$$\int\int_A f_{X,Y}(x, y)\mathrm{d}x\mathrm{d}y. \tag{A.17}$$

The probability or density functions associated with the component random variables are obtained from the corresponding joint function as the so-called *marginal* distributions, namely

$$f_X(k) = \sum_{n=0}^{\infty} f_{X,Y}(k, n) \quad \text{or} \quad f_X(x) = \int_0^{\infty} f_{X,Y}(x, y)\mathrm{d}y, \tag{A.18}$$

according as the random variables are both discrete or both continuous.

As in the single random variable case, we assume that joint density functions exist unless otherwise indicated, but readers should be cautioned that for two continuous distributions, X, and Y, a joint density function need not exist even when both f_X and f_Y do. For example, if $X = Y$, then (X, Y) lies in the region $A = \{(x, y) : x = y\}$ with probability 1, but the double integral of any two-variable function over A will equal 0.

For a function g of two variables defined on a set that contains all the points $\big(X(\omega), Y(\omega)\big)$ where ω is in the sample space Ω, it can be shown, analogously to (A.7), that

$$E[g(X, Y)] = \sum_{k=0}^{\infty}\sum_{n=0}^{\infty} g(k, n)f_{X,Y}(k, n) \tag{A.19}$$

in the discrete case, or

$$E[g(X, Y)] = \int_0^{\infty}\int_0^{\infty} g(x, y)f_{X,Y}(x, y)\mathrm{d}x\mathrm{d}y \tag{A.20}$$

in the continuous case. An important example is the function $g(xy) = xy$, which is used to define the *covariance* of X and Y given by

$$\mathrm{Cov}(X, Y) = E[(X - E(X))(Y - E(Y))] = E(XY) - E(X)E(Y). \tag{A.21}$$

When $\mathrm{Cov}(X, Y) > 0$, it means that the two random variables move in the same direction. High (low) values of one will tend to imply high (low) values of the other. Cov $(X, Y) < 0$ means the two random variables move in opposite directions. When $\mathrm{Cov}(X, Y) = 0$, we say that X and Y are *uncorrelated.*

For another frequently used example, take $g(x, y) = x + y$. Then

$$E(X + Y) = \int_0^{\infty}\int_0^{\infty} (x + y)f_{X,Y}(x, y)\mathrm{d}x\mathrm{d}y = \int_0^{\infty}\int_0^{\infty} xf_{X,Y}(x, y)\mathrm{d}x\mathrm{d}y$$
$$+ \int_0^{\infty}\int_0^{\infty} yf_{X,Y}(x, y)\mathrm{d}y\mathrm{d}x.$$

(Note the interchange of variables in the second integral.) Now using (A.20) we derive the often used result that

$$E(X+Y)=E(X)+E(Y). \tag{A.22}$$

We need not confine ourselves to just two random variables, and may want to consider functions of n random variables defined on the same sample space, where n is any positive integer. In the case of a sum, formula (A.22) extends by induction to

$$E\left[\sum_{i=1}^{n} X_i\right]=\sum_{i=1}^{n} E(X_i). \tag{A.23}$$

Another key formula involving a sum of random variables is

$$\text{Var}\left[\sum_{i=1}^{n} X_i\right]=\sum_{i=1}^{n} \text{Var}(X_i)+2\sum_{i<j} \text{Cov}(X_i,X_j). \tag{A.24}$$

A.8 Conditioning and independence for random variables

The notions of independence and conditioning can be extended from events to random variables. We say that two random variables X and Y are *independent* if their associated events are independent. That is, for any two Borel sets A and B,

$$P(X\in A \text{ and } Y\in B)=P(X\in A)P(Y\in B).$$

As in (A.4), we can extend this definition to any collection of random variables by requiring that, given any finite number of random variables $X_1, X_2, \ldots, X_n$ from this collection and any Borel sets $A_1, A_2, \ldots, A_n$, we have

$$P(X_1\in A_1, X_2\in A_2, \ldots, X_n\in A_n)=P(X_1\in A_1)P(X_2\in A_2)\cdots P(X_n\in A_n).$$

Equivalent formulations can be made in terms of the joint probability or density, or distribution functions. For example, in the bivariate case, the random variables X and Y are independent if and only if for all points (x,y),

$$f_{X,Y}(x,y)=f_X(x)f_Y(y), \tag{A.25}$$

or

$$F_{X,Y}(x,y)=F_X(x)F_Y(y). \tag{A.26}$$

This leads immediately to the fact that

$$E(XY)=E(X)E(Y) \quad \text{if } X \text{ and } Y \text{are independent.} \tag{A.27}$$

As we would expect, two independent random variables are uncorrelated. However, X and Y can be uncorrelated without being independent (see Section 15.7).

If X is a random variable and B an event with $P(B) > 0$ we can speak of the random variable $X|B$. This is just the restriction of X to the sample space B. Then

$$E(X|B) = \text{ the expectation of } X|B \tag{A.28}$$

with respect to the probability measure $P(\cdot|B)$.

Conditioning is often a useful tool in computing an expectation, since we can break up the calculation into various cases. Formally we have what is known as the *law of total expectation*, which states that given a partition of Ω, that is, a collection of pairwise disjoint sets $B_1, B_2, \ldots, B_n$ with union Ω,

$$E(X) = \sum_{i=1}^{n} E(X|B_i)P(B_i). \tag{A.29}$$

Taking X to be I_A for an event A gives the *law of total probability*

$$P(A) = \sum_{i=1}^{n} P(A|B_i)P(B_i). \tag{A.30}$$

A.9 Moment generating functions

Given any random variable X, the *moment generating function* (m.g.f.) is a function M_X of a real variable t defined by

$$M_X(t) = E[\mathrm{e}^{tX}], \tag{A.31}$$

provided that $E[\mathrm{e}^{tX}]$ exists in some neighbourhood of 0. (There are distributions for which $E(\mathrm{e}^{tX}) = \infty$ for all positive values of t, in which case the m.g.f. will not exist in any neighbourhood of 0.)

Recalling the series expansion

$$\mathrm{e}^x = 1 + x + \frac{x^2}{2!} + \frac{x^3}{3!} + \cdots,$$

so that

$$\mathrm{e}^{tX} = 1 + tX + \frac{t^2X^2}{2!} + \frac{t^3X^3}{3!} + \cdots,$$

we see that the m.g.f. has the series expansion

$$M_X(t) = 1 + tE(X) + \frac{t^2E(X^2)}{2!} + \frac{t^3E(X^3)}{3!} + \cdots,$$

from which we obtain

$$E(X^n) = \text{the } n\text{th derivative of } M_X(t) \text{ evaluated at } t = 0. \tag{A.32}$$

Some important facts about m.g.f.'s are as follows

Theorem A.1 *For X and Y independent*

$$M_{X+Y}(t) = M_X(t)M_Y(t).$$

Proof. Using A.26 we have

$$M_{X+Y}(t) = E(e^{t(X+Y)}) = E(e^{tX}e^{tY}) = E(e^{tX})E(e^{tY}) = M_X(t)M_Y(t),$$

where we invoke the fact that functions of independent random variables are themselves independent to justify the third equality.

The second theorem, which we state without proof, tells us that the m.g.f. determines the distribution when it exists. □

Theorem A.2 (The uniqueness theorem) *Suppose that, for two random variables X and Y, $M_X(t)$ and $M_Y(t)$ are equal in some neighbourhood of zero. Then $X \sim Y$.*

This theorem does not say that one can easily compute the distribution function or density function from the m.g.f. The idea is rather that if one recognizes an m.g.f. as being that of a certain known distribution, then the distribution in question must be that distribution. An example of its use will appear in Section A.11.2 below.

A.10 Probability generating functions

For discrete distributions that take nonnegative integer values, it is often more convenient to use another function closely related to the m.g.f. This is known as the *probability generating function* (p.g.f.) and is defined by

$$P_X(t) = \sum_{k=0}^{\infty} t^k f_X(k). \tag{A.33}$$

This should not be confused with the probability function as defined above.

Given the p.g.f., we can immediately find the complete distribution by

$$f_X(k) = \frac{P_X^{(k)}(0)}{k!},$$

where the superscript (k) denotes the kth derivative with respect to t.

The p.g.f. is related to the m.g.f. because we can write it as

$$P_X(t) = E(t^X) = E(e^{\log tX}) = M_X(\log t), \tag{A.34}$$

and it follows that

$$M_X(t) = P_X(e^t). \tag{A.35}$$

If we find the p.g.f. of a discrete distribution, (A.35) immediately gives us the m.g.f. as well.

The following technique will illustrate a good way to remember the p.g.f.'s of familiar distributions. Suppose we start with any function g whose power series expansion about zero has all nonnegative coefficients (i.e., all its derivatives are nonnegative at 0) and whose radius of convergence is greater than 1. Then

$$g(t) = a_0 + a_1 t + a_2 t^2 + \cdots,$$

where $a_n \geq 0$ for all n. Since

$$g(1) = a_0 + a_1 + \cdots,$$

such an expansion immediately gives rise to a probability distribution with probability function

$$f(k) = \frac{a_k}{g(1)}, \quad k = 0, 1, \ldots. \tag{A.36}$$

The resulting p.g.f. is immediate since, for any random variable X with this distribution,

$$P_X(t) = \frac{1}{g(1)} \sum_{k=0}^{\infty} a_k t^k = \frac{g(t)}{g(1)}. \tag{A.37}$$

Particular applications of this idea are found in Sections A.11.2 and A.11.3 below.

To calculate moments given the p.g.f., we could find the m.g.f. by (A.35) and then use (A.32), but we can also use (A.33) directly. Notice, for example, that

$$P'_X(1) = E(X), \tag{A.38}$$

and similarly

$$P''_X(1) = E[X(X-1)],$$

from which we can derive

$$\mathrm{Var}(X) = E[X(X-1)] + E(X) - E(X)^2 = P''_X(1) + P'_X(1) - P'_X(1)^2. \tag{A.39}$$

We close this section by noting that (A.34) and (A.35) easily imply that the analogue of Theorem A.1 holds for the p.g.f. That is, if Z is an independent sum of X and Y, then

$$P_Z(t) = P_X(t) P_Y(t). \tag{A.40}$$

A.11 Some standard distributions

We review here the main features of some standard distributions.

A.11.1 The binomial distribution

Suppose we take m independent trials of a random experiment where the outcomes are classified as either 'success' or 'failure', and the probability of a success is p. The number of successes is a random variable with probability function

$$p(k) = \binom{m}{k} p^k (1-p)^{m-k}, \quad k = 0, 1, \ldots, m.$$

This is known as a binomial random variable with parameters m and p, and we will denote it by $\text{Bin}(m, p)$. It can be generated, as in (A.36), by the function

$$g(t) = (1 - p + pt)^m$$

(a binomial expansion, which is the source of the name). It follow from (A.37) to (A.39) that

$$P_{\text{Bin}(m,p)}(t) = (1 - p + pt)^m, \quad E[\text{Bin}(m,p)] = mp, \quad \text{Var}[\text{Bin}(m,p)] = mp(1-p). \quad \text{(A.41)}$$

A.11.2 The Poisson distribution

The Poisson distribution arises as the limit of Binomials where the mean is held constant. Precisely, for any $\lambda > 0$ the $\text{Poisson}(\lambda)$ distribution is the limit of $\text{Bin}(m, \lambda/m)$ as m goes to ∞. It will have a probability function given by

$$\begin{aligned} p(k) &= \lim_{m\to\infty} \frac{m!}{k!(m-k)!} \left(\frac{\lambda}{m}\right)^k \left(1 - \frac{\lambda}{m}\right)^{m-k} \\ &= \frac{\lambda^k}{k!} \lim_{m\to\infty} \left[\left(1 - \frac{\lambda}{m}\right)^m \left(1 - \frac{\lambda}{m}\right)^{-k} \left(\frac{m(m-1)(m-2)\cdots(m-k+1)}{m^k}\right)\right]. \end{aligned}$$

The first factor in the square brackets approaches $e^{-\lambda}$, and the other two each approach 1, so we are left with

$$p(k) = e^{-\lambda} \frac{\lambda^k}{k!}, \quad k = 0, 1, 2 \ldots,$$

This is generated as in (A.36) from the function $g(t) = e^{-\lambda t}$ so we can deduce from (A.37) to (A.39) that

$$P_{\text{Poisson}(\lambda)}(t) = e^{\lambda(t-1)}, \quad E[\text{Poisson}(\lambda)] = \lambda, \quad \text{Var}[\text{Poisson}(\lambda)] = \lambda^2 + \lambda - \lambda^2 = \lambda. \quad \text{(A.42)}$$

The m.g.f. can be calculated directly from the definition, but it easier to derive it from (A.35).

$$M_{\text{Poisson}(\lambda)}(t) = P_{\text{Poisson}(\lambda)}(e^t) = e^{\lambda(e^t-1)}. \tag{A.43}$$

Here is an example to illustrate the use of m.g.f.'s in deducting distributions.

Example A.1 If X has a Poisson(λ) distribution and Y has a Poisson (μ), distribution. Find the distribution of an independent sum $Z = X + Y$.

Solution. Directly from (A.43) and Theorem A.1

$$M_Z(t) = M_X(t)M_Y(t) = e^{(\lambda+\mu)(e^t-1)}$$

and we conclude that Z has a Poisson($\lambda + \mu$) distribution.

This example illustrates a familiar phenomenon. For many common distributions, the sum of an independent collection is just another distribution of the same type but with a different parameter. In the Poisson case, since the parameter equals the mean, it is clear that the parameter for the sum of the random variables must be the sum of the parameters.

A.11.3 The negative binomial and geometric distributions

Let us again take repeated trials of an experiment where the probability of a 'success' is p. Now, however, instead of taking a fixed number of repetitions, we continue until we get a 'failure'. We then count the number of successes. This is a random variable with

$$p(k) = p^k(1-p), \quad k = 0, 1, \ldots .$$

It is known as the *geometric distribution* with parameter p (the name is chosen to reflect the fact the values of the probability function form a geometric progression) and will be denoted by Geom(p). If N is such a random variable,

$$P(N \geq m) = (1-p)\sum_{k=m}^{\infty} p^k = p^m.$$

It follows that, for positive integers r and t,

$$P(N \geq t + r | N \geq r) = \frac{p^{t+r}}{p^r} = p^t = P(N \geq t). \tag{A.44}$$

This says that at any point, the probability that you will continue for a further s repetitions is independent of the number of repetitions you have already taken, sometimes referred to as a *memoryless* feature of the distribution.

Suppose we repeat trials as above, except that instead of continuing until we get a single failure we continue until we get r failures. The probability function for the number of successes is now given by

$$p(k) = \binom{r+k-1}{k} p^k(1-p)^r$$
$$= \frac{r(r+1)\cdots(r+k-1)}{k!} p^k(1-p)^r, \quad k = 0, 1, \ldots. \tag{A.45}$$

It is clear from the second expression that r can be any positive number and not just an integer. We no longer get the interpretation of repeated trials as given but we still have a perfectly valid distribution. It is known as the *negative binomial* distribution with parameters r and p, and we denote it by Negbin (r, p). It is generated as in (A.36) from the function $g(t) = (1 - pt)^{-r}$, which is the source of the name of the distribution. From (A.37) to (A.39) we have that

$$P_{\text{Negbin}(r,p)}(t) = \left(\frac{1-p}{1-pt}\right)^r, \quad E[\text{Negbin}(r,p)] = \frac{rp}{1-p}, \quad \text{Var}[\text{Negbin}(r,p)] = \frac{rp}{(1-p)^2}. \tag{A.46}$$

The above formulas give the same quantities for the geometric distribution, simply by taking $r = 1$.

There is another parameterization of the negative binomial that is convenient for certain purposes. In place of p, we use the quantity $\alpha = p/(1 - p)$. Then $1/(1 - p) = 1 + \alpha$. In terms of this parameter, we can write

$$E[\text{ Negbin}(r,p)] = r\alpha, \qquad \text{Var}[\text{Negbin}(r,p)] = r\alpha(1+\alpha).$$

A.11.4 The continuous uniform distribution

Suppose we choose a point at random in an interval (a, b). If we decide that any interval of a given length h is equally likely regardless of where the interval starts, then we have a uniform distribution U. It has a continuous density function given by

$$f_U(t) = \frac{t}{b-a}, \quad a < t < b,$$

and direct integration shows that

$$E(U) = \frac{b-a}{2}, \qquad \text{Var}(U) = \frac{(b-a)^2}{12}. \tag{A.47}$$

A.11.5 The normal distribution

One of the most widely used distributions in probability and statistics is the normal. The *standard normal* distribution is that with density function given by

$$f(z) = \frac{1}{\sqrt{2\pi}} e^{-z^2/2}, \quad -\infty < x < \infty,$$

whose graph is the familiar bell-shaped curve. This is one instance where we have a random variable that takes negative values. For a standard normal random variable Z

$$E(Z) = 0, \qquad \mathrm{Var}(Z) = 1,$$

as shown by (A.32) and (A.49) below. The complete family of normal random variables consists of those of the form.

$$\mu + \sigma Z$$

where μ is any constant $\sigma > 0$ and Z has a standard normal distribution. Such a random variable will have mean μ and variance σ^2.

The importance of normal distributions comes from the famous *central limit theorem.* Suppose we have an independent sequence $X_1, X_2, \ldots$ of random variables, each with the same distribution. Then the averages $S_n = (X_1, X_2, \ldots, X_n)/n$ converges in some sense to a normal distribution. (We will not give precise details here.) As shown below in Section A.12 it can be difficult to compute the distribution of a sum of random variables. In many cases one invokes the central limit theorem to approximate the distribution of a sum by assuming it is normal, and then only the mean and variance need be known.

Let Φ denote the cumulative distribution function of the standard normal. The integration to compute this from the density function must be done by some numerical method. It used to be customary to publish tables of various values of Φ but now it can be done by various computer programs. In Excel®, the formula = NORMSDIST(t) returns the value $\Phi(t)$. For example =NORMSDIST (0.5) returns 0, which is obvious since the distribution is symmetric about the mean of 0. For another example, =NORMSDIST (1.645) returns 0.95, meaning that with probability 0.95 a standard normal takes a value less than or equal to 1.645.

In most applications we are interested in the inverse calculation. Given α we we want to find the α-quantile of the distribution, that is the point t so that the probability of being less than t is α. This is $\Phi^{-1}(\alpha)$, which in Excel® is denoted by NORMSINV(α). For example, taking $\alpha = 0.95$, the formula = NORMSINV(0.95) returns 1.645, as we have verified.

For the case of a general normal $X = \mu + \sigma(Z)$ we have

$$X \leq x \quad \text{if and only if} \quad \mu + \sigma(Z) \leq x \quad \text{if and only if} \quad Z \leq (x - \mu)/\sigma,$$

from which it follows that

$$F_X(x) = \Phi\left(\frac{x-\mu}{\sigma}\right), \quad f_X(x) = \frac{1}{\sigma} f_Z\left(\frac{x-\mu}{\sigma}\right) \tag{A.48}$$

where we invoke (A.6) for the second equation. It follows moreover that the α-quantile of X is given by $\mu + \Phi^{-1}(\alpha)\sigma$.

To calculate the m.g.f. we start with Z, the standard normal. Then

$$M_Z(t) = \frac{1}{\sqrt{2\pi}} \int_{-\infty}^{\infty} \mathrm{e}^{tx} \mathrm{e}^{-x^2/2} \mathrm{d}x.$$

We handle this by the familiar 'completing the square' trick. The exponent in the integrand is $-\frac{1}{2}[x^2 - 2tx + t^2 - t^2] = -\frac{1}{2}[(x-t)^2 - t^2]$. Since x is the variable of integration here, the terms involving t can be factored out, and we write

$$M_Z(t) = \mathrm{e}^{t^2/2}\left[\frac{1}{\sqrt{2\pi}}\int_{-\infty}^{\infty}\mathrm{e}^{-(x-t)^2/2}\mathrm{d}x\right].$$

Now the term inside the bracket is just the integral over the entire range of the density function of the normal distribution with mean t and standard deviation 1, as we see from (A.48) so it must equal 1. We conclude that

$$M_Z(t) = \mathrm{e}^{t^2/2}. \tag{A.49}$$

To handle the general case, we will use some formulas involving modifications by constants. For any constant a,

$$M_{X+a}(t) = E[\mathrm{e}^{t(X+a)}] = \mathrm{e}^{ta}E[\mathrm{e}^{tX}] = \mathrm{e}^{ta}M_X(t). \tag{A.50}$$

For any constant b,

$$M_{bX}(t) = E[\mathrm{e}^{tbX}] = M_X(bt). \tag{A.51}$$

The general normal random variable X with mean μ and standard deviation σ is distributed as $\mu + \sigma Z$ that by combining (A.49)–(A.51), we obtain

$$M_X(t) = \mathrm{e}^{t\mu+\sigma^2 t^2/2}. \tag{A.52}$$

A.11.6 The gamma and exponential distributions

The gamma distribution is produced from the gamma function, defined by

$$\Gamma(\alpha) = \int_0^{\infty} x^{\alpha-1}\mathrm{e}^{-x}\mathrm{d}x, \quad \alpha > 0.$$

We cannot evaluate this integral in terms of elementary functions. We can, however, get some information by integrating by parts. For $\alpha > 1$,

$$\Gamma(\alpha) = -x^{\alpha-1}\,\mathrm{e}^{-x}\Big|_0^{\infty} + (\alpha-1)\int_0^{\infty} x^{\alpha-2}\mathrm{e}^{-x}\mathrm{d}x = (\alpha-1)\Gamma(\alpha-1). \tag{A.53}$$

Since $\Gamma(1) = 1$, it follows by induction that for any positive integer n,

$$\Gamma(n) = (n-1)!.$$

Choose any $\beta > 0$ and consider a more general integral,

$$\int_0^\infty x^{\alpha-1}\,\mathrm{e}^{-\beta x}\mathrm{d}x = \frac{\Gamma(\alpha)}{\beta^\alpha},$$

which follows by making a change of variable from x to βx. From this, we define a continuous positive-valued random variable with the density function

$$f(x) = \frac{\beta^\alpha}{\Gamma(\alpha)}x^{\alpha-1}\mathrm{e}^{-\beta x}, \quad x > 0.$$

This is known as a Gamma distribution with parameters α and β and will be denoted by Gamma(α, β). For such a distribution X

$$M_X(t) = \frac{\beta^\alpha}{\Gamma(\alpha)}\int_0^\infty x^{\alpha-1}\,\mathrm{e}^{-(\beta-t)x}\,\mathrm{d}x = \frac{\beta^\alpha}{\Gamma(\alpha)}\frac{\Gamma(\alpha)}{(\beta-t)^\alpha} = \left(\frac{\beta}{\beta-t}\right)^\alpha. \tag{A.54}$$

Evaluating derivatives at 0 then gives

$$E(X) = \frac{\alpha}{\beta}, \qquad E(X^2) = \frac{\alpha(\alpha+1)}{\beta^2}, \qquad \mathrm{Var}(X) = \frac{\alpha}{\beta^2}. \tag{A.55}$$

The special case of the gamma when $\alpha = 1$ arises frequently. It is known as an *exponential distribution* and will be denoted by Exp(β). For such a distribution X we have from above that

$$f_X(t) = \beta e^{-\beta t}, \quad s_X(t) = \mathrm{e}^{-\beta t}, \quad M_X(t) = \frac{\beta}{\beta - t}, \qquad E(X) = \frac{1}{\beta}, \qquad \mathrm{Var}(X) = \frac{1}{\beta^2}. \tag{A.56}$$

This is the unique continuous distribution which has the same “memoryless” property that we observed in the discrete case for the geometric distribution, namely, for any nonegative t, and r

$$\mathrm{P(X > t + r|X > r) = P(X > t)}. \tag{A.57}$$

A.11.7 The lognormal distribution

Another commonly used positive-valued distribution is the *lognormal*. A random variable Y is said to have a lognormal distribution if $Y = \mathrm{e}^X$ where X has a normal distribution. That is, just as the name suggests, the logarithm of Y is normal. We can calculate the moments of this distribution from (A.52). Namely, if $X \sim N(\mu, \sigma^2)$ (the normal distribution with mean μ and variance σ^2), then

$$E(Y) = E(\mathrm{e}^X) = M_X(1) = \mathrm{e}^{\mu+\sigma^2/2}, \tag{A.58}$$

and

$$E(Y^2) = M_X(2) = \mathrm{e}^{2\mu+2\sigma^2}, \tag{A.59}$$

from which we obtain

$$\mathrm{Var}(Y) = E(Y^2) - (EY)^2 = \mathrm{e}^{2(\mu+\sigma^2)}(\mathrm{e}^{\sigma^2} - 1).$$

A.11.8 The Pareto distribution

The probability density function of the Pareto (θ, α) distribution is proportional to $(x + \theta)^{-(\alpha+1)}$. Since

$$\int_0^\infty (x + \theta)^{-(\alpha+1)}\,\mathrm{d}x = \frac{\theta^{-\alpha}}{\alpha},$$

this density function is given by

$$f(x) = \frac{\alpha\theta^\alpha}{(x + \theta)^{\alpha+1}},$$

For this distribution,

$$s(x) = \alpha\theta^\alpha \int_x^\infty (y + \theta)^{-(\alpha+1)}\,\mathrm{d}y = \left(\frac{\theta}{x + \theta}\right)^\alpha$$

and, provided $\alpha > 1$,

$$E(X) = \int_0^\infty s(t)\,\mathrm{d}t = \frac{\theta}{\alpha - 1}.$$

Further integration leads to

$$E(X^2) = \frac{2\theta^2}{(\alpha - 1)(\alpha - 2)}, \quad \text{provided } \alpha > 2.$$

In general, $E(X^k)$ becomes infinite for $k \geq \alpha$.

A.12 Convolution

A.12.1 The discrete case

Convolution is a tool for finding the distribution of a sum of random variables. Consider a simple example. Suppose X takes the values 2, 5 and 8 with probabilities 0.5, 0.3 and 0.2, respectively, while Y takes the values 3, 6 and 9, with probabilities 0.6, 0.3 and 0.1, respectively, and that X and Y are independent. Let $Z = X + Y$. What is the probability that $Z = 5$? This can happen if and only if $X = 2$ and $Y = 3$. By the independence assumption, we multiply to get the probability of both these occurrences, and we see that

$$f_Z(5) = 0.5 \times 0.6 = 0.30.$$

What is the probability that $Z = 8$? There are two mutually exclusive ways that this can occur, namely, $X = 2, Y = 6$ and $X = 5, Y = 3$. We deduce that

$$f_Z(8) = f_X(2)f_Y(6) + f_X(5)f_Y(3) = 0.5 \times 0.3 + 0.3 \times 0.6 = 0.33.$$

For the probability that $Z = 11$, we must take three terms:

$$f_Z(11) = f_X(2)f_Y(9) + f_X(5)f_y(6) + f_X(8)f_Y(3) = 0.5 \times 0.1 + 0.3 \times 0.3 + 0.2 \times 0.6 = 0.26.$$

Proceeding similarly, we can calculate $f_Z(14) = 0.09$ and $f_Z(17) = 0.02$, completing the distribution.

The general rule can be written as follows. Let X and Y be independent random variables, with nonnegative integers as values, and let $Z = X + Y$. Then

$$f_Z(n) = \sum f_X(k)f_Y(m), \tag{A.60}$$

where the sum is taken over all pairs (k, m) for which $k + m = n$.

Formula (A.60) is fine for small examples like this, but in cases where the random variables take on many more values (perhaps even an infinite number) we want to write the formula in a more systematic way, which will ensure that we do not miss any combinations. In order for a sum of nonnegative integers $k + m$ to sum to a nonnegative integer n, we can have k take any value from 0 to n and then m must take the value $n - k$. We can then write (A.60) as

$$f_Z(n) = \sum_{k=0}^{n} f_X(k)f_Y(n-k). \tag{A.61}$$

Equation (A.61) is the general so-called convolution formula. Note, however, that for small problems like the one above, which we want to do by hand calculation, it can be inefficient compared to (A.60) since we will be adding up many terms of 0.

Suppose we want the probability that $Z \leq n$. We can reason in the same way. We need X to take a value k from 0 to n and now Y must take a value less than or equal to $n - k$, so we have

$$F_Z(n) = \sum_{k=0}^{n} f_X(k)F_Y(n-k). \tag{A.62}$$

It is not hard to verify that we can interchange F and f also write (A.28) as

$$F_Z(n) = \sum_{k=0}^{n} F_X(k)f_Y(n-k). \tag{A.63}$$

This follows simply by changing the variable of summation from k to $n - k$. The reader is warned that we *cannot* write $F_Z(n) = \sum_{k=0}^{n} F_X(k)F_Y(n-k)$.

As an illustrate of convolution, we will redo Example A.1, which we did before by p.g.f.'s, now using convolution.

From (A.61),

$$f_Z(n) = \sum_{k=0}^{n} e^{-\lambda} e^{-\mu} \frac{\lambda^k \mu^{n-k}}{k!(n-k)!} = \frac{e^{-(\lambda+\mu)}(\lambda+\mu)^n}{n!} \sum_{k=0}^{n} \binom{n}{k} \left(\frac{\lambda}{\lambda+\mu}\right)^k \left(\frac{\mu}{\lambda+\mu}\right)^{n-k}.$$

The summation on the right hand side is just a binomial expansion

$$\left[\left(\frac{\lambda}{\lambda+\mu}\right) + \left(\frac{\mu}{\lambda+\mu}\right)\right]^n = 1,$$

and we conclude that Z has a Poisson$(\lambda + \mu)$ distribution.

Suppose we want the sum of more than two discrete random variables. For example, given independent random variables X, Y, W, find the distribution of $V = X + Y + W$. For cases where there are few nonzero values, we can proceed just as in (A.60) and write

$$f_V(n) = \sum f_X(k) f_Y(m) f_W(p),$$

where the sum is taken over all ordered triples (k, m, p) such that $k + m + p = n$. For the general case, however, it may not be so easy to pick out all such triples, and we have to proceed more systematically. The basic procedure is to iterate the calculation. We first apply (A.60) or (A.61) to find the distribution of $Z = X + Y$ and then apply it again to find the distribution of $V = Z + W$. For the general case of a sum of n independent random variables, we just iterate this $n - 1$ times.

Example A.2 Let $Z = \sum_{i=1}^{5} X_i$, where the X_i's are independent random variables, each taking the value 0 with probability 0.7, 1 with probability 0.2 and 2 with probability 0.1. Find the probability that $Z = 2$.

Solution. We can derive this single number by inspection, without having to do the four iterations. One way for the five values to add up to 2 is that one of them is 2 and the rest are zero. Since there are five possibilities for the 2, the probability of this is $5 \times 0.7^4 \times 0.1$. The only other way is for three of the values to be zero and the other two to be 1. The probability of this $10 \times 0.7^3 \times 0.2^2$. So

$$P(Z = 2) = 5 \times 0.7^4 \times 0.1 + 10 \times 0.7^3 \times 0.2^2 = 0.2573.$$

A.12.2 The continuous case

We now suppose that X and Y are independent, *continuous*, nonnegative random variables, and we want the distribution of $Z = X + Y$. If X takes the value x and Y takes the value y, the sum will be less than or equal to s if and only if the point (x, y) lies in the region bounded by the x and y axes, and the line $y = s - x$. Integrating the joint density function over this region gives

$$F_Z(s) = \int_0^s \int_0^{s-x} f_{X,Y}(x, y) \mathrm{d}y \mathrm{d}x. \tag{A.64}$$

Equation (A.64) is in fact true without the independence assumption, but with the assumption of independence the integral simplifies considerably. The joint density function factors as $f_X(x) f_Y(y)$, the inner integral becomes

$$\int_0^{s-x} f_Y(y) \mathrm{d}y = F_Y(s - x)$$

and we can write

$$F_Z(s) = \int_0^s f_X(x) F_Y(s - x) \mathrm{d}x, \tag{A.65}$$

a direct analogue of (A.62).

Using Leibniz's rule for differentiating integrals and the fact that $F_Y(0) = 0$, we have

$$f_Z(s) = \int_0^s f_X(x) f_Y(s - x) \mathrm{d}x, \tag{A.66}$$

a direct analogue of (A.61).

The following example serves to indicate that calculating convolutions for continuous distributions can be a very involved procedure, even in the simplest cases.

Example A.3 Find the distribution of an independent sum $Z = X + Y$, where X has a uniform distribution on the interval [0, 2] and Y has a uniform distribution on the interval [0, 3].

Solution. Whether we use (A.65) or (A.66) depends on the particular example. In this case, it is easier to use (A.66). See Figure A.1, and notice that it depicts a region of the (x, s) plane (rather than the (x, y) place. It is that portion of g the positive quadrant in the (x, s) plane given by $0 \leq s \leq 5, 0 \leq x \leq s$. A distribution that is uniform on an interval $[a, b]$ has a density function that is a constant $(b - a)^{-1}$ on this interval and zero elsewhere. Consequently, the integrand in (A.66) takes the value 1/6 when $0 \leq x \leq 2$ and $0 \leq s - x \leq 3$, as indicated by the union of regions R_1, R_2, R_3, and it take the value of 0 elsewhere, as indicated by the union of regions S_1, S_2, S_3. Now we consider in turn all possible values of s and the value of the integral in (A.66).

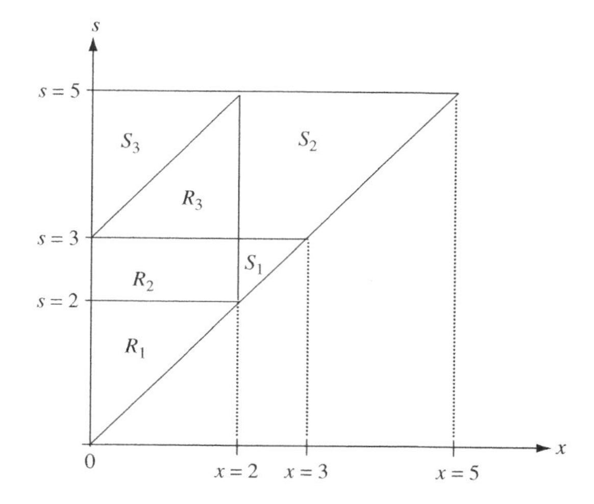

Figure A.1 The region of integration in Example A.3

If $0 \leq s \leq 2$, the value of x in R_1 will vary from 0 to s, so that the integrand will be 1/6 for all x. Therefore,

$$f_Z(s) = \frac{s}{6}, \quad 0 < s \leq 2.$$

If $2 \leq s \leq 3$, the value of x in R_2 will vary from 0 to 2, so that the integrand will be 1/6 when $x \leq 2$ and zero when $2 < x \leq s$. Therefore,

$$f_Z(s) = \frac{1}{3}, \quad 2 \leq s \leq 3.$$

If $3 \leq s \leq 5$, then the value of X in R_3 will vary from $s - 3$ to 2, so that the integrand will be 1/6 when $s - 3 \leq x \leq 2$, but zero when $0 \leq x < s - 3$ or when $2 \leq x$. Therefore,

$$f_Z(s) = \frac{5 - s}{6}, \quad 3 \leq s < 5.$$

Of course, since Z varies from 0 to 5, we know that

$$f_Z(s) = 0, \quad s < 0 \text{ or } s > 5.$$

A.12.3 Notation and remarks

Suppose that X has probability (or density) function f, and distribution function F, while Y has probability (or density) function g and distribution function G.

The probability (or density) function given in (A.61) (or (A.66)) is denoted by $f * g$. The distribution function given in (A.62) (or (A.65)) is denoted by $F * G$. One must take care with this latter notation as it could induce the error that the reader was warned about above, of using the c.d.f. in both places when calculating the c.d.f. of the sum. Similarly, given a sum of n independent random variables with the same distribution, we use f^{*n} for the probability (density) function of the sum and F^{*n} for the distribution function of the sum. To illustrate, in Example A.2 we would denoted the desired answer as $f^{*5}(2)$, where f was the given probability function.

As a final remark in this section, we note that all of the above formulas hold for random variables that take negative values, with the exception that we must replace the lower limit on the integrals and sums by $-\infty$.

A.13 Mixtures

Suppose you are faced with a choice of two games to play. Game 1 has a return of either 2 or -2 each with probability 1/2, while game 2 has a return of 4 with probability 1/4, 2 with probability 1/4 or -3 with probability 1/2. Racked with indecision, you flip a coin, intending to pay game 1 if a head turns up, or game 2 if a tails comes up. What you are really playing is a *mixture* of game 1 and game 2 with equal weights. The resulting return is easily calculated to be 4 with probability 1/8, 2 with probability 3/8, -2 with probability 2/8 and -3 with probability 2/8. More generally, we could have n random variables $X_1, X_2, \dots, X_n$ and a probability distribution on $\{1, 2, \dots, n\}$, called the mixing distribution. The resulting mixture is a random variable X that takes a value from X_i with probability $p(i)$, where p is the probability function of the mixing distribution. (Our initial example had $p(1) = p(2) = 1/2$.) Calculating quantities with respect to the mixed distribution presents no problems as everything follows the same convex combination. We have, for example,

$$f_X(s) = \sum_{k=1}^{n} p(k) f_{X_k}(s), \qquad F_X(s) = \sum_{k=1}^{n} p(k) F_{X_k}(s), \qquad M_X(s) = \sum_{k=1}^{n} p(k) M_{X_k}(s). \quad \text{(A.67)}$$

Let us consider a slightly trickier situation, in which we have continuous mixing. Suppose now that we have a whole family of random variables X_t, either all discrete or all continuous, indexed on all the nonnegative reals $[0, \infty)$ instead of just the integers. For a mixing distribution, we take a continuous nonnegative random variable with density function p. The resulting mixture is a random variable X with density function

$$f_X(s) = \int_0^{\infty} p(t) f_{X_t}(s) \mathrm{d}t. \quad \text{(A.68)}$$

References

Arrow, K.J. (1963). Uncertainty and the welfare of medical care. *American Economic Review* 53, 941–973.

Artzner, P., Delbaen, F., Eber, J.M. and Heath, D. (1999). Coherent measures of risk. *Math Finance* 9, 203–228.

Björk, T. (2009). *Arbitrage Theory in Continuous Time*, 3rd edn. New York, NY: Oxford University Press.

Bowers, N., Gerber, H., Hickman, J., Jones, D. and Nesbitt, C. (1997). *Actuarial Mathematics*, 2nd edn. Schaumburg, IL: Society of Actuaries.

Brillinger, D.R. (1961). A justification of some common laws of mortality. *Transactions of the Society of Actuaries* XIII, 116–119.

Broverman, S. (2010). *Mathematics of Investment and Credit*, 5th edn. Winstead CT: ACTEX.

Bülhman, H. and Gisler, A. (2005). *A Course in Credibility Theory*. Berlin-Heidelberg: Springer-Verlag.

Carrière, J. (1994a). An investigation of the Gompertz law of mortality. *Actuarial Research Clearing House*, 2.

Carrière, J. (1994b). Dependent decrement theory. *Transactions of the Society of Actuaries* XLVI, 45–65.

Daniel, J.W. and Vaaler, L.J.F. (2009). *Mathematical Interest Theory*, 2nd edn. Mathematical Association of America.

Dickson, D., Hardy, M. and Waters, H. (2013). *Actuarial Mathematics for Life Contingent Risks*, Cambridge: Cambridge University Press.

Etheridge, A. (2002). *A Course in Financial Calculus*, Cambridge: Cambridge University Press.

Frees, E., Carrière, J. and Valdez, E. (1996). Annuity valuation with dependent mortality. *Journal of Risk and Insurance* 63, 229–261.

Frees, E. and Valdez E. (1998). Understanding relationships using copulas. *North American Actuarial Journal* 2, 1–25.

Gerber, H. and Shiu, E.S. (1998). On the time value of ruin. *North American Actuarial Journal* 2, 48–78.

Grandel, J. (1991). *Aspects of Risk Theory*. New York, NY: Springer-Verlag.

Herzog, T.N. (1999). *Introduction to Credibility Theory*, 3rd edn. Winstead, CT: ACTEX.

Hoel, P., Port, S. and Stone, C. (1972). *Introduction to Stochastic Processes*, Boston: Houghton Mifflin.

Fundamentals of Actuarial Mathematics, Third Edition. S. David Promislow.

Companion website: http://www.wiley.com/go/actuarial

Hull, J.C. (2014). *Options, Futures, and other Derivative Securities*, 9th edn, Cambridge, Upper Saddle River, NJ: Prentice Hall.

Jones, B.L. (1997). Stochastic models for continuing care retirement communities. *North American Actuarial Journal* 1, 50–73.

Jordan, C.W. (1967). *Life Contingencies*, 2nd edn. Schaumburg, IL: Society of Actuaries.

Kaas, R., Goovaerts, Dhaene, M.J. and Denuit, M. (2008). *Modern Actuarial Risk Theory*, 2nd edn. Springer-Verlag.

Kemeny, J.G. and Snell, J.L. (1963). *Finite Markov Chains*. Princeton, NJ: D. Van Nostrand.

Klugman, S., Panjer, H. and Willmot, G. (2012). *Loss Models*, 4th edn. Hoboken, NJ: John Wiley & Sons.

London, D. (1997). *Survival Models and their Estimation*, 3rd edn. Winstead CT: ACTEX.

McDonald, R.L. (2012). *Derivatives Markets*, 3rd edn. Upper Saddle River, NJ: Prentice Hall.

Mikosch, T. (1998). *Elementary Stochastic Calculus-with Finance in View*, River Edge, NJ: World Scientific.

Nelsen, R.B. (1999). *An Introduction to Copulas*, Lecture Notes in Mathematics 139. New York, NY: Springer-Verlag.

Norberg, R. (2008). Multi-state models for life insurance mathematics. In: E. Melnick and B. Everitt (eds.), *Encyclopedia of Quantitative Risk Analysis and Assessment*. Chichester: Wiley-Blackwell.

Powers, M.R. (1995). A theory of risk, return, and solvency. *Insurance: Mathematics and Economics* 17 , 101–118.

Promislow, S.D. (1980). A new approach to the theory of interest. *Transactions of the Society of Actuaries* XXXII, 53–92.

Promislow, S.D. (1991). Select and ultimate models in multiple decrement theory. *Transactions of the Society of Actuaries* XLIII, 281–300.

Promislow, S.D. (1997). Classification of usurious loans. In: M. Sherris (ed.), *Proceedings of the 7th International AFIR Colloquium*. Institute of Actuaries of Australia.

Ross, S.M. (2010). *Introduction to Probability Models*, 10th edn. San Diego, CA: Academic Press.

Rothschild, M. and Stiglitz, J. (1970). Increasing risk: A definition. *Journal of Economic Theory* 2, 225–243.

Royden, H.L.(1988). *Real Analysis*, 3rd edn. New York, NY: Macmillan.

Rudin, W.R. (1976). *Principles of Mathematical Analysis*, 3rd edn. New York, NY: McGraw-Hill.

Shaked, M. and Shantikumar, J. G. (2007). *Stochastic Orders*. Springer.

Steland, A. (2012). *Financial Statistics and Mathematical Finance*, Chichester: John Wiley & Sons.

Teichroew, D., Robichek, A.A. and Montalbano, M. (1965a). Mathematical analysis of rates of return under certainty. *Management Science* 11, 395–403.

Teichroew, D., Robichek, A.A. and Montalbano, M. (1965b). An analysis of criteria for investment and financing decisions under certainty. *Management Science* 12, 151–179.

Tennenbein, A. and Vanderhoof, I. (1980). New mathematical laws of select and ultimate mortality. *Transactions of the Society of Actuaries* XLIV, 509–538.

Young, V.R. (2004). Premium principles. In: J.F. Teugels and B. Sundt (eds.), *Encyclopedia of Actuarial Science*. Chichester: John Wiley & Sons.

Notation index

The following is a list of the major symbols which are used in the book. For the most part, the first page they appear on is listed. An exception is the notation in Appendix A, where the first appearance in that chapter is noted. This list excludes that part of the standard actuarial notation which is not used in the main body of the text. The latter can be found in the appropriate sections of Chapters 2–6, 8 and 10 entitled 'Standard Notation and Terminology'.

Chapter 2

$\ddot{a}(\mathbf{c}; v)$ 15
$\mathbf{a} * \mathbf{b}$
$B_k(\mathbf{c}; v)$ 21
$\tilde{B}_k(\mathbf{c})$ 25
${}_k\mathbf{c}$ 21
${}^k\mathbf{c}$ 21
$\mathbf{c} \circ k$ 25
d_k 12
$\mathbf{e}^k$ 17
i_k 12
${}_kV(\mathbf{c}; v)$ 21
$v(k)$ 11
$v(k, n)$ 10
$v \circ k$ 25
$\mathrm{Val}_n(\mathbf{c}; v)$ 26
$\Delta\mathbf{b}$ 17
$\nabla\mathbf{b}$ 18

Chapter 3

d_x 39
e_x 42
$\mathring{e}_x$ 42
${}_np_x$ 40
${}_nq_x$ 40
p_x 40
q_x 40
ℓ_x 39
ω 40

Chapter 4

$\ddot{a}_x(\mathbf{c})$ 49
(1_∞) 51
$y_x(k)$ 50
$\ddot{a}_{[x]+k}(\mathbf{c})$ 52

Chapter 5

$A_x(\mathbf{b})$ 62
$w_x(k)$ 69
$\ddot{A}_{[x]+k}(\mathbf{c})$ 70

Chapter 6

η_k 81
${}_kV$ 78

Fundamentals of Actuarial Mathematics, Third Edition. S. David Promislow.

Companion website: http://www.wiley.com/go/actuarial

Chapter 7

$\ddot{a}^{(m)}(\mathbf{c}; y)$ 99
$a^{(m)}(\mathbf{c}; y)$ 104
$\ddot{a}_x^{(m)}(\mathbf{c})$ 101
$d^{(m)}$ 100
$i^{(m)}$ 100
$\alpha(m)$ 102
$\beta(m)$ 102

Chapter 8

$\bar{a}(c; y)$ 113
$\bar{a}_x(c)$ 115
$\bar{A}_x(\mathbf{b})$ 119
$\alpha(\infty)$ 116
$\beta(\infty)$ 116
$\delta_y(t)$ 113
δ 114
$\lambda_x(t)$ 119
$\mu(x)$ 118
$\mu_x(t)$ 118

Chapter 9

$\ell_{[x]+k}$ 139
${}_sp_{[x]+t}$ 138
${}_sq_{[x]+t}$ 137
$q_{[x]+t}$ 138

Chapter 10

$\ddot{a}_{xy}(\mathbf{c})$ 145
$\bar{a}_{xy}(c)$ 145
$\ddot{a}_{\overline{xy}}(\mathbf{c})$ 147
$A_{xy}(\mathbf{c})$ 148
$\bar{A}_{xy}(c)$
$A_{\overline{xy}}(\mathbf{c})$
$\bar{A}_{\overline{xy}}(c)$ 148
$A^1_{xy}(b)$ 152
$A^2_{xy}(b)$ 152
$\bar{A}^1_{xy}(b)$ 153
$\bar{A}^2_{xy}(b)$ 153
${}_np_{xy}$ 143
${}_np_{\overline{xy}}$ 146
${}_nq_{xy}$ 146
${}_nq_{\overline{xy}}$ 153
${}_sq^1_{xy}$ 147
$\mu_{xy}(t)$ 147

Chapter 11

$\bar{A}_x^{(j)}(\mathbf{b})$ 166
$d_x^{(j)}$ 163
$d_x^{(\tau)}$ 163
$\ell_x^{(\tau)}$ 163
$\ell_x^{(j)}$ 164
${}_np_x^{(j)}$ 165
${}_np_x^{(\tau)}$ 160
${}_nq_x^{(j)}$ 164
$q_x^{(j)}$ 164
$q_x^{\prime(j)}$ 171
$\mu_x^{(j)}(t)$ 165

Chapter 12

AS_k 187
Pr_k 190
Π_k 194

Chapter 13

AV_k 201
COI_k 202
S_x 205

Chapter 14

$f_x(t)$ 220
$s_x(t)$ 220
$T \circ u$ 216
$\tilde{T}$ 217

Chapter 15

$\ddot{a}_{\tilde{T}}(\mathbf{c}; v)$ 229
$\bar{a}_T(c, v)$ 231
$A_{\tilde{T}}(\mathbf{b}, v)$ 225
$\bar{A}_T(b; v)$ 226
$CV(X)$ 238
${}_rL$ 234
L 235

Chapter 16

$\hat{F}_X$ 252
$\hat{f}_X$ 252

Chapter 17

$f_{T_1,T_2,\ldots,T_M}(t_1, t_2, \ldots, t_m)$ 260
$F_{T_1,T_2,\ldots,T_M}(t_1, t_2, \ldots, t_m)$ 260

$s_{T_1,T_2,\ldots,T_M}(t_1, t_2, \ldots, t_m)$ 260
$F_{T,J}(t,j)$ 262
$f_{T,J}(t,j)$ 264
$s_{T,J}(t,j)$ 264
$\mu_{T,J}(t,j)$ 266

Chapter 18

$p_{xy}(k,n)$ 284
p_{xy} 284
$o(h)$ 294
$\sim$ 284

Chapter 19

$p_{ii}(s,t)$ 311
$p_{\overline{ii}}(s,t)$ 314
$\mu_{ij}(t)$ 312

Chapter 20

$E_k(W)$ 350
$E_{\Pi}(W)$ 366

Chapter 21

$\langle N, X \rangle$ 379
S^+ 384
$(X-d)_+$ 388
$X \wedge d$ 388

Chapter 22

VaR 413
TVaR_α 413

Chapter 23

$D(u)$ 423
$J(u)$ 433
$\mathcal{L}$ 440
$\psi(u)$ 420
$\psi_k(u)$ 436
$\psi(u,t)$ 420

Chapter 24

$E(X|Y)$ 453
$\text{Var}(X|Y)$ 455
$\propto$ 459

Appendix A

$P(A)$ 477
$P(A|B)$ 479
$f_X(x)$ 480
$F_X(x)$ 480
$E(X)$ 481
$\text{Var}(X)$ 481
$s(t)$ 482
$\text{Cov}(X,Y)$ 484
$E(X|B)$ 486
$f * g$ 499
$F * G$ 499
f^{*n} 499
F^{*n} 499
$M_X(t)$ 486
$P_X(t)$ 487

Index

accumulated value 14, 34
actuarial equivalence 13–15
actuarial present value 225
adjustment coefficient 429–31, 440, 445
aggregate mortality 139, 219
american option 338, 346–8
amortization 25
annuity
  cash refund 69
  certain 58
  continuous 112–13, 232–3
  deferred 33, 49, 55–6
  due 33, 104
  guaranteed 54
  immediate 33, 104
  instalment refund 60, 69
  joint-life 146
  last survivor 147–8
  life 47–8, 150–1, 229–31
  (m)thly 98, 101–4
  reversionary 152, 158
  temporary 49, 160, 248
  variable 203–4
  whole life 49, 51, 57
arbitrage 31, 334
arbitrage-free market 334–7, 342, 343, 351, 367–8
Arrow's optimal insurance theorem 411
asset share 188, 191
associated single decrement table 175–81

Balducci hypothesis 109
binomial tree 343, 348
Black-Scholes-Merton formula 361–4
Brownian motion 295–9, 362

call option 338, 344, 346, 362
cash surrender value 88
cash flow vector 7, 13, 78
central limit theorem 492
central rate of mortality 134
Chapman-Kolmogorov equations 313
Chebyshev's inequality 432, 482
collective risk model 377
comonotonic 415
complete market 359–61
common shock model 271–3, 321–2
compound distribution 377–98
compound Poisson
  distribution 379
  process 438–40
concave function 404, 406–7
conditional expectation 366, 453–5
conditional probability 450, 479
conditional variance 453–4
conditional tail expectation (CTE) 414
conjugate prior 467
constant force assumption 132
constant force of mortality 126–7, 215
contingency loading 239
contingent insurances 153–5

Fundamentals of Actuarial Mathematics, Third Edition. S. David Promislow.

Companion website: http://www.wiley.com/go/actuarial

convex
  function 405
  order 417
  set 353
convolution 381, 394, 495–500
cost of insurance 200, 201
copula 273–6
counting process 293
covariance 484
credibility
  Bayesian 457–9
  Bühlman 463–4
  Bühlman – Straub 464–5
  exact 465–8
cumulative distribution function 480

deductible 388–92
deferred
  annuity 33, 49, 55–6
  contract 233
  insurance 73, 255
deficit at ruin 422, 432
defined benefit plan (DB) 57, 204
defined contribution plan (DC) 57, 206–7
density function 480
Demoivre's law 127
difference formula 24
differential equation 124–5, 313, 324
discount
  force of 113
  function 9–11
  rate 12
distortion risk measure 417
distribution
  beta 474
  binomial 382, 489
  exponential 215, 494
  gamma 383, 398, 493–4
  geometric 291, 490
  Gompertz 215
  lognormal 494
  Makeham 216
  negative binomial 382–3, 398, 490–1
  normal 238, 296, 491
  Pareto 384, 495
  Poisson 294, 382, 489
  uniform 215, 491
  Weibull 223
distribution function 480
dividends 85

endowment
  identity 71
  insurance 63–4, 248
  pure 48
Euler's method 125, 317
European option 338
expectation 481
expenses 88, 184–6
expense-augmented premium 185
expense-augmented reserve 186
expiration date 338

Fackler reserve formula 83
failure time 211
first-death insurance 153
force
  of decrement 170
  of discount 113
  of failure 149
  of interest 114
  of mortality 118
  of transition 312
forward
  contract 30–1
  interest rates 32, 370
  prices 30–2, 370
frequency 378, 381–3
full preliminary term 187
fundamental theorem of asset pricing 352, 357–8

gains and losses 83–5, 191–3
gambler's ruin 427
generational annuity table 142
Geometric Brownian motion 299, 362
Gompertz's law 222
gross premium 56, 185
gross premium reserve 188

Hattendorf's theorem 241
hazard rate 212

increment
  independent 293
  stationary 293
independence 479, 485
individual risk model 378
insurance
  casualty 377
  deferred 73, 254
  endowment 63–4, 248
  life 61–74, 225–9
  term 64, 278
  universal life 199–202
  whole life 64
intensity function 192
intensity matrix 313
interest
  constant 12–13
  force of 114
  nominal rate 100
  rate 12, 372
interest and survivorship 50–2
internal rate of return 29

Jensen's inequality 407–8
joint density function 483
joint distribution 483–4
joint distribution function 483–4
joint-life status 144–6, 171
joint survival function 260

Kolmogorov equations 312, 317, 318, 330

lapse 88
last survivor status 147
life annuity 47–8, 150–1, 229–31
life expectancy 42–3, 117–18, 221
  complete 42
  curtate 42
  temporary 43, 117
life insurance 61–74, 225–9
life table 39–45
loss elimination ratio 401
Lundberg's inequality 445

Makeham's law 222
Markov chain 282–4
  finite state 287–92
  limiting distribution 289
  models for insurance and annuities 304
  non-stationary 305
  periodic 289
  reducible 289
martingale 286–7, 352, 424
maximal aggregate loss 441–3
mean 469, 481
minimum random variable 259
mixtures 500
mode 215
modified reserve system 187
moments 481
moment generating function 486–7, 488
Monte Hall problem 451–2
mortality
  force of 113
  rate 40
  table 39
multi-state models 304–31
multiple decrement
  models 166–83
  table 167
multiplication rule 41

net amount at risk 83
net annual premium 53, 64
net single premium 51
nonforfeiture 88
nonhomogeneous Poisson process 295
nonidentifiability 268

options 337–9
  American 338, 346–8
  call 338
  European 338
  embedded 333
  lookback 343
  put 338
optional stopping theorem 426–7

paid up reserve formula formula 91
pension plans 57, 204–7
periodic Markov chain 289
Poisson process 293–5

premium 4, 47, 61, 122, 235–40
  annual 53, 64, 255
  expense-augmented 185
  equivalence principle 23
  gross 56, 185
  net 56
  pattern vector 55
  percentile 236–7, 413
posterior distribution 458
premium difference reserve formula 90
premium principle 240–1
present value 13, 51
prior distribution 458
probability density function 480
probability function 480
probability generating function 487–8
probability mass function 480
probability measure 477–8
profit margin 195
profit signature 195
profit testing 193–5
prospective loss 234, 249
prospective method 23
pure endowment 48, 65
put-call parity 342

random variable 479
  continuous 480
  discrete 480
random walk 284, 293
recurrent state 290–3
recursion formulas 24
  aggregate claims 199
  balances 25
  life expectancy 43
  reserves 24, 82, 188, 326
  ruin probabilities 434–8
reserve 76–96, 187–9, 324–7
  definition of 21
  differential equation 124–5, 327
  expense-augmented 186
  at fractional durations 107
  gross premium 188
  initial 83
  modified 187
  net premium 186
  prospective 23
  retrospective 23
  terminal 83
  Zilmerized 186
retrospective method 21
risk averse 404, 408
risk free
  bond 31
  rate of interest 334
risk comparison 408–12
risk loading 239
risk measures 412–17
risk-neutral 340
risk portion of premium 86
Rothschild-Stiglitz 408
ruin 420–48
  functional equation approach 422
  martingale approach 424
  recursion formula 43–8
  time of 420

salary scales 205
sample space 477
savings portion of premium 86
second-death insurance 148
select and ultimate 139
select mortality 137–40
select period 138
self-financing trading strategy 335, 337, 364
semi-Markov process 328
severity 378, 383
short selling 31
sojourn probabilities 314–15
spot rate of interest 32
standard deviation 240, 482
stationary increment 293
stochastic process 281–303
  discrete-time 281
  continuous-time 293
  realization 282
stop-loss reinsurance 392
stopping time 424–6
strike price 338
St. Petersburg paradox 404
submartingale 286
supermartingale 287

surplus process
  compound Poisson 438–40
  discrete 421
survival
  distribution 211–16
  function 212, 214

tail value at risk (TVaR) 413–17
term structure of interest rates 32
Thiele's differential equation 125, 317, 327
time value of money 8
total probability, law of 486
total expectation, law of 486
trading strategy 334
transient state 290–3
transition
  matrix 287
  probability 287

unearned premium 108
uniform distribution of deaths (UDD) 101–2
uniform seniority 223
universal life 199–203
utility 403–6
  exponential 417
  function 404
  power 405

valuation 78
  premium 85
value at time *n* 14
value at risk 413
variable annuity 203–4
variance 224, 481
volatility 362

waiting times 295
Woolhouse's formulas 106, 129–31, 132

yield 29
yield curve 32

zero-coupon bonds 30

Printed and bound by CPI Group (UK) Ltd, Croydon, CR0 4YY

12/01/2025

14624503-0004